VOLUME FIVE HUNDRED AND SIXTY SIX

METHODS IN ENZYMOLOGY

Isotope Labeling of Biomolecules – Applications

METHODS IN ENZYMOLOGY

VOLUME FIVE HUNDRED AND SIXTY SIX

METHODS IN ENZYMOLOGY

Isotope Labeling of Biomolecules – Applications

Edited by

ZVI KELMAN

Biomolecular Labeling Laboratory, and Biomolecular Structure and Function Group, Institute for Bioscience and Biotechnology Research, National Institute of Standards and Technology and the University of Maryland, Rockville, Maryland, USA

ELSEVIER

AMSTERDAM • BOSTON • HEIDELBERG • LONDON
NEW YORK • OXFORD • PARIS • SAN DIEGO
SAN FRANCISCO • SINGAPORE • SYDNEY • TOKYO
Academic Press is an imprint of Elsevier

Academic Press is an imprint of Elsevier
50 Hampshire Street, 5th Floor, Cambridge, MA 02139, USA
525 B Street, Suite 1800, San Diego, CA 92101–4495, USA
The Boulevard, Langford Lane, Kidlington, Oxford OX5 1GB, UK
125 London Wall, London, EC2Y 5AS, UK

First edition 2016

ISBN: 978-0-12-803049-3
ISSN: 0076-6879

For information on all Academic Press publications visit our website at http://store.elsevier.com/

CONTENTS

Section III
Mass Spectrometry

CONTRIBUTORS

Luke W. Arbogast
Institute for Bioscience and Biotechnology Research, National Institute of Standards and Technology and the University of Maryland, Rockville, Maryland, USA

Robert G. Brinson
Institute for Bioscience and Biotechnology Research, National Institute of Standards and Technology and the University of Maryland, Rockville, Maryland, USA

Zimei Bu
Department of Chemistry and Biochemistry, City College of New York, City University of New York, New York, USA

David J.E. Callaway
Department of Chemistry and Biochemistry, City College of New York, City University of New York, New York, USA

Luke A. Clifton
ISIS Neutron Facility, STFC Rutherford Appleton Laboratory, Didcot, Oxfordshire, United Kingdom

Erdem Coskun
Biochemical Measurement Division, National Institute of Standards and Technology, Gaithersburg, Maryland, USA

Joseph E. Curtis
National Institute of Standards and Technology, Gaithersburg, Maryland, USA

Juliette M. Devos
Life Sciences Group, Institut Laue-Langevin, Grenoble Cedex 9, France

Miral Dizdaroglu
Biochemical Measurement Division, National Institute of Standards and Technology, Gaithersburg, Maryland, USA

Orla Dunne
Life Sciences Group, Institut Laue-Langevin, Grenoble Cedex 9, France, and Department of Structural and Molecular Biology, University College London, London, United Kingdom

Karen G. Fleming
T.C. Jenkins Department of Biophysics, Johns Hopkins University, Baltimore, Maryland, USA

Patrick J. Fleming
T.C. Jenkins Department of Biophysics, Johns Hopkins University, Baltimore, Maryland, USA

Elyssia S. Gallagher
Institute for Bioscience and Biotechnology Research, National Institute of Standards and Technology and the University of Maryland, Rockville, Maryland, USA

Miklos Guttman
Department of Medicinal Chemistry, University of Washington, Seattle, Washington, USA

Michael Haertlein
Life Sciences Group, Institut Laue-Langevin, Grenoble Cedex 9, France

Frank Heinrich
Department of Physics, Carnegie Mellon University, Pittsburgh, Pennsylvania, and Center for Neutron Research, National Institute of Standards and Technology, Gaithersburg, Maryland, USA

Thomas Helleday
Science for Life Laboratory, Division of Translational Medicine and Chemical Biology, Department of Medical Biochemistry and Biophysics, Karolinska Institutet, Stockholm, Sweden

Mirko Hennig
Nutrition Research Institute, University of North Carolina at Chapel Hill, Kannapolis, North Carolina, USA

Peter J. Holden
National Deuteration Facility, Bragg Institute, Australian Nuclear Science and Technology Organization, Lucas Heights, New South Wales, Australia

Stephen A. Holt
Bragg Institute, Australian Nuclear Science and Technology Organisation, Lucas Heights, New South Wales Australia

James T. Hoopes
Institute for Bioscience and Biotechnology Research, University of Maryland, Rockville, Maryland, USA

Jeffrey W. Hudgens
Institute for Bioscience and Biotechnology Research, and Biomolecular Measurement Division, National Institute of Standards and Technology, Rockville, Maryland, USA

Pawel Jaruga
Biochemical Measurement Division, National Institute of Standards and Technology, Gaithersburg, Maryland, USA

Ann-Sofie Jemth
Science for Life Laboratory, Division of Translational Medicine and Chemical Biology, Department of Medical Biochemistry and Biophysics, Karolinska Institutet, Stockholm, Sweden

Jason D. Kahn
Department of Chemistry and Biochemistry, University of Maryland, College Park, Maryland, USA

Eric L. Kilpatrick
Bioanalytical Science Group, Biomolecular Measurement Division, Material Measurement Laboratory, National Institute of Standards and Technology, Gaithersburg, Maryland, USA

Jonggul Kim
Department of Chemistry, University of Minnesota, Minneapolis, Minnesota, USA

Susan Krueger
National Institute of Standards and Technology, Gaithersburg, Maryland, USA

Jeremy H. Lakey
Institute for Cell and Molecular Biosciences, Newcastle University, Framlington Place, Newcastle upon Tyne, United Kingdom

Valerie Laux
Life Sciences Group, Institut Laue-Langevin, Grenoble Cedex 9, France

Anton P. Le Brun
National Deuteration Facility, Bragg Institute, Australian Nuclear Science and Technology Organization, Lucas Heights, New South Wales, Australia

Stuart F.J. Le Grice
Basic Research Laboratory, NCI-Frederick National Institutes of Health, Frederick, Maryland, USA

Kelly K. Lee
Department of Medicinal Chemistry, University of Washington, Seattle, Washington, USA

Geoffrey Li
Department of Chemistry, University of Minnesota, Minneapolis, Minnesota, USA

Olga Loseva
Science for Life Laboratory, Division of Translational Medicine and Chemical Biology, Department of Medical Biochemistry and Biophysics, Karolinska Institutet, Stockholm, Sweden

Mark S. Lowenthal
Biochemical Measurement Division, National Institute of Standards and Technology, Gaithersburg, Maryland, USA

John P. Marino
Institute for Bioscience and Biotechnology Research, National Institute of Standards and Technology and the University of Maryland, Rockville, Maryland, USA

Leland Mayne
Johnson Research Foundation, Department of Biochemistry and Biophysics, Perelman School of Medicine, The University of Pennsylvania, Philadelphia, Pennsylvania, USA

Jennifer T. Miller
Basic Research Laboratory, NCI-Frederick National Institutes of Health, Frederick, Maryland, USA

Martine Moulin
Life Sciences Group, Institut Laue-Langevin, Grenoble Cedex 9, France

Bryant C. Nelson
Biosystems and Biomaterials Division, National Institute of Standards and Technology, Gaithersburg, Maryland, USA

Karen W. Phinney
Biomolecular Measurement Division, National Institute of Standards and Technology, Gaithersburg, Maryland, USA

Prasad T. Reddy
Institute for Bioscience and Biotechnology Research, National Institute of Standards and Technology and the University of Maryland, Rockville, Maryland, USA

Clifford W. Sandlin
T.C. Jenkins Department of Biophysics, Johns Hopkins University, Baltimore, Maryland, USA

Kerry Bauer Scott
Institute for Bioscience and Biotechnology Research, National Institute of Standards and Technology and the University of Maryland, Rockville, Maryland, USA

Lincoln G. Scott
Cassia LLC, San Diego, California, USA

V. Trevor Forsyth
Life Sciences Group, Institut Laue-Langevin, Grenoble Cedex 9, France, and Macromolecular Structure Group, Keele University, Staffordshire, United Kingdom

Illarion V. Turko
Institute for Bioscience and Biotechnology Research, National Institute of Standards and Technology and the University of Maryland, Rockville, Maryland, USA

Gianluigi Veglia
Department of Chemistry, and Department of Biochemistry, Molecular Biology, and Biophysics, University of Minnesota, Minneapolis, Minnesota, USA

Yingjie Wang
Department of Chemistry, University of Minnesota, Minneapolis, Minnesota, USA

Nathan R. Zaccai
T.C. Jenkins Department of Biophysics, Johns Hopkins University, Baltimore, Maryland, USA

PREFACE

A number of structural biology techniques such as nuclear magnetic resonance (NMR), neutron reflectometry (NR), and small-angle neutron scattering (SANS) can be performed on biomolecules with nuclei at natural isotope abundance. However, the use of biomolecules labeled with stable isotopes (mainly ^{2}H, ^{13}C, or ^{15}N) allows for additional experimental approaches. Labeled biomolecules are also important in quantitative mass spectrometry (MS), and deuteration of exchangeable protons is required for hydrogen–deuterium exchange mass spectrometry (HDX-MS).

The aim of this volume of *Methods in Enzymology* is to provide methodologies for the use of labeled biomolecules in different structural and biophysical applications. The accompanying volume describes the production and purification of labeled biomolecules.

Structural characterization of biomolecules by NMR has utilized stable isotopic enrichment to allow for the use of the NMR active spin ½ of ^{13}C and ^{15}N. Correlations to these heteronuclei alleviate ^{1}H resonance overlap. And perdeuteration of biomolecules increases the accessibility of large proteins for NMR analysis. In Chapter 1, Arbogast and coworkers describe the use of 2D spectra collected at natural isotopic abundance of ^{13}C and ^{15}N in the structural analysis of protein therapeutics. Chapter 2, from the laboratory of Gianluigi Veglia, describes an approach to assign the resonances of labeled methyl group side chains in large proteins.

Fluorine (^{19}F) nuclei, which have 100% natural abundance, are spin ½, have a high gyromagnetic ratio, and provide NMR signals that can serve as sensitive and specific reporters of the structural and chemical environment around the site of substitution since ^{19}F is absent from all naturally occurring biomolecules. Two chapters describe the use of ^{19}F-labeled nucleotides for structural analysis by NMR. Scott and Hennig describe the use of ^{19}F-labeled nucleotides for nucleic acid structural analysis (Chapter 3), while Brinson and coworkers describe the use of these molecules in the structural and dynamic analysis of homopyrimidine/homopurine tracts (Chapter 4).

In addition to NMR, deuterated proteins are used in structural determination by scattering techniques. One of the advantages of using deuterated material is the ability to perform contrast variation experiments, in which molecules containing hydrogen and those containing deuterium can be matched out against the isotopic composition of the solvent. Chapter 5, from the laboratory of Trevor Forsyth, provides an overview of the methods used for deuteration of biomolecules, the type of labeling (perdeuterated, fractional, etc.), and the assays that use deuterated material. Zaccai et al. describe the use of deuterium labeling and SANS contrast variation in the study of protein–protein interactions (Chapter 6). In Chapter 7, Heinrich provides general applications of deuterated material in biological NR studies, while Le Brun and coworkers describe the use of NR and deuterium labeling to study the bacterial outer membrane (Chapter 8). Callaway and Bu describe the strategies and general considerations for analyzing protein dynamics by neutron spin echo spectroscopy in Chapter 9.

Stable isotopes are also commonly used in MS as standards for protein quantitation. Kilpatrick describes a computer program to calculate the percentage of ^{15}N incorporation in peptides and proteins (Chapter 10). In Chapter 11, Scott et al. describe the design, expression, and preparation of labeled concatenated peptides (QconCAT) and their use as internal standards for absolute or relative protein quantification. Reddy et al. (Chapter 12) describe the use of ^{15}N-labeled DNA repair proteins as internal standards for MS.

HDX-MS is used to study protein structure and dynamics, protein–protein interactions, and protein–ligand interactions. In Chapter 13, Mayne discusses the principle behind HDX-MS and the considerations needed when using this technique. Gallagher and Hudgens describe the use of HDX-MS in the study of protein–ligand interactions (Chapter 14), while Guttman and Lee describe the use of the technique for the study of viruses (Chapter 15).

I hope that this and the accompanying volume prove useful for the scientific community by providing descriptions of techniques for biomolecule labeling and their use in structural analysis.

I would like to acknowledge the help of Dr. Lori M. Kelman during the course of editing these volumes.

Zvi Kelman

SECTION I

Nuclear Magnetic Resonance Spectroscopy

CHAPTER ONE

Application of Natural Isotopic Abundance ^1H–^{13}C- and ^1H–^{15}N-Correlated Two-Dimensional NMR for Evaluation of the Structure of Protein Therapeutics

Luke W. Arbogast, Robert G. Brinson, John P. Marino[1]

Institute for Bioscience and Biotechnology Research, National Institute of Standards and Technology and the University of Maryland, Rockville, Maryland, USA

[1]Corresponding author: e-mail address: john.marino@nist.gov

Contents

Abstract

Methods for characterizing the higher-order structure of protein therapeutics are in great demand for establishing consistency in drug manufacturing, for detecting drug product variations resulting from modifications in the manufacturing process, and for comparing a biosimilar to an innovator reference product. In principle, solution

Methods in Enzymology, Volume 566
ISSN 0076-6879
http://dx.doi.org/10.1016/bs.mie.2015.09.037

NMR can provide a robust approach for characterization of the conformation(s) of protein therapeutics in formulation at atomic resolution. However, molecular weight limitations and the perceived need for stable isotope labeling have to date limited its practical applications in the biopharmaceutical industry. Advances in NMR magnet and console technologies, cryogenically cooled probes, and new rapid acquisition methodologies, particularly selective optimized flip-angle short transient pulse schemes and nonuniform sampling, have greatly ameliorated these limitations. Here, we describe experimental methods for the collection and analysis of 2D $^{1}H^{N}$–^{15}N-amide- and ^{1}H–^{13}C-methyl-correlated spectra applied to protein drug products at natural isotopic abundance, including representatives from the rapidly growing class of monoclonal antibody (mAb) therapeutics. Practical aspects of experimental setup and data acquisition for both standard and rapid acquisition NMR techniques are described. Furthermore, strategies for the statistical comparison of 2D $^{1}H^{N}$–^{15}N-amide- and ^{1}H–^{13}C-methyl-correlated spectra are detailed.

1. INTRODUCTION

Over the past two decades, there has been a tremendous growth in the number of clinically approved biopharmaceuticals, particularly driven by the growth in FDA-approved monoclonal antibodies (mAbs) therapeutics. This success has brought with it a need to assess the folded structure(s) of protein therapeutics, which may vary from batch to batch in productions, from process and formulation changes, and in particular between an originator biologic and a potential biosimilar product. Facile methods that can characterize the secondary, tertiary, and/or quaternary structure (collectively the higher-order structure (HOS)) of protein therapeutics and distinguish structural variants are therefore of high value (Berkowitz, Engen, Mazzeo, & Jones, 2012). Currently, a number of spectroscopic techniques are employed for such HOS assessment, including conventional and vibrational circular dichroism, Raman optical activity, and Fourier transform (FT) infrared spectroscopies (Bertucci, Pistolozzi, & De Simone, 2011; Jiang et al., 2011; Yamamoto, 2012). While these methods may yield relatively rapid spectral markers or "fingerprints" that are related to protein structure, aggregation, and conformational dynamics, none afford the ability to directly correlate spectral features to the structural state(s) and dynamics of a given protein at atomic resolution (Barth, 2007). To facilitate a more precise definition of a protein drug structure and gain a greater confidence in the similarity or dissimilarity between two protein drug species, methods that provide assignment of spectral signals to residue level or better resolution are therefore desirable.

In principle, solution nuclear magnetic resonance (NMR) spectroscopy is ideally suited to the task. Indeed, atomic-level characterization of structure, dynamics, and interactions of small to modest size proteins (<25 kDa) by NMR has become routine since the advent of recombinant protein expression. This success, however, has been largely driven by the use of stable isotope (^{13}C, ^{15}N, ^{2}H) enrichment of protein samples, which allows the complex spectra of proteins to be resolved through multidimensional heteronuclear editing and correlation experiments (Wang et al., 2014). However, stable isotope enrichment of protein therapeutics is neither practical in a pharmaceutical research and development setting, nor is it desirable since it represents a shift away from an approved therapeutic product. Due to the high ^{1}H natural abundance (99.9%) and sensitivity, one- and two-dimensional proton-based methods for the characterization of HOS in biotherapeutic drug products have received consideration and were among the first successful demonstrations of the applicability of NMR for such purposes (Freedberg, 2005; Poppe, Jordan, & Lawson, 2013; Sörgel, Lerch, & Lauber, 2010; Zuperl, Pristovšek, Menart, Gaberc-Porekar, & Novic, 2007). Solid-state methods have also been explored, including ^{14}N quadrupolar correlation spectroscopy (Jarvis, Haies, Williamson, & Carravetta, 2013) and the freezing rotational diffusion of protein solutions at low temperature and high viscosity magic angle spinning technique (FROSTY-MAS), which shows promise for characterization of protein therapeutics in highly concentrated formulation, though it has yet to be fully demonstrated in a pharmaceutical context (Mainz, Jehle, van Rossum, Oschkinat, & Reif, 2009).

Solution methods that are sensitive, well resolved, and report directly on HOS of protein therapeutics in formulation remain desirable. While 2D ^{1}H,^{13}C-methyl and ^{1}H^{N},^{15}N-amide chemical shifts are well established as sensitive reporters of HOS (Amezcua & Szabo, 2013; Panjwani, Hodgson, Sauvé, & Aubin, 2010), ^{13}C and ^{15}N natural abundance spectroscopy has historically been seen as too challenging and impractical for widespread use in the biopharmaceutical setting (Berkowitz et al., 2012). However, with the continuing improvements in ultra-high-field spectrometer strengths, cryogenically cooled probe hardware, and console electronics (Wishart, 2013), the feasibility of performing 2D ^{15}N and ^{13}C correlation spectroscopy on natural isotopic abundance protein therapeutic samples of modest molecular weights (<25 kDa) has been demonstrated (Aubin, Gingras, & Sauvé, 2008; Panjwani et al., 2010). In combination with state-of-the-art rapid acquisition techniques such as selective optimized flip-angle short transient (SOFAST) (Schanda, Kupce, & Brutscher, 2005) and nonuniform sampling (NUS) of the indirect domain, 2D ^{1}H–^{13}C-methyl and $^{1}H^{N}$–^{15}N-amide correlation spectra on protein therapeutic drug products, including mAbs, have also now been shown to be acquired in a time frame of hours to days instead of days to weeks (Arbogast, Brinson, & Marino, 2015; Marino et al., 2015). Here, we outline general consideration for the collection of standard and rapid acquisition ^{13}C and ^{15}N 2D correlation NMR experiments on natural abundance protein therapeutic samples followed by a detailed description of tools for statistical comparability. We illustrate these methods using granulocyte-colony stimulating factor (metG-CSF, known as Filgrastim when expressed from *Escherichia coli*) and enzymatically prepared Fab and Fc domains from the NISTmAb, an IgG1κ, drug-like substance (Arbogast, Brinson, Formolo, Hoopes, & Marino, 2015; Arbogast, Brinson, & Marino, 2015).

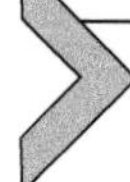

2. PREPARATION OF PROTEIN THERAPEUTIC SAMPLES

2.1 Considerations for Sample Conditions

Ideally, NMR characterization of protein therapeutics would be conducted under native formulation conditions. However, for this to be practical, it is necessary that the protein be at high enough concentration (ca. ≥200 μ*M*, 30 mg/mL for a 150-kDa protein) to achieve adequate sensitivity at natural abundance, although not so concentrated (ca. <50 mg/mL) that high solution viscosity leads to decreased molecular diffusion. The effect of viscosity on an NMR spectrum is clearly visualized in Fig. 1A and B, in which an intact mAb at 100 mg/mL exhibits very broad lines in a ^{1}H–^{15}N SOFAST-heteronuclear multiple quantum coherence (HMQC) experiment, while the same mAb at 30 mg/mL yielded a spectrum with better defined peaks (Marino et al., 2015). For the Fab and Fc domains, better rotational diffusion properties in general allow for the use at higher concentrations, although for the experiments described in this chapter, concentrations were kept between 20 and 40 mg/mL.

The formulation of a mAb may also impact the quality of a spectrum. Many common proton-containing excipients can interfere with the spectral region of interest, leading to obscuration of signals under ridges arising from t_1 noise (Fig. 1B). Worse still, these signals can overload the analog-to-digital converter, leading to clipping of the free induction decay and accompanying spectral distortions following FT of the time-domain signal. Some success has been found in employing diffusion-ordered spectroscopy-type diffusion filters to diminish the effects of excipients (Poppe et al., 2013). For the case

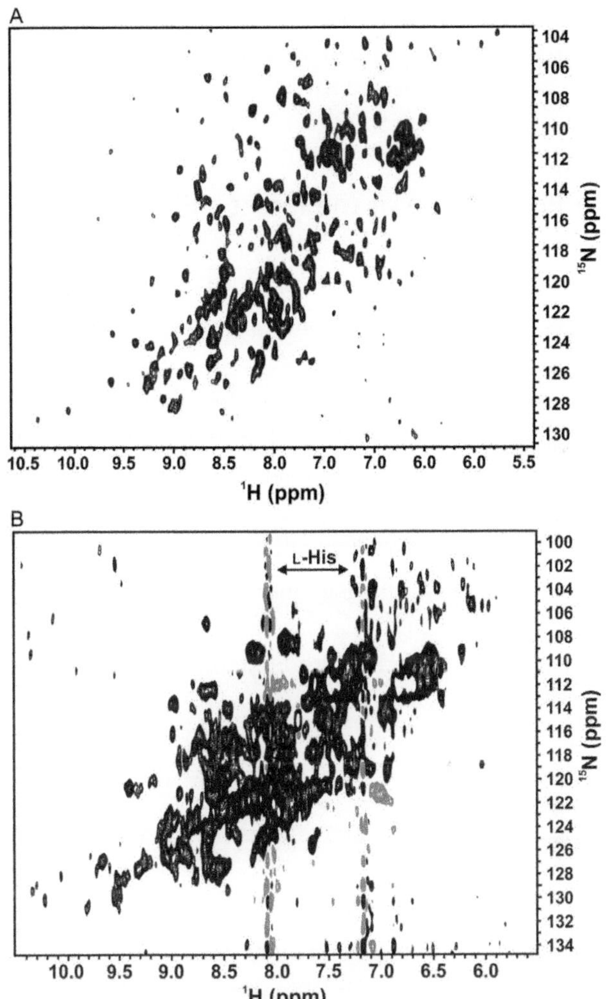

Figure 1 2D ^{1}H–^{15}N SOFAST-HMQC spectra of (A) the intact NISTmAb at 30 mg/mL in 20 m*M* bis–tris-d_{19}, pH 6.0. (B) The intact NISTmAb in a formulation at 100 mg/mL in 25 m*M* L-histidine, pH 6.0. The ridges at 7.2 and 8.1 ppm in ^{1}H in panel B are from the aromatic protons of the histidine buffer. Positive contours are black and negative contours gray. Both spectra were collected at 900 MHz and 50 °C. *Panel A is adapted from published data (Marino et al., 2015).*

of ^{15}N,^{1}H correlations, fewer excipients interfere with the amide region (^{1}H, 6.5–11 ppm), since many are hydrocarbon-based molecules (i.e., polysorbate, sucrose). However, these excipients potentially adversely impact collection of ^{13}C,^{1}H correlations, since protein aliphatic carbons

(^{1}H, −0.5 to 5.5 ppm) fall in approximately the same region as the excipient carbons (^{1}H, 1.0–5.5 ppm). If a mAb molecule can be exchanged into a deuterated analog of the excipient, many of these problems can be greatly ameliorated.

2.2 Sample Preparation of the Intact NISTmAb

The NISTmAb, an IgG1κ antibody that is a drug-like substance donated to NIST, was originally formulated at 100 mg/mL in 25 m*M* L-histidine at pH 6.0. The intact NISTmAb was buffer exchanged twice into 20 m*M* L-histidine-d3 (α-d1; imidazole-2,5-d2) or 20 m*M* bis–tris-d19, both at pH 6.0, using a 100 kDa molecular weight cut-off (MWCO) Amicon (Millipore, Inc.) Ultra-4 centrifugal filter. The sample was concentrated to a final volume of 300 μL (ca. 250 μ*M* NISTmAb) in 3% (9 μL) D_2O.

2.3 Enzymatic Digest of mAbs into Fab and Fc Domains

Cleavage of the NISTmAb into Fab and Fc fragments using immobilized papain (Moorhouse et al., 1997) was carried out with the Pierce Fab Preparation Kit (Thermo Scientific) to produce the Fab and Fc fragments with high fidelity and efficiency. With this approach, two moles of Fab domain is generated for every mole of Fc fragment. The procedure is well documented by the vendor and described elsewhere (Coulter & Harris, 1983; Goding, 1978; Rousseaux, Rousseaux-Prévost, & Bazin, 1983). For typical IgG1κs, the enzymatic cleavage is allowed to proceed for 4–6 h, and the cleaved domains are purified together by removal of immobilized papain using centrifugal filtration. Purification of the individual domains was accomplished using protein A affinity chromatography as described elsewhere (see above references). The Fab fraction was further purified from light chain contaminant produced during cleavage, concentrated and exchanged into the deuterated formulation buffer using a 30-kDa MWCO Amicon filter. Similarly, the Fc fraction was passed through a 100-kDa MWCO Amicon centrifugal filter to remove undigested intact mAb and other higher-order cleavage products followed by concentration and reconstitution in a deuterated formulation buffer using a 30-kDa MWCO Amicon filter. Fidelity of cleavage was confirmed by mass spectrometry (Arbogast, Brinson, Formolo, et al., 2015). All mAb samples (intact and fragments) were loaded into Shaped Tubes (300 μL, 0.25-mm-thick walls; Bruker Biospin) to deliver maximum signal-to-noise (S/N) on AVANCE III Bruker TCI-cryoprobe 900 MHz and 600 MHz platforms, hereafter referred to as cryo900 and cryo600.

Additionally, samples were loaded in 5 mm D_2O susceptibility-matched symmetrical microtubes (300 μL; Shigemi, Inc.) for use on Bruker DMX 500 MHz (RT500) and 600 MHz AVANCE I (RT600) NMR platforms, both equipped with TXI probes to access data quality on room temperature probes. Buffer conditions and concentration of each protein sample are given in figure captions.

2.4 Sample Preparation of Filgrastim

Unlabeled and ^{15}N-labeled recombinant granulocyte-colony stimulating factor (metG-CSF), otherwise known as Filgrastim, was expressed in inclusion bodies in *E. coli* (BL21 DE3 cells; New England BioLabs) that contained the native metG-CSF sequence in pET-21a(+) (Arbogast, Brinson, Formolo, et al., 2015). metG-CSF was refolded from 6 *M* guanidine hydrochloride on a 200-mL G-25 column (GE Healthcare) followed by the equilibration in 20 m*M* sodium acetate, 150 m*M* NaCl, at pH 4.0. metG-CSF was buffer exchanged into 50 m*M* sodium chloride and 10 m*M* sodium acetate-d3, pH 4.0 using a 10-kDa centrifugal filter. Final NMR concentrations were 0.1 and 1.0 m*M*, respectively, for ^{15}N-labeled and unlabeled metG-CSF.

3. ACQUISITION OF 2D SPECTRA

3.1 Amide ^{1}H–^{15}N-Correlated Spectra

2D $^1H^N$,^{15}N-amide correlation spectra are generally considered the "gold standard" for HOS characterization of proteins. Each nonproline residue in a protein gives rise to a cross-peak in the 2D spectrum that reports directly on the chemical environment at the backbone of a given residue. Because of the exquisite sensitivity of the NH chemical shifts to the HOS of proteins, a 2D $^1H^N$,^{15}N-amide correlation plot can serve as a unique marker, or "fingerprint" for a given structural state of a protein (Aubin, Jones, & Freedberg, 2010). These correlation spectra are commonly recorded using heteronuclear single quantum coherence (HSQC) experiments. When combined with pulsed-field gradient coherence selection and sensitivity-enhanced coherence transfer, these HSQC experiments can deliver artifact and phase distortion-free spectra with the highest possible sensitivity (Kay, Keifer, & Saarinen, 1992).

The HSQC experiment has become a mainstay of protein NMR, and both phase-cycled and gradient-selected versions come precoded in current

Bruker pulse sequence 2.0 and higher libraries (*hsqcfpf3gpphwg, hsqcetf3pgsi*). Their implementation has become nearly automated with only minimal optimization of parameters (i.e., proton carrier frequency, pulse widths, and indirect spectral widths) required prior to acquisition. In general, they can readily be acquired and processed by the nonexpert. To compensate for the very low ^{15}N natural abundance (0.37%), experiments must be signal averaged over thousands of scans, and even with a modest indirect acquisition time of 10 ms, spectra can take multiple days to collect with adequate sensitivity. For relatively small protein therapeutics like metG-CSF (18.8 kDa), these standard methods can be readily employed and high-quality spectra can be collected at even modest field strengths (≥500 MHz) (Ghasriani et al., in press) and without the necessity of cryo-probe platforms. For example, on RT600 platform, we have collected a high-sensitivity ^{1}H–^{15}N gsHSQC spectrum of a 1-m*M* unlabeled metG-CSF sample using 4096 scans with acquisition times of 100 and 33 ms over spectral widths of 14 and 32 ppm (1676 × 128) points in the t_2 (^{1}H) and t_1 (^{15}N) domains, respectively, and with a recycle delay of 1 s in 6.5 days. On the cryo900 platform, a spectrum with the same effective resolution can be collected in just 2.5 days using 1024 scans while keeping other acquisition parameters the same (Table 1).

For larger proteins (ca. >30 kDa), however, efficient T_2 transverse nuclear spin relaxation processes diminish resolution and sensitivity of ^{1}H^{N}–^{15}N-amide correlation experiments. For isotopically labeled proteins, specialized transverse relaxation-optimized spectroscopy (TROSY) pulse sequences have been developed to take advantage of destructive interference between different cross-correlated relaxation pathways in the ^{1}H^{N}–^{15}N spin pair, greatly extending the molecular weight limit of NMR spectroscopy (Pervushin, Riek, Wider, & Wüthrich, 1997). These gains from improvement of relaxation characteristics come at the expense of eliminating half of the initial magnetization, a tradeoff that is generally considered necessary for large proteins. However, aliphatic protons drive alternate relaxation pathways which degrade the cross-correlated interference, thereby diminishing the gains from the TROSY effect (Kontaxis, Clore, & Bax, 2000). The efficient relaxation pathway afforded from the aliphatic protons is overcome by perdeuteration of nonlabile protons sites in order to achieve isolated ^{1}H^{N}–^{15}N spin pairs. In a biopharmaceutical setting, with the goal of measuring the HOS of protein therapeutics at natural isotopic abundance, perdeuteration is not a practical solution. While TROSY methods can yield sharper lines for protonated high molecular weight molecules, these

Table 1 Acquisition Parameters for 2D $^{1}H,^{15}N$ Natural Abundance Spectra

Sample	Experiment	Platform	Scans	Recycle Delay (s)	^{1}H Acquisition pts/ Time(ms)/sw (ppm)	^{15}N Acquisition pts/ Time(ms)/sw (ppm)	Experimental Time
Intact mAb	sfHMQC	cryo900	8192	0.4	1260/50/14	60/10/32	2d 16.5 h
Fab	sfHMQC	cryo600	8192	0.4	1536/80/16	52/15/28	2d 11.5 h
Fab	sfHMQC	cryo900	4096	0.4	1260/50/14	62/10/33.5	1d 16 h
Fab	gsHSQC	cryo900	4096	2	2018/80/14	54/10/29.5	5d 8 h
Fc	gsHMQC	cryo900	4096	1.6	1260/50/14	54/10/29.5	4d 7 h
Fc	sfHMQC	cryo600	8192	0.4	1536/80/16	52/15/28	2d 11.5 h
Fc	sfHMQC	cryo900	4096	0.4	1260/50/14	54/10/29.5	1d 5 h
metG-CSF	gsHSQC	RT600	4096	1.0	1676/100/14	128/33/32	6d 20 h
metG-CSF	gsHSQC	RT600	4096	1.0	1676/100/14	128/33/32	6d 20 h
metG-CSF	gsHSQC	cryo600	1536	1.0	2728/162/14	128/33/32	2d 18 h
metG-CSF	gsHSQC	cryo900	1024	1.5	3534/140/14	128/22/32	2d 13 h
metG-CSF	sfHMQC	cryo600	1024	0.4	840/50/14	128/33/32	17.5 h
metG-CSF	sfHMQC	cryo900	1024	0.4	1024/40/14	128/22/32	17 h

methods tend to deliver lower sensitivity and inferior sequence coverage than HMQC/HSQC-type methods.

Similarly, for larger natural abundance proteins, the performance of gradient selection and sensitivity-enhanced (SE) methods (Schleucher et al., 1994) is compromised, since the additional pulses and evolution delays of the SE coherence transfer result in transverse relaxation losses that are likely to overwhelm gains from this approach. While gradient selection best suppresses signals arising from ^{1}H–^{14}N spin pairs and excipients, it comes at a cost of square root of 2 in S/N in the absence of SE coherence transfer. Therefore, phase-cycled coherence selection is generally preferable under these circumstances. Unlike the case with metG-CSF, these larger proteins require higher field strength spectrometers ($\geq$600 MHz) equipped with cryoprobe platforms to overcome the poorer resolution and sensitively needed to achieve useful ^{1}H^N–^{15}N-amide correlation spectra. For example, using the cryo900 platform, we have collected the spectrum of a 500 μM Fab domain of the NISTmAb with 4096 scans and acquisition times of 80 and 10 ms over spectral widths of 14 and 32 ppm (2522 $\times$ 60) pts in the t_2 (^{1}H) and t_1 (^{15}N) dimensions, respectively, with a recycle delay of 2 s in 5 days (Table 2).

To facilitate a more rapid acquisition of 2D ^{1}H^N–^{15}N correlation spectra of larger proteins at natural isotopic abundance, use of specialized pulse sequences designed for protonated proteins such as the SOFAST-HMQC can be employed. By selectively exciting only the amide spectral region of interest, the large reservoir of aliphatic protons remains unperturbed leading to highly efficient return of the ^{1}H^N spin system to equilibrium by T_1 relaxation, thus allowing for a reduction of the interscan recycle delay by a factor of two- to fivefold (Schanda et al., 2005). Although the ^{1}H^N–^{15}N SOFAST-HMQC comes precoded in the current Bruker Topspin 3.0 or greater pulse sequence library (*sfhmqcf3gpph*), it requires additional optimization relative to standard HSQC/HMQC sequences. The default selective pulses are the PC-9 excitation pulse (Kupce & Freeman, 1994) and the r-Snob refocusing pulse (Kupce, Boyd, & Campbell, 1995). For the PC-9 excitation pulse, the flip-angle for the chosen recycle delay must be optimized based upon the appropriate Ernst-angle to deliver maximum sensitivity (Schanda et al., 2005). Additionally, we found the use of a Reburp refocusing pulse (Geen & Freeman, 1991) afforded greater sensitivity in comparison to the r-Snob pulse. After normal optimization of ^{1}H carrier frequency and 90° pulse width, the selective pulse bandwidths and RF-carrier offsets must be determined to adequately cover the spectral region of interest while leaving the water and aliphatic resonances

Table 2 Acquisition Parameters for 2D $^{1}H,^{13}C$ Natural Abundance Spectra

Sample	Experiment	Platform	Scans	Recycle Delay (s)	^{1}H Acquisition pts/Time (ms)/sw (ppm)	^{13}C Acquisition pts/Time (ms)/sw (ppm)	Experimental Time (h)
mAb	gsHSQC	cryo600	128	2	1344/80/14	84/10/28	6.5
mAb	gsHSQC	cryo900	128	2	1766/70/14	102/8/28	7.5
Fc	gsHSQC	RT500	512	2	1398/100/14	72/10/28	22
Fc	gsHSQC	RT600	512	1.5	1340/80/14	84/10/28	19.5
Fc	gsHSQC	cryo600	128	2	1344/80/14	84/10/28	6.5
Fc	gsHSQC	cryo900	64	2	2018/80/14	116/8/32	4.5
Fc	sfHMQC	cryo900	64	0.5	1260/50/14	102/8/28	1
Fab	gsHSQC	cryo600	128	2	2018/120/14	84/10/28	6.5
Fab	gsHSQC	cryo900	64	2	2048/80/14	116/8/32	4.5
Fab	sfHMQC	cryo900	64	0.5	1260/50/14	102/8/28	1

unaffected. Optimization can be performed using a ^{15}N-labeled standard sample and optimized parameters imported into experiments used for natural abundance samples. For the measurements reported herein, typical SOFAST-HMSQC experiments applied a PC-9 90° excitation and Reburp 180° refocusing pulses with bandwidths of 5.5 ppm and the carrier frequency set to 8.25 ppm.

Due of the shortened recycle delays employed in SOFAST experiments, direct acquisition times must be limited when heteronuclear decoupling is employed in order to satisfy cryoprobe duty cycle requirements. For larger proteins with short transverse relaxation time, the acquisition time may be sufficient. However, longer acquisition times may be utilized by switching to an in-phase/antiphase (IP/AP) virtual decoupling scheme rather than a broadband decoupling sequence (Kern, Schanda, & Brutscher, 2008). Such schemes, however, require an IP/AP filter evolution period that extends the pulse sequence, and for large natural abundance proteins, additional relaxation loses during this period may compromise sensitivity.

Representative spectra on the intact NISTmAb and the papain-cleaved Fab and Fc domains were collected using the SOFAST-HMQC approach on the cryo900 and cryo600 spectrometers (Figs. 1A and 2). Experimental parameters are given in Table 1. Although a spectrum of the intact mAb can be recorded using a SOFAST approach in just over 2 days on the cryo900 platform, the unfavorable relaxation characteristics cannot be completely overcome and the spectrum is characterized by poor resolution and is inadequate for fingerprinting purposes (Arbogast, Brinson, Formolo, et al., 2015). However, spectra of the individual domains can be collected in just over 1 day and display reasonable linewidths and many well-resolved peaks, especially in the downfield regions of the spectra. Indeed, for the Fc spectrum, 81% of expected number of peaks can be readily identified overall, while in the well-resolved downfield region this increases to 92% (Marino et al., 2015). On the cryo600 platform, adequate sensitivity can be maintained by collecting domain spectra with double the number of scans (~75% S/N at 900 MHz), and while the expected loss of resolution is apparent, especially in the core region of the spectrum, well-resolved distinguishing features of the spectra are still present, allowing the spectra to serve as benchmarks of HOS.

3.2 Methyl ^{1}H–^{13}C-Correlated Spectra

2D ^{1}H–^{13}C-methyl correlation spectra can provide a second spectral fingerprint for HOS characterization of protein therapeutics. Use of methyl

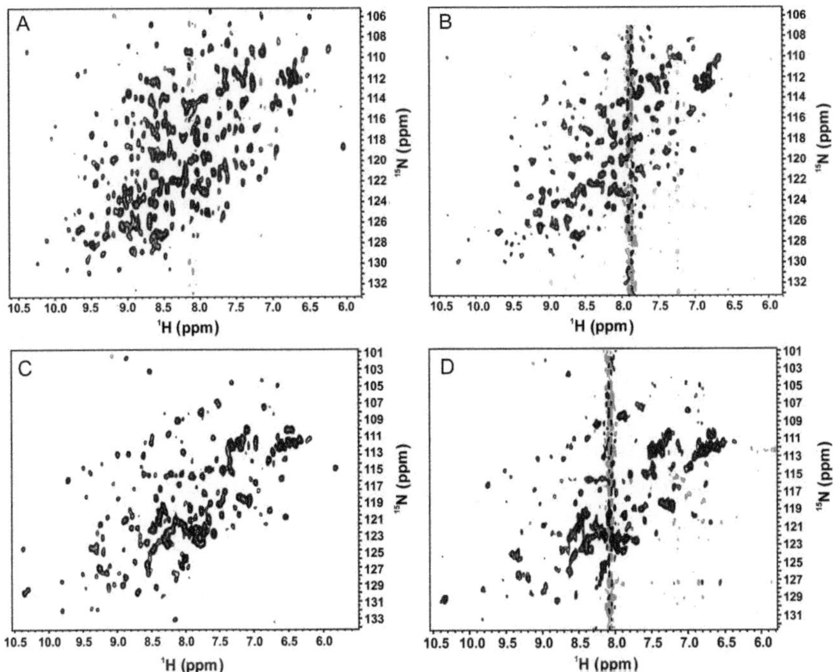

Figure 2 Representative NISTmAb Fab and Fc domain ^{1}H–^{15}N SOFAST-HMQC spectra. (A) Fab spectrum collected at 50 °C with the cryo900. (B) Fab spectrum collected at 45 °C with the cryo600. (C) Fc spectrum collected at 50 °C with the cryo900. (D) Fab spectrum collected at 45 °C with the cryo600. All samples were in 25 m*M* L-histidine-d3, pH 6.0. In panels B and D, the ridges at 7.2 and 8.1 ppm in ^{1}H dimension are from back exchange of the aromatic protons from the histidine buffer. Positive contours are black and negative contours gray. The lower temperature on the cryo600 platform was due to hardware limitations of the cryoprobe.

fingerprints has several distinct advantages over use of $^1H^N$–^{15}N-amide correlations at natural isotopic abundance. Though only six residues bear methyl groups (alanine, leucine, isoleucine, valine, threonine, and methionine), they tend to be well dispersed throughout primary structures and directly report on the nature of the hydrophobic pockets, making methyl groups sensitive reporters of HOS (Rosen et al., 1996). The larger natural abundance of ^{13}C (1.1%) to ^{15}N (0.37%), greater proton-driven polarization in HSQC/HMQC experiments, rapid rotation about the methyl symmetry axis that gives rise to favorable relaxation properties, and lack of exchangeability with solvent make methyl groups a much more sensitive reporter of HOS in natural abundance samples than amide groups (Serber et al., 2004). Furthermore, the increased dispersion and lowered complexity of the methyl spectrum relative that of the amide groups help negate the effects of limited acquisition time in the indirect dimension.

In general, because of the superior spectral properties of methyl groups relative to amide groups, gradient-selected sensitivity-enhanced HSQC (gsHSQC) sequences may be used to acquire high sensitivity and artifact-free spectra with experimental durations of only several hours. Because of the low ^{13}C natural abundance, ^{13}C occupancy at neighboring sites is statistically unlikely, so $^{1}J_{CC}$ couplings are not active, removing the need for constant time ^{13}C virtual homonuclear decoupling schemes. Selection of the methyl groups can be achieved passively by optimization of the INEPT transfer period to maximize transfer efficiency of CH_3 groups ($^{1}J_{CH} \approx 125$ Hz) over CH_2 and CH groups ($^{1}J_{CH} \approx 140$–150 Hz). Less efficient polarization transfer and more efficient relaxation of the latter groups lead to loss of intensity. Still residual signal may still remain slightly downfield of the methyl resonances in one or both dimensions, especially for smaller proteins, and it is necessary to set indirect spectral widths adequately large (ca. 30 ppm) to avoid folding these signals into the methyl region. Though broad and overlapped, these signals may still be used as part of the spectral fingerprint. However, if their removal is desired, either because of overlap with a methyl resonance of interest or to decrease the indirect spectral width to accommodate greater indirect resolution/unit time, multiplicity-edited and separated sequences may be used to actively suppress or filter signals from CH and CH_2 groups (Chen, Freedberg, & Keire, 2015). These sequences require additional evolution periods which may compromise sensitivity and require additional signal averaging, especially for larger proteins. The standard ^{1}H–^{13}C gsHSQC sequences come precoded in the current Bruker Topspin pulse sequence library (*hsqcetfpgpsi2*). In addition to calibration of ^{1}H carrier frequencies and pulse widths, optimization of the INEPT transfer delay to select for methyl groups is required and can be accomplished using a ^{13}C-labeled standard sample.

The superior NMR properties of methyl groups allow for the successful collection of ^{1}H–^{13}C correlation spectra of intact antibodies (Fig. 3). On the cryo900 platform, we have recorded a ^{1}H–^{13}C gsHSQC spectra using 128 scans with acquisition times of 100 and 10 ms over spectral widths of 14 and 27 ppm (22,706 × 124) pts in the t_2 (^{1}H) and t_1 (^{13}C) domains, respectively, with a recycle delay of 2 s in 9.5 h (Arbogast, Brinson, & Marino, 2015). On the other hand, Fab and Fc domain methyl spectra can be collected at fields as low as 500 MHz using room temperature probes (Fig. 3F). Employing the RT600 platform, a ^{1}H–^{13}C gsHSQC spectra of the Fc domain have been

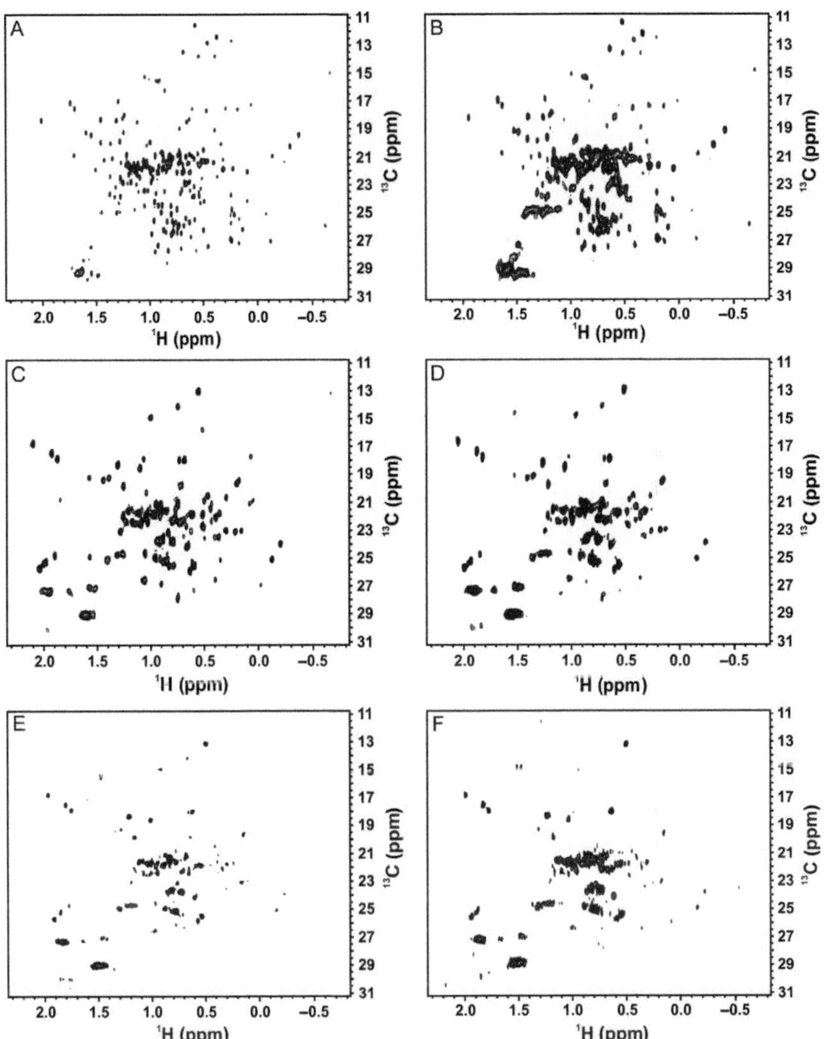

Figure 3 Representative NISTmAb Fab and Fc domain ^{1}H–^{13}C-methyl gsHSQC spectra. (A) Fab spectrum collected at 50 °C with the cryo900. (B) Fab spectrum collected at 45 °C with the cryo600. (C) Fc spectrum collected at 50 °C with the cryo900. (D) Fc spectrum collected at 45 °C with the cryo600. (E) Fc spectrum collected at 50 °C with the RT600. (F) Fc spectrum collected at 50 °C with the RT500. Panels A–D are in 25 m*M* L-histidine-d3, pH 6.0, while panels E and F are in 20 m*M* bis–tris-d_{19}, pH 6.0. The lower temperature on the cryo600 platform was due to hardware limitations of the cryoprobe.

recorded using 512 scans with acquisition times of 100 and 10 ms over spectral widths of 14 and 27 ppm (1436 × 82) pts in the t_2 (^{1}H) and t_1 (^{13}C) domains, respectively, with a recycle delay of 2 s in 18.5 h (Fig. 3E). Using the cryo600 platform, the same spectrum can be collected with only 128 scans in 6 h, while on the cryo900 platform with 64 scans in 4.5 h (Fig. 3D, Table 2).

Like amide groups, methyl groups may also be specifically measured using a SOFAST-HMQC sequence, resulting in further decrease in experimental duration. However, because of the inherently shorter T_1 relaxation time of methyl groups and because amide protons are poorer drivers of relaxation than methyl groups, gains from this approach are generally less than for amides. ^{1}H–^{13}C SOFAST-HMQC sequences are not currently precoded on Bruker Topspin pulse sequence libraries and must be recoded from the ^{1}H–^{15}N SOFAST-HMQC sequences. This can be accomplished by changing the channel of the applied RF of the hard pulses in the indirect dimension from ^{15}N to ^{13}C (i.e., from f3 to f2) in the pulse sequence code and then adjusting the pulse widths and power levels of the indirect pulses for optimized ^{13}C values. Additionally, selective pulses must be optimized for the methyl region in a manner similar to the ^{15}N SOFAST experiments. The selective PC-9 90° excitation and Reburp 180° refocusing pulses were applied at 1.5 ppm with a bandwidth of 5 ppm. Using this approach, we have demonstrated that for mAb Fab and Fc domain, methyl spectra may be recorded with a similar S/N as experiments described above in as little as 1 h using a SOFAST approach (Arbogast, Brinson, & Marino, 2015).

3.3 Nonuniform Sampling Method

While SOFAST methods may be used to reduce experimental duration by speeding up the effective T_1 relaxation rate and thus shortening the necessary interscan delay, increases in S/N per unit time can also be accomplished by employing NUS methods in the indirect evolution domain. In contrast to typical uniform sampling (US) where all incremented points in the indirect domain are sampled in a linear manner, in NUS implementations only a fraction of points in the indirect domain are sampled as a function of a specific nonlinear sample schedule. While a number of sampling schedules have been proposed for NUS, systematic characterization and comparison of their relative performance have only begun to be considered (for a review of NUS strategies, see Maciejewski, Mobli, Schuyler, Stern, & Hoch, 2012 and

Mobli & Hoch, 2014). In the context of natural isotopic abundance, besides our work (Arbogast, Brinson, & Marino, 2015), only one other study employing NUS has yet been reported in the literature, also employing Poisson Gap sinusoidal weighted sampling (Palmer, Wenrich, Stahlfeld, & Rovnyak, 2014). For our part, this schedule was chosen for ease of implementation. Sampling schedules can be generated using the istHMS Poisson Gap Generator applet, and detailed instruction for implementation and processing (see below) on various common spectrometer platforms can be found at http://gwagner.med.harvard.edu/intranet/hmsIST/ (Hyberts, Takeuchi, & Wagner, 2010). For Bruker platforms running Topspin 3.0 or higher, built-in NUS capabilities allow for easy implementation without changes to pulse sequences or other programming requirements. After generation of the sampling schedule using the istHMS applet, the list of points to be sampled can be copied into the topspin "vc" directory as a text file. In the AcquPars window, the FnType must be switched from "traditional (planes)" to "nonuniform sampling" and the sample schedule file name along with sampling density entered under the NUS parameter heading in the AcquPars tab prior to acquisition.

Sampling densities as low as 20% for a $^{1}H^{N}$–^{15}N HSQC have been successfully demonstrated using isotopically labeled samples (Hyberts, Arthanari, & Wagner, 2012). However, for natural abundance samples, it is expected that sampling densities will need to be much higher. For mAb domain samples, we ran gsHSQC experiments from 10% to 50% sampling densities and compared the spectra against that of a uniformly sampled reference spectrum (Fig. 4) (Arbogast, Brinson, & Marino, 2015). From these results, we have found that sampling densities on the order of 30–40% are needed produce a spectrum within reasonable agreement with the uniformly sampled reference spectrum, and 50% NUS maintains full sequence coverage and spectral fidelity without introduction of artifacts (Fig. 4). Using a 50% NUS implementation of a ^{1}H–^{13}C-methyl SOFAST-HMQC, we have shown that a spectrum of a 50 kDa Fc or Fab fragment can be collected in as little as ca. 30 min (Arbogast, Brinson, & Marino, 2015). For ^{1}H–^{15}N-amide SOFAST-HMQC spectra, experimental duration may be shortened to ca. 15 h using 50% NUS. However, given the inherently lower spectral quality of the amide spectra, it may be more advantageous to use NUS to increase sensitivity and/or indirect acquisition times in a time equivalent manner (Arbogast, Brinson, Formolo, et al., 2015).

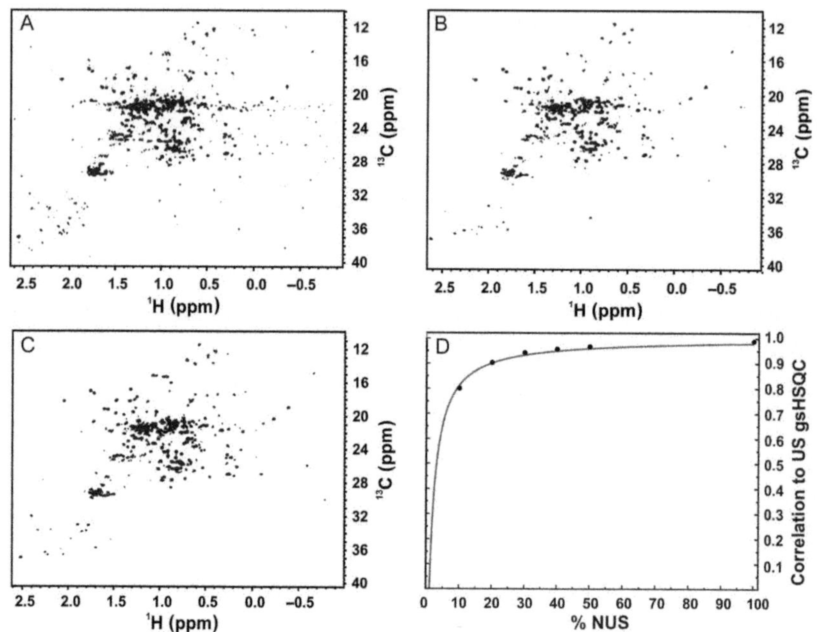

Figure 4 NISTmAb Fab domain ^{1}H–^{13}C-methyl gsHSQC spectra sampled various densities. (A) 30% NUS, (B) 40% NUS, (C) 50% NUS. (D) Correlation of NUS data sets to a uniformly sampled (US) gsHSQC reference experiment as a function of NUS sampling density for IgG1κ mAb Fc and Fab domains. The correlation coefficient, *R*, between the US and NUS data sets can then be plotted and fit to the equation $R = \frac{A[\%NUS-1]}{1+A[\%NUS-1]}$, where *A* is dependent on the experimental precision and is a fit parameter (Bershad & Rockmore, 1974). Data were collected in 25 m*M* L-histidine-d3, pH 6.0 at 37 °C with the cryo900.

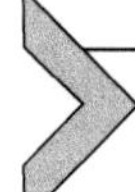

4. PROCESSING 2D-CORRELATED SPECTRA

4.1 Processing Uniformly Sampled Data

Natural abundance spectra can generally be processed in a similar manner to those collected on isotopically labeled samples. However, because of the necessity to impose limited acquisition times in the indirect domain of natural abundance spectra, the choice of data postprocessing parameters is of concern since it will determine linewidths and shapes in the final Fourier transformed spectrum. At minimum, solvent subtraction, baseline correction, apodization, and zero filling in both dimensions are necessary to minimize truncation distortions and artifacts. These can all be applied using precompiled scripts or manually by the nonexpert either using the spectrometer console software supplied by the instrument vendor or using many common third-party software packages, like NMRPipe (Delaglio et al., 1995). In-depth theoretical and practical considerations of these standard processing algorithms have been detailed extensively, and for more information, the reader is directed to these sources (Hoch & Stern, 1996; Rule & Hitchens, 2006). More stringent processing can be applied to increase apparent spectral resolution, which may be advantageous for purposes of tracking peaks of interest in crowded spectral regions or more generally for aesthetic spectral presentation. Forward–backward linear prediction (LP) can be readily employed by the nonexpert on common platforms to double the effective number of points in the indirect dimension to deliver significant gains in spectral resolution (Fig. 5). For data collected using a SOFAST implementation where direct acquisition time is also limited, LP in both domains can be especially useful. Even further gains can be delivered by employing a two-dimensional maximum entropy (ME)

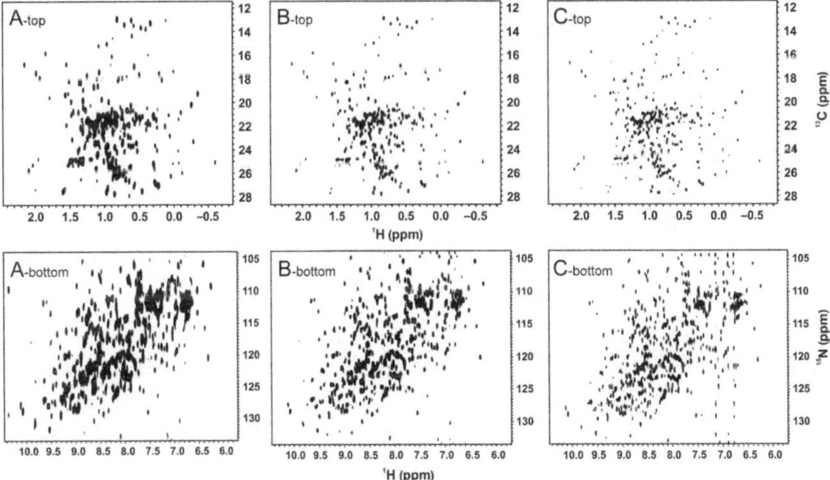

Figure 5 Comparison of the performance of resolution enhancement processing techniques on methyl and amide spectra of the combined Fab/Fc (2:1) domains of the NISTmAb. (A) Zero-filling only, (B) linear prediction, (C) maximum entropy reconstruction. Methyl spectra are labeled as 'top' and amide spectra as 'bottom.' All spectra were processed with polynomial baseline correction and zero filling in both dimensions. Spectra with no resolution enhancement and linear prediction were apodized with shifted sine-squared window function. Linear predicted spectra had time domain doubling in both direction by forward linear prediction. Maximum entropy reconstructed spectra were fit with an exponentially modulated window function with 10-Hz line broadening in the ^{1}H dimension and a shifted sine-squared bell in the indirect dimension.

```
nmrPipe -in test.fid \
| nmrPipe -fn SOL -poly \
| nmrPipe -fn LP -fb -x1 1 -xn SIZE -ord 8 \
| nmrPipe -fn SP -off 0.4 -end 0.98 -pow 2 -c 0.5 \
| nmrPipe -fn ZF -zf 2 -auto \
| nmrPipe -fn FT -auto \
| nmrPipe -fn PS -p0 $1 -p1 $2 -di -verb \
| nmrPipe -fn EXT -x1 -1ppm -xn 2.6ppm -sw \
| nmrPipe -fn TP \
| nmrPipe -fn LP -fb -x1 1 -xn SIZE -ord 8 \
| nmrPipe -fn SP -off 0.4 -end 0.98 -pow 2 -c 0.5 \
| nmrPipe -fn ZF -zf 2 -auto \
| nmrPipe -fn FT \
| nmrPipe -fn PS -p0 $3 -p1 $4 -di -verb \
| nmrPipe -fn EXT -x1 10ppm -xn 38ppm -sw \
| nmrPipe -fn POLY -auto -ord 0 \
| nmrPipe -fn TP \
| nmrPipe -fn POLY -auto -ord 0 \
   -ov -out test.ft2
```

Scheme 1 Example of an NMRPipe processing script which applies linear prediction (bolded) in both dimensions.

reconstruction scheme on the indirect frequency domain signal (Fig. 5). However, more complicated processing algorithms are necessary and the possibility of introducing artifacts is greater; therefore, ME cannot generally be accomplished by the nonexpert. While both LP and ME can deliver significant improvements in resolution and line shape, they are not without their pitfalls; both can introduce errors in line position and intensity, exacerbate the effects of excipient t_1 ridges, and diminish effective S/N and dynamic range. Therefore, their application is generally not recommended for more rigorous chemometric analysis of spectra and should be restricted to qualitative and semi-quantitative analytical applications only. Typical NMRPipe processing script for methyl data using LP and ME reconstruction are given in Schemes 1 and 2.

4.2 Processing Nonuniformly Sampled Data

For data collected using NUS, the choice of reconstruction algorithm will influence the magnitude and extent of artifacts present. In a manner similar to the sampling schedules, a number of reconstruction algorithms have been proposed, although a rigorous comparison of their relative performance has yet to be reported. The current generation of Topspin 3.0 software

```
nmrPipe -in test.fid \
| nmrPipe -fn SOL -poly \
| nmrPipe -fn ZF -zf 2 -auto \
| nmrPipe -fn FT -auto \
| nmrPipe -fn PS -p0 $1 -p1 $2 -di -verb \
| nmrPipe -fn EXT -x1 -1ppm -xn 2.6ppm -sw \
| nmrPipe -fn TP \
| nmrPipe -fn ZF -zf 2 -auto \
| nmrPipe -fn FT - \
| nmrPipe -fn PS -p0 $3 -p1 $4 -di -verb \
| nmrPipe -fn MEM -ndim 2 -pos -sigma $5 -alpha 1e-4 -itmax 100
      -xconv EM -xcQ1 20 -yconv SP -ycQ1 0.5 -ycQ2 0.98 -ycQ3 2 -report 2 \
| nmrPipe -fn EXT -x1 10ppm -xn 38ppm -sw \
| nmrPipe -fn POLY -auto -ord 0 \
| nmrPipe -fn TP \
| nmrPipe -fn POLY -auto -ord 0 \
  -ov -out test.ft2
```

Scheme 2 An example of an NMRPipe processing script which applies maximum entropy (bolded).

incorporates a NUS processing module, although it requires additional software licensing. However, several reconstruction methods can be accomplished using NMRPipe processing scripts. For the purpose of employing NUS to natural abundance 2D NMR experiments, we have found that iterative soft threshold-compressed sensing delivers computationally efficient reconstruction in both the low S/N and high dynamic range regimes of natural abundance protein samples compared to other methods such as ME which is more computationally demanding and tends to amplify t_1 noise and other sampling artifacts at the expense of real, low-intensity threshold peaks (Hyberts, Milbradt, Wagner, Arthanari, & Wagner, 2012). Using the NMRPipe istHMS processing scripts, reconstruction has been easily incorporated into the standard processing pipe routine and can be accomplished by users with a rudimentary understanding of NMRPipe processing. More information about the istHMS processing scripts can be found at the istHMS Website given above.

5. COMPARATIVE ANALYSIS OF 2D-CORRELATED SPECTRA

Ultimately, the goal of many HOS characterization studies of protein therapeutics is the comparison of one of more spectra against a reference spectrum. The simplest and first level analysis is typically a direct visual inspection of the analyte spectrum, generally by overlay with the reference spectrum of interest. If there are differences in peak position that can be discerned by eye after correcting for any referencing offset, then it can be said that the two samples are definitively different in some way, although the interpretation of the difference in terms of structure is not likely to be possible in the absence of resonance assignments. In such comparability exercises, caution needs to be taken to match pH and temperature conditions before assigning any structural basis to differences observed between two spectra. Differences in salt and excipient concentrations may also affect the spectrum; however, these differences generally need to be quite large in magnitude before they visibly alter the spectrum (Panjwani et al., 2010).

While the case for two spectra that are visibly different is somewhat trivial to deal with, spectra that overlay well with one another may require additional analysis, since subtle changes in peak intensity and position that are above experimental imprecision but below easy visual discrimination may be present. Furthermore, choice of a high contour level may mask differences between spectra, and a more thorough analysis may be necessary for an objective review of the data. Because of both the poor expected resolution of natural abundance spectra as a result of the large number of peaks, limited indirect acquisition times, as well as the potential lack of resonance assignments for protein therapeutics targets, methods which do not require the identification of individual peaks are advantageous. One simple and rather elegant method for comparison of 2D methyl spectra of natural abundance protein therapeutics, so-called easy comparison of high-order structure (ECHOS), has recently been proposed (Amezcua & Szabo, 2013), and we have extended a slightly modified ECHOS method to $^{1}H^{N}$–^{15}N-amide correlations and propose the workflow given in Fig. 6 (Arbogast, Brinson, Formolo, et al., 2015).

After spectral processing (Section 4), the first goal is to reduce the 2D spectral matrix, defined by the spectral region of interest in $t_2 \times t_1$ points, into a 1D intensity vector (Fig. 6, point a; Fig. 7; Scheme 3). The NMRPipe "pipe2txt.tcl" script accomplishes this by creating a spectral intensities vector binned by two-dimensional frequency according to the digital resolution of the NMRPipe processed spectrum. The output spectral intensity vector text file can then be further processed in vector manipulation program such as Matlab to facilitate spectral comparison. While the ECHOS method binned intensity vectors to widths greater than the digital resolution of the spectrum to remove the effects of minor solvent composition differences, this

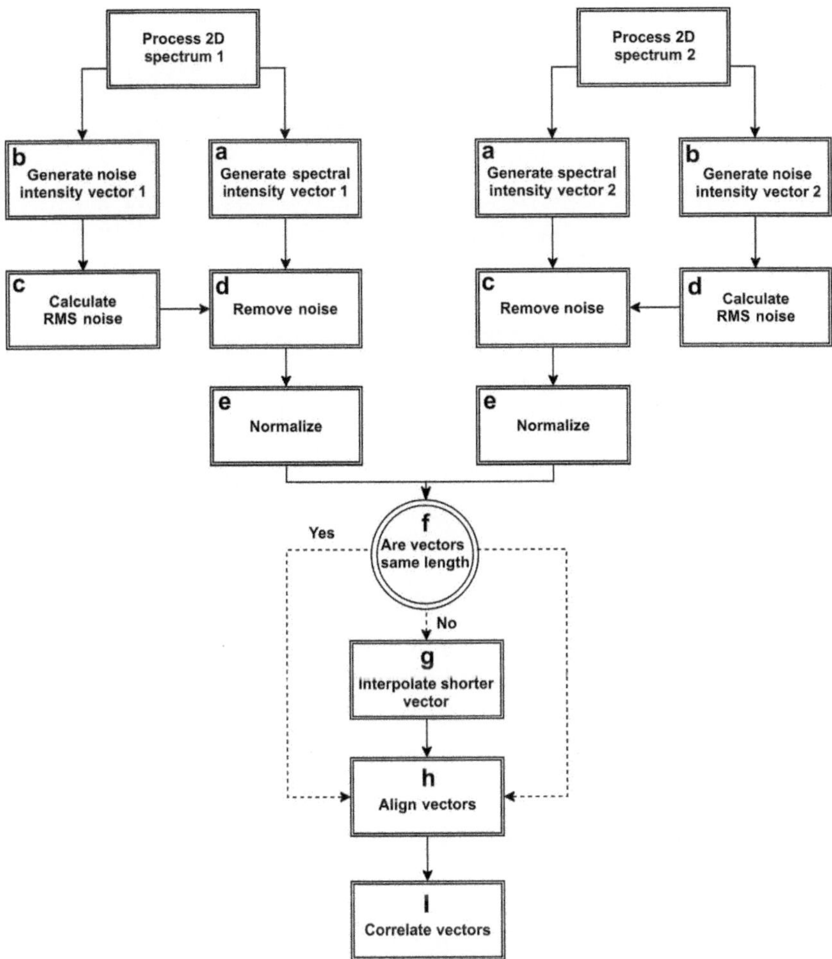

Figure 6 Process flow chart for generating spectral correlations.

effectively discards data by lowering the actual spectral resolution. If a lower resolution analysis is sufficient, acquiring the data at the desired resolution would be more economical and efficient by shorter acquisition times in one or more dimensions. Similarly, postprocessing resolution enhancement (see Section 4) should be limited to avoid introducing additional errors in peak position/intensity and unnecessarily accentuating subtle and structurally meaningless spectral differences. Once the spectral intensity vector is obtained, the noise contribution can be removed via thresholding (Fig. 6, points b–d). To avoid removing significant but low-intensity peaks, conservative application of noise is encouraged. Noise threshold levels can be

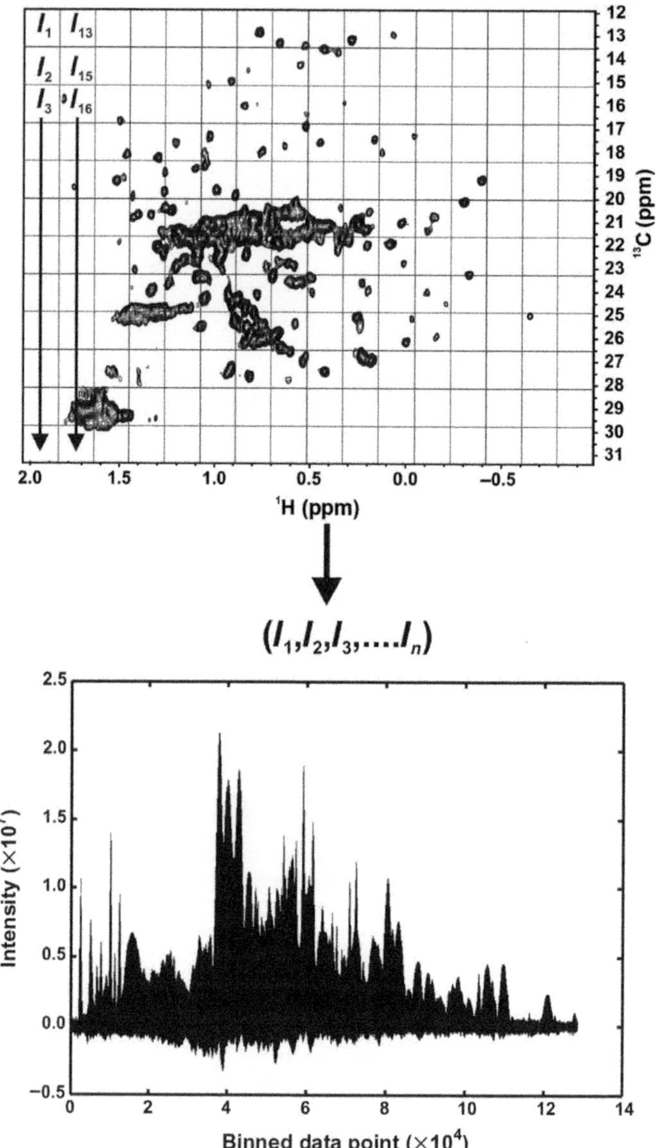

Figure 7 Conversion of a 2D HSQC spectrum into pseudo 1D intensity vector. The maximum intensity of each bin of the 2D grid shown above is taken and arranged as a vector of end-to-end strip plots across the indirect dimension (in this case, ^{13}C) to produce a pseudo 1D intensity spectrum. Creating the pseudo 1D intensity vectors from the direct (^{1}H) dimension does not change the statistical result. For clarity, 2D bins are illustrated at a much lower resolution. In practice, the data was binned at the digital resolution of the spectrum, which was 0.003 ppm for ^{1}H and 0.125 ppm for ^{13}C. The negative intensity is in large part from stochastic noise.

```
% Spec_Cor.m v.2.0 Correlate ASCII pipe2txt file spectral intensity
vectors (SIVs)

% Work Flow outline:
% 1. Import ASCII SIV data from pipe2txt.tcl
% 2. Import ASCII noise vector and remove noise form SIVs
% 3. Normalize SIVs
% 4. Interpolate SIVs
% 5. Align SIVS
% 6. Correlate SIVs and plot data
% 7. Write processed data to ASCII file (optional)

% Begin Program

clear

% 1. Prompt for and import Spectral Intensity Vector data files

SIV1 = importdata(input('Name of 1st SIV file? ')); % prompt for name
of and read 1stSIV file
SIV1=abs(SIV1(:,3));  % remove indexing and take abs value (to
protect negative aliased peaks)

SIV2 = importdata(input('Name of 2nd SIV file? '));
SIV2=abs(SIV2(:,3));

AS1=SIV1;                 % create dummy SIV for S/N calculation
AS2=SIV2;

% 2. Calculate RMS noise and remove from SIVs

NV1=importdata(input('Enter noise file for 1st spectrum ')); % prompt
for noise file for spectrum 1
NV1=NV1(:,3);                   % remove indexing for NV1

avgN1=mean(NV1);                % calculate average NV1 amplitude
noise1=NV1-avgN1;               % mean center noise to correct for any
                                baseline offset
rmsnoise1=rms(noise1);          % calculate RMS noise for NV1

AS1(AS1<rmsnoise1)=[];          % remove noise contribution from SIV1
RMS1=rms(AS1);                  % calculate RMS signal
SN1=RMS1/rmsnoise1              % Calculate and display S/N for spectrum 1


NV2=importdata(input('Enter noise file for 2nd spectrum '));
NV2=NV2(:,3);

avgN2=mean(NV2);
noise2=NV2-avgN2;
rmsnoise2=rms(noise2);

AS2(AS2<rmsnoise2)=[];
RMS2=rms(AS2);
SN2=RMS2/rmsnoise2

SIV1(SIV1<rmsnoise1)=0;         % remove noise by replacing SIV values
                                < RMS noise w/0
SIV2(SIV2<rmsnoise2)=0;

SN_avg=(SN1+SN2)/2              % Calculate average S/N for both SIVs

clear avgN1 avgN2 NV1 NV2 noise1 noise2 % comment out for debug

%3 Normalize vectors

max1=max(SIV1);                 % find maximum intensity of SIV
max2=max(SIV2);
```

Scheme 3 Matlab script for correlating spectral intensity vectors.

(Continued)

```
SIV1n=SIV1/max1;                % normalize intensity vectors
SIV2n=SIV2/max2;

clear SIV1 SIV2 max1 max2       % comment out for debug

% 4. Interpolate smaller file so both are same size

size1=length(SIV1n);            % determine # of data points in file 1
size2=length(SIV2n);

if size1>size2;                 % determine vector with smallest length
  ratio=size2/size1;            % find ratio of SIV lengths
  range_old=transpose(1:1:size2);  % create old index vector for
                                     points in smaller vector
  range_new=transpose(0:ratio:size2); % create new index vector of
                                        length of larger vector (+1)
  range_new(1)=[];              % remove initial zero point in new
                                index vector
  SIV2i=interp1(range_old,SIV2n,range_new,'linear','extrap');  % perform
                                                   linear interpolation
  SIV2i(isnan(SIV2i))=0;        % replace accidental NAN values with
                                zero (not needed with 'extrap' )
  SIV1i=SIV1n;                  % rename larger SIV to match new
                                naming convention
end

if size2>size1;
  ratio=size1/size2;
  range_old=transpose(1:1:size1);
  range_new=transpose(0:ratio:size1);
  range_new(1)=[];
  SIV1i=interp1(range_old,SIV1n,range_new,'linear','extrap');
  SIV1i(isnan(SIV1i))=0;
  SIV2i=SIV2n;
end

if size1 == size2               % if vector lengths are identical no
                                interpolation needed
  SIV2i=SIV2n;                  % rename files to match new naming
                                convention
  SIV1i=SIV1n;
end

clear size1 size2 ratio range_old range_new SIV1n SIV2n

%5. Align input vectors to remove referencing offset

disp('Aligning vectors');

        % align signals using cross-correlation vector, truncate
        non-overlapping regions

[SIV1i,SIV2i]=alignsignals(SIV1i,SIV2i,[],'truncate');

% 6.1 Calculate and display correlation coefficient between input files

R=corr(SIV1i,SIV2i)
```

Scheme 3—Cont'd

```
% 6.2 Plot correlation and overlay

Cor=figure ;
scatter(SIV1i,SIV2i,50,'fill');
hold on
refline(1,0)                    % add 1:1 reference line
axis([0.01,1.1,0.01,1.1])   % set corr plot axis
title(['R=' num2str(R)])      % display correlation coefficient as title
xlabel('SIV1');
ylabel('SIV2');

overlay=[SIV1i SIV2i];         % Plot pseudo 1D overlay of SVIs
ol=figure;
plot(overlay);
hold on
legend('SIV1', 'SIV2');
xlabel('Binned data points (#)');
ylabel('Normalized intensity');

% 7. Write output files of binned data/print figures

figs=input('Print figures? 1=yes 2=no (default) ');
if figs==1
  [figname1]=input('overlay plot file name ');
  saveas(ol,figname1,'eps');
  [figname2]=input('cor plot file name ');
  saveas(Cor,figname2,'eps');

  clear figs prompt figname1 figname2
end

out=input('Write output bin files? 1=yes 2=no (default) ') ;
if out == 1
  [outname1]=input('SIV1 output name ');
  dlmwrite(outname1,SIV1i);
  [outname2]=input('SIV2 output name ');
  dlmwrite(outname2,SIV2i);
clear outname1 outname2
end
```

Scheme 3—Cont'd

obtained by generating a spectral intensity vector for a spectral region only containing noise and computing the RMS amplitude of the noise vector. Spectral intensity vector values between ± the RMS noise amplitude are then replaced by zero values. After noise subtraction, effects of potential sample concentration differences are minimized by normalizing all intensity vectors to the maximum vector intensity (Fig. 6, point e). Once spectral intensity vectors are processed for spectrum 1 and spectrum 2, the total number of points in each 1D vector must be compared to assure they have identical lengths (Fig. 6, point f). Ideally, the acquisition and processing parameters of the spectra to be compared would be identical, and no correction would be necessary. However, if data were collected at different acquisition times on different instruments, parameters may not be identically matched and vector lengths may be slightly different. In this case, interpolation of the shorter vector using a 2-point nearest neighbor fit to increase the length of the vector while maintaining its range can be used as a correction (Fig. 6, point g). After linear interpolation has equalized the length of the vectors, they must be aligned to remove referencing offsets, which may be accomplished by maximization of the two vectors cross-correlation coefficient (Fig. 6, point h). Finally, the vectors may be correlated by linear regression to give a Pearson's correlation coefficient (Fig. 6, point i). Examples of these plots are given in Fig. 8.

In order to derive meaningful analysis from the correlation coefficient, it is necessary to calibrate the expectation in terms of S/N, which dictates the maximum possible correlation between species and resolution, which dictates the minimum observable correlation. For a given experiment and conditions (i.e., SOFAST-HMQC, gsHSQC, and US, NUS), the expectation in terms of S/N can be calibrated using an isotopically labeled reference standard by collecting a series of duplicate data sets where the number of scans is incremented as described in our previous report (Arbogast, Brinson, Formolo, et al., 2015). The tolerance for statistical precision and the S/N, resolution, and total experimental time will depend on exact need of a given product application.

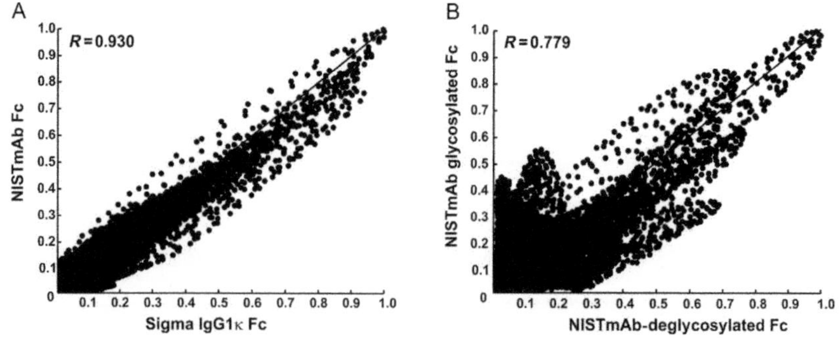

Figure 8 Total spectral correlation plots between (A) the NISTmAb Fc domain and another IgG1κ Fc domain from a polyclonal antibody supplied by Sigma-Aldrich and (B) the fully glycosylated NISTmAb Fc domain and the enzymatically deglycosylated NISTmAb Fc domain. All data were obtained from spectra collected at 50 °C with the cryo900. *Total spectral correlation plots are adapted from Arbogast et al. (2015).*

Although we have outlined a general procedure for the comparison of two spectra, if a comparison of a larger sample set is desired, the same methods can be used to produce and process spectral intensity vectors of many spectra, which may then be subjected to higher-order statistical analyses, such as principal component analysis (Ghasriani et al., in press). These details are beyond the scope of this chapter.

6. PERSPECTIVES

The ability to perform 2D ^{1}H–^{13}C-methyl and ^{1}H–^{15}N-amide NMR on natural abundance protein therapeutic samples has now been firmly established for proteins as large as intact mAbs using state-of-the-art spectrometer platforms and spectroscopic methodologies, while proteins of a more moderate size can be routinely analyzed using more conventional platforms. While we have outlined here considerations and methods employed for the successful collection and analysis of the spectra of mAbs and their constituent domains using a variety of techniques and spectrometer setups, we anticipate the further refinement of the methodologies as their usage become more widespread and consensus on best practices can be reached. In particular, it is expected that there will be continued development of methods for the collection of spectra with higher sensitivity and resolution in shorter time frames that will allow for even greater adoption of 2D NMR for HOS characterization of biotherapeutics as well as application to other protein classes and targets. Similarly, adaptation and development of chemometric methods for quantitative analysis and comparison, already well established in the metabolomics NMR community, will allow 2D NMR to become a more routine and indispensable tool in the analytical chemist's tool box for HOS characterization of protein therapeutics.

Note: Certain commercial equipment, instruments, and materials are identified in this chapter in order to specify the experimental procedure. Such identification does not imply recommendation or endorsement by the National Institute of Standards and Technology, nor does it imply that the material or equipment identified is necessarily the best available for the purpose.

ACKNOWLEDGMENTS

We acknowledge the support by NIST Biomanufacturing Initiative, NIST, and W.M. Keck for support of Biomolecular NMR instrumentation. L.W.A. acknowledges a National Research Council postdoctoral fellowship.

REFERENCES

Amezcua, C., & Szabo, C. (2013). Assessment of higher order structure comparability in therapeutic proteins using nuclear magnetic resonance spectroscopy. *Journal of Pharmaceutical Sciences, 102*, 1724–1733.

Arbogast, L. W., Brinson, R. G., Formolo, T., Hoopes, J. T., & Marino, J. P. (2015). 2D 1HN, 15N correlated NMR methods at natural abundance for obtaining structural maps and statistical comparison of monoclonal antibodies. *Pharmaceutical Research*. http://dx.doi.org/10.1007/s11095-015-1802-3.

Arbogast, L. W., Brinson, R. G., & Marino, J. P. (2015). Mapping monoclonal antibody structure by 2D 13C NMR at natural abundance. *Analytical Chemistry, 87*, 3556–3561.

Aubin, Y., Gingras, G., & Sauvé, S. (2008). Assessment of the three-dimensional structure of recombinant protein therapeutics by NMR fingerprinting: Demonstration on recombinant human granulocyte macrophage-colony stimulation factor. *Analytical Chemistry, 80*, 2623–2627.

Aubin, Y., Jones, C., & Freedberg, D. I. (2010). Using NMR spectroscopy to obtain the higher order structure of biopharmaceutical products: Simple methods can characterize polysaccharide vaccines and recombinant cytokines at high resolution. *Biopharm International*, 28–34.

Barth, A. (2007). Infrared spectroscopy of proteins. *Biochimica et Biophysica Acta, 1767*, 1073–1101.

Berkowitz, S. A., Engen, J. R., Mazzeo, J. R., & Jones, G. B. (2012). Analytical tools for characterizing biopharmaceuticals and the implications for biosimilars. *Nature Reviews. Drug Discovery, 11*, 527–540.

Bershad, N., & Rockmore, A. (1974). On estimating signal-to-noise ratio using the sample correlation coefficient (Corresp.). *IEEE Transactions on Information Theory, 20*, 112–113.

Bertucci, C., Pistolozzi, M., & De Simone, A. (2011). Structural characterization of recombinant therapeutic proteins by circular dichroism. *Current Pharmaceutical Biotechnology, 12*, 1508–1516.

Chen, K., Freedberg, D. I., & Keire, D. A. (2015). NMR profiling of biomolecules at natural abundance using 2D 1H–15N and 1H–13C multiplicity-separated (MS) HSQC spectra. *Journal of Magnetic Resonance, 251*, 65–70.

Coulter, A., & Harris, R. (1983). Simplified preparation of rabbit Fab fragments. *Journal of Immunological Methods, 59*, 199–203.

Delaglio, F., Grzesiek, S., Vuister, G. W., Zhu, G., Pfeifer, J., & Bax, A. (1995). NMRPipe: A multidimensional spectral processing system based on UNIX pipes. *Journal of Biomolecular NMR, 6*, 277–293.

Freedberg, D. I. (2005). Using nuclear magnetic resonance spectroscopy to characterize biologicals. *Developments in Biologicals, 122*, 77–83.

Geen, H., & Freeman, R. (1991). Band-selective radiofrequency pulses. *Journal of Magnetic Resonance, 93*, 93–141.

Ghasriani, H., Hodgson, D. J., Brinson, R. G., McEwen, I., Buhse, L., Kozlowski, S., et al. (in press). Demonstration of the precision and robustness of 2D-NMR for higher-order structure assessment of filgrastim. *Nature Biotechnology*.

Goding, J. W. (1978). Use of staphylococcal protein A as an immunological reagent. *Journal of Immunological Methods, 20*, 241–253.

Hoch, J. C., & Stern, A. S. (1996). *NMR data processing*. New York: John Wiley & Sons, Inc.

Hyberts, S. G., Arthanari, H., & Wagner, G. (2012). Applications of non-uniform sampling and processing. *Topics in Current Chemistry, 316*, 125–148.

Hyberts, S. G., Milbradt, A. G., Wagner, A. B., Arthanari, H., & Wagner, G. (2012). Application of iterative soft thresholding for fast reconstruction of NMR data non-uniformly sampled with multidimensional Poisson Gap scheduling. *Journal of Biomolecular NMR*, *52*, 315–327.

Hyberts, S. G., Takeuchi, K., & Wagner, G. (2010). Poisson-gap sampling and forward maximum entropy reconstruction for enhancing the resolution and sensitivity of protein NMR data. *Journal of the American Chemical Society*, *132*, 2145–2147.

Jarvis, J. A., Haies, I. M., Williamson, P. T. F., & Carravetta, M. (2013). An efficient NMR method for the characterisation of 14N sites through indirect 13C detection. *Physical Chemistry Chemical Physics*, *15*, 7613–7620.

Jiang, Y., Li, C., Nguyen, X., Muzammil, S., Towers, E., Gabrielson, J., et al. (2011). Qualification of FTIR spectroscopic method for protein secondary structural analysis. *Journal of Pharmaceutical Sciences*, *100*, 4631–4641.

Kay, L. E., Keifer, P., & Saarinen, T. (1992). Pure absorption gradient enhanced heteronuclear single quantum correlation spectroscopy with improved sensitivity. *Journal of the American Chemical Society*, *114*, 10663–10665.

Kern, T., Schanda, P., & Brutscher, B. (2008). Sensitivity-enhanced IPAP-SOFAST-HMQC for fast-pulsing 2D NMR with reduced radiofrequency load. *Journal of Magnetic Resonance*, *190*, 333–338.

Kontaxis, G., Clore, G. M., & Bax, A. (2000). Evaluation of cross-correlation effects and measurement of one-bond couplings in proteins with short transverse relaxation times. *Journal of Magnetic Resonance*, *143*, 184–196.

Kupce, E., Boyd, J., & Campbell, I. D. (1995). Short selective pulses for biochemical applications. *Journal of Magnetic Resonance, Series B*, *106*, 300–303.

Kupce, E., & Freeman, R. (1994). Wideband excitation with polychromatic pulses. *Journal of Magnetic Resonance, Series A*, *108*, 268–273.

Maciejewski, M. W., Mobli, M., Schuyler, A. D., Stern, A. S., & Hoch, J. C. (2012). Data sampling in multidimensional NMR: Fundamentals and strategies. *Topics in Current Chemistry*, *316*, 49–77.

Mainz, A., Jehle, S., van Rossum, B. J., Oschkinat, H., & Reif, B. (2009). Large protein complexes with extreme rotational correlation times investigated in solution by magic-angle-spinning NMR spectroscopy. *Journal of the American Chemical Society*, *131*, 15968–15969.

Marino, J. P., Brinson, R. G., Hudgens, J. W., Ladner, J. E., Gallagher, D. T., Gallagher, E. S., et al. (2015). Emerging technologies to assess the higher-order structure of monoclonal antibodies. In J. E. Schiel, D. D. Davis, & O. V. Borisov (Eds.), *State-of-the-art and emerging technologies for therapeutic monoclonal antibody characterization: Vol. 3.* (pp. 17–43). WashingtonDC: American Chemical Society. Chapter 2.

Mobli, M., & Hoch, J. C. (2014). Nonuniform sampling and non-Fourier signal processing methods in multidimensional NMR. *Progress in Nuclear Magnetic Resonance Spectroscopy*, *83*, 21–41.

Moorhouse, K. G., Nashabeh, W., Deveney, J., Bjork, N. S., Mulkerrin, M. G., & Ryskamp, T. (1997). Validation of an HPLC method for the analysis of the charge heterogeneity of the recombinant monoclonal antibody IDEC-C2B8 after papain digestion. *Journal of Pharmaceutical and Biomedical Analysis*, *16*, 593–603.

Palmer, M. R., Wenrich, B. R., Stahlfeld, P., & Rovnyak, D. (2014). Performance tuning non-uniform sampling for sensitivity enhancement of signal-limited biological NMR. *Journal of Biomolecular NMR*, *58*, 303–314.

Panjwani, N., Hodgson, D. J., Sauvé, S., & Aubin, Y. (2010). Assessment of the effects of pH, formulation and deformulation on the conformation of interferon alpha-2 by NMR. *Journal of Pharmaceutical Sciences*, *99*, 3334–3342.

Pervushin, K. V., Riek, R., Wider, G., & Wüthrich, K. (1997). Attenuated T2 relaxation by mutual cancellation of dipole-dipole coupling and chemical shift anisotropy indicates an avenue to NMR structures of very large biological macromolecules in solution. *Proceedings of the National Academy of Sciences of the United States of America*, *94*, 12366–12371.

Poppe, L., Jordan, J., & Lawson, K. (2013). Profiling formulated monoclonal antibodies by 1H NMR spectroscopy. *Analytical Chemistry, 85*, 9623–9629.

Rosen, M. K., Gardner, K. H., Willis, R. C., Parris, W. E., Pawson, T., & Kay, L. E. (1996). Selective methyl group protonation of perdeuterated proteins. *Journal of Molecular Biology, 263*, 627–636.

Rousseaux, J., Rousseaux-Prévost, R., & Bazin, H. (1983). Optimal conditions for the preparation of Fab and F(ab')2 fragments from monoclonal IgG of different rat IgG subclasses. *Journal of Immunological Methods, 64*, 141–146.

Rule, G. S., & Hitchens, T. K. (2006). *Fundamentals of protein NMR spectroscopy*. Dordrecht, Netherlands: Springer.

Schanda, P., Kupce, E., & Brutscher, B. (2005). SOFAST-HMQC experiments for recording two-dimensional heteronuclear correlation spectra of proteins within a few seconds. *Journal of Biomolecular NMR, 33*, 199–211.

Schleucher, J., Schwendinger, M., Sattler, M., Schmidt, P., Schedletzky, O., Glaser, S. J., et al. (1994). A general enhancement scheme in heteronuclear multidimensional NMR employing pulse field gradients. *Journal of Biomolecular NMR, 4*, 301–306.

Serber, Z., Straub, W., Corsini, L., Nomura, A. M., Shimba, N., Craik, C. S., et al. (2004). Methyl groups as probes for proteins and complexes in in-cell NMR experiments. *Journal of the American Chemical Society, 126*, 7119–7125.

Sörgel, F., Lerch, H., & Lauber, T. (2010). Physicochemical and biologic comparability of a biosimilar granulocyte colony-stimulating factor with its reference product. *BioDrugs: Clinical Immunotherapeutics, Biopharmaceuticals and Gene Therapy, 24*, 347–357.

Wang, G., Zhang, Z.-T., Jiang, B., Zhang, X., Li, C., & Liu, M. (2014). Recent advances in protein NMR spectroscopy and their implications in protein therapeutics research. *Analytical and Bioanalytical Chemistry, 406*, 2279–2288.

Wishart, D. S. (2013). Characterization of biopharmaceuticals by NMR spectroscopy. *TrAC Trends in Analytical Chemistry, 48*, 96–111.

Yamamoto, S. (2012). Conformational analyses of peptides and proteins by vibrational Raman optical activity. *Analytical and Bioanalytical Chemistry, 403*, 2203–2212.

Zuperl, Š., Pristovšek, P., Menart, V., Gaberc-Porekar, V., & Novic, M. (2007). Chemometric approach in quantification of structural identity/similarity of proteins in biopharmaceuticals. *Journal of Chemical Information and Modeling, 47*, 737–743.

CHAPTER TWO

A Semiautomated Assignment Protocol for Methyl Group Side Chains in Large Proteins

Jonggul Kim*, Yingjie Wang*, Geoffrey Li*, Gianluigi Veglia*,†,1

*Department of Chemistry, University of Minnesota, Minneapolis, Minnesota, USA

†Department of Biochemistry, Molecular Biology, and Biophysics, University of Minnesota, Minneapolis, Minnesota, USA

[1]Corresponding author: e-mail address: vegli001@umn.edu

Contents

Abstract

The developments of biosynthetic specific labeling strategies for side-chain methyl groups have allowed structural and dynamic characterization of very large proteins and protein complexes. However, the assignment of the methyl-group resonances remains an Achilles' heel for NMR, as the experiments designed to correlate side chains to the protein backbone become rather insensitive with the increase of the transverse relaxation rates. In this chapter, we outline a semiempirical approach to assign the resonances of methyl-group side chains in large proteins. This method requires a crystal structure or an NMR ensemble of conformers as an input, together with NMR data sets such as nuclear Overhauser effects (NOEs) and paramagnetic relaxation enhancements (PREs), to be implemented in a computational protocol that provides a probabilistic

Methods in Enzymology, Volume 566
ISSN 0076-6879
http://dx.doi.org/10.1016/bs.mie.2015.08.033

assignment of methyl-group resonances. As an example, we report the protocol used in our laboratory to assign the side chains of the 42-kDa catalytic subunit of the cAMP-dependent protein kinase A. Although we emphasize the labeling of isoleucine, leucine, and valine residues, this method is applicable to other methyl group side chains such as those of alanine, methionine, and threonine, as well as reductively methylated cysteine side chains.

1. INTRODUCTION

Isotopic labeling is at the heart of NMR spectroscopy. After the pioneering work in the early 1960s by several research groups (Crespi & Katz, 1969; Crespi, Rosenberg, & Katz, 1968; Markley, Putter, & Jardetzky, 1968; Putter, Barreto, Markley, & Jardetzky, 1969), more complex labeling schemes have been used to attenuate transverse relaxation and push the boundaries of NMR analysis of large macromolecular complexes (Kainosho et al., 2006; Kasinath, Valentine, & Wand, 2013; Meissner & Sorensen, 1999). A significant step forward was made when nonlabile protons in proteins were substituted by deuterons, ameliorating the dipolar relaxation due to ^{1}H–^{13}C, ^{1}H–^{15}N, and ^{1}H–^{1}H interactions, increasing sensitivity and resolution, thereby rendering proteins larger than 20 kDa amenable to structural NMR studies (Nietlispach et al., 1996; Venters et al., 1995). Although complete deuteration has been used for backbone assignments in conjunction with transverse relaxation optimized spectroscopy (TROSY) (Pervushin, Riek, Wider, & Wüthrich, 1997), NOESY-based experiments benefit from incomplete (or fractional) deuteration, which preserves some protons for distance measurements (Nietlispach et al., 1996). Another quantum leap in the NMR structure determination of large complexes was made by Kay and coworkers, who developed a biosynthetic strategy for ^{1}H,^{13}C methyl group labeling in a highly deuterated background (Gardner & Kay, 1997; Goto, Gardner, Mueller, Willis, & Kay, 1999). Since then, this labeling strategy has expanded to include selective methyl group labeling on isoleucine, valine, leucine, methionine (Gelis et al., 2007), threonine (Velyvis, Ruschak, & Kay, 2012), and alanine (Ayala, Sounier, Use, Gans, & Boisbouvier, 2009). Although selective labeling schemes for aromatic side chains will certainly have an impact both in structure determination and characterization of dynamics large proteins (Kasinath et al., 2013), here we will focus on the use of methyl groups for the spectroscopy of large systems (Tugarinov, Hwang, Ollerenshaw, & Kay, 2003) akin to the TROSY method developed for backbone amides (Pervushin et al., 1997). Following these methyl labeling schemes, quantitative studies on binding, structure, and conformational dynamics of proteins of several hundred kDa are emerging, such as large molecular machinery (Religa, Sprangers, & Kay, 2010; Ruschak, Religa, Breuer, Witt, & Kay, 2010; Sprangers & Kay, 2007), allosteric enzymes (Shi & Kay, 2014), chaperones (Saio, Guan, Rossi, Economou, & Kalodimos, 2014), and protein thermodynamics (Tzeng & Kalodimos, 2012).

2. LABELING OF SIDE CHAIN METHYL GROUPS FOR LARGE PROTEINS

The first methyl labeling scheme was published in 1997 by Gardner and Kay (1997), who obtained selective protonation of amino acids in which the direct biosynthetic precursor to isoleucine, 2-ketobutyrate, was enzymatically and chemically prepared with selective $^{13}CH_3$ labeling in a highly deuterated background. Exploiting the *Escherichia coli* biosynthetic pathway, this protocol enables the specific $^{13}CH_3$ labeling of the Cδ of Ile, while the remaining nonlabile protons are replaced by deuterons. The incorporation of 2-ketoisovalerate, in a similar fashion, leads to selective labeling on the Cδ and Cγ of leucine and valine, respectively (Goto et al., 1999), resulting in the so-called "ILV labeling" scheme (Fig. 1). Nowadays, these biosynthetic precursors are commercially available in a variety of labeling schemes for assignment, structure determination, and dynamics studies. Selective labeling of the Cγ methyl of isoleucine has also been devised (Ruschak, Velyvis, & Kay, 2010b), but is uncommon due to the superior spectral qualities of the Cδ methyl group. Usually, selective labeling of other amino acids is achieved through direct addition of the amino acid to the growth medium during cell growth. For alanine and methionine, this is most commonly accomplished by directly incorporating the amino acid prior to induction (30 min to 1 h) at final concentrations of 100–250 mg/L (Saio et al., 2014; Tzeng & Kalodimos, 2012) for Met and from 100–800 mg/L for alanine (Ayala et al., 2009; Tzeng & Kalodimos, 2012). An alternate approach is to engineer cysteine and react with ^{13}C-methyl-methanethiosulfonate (^{13}C-MMTS) to produce a methionine mimic (Religa, Ruschak, Rosenzweig, & Kay, 2011). Kay and coworkers have also developed an *in vitro* biosynthetic method to produce U-[^{2}H],Thr-γ2[$^{13}CH_3$] ($^{13}CH_3$ labeled threonine in a deuterated background) (Velyvis et al., 2012). Threonine is a direct biosynthetic precursor to 2-ketobutyrate, leading to isoleucine methyl (Cδ) labeling. Since

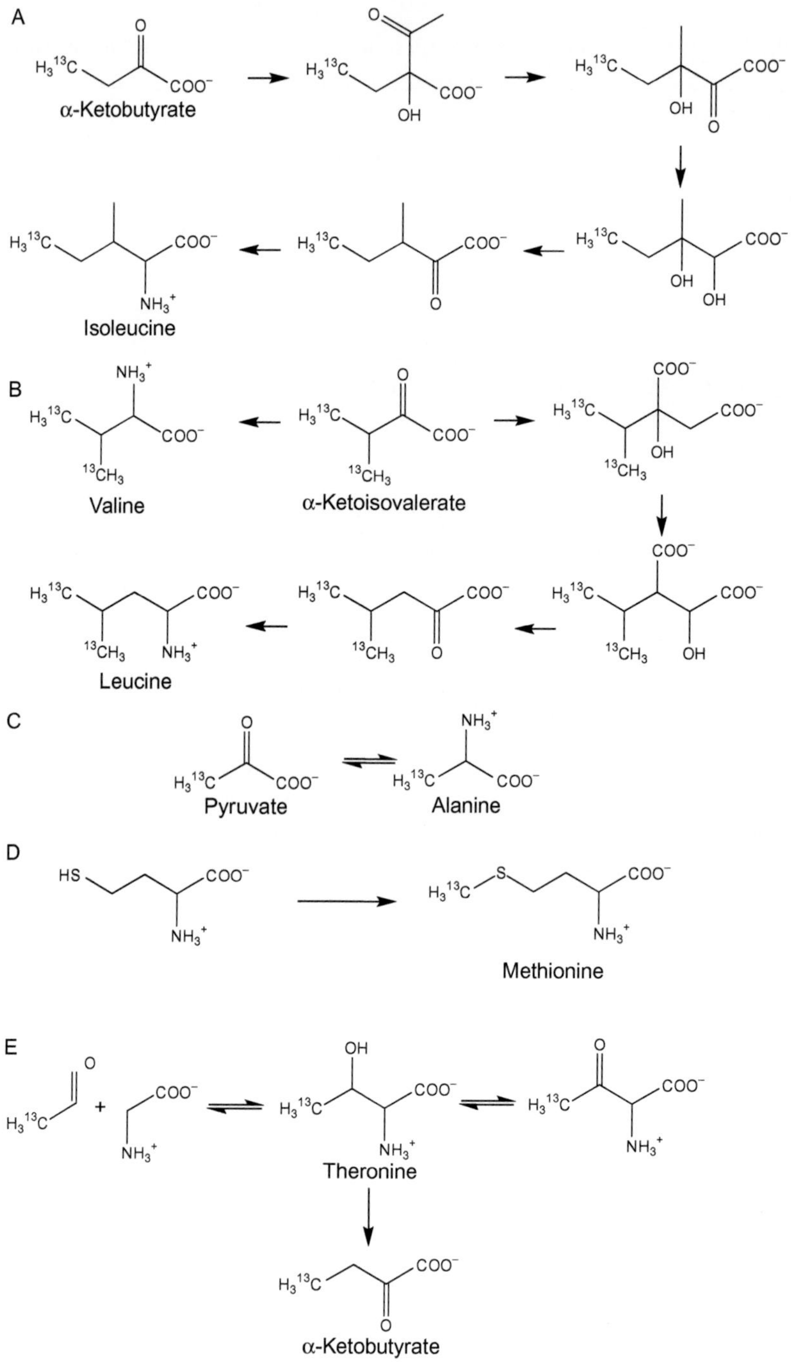

A

α-Ketobutyrate

Isoleucine

B

Valine

α-Ketoisovalerate

Leucine

C

Pyruvate

Alanine

D

Methionine

E

Theronine

α-Ketobutyrate

Figure 1 Biosynthetic pathways for the specific labeling of (A) isoleucine, (B) leucine and valine, (C) alanine, (D) methionine, and (E) threonine. Note the off-pathway dilution of alanine and threonine amino acids.

threonine is converted to glycine either from threonine aldolase or from threonine dehydrogenase/2-amino-3-ketobutyrate ligase, the authors used 100 mg/L of d_5-glycine along with 50 mg/L of U-[^{2}H], Thr-γ2[$^{13}CH_3$] and 50 mg/L of 2-ketobutyrate for specific threonine and isoleucine labeling, respectively (Fig. 1).

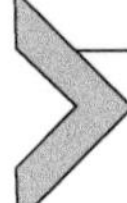

3. METHYL LABELING PROTOCOL FOR THE cAMP-DEPENDENT PROTEIN KINASE A

In this outline, we highlight a method to label the ^{13}C isoleucine (Cδ), leucine, and valine methyl groups for the catalytic subunit of cAMP-dependent protein kinase A (PKA-C) (Fig. 2). Specific labeling of side-chain methyl groups is achieved using the M9 minimal medium typically used for ^{15}N labeling of proteins with the addition of appropriate amounts of methyl labeling precursors. As the protocols for overexpression and purification of PKA-C have been described extensively (Hemmer, McGlone, & Taylor, 1997; Narayana, Cox, Shaltiel, Taylor, & Xuong, 1997; Yonemoto et al., 1991), here we report only the main steps of the protocol (Fig. 3). Note that for binding titrations studies we used fully protonated kinase and 2D NMR experiments; however, for methyl group assignment experiments (i.e., 3D [^{13}C,^{13}C]-methyl NOESY experiments) we utilized fractionally deuterated kinase expressed in 80% D_2O. In our hands, the utilization of fully per-deuterated media resulted in a scarce expression of the kinase. Nonetheless, for larger proteins with molecular weights greater than 100 kDa, high levels of deuteration are often necessary. In the latter case, overexpression of proteins in 100% D_2O is highly strain-, vector-, and protein-dependent

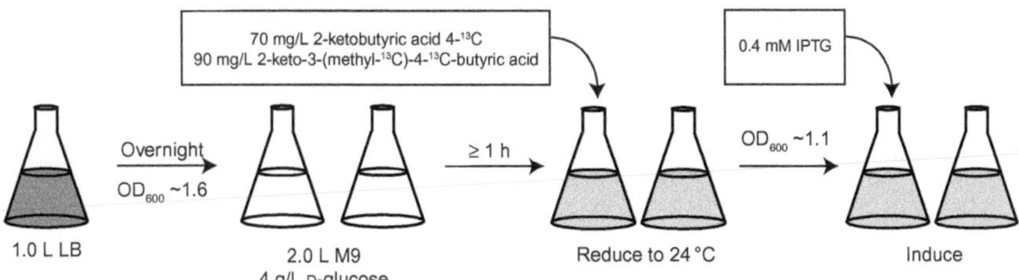

Figure 2 General growth scheme used to express ^{15}N ^{13}C-ILV-labeled PKA-C. All cultures are grown in 32 °C with induction occurring at 24 °C. This is to prevent the formation of inclusion bodies during expression (Yonemoto, McGlone, Slice, & Taylor, 1991). In M9 media, $^{15}NH_4Cl$ (1 g/L) and D-glucose (4 g/L) are used as the only nitrogen and carbon sources, respectively.

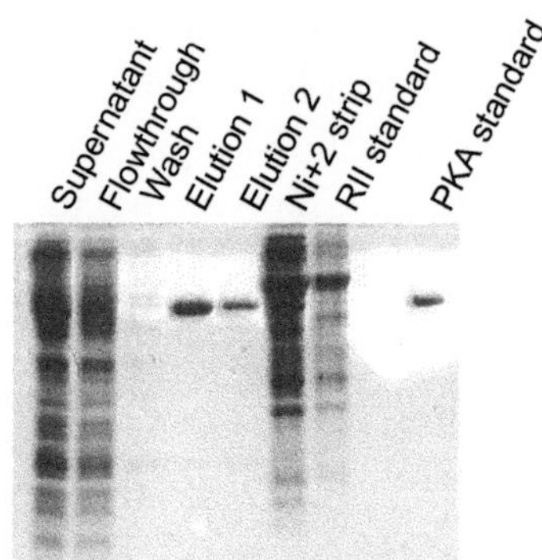

Figure 3 SDS-PAGE of the purification of PKA-C using a His_6-RIIα immobilized construct.

and requires optimization for each of these variables. Several tips for the optimization of perdeuterated growths have been provided by Gardner, Zhang, Gehring, and Kay (1998).

3.1 Materials

Note: All stocks should be sterile before usage.

LB media (1 L)

10.0 g Tryptone
5.0 g Bacto-Yeast
10.0 g NaCl

M9 Media (1 L)

3.0 g KH_2PO_4
12.8 g $Na_2HPO_4 \cdot 7H_2O$
1.0 g NH_4Cl (for ^{15}N labeling use $^{15}NH_4Cl$)
0.5 g NaCl

Vitamins (50 mL)

5.0 mg Ca D(+)-panthothenate
5.0 mg Biotin
5.0 mg Folic acid
5.0 mg Niacinamide USP
5.0 mg Pyridoxal 5-phosphate monohydrate
1000 mg Thiamin

Trace metals (1×, 100 mL)

50 m*M* $FeCl_3$
20 m*M* $CaCl_2$
10 m*M* $MnCl_2$

10 m*M* $ZnSO_4$
2 m*M* $CoCl_2$
2 m*M* $CuCl_2$
2 m*M* $NiCl_2$
2 m*M* Na_2MoO_4
2 m*M* Na_2SeO_3
2 m*M* H_3BO_3
Note: More details in the preparation of Trace Metal solution is provided by Studier (2005)

Keto-acid stock (10 mL, water)
2-Ketobutyric acid-4-^{13}C sodium salt (70 mg/L of media)
2-Keto-3-(methyl-^{13}C)-butyric acid-4-^{13}C sodium salt (90 mg/L of media)

Calcium chloride
Magnesium sulfate
D-glucose
Ampicillin
Isopropyl β-D-1-thiogalactopyranoside (IPTG)

1. Place 500 mL of M9 media into baffled flasks and autoclave.
2. Grow BL21 (DE3) cells containing the pRSET plasmid with PKA-C overnight in LB media (with Ampicillin).
3. In the morning, the culture should be in log phase growth. Then, pellet the BL21 (DE3) cells under 2200 × *g* for 10 min. During this time add the following per 500 mL flask containing M9 media:
   a. 10 mL of 20% (w/v) glucose
   b. 5 mL of Vitamin stock
   c. 1 mL of 1 *M* $MgSO_4$
   d. 1 mL of 50 m*M* $CaCl_2$
   e. 500 μL of 1000 × Trace Metals
   f. 500 μL of 10% (w/v) Ampicillin stock
4. Resuspend the cell pellet in M9 media and allow to grow for at least an hour to allow for the culture to adapt to the adjusted metabolism. We usually have an initial OD_{600} of ~0.8.
5. Once the cells reach log phase growth, lower the temperature to 24 °C and add in the appropriate concentration of keto-acids into each flask (70 mg/L of 2-ketobutyric acid-4-^{13}C sodium salt and 90 mg/L of 2-keto-3-(methyl-^{13}C)-butyric acid-4-^{13}C sodium salt).
6. After 1 h, induce with 0.4 m*M* IPTG for at least 5 h. Harvest under 8000 × *g* for 25 min at 4 °C.

3.2 Purification of PKA-C

Buffer A (1 L)

– 30 m*M* MOPS (3-(*N*-morpholino)propanesulfonic acid)	(6.28 g)
– 15 m*M* $MgCl_2$	(3.05 g)
– 5 m*M* β-mercaptoethanol	(352 μL/L of buffer)
– pH 8.0	

Buffer B (1 L)

– 30 m*M* MOPS	(6.28 g)
– 15 m*M* $MgCl_2$	(3.05 g)
– 25 m*M* KCl	(1.86 g)
– 5 m*M* β-mercaptoethanol	(352 μL/L of buffer)
– pH 8.0	

Elution Buffer (100 mL)

– 30 m*M* MOPS	(628 mg)
– 15 m*M* $MgCl_2$	(305 mg)
– 25 m*M* KCl	(186 mg)
– 1 m*M* cAMP	(33 mg)
– 5 m*M* β-mercaptoethanol	(35.2 μL/100 mL of buffer)
– pH 8.0	

Cell lysis

(1) Resuspend PKA and the His_6-RIIα(R213K) overexpressed cells in Buffer A. Make sure the pellet containing RIIα was grown in at least 1/3 of the volume of your PKA-C.

(2) Grind the cells in ice for approximately 10 min then lyse using a French press using 1000 psi of pressure.

(3) Spin down the solution at 20,000 rpm for 40 min at 4 °C. Mix with 10 mL of NiNTA slurry and batch bind at 4 °C for a minimum of 3 h.

Elution

(1) Run the slurry through the column (make sure the resin does not dry).

(2) Wash with Buffer B (~200 mL). If preferred the wash buffer can also include 20 m*M* Imidazole to remove some nonspecific binding proteins. This will not affect the PKA-C:RIIα interaction.

(3) Elute PKA-C using 150 mL of elution buffer.

(4) To remove RIIα wash with ~50 mL of Buffer B with 250 m*M* imidazole.

Although mass spectrometry is a viable method to assess the labeling of proteins with stable isotopes, the most straightforward way to determine the correct isotopic incorporation is to acquire the [^{1}H,^{13}C]-HMQC spectra of the system of interest. The characteristic chemical shift patterns allow one to easily identify isoleucine (Cδ), leucine, and valine methyl groups as they resonate in distinct regions of the ^{1}H/^{13}C correlated spectra. Due to the high sensitivity of methyl groups, 2D spectra of large proteins (~40 kDa) with reasonable sensitivity (*S*/*N* ratio 200–400) can be measured with modern spectrometers (700–900 MHz) using relatively modest concentrations (~200 μ*M*). A [^{1}H,^{13}C]-HMQC spectrum of apo PKA-C shows the labeling and spectral quality of ILV-methyl group labeling (Fig. 4). Below, we outline the NMR sample preparation that we use for *apo* form of PKA-C.

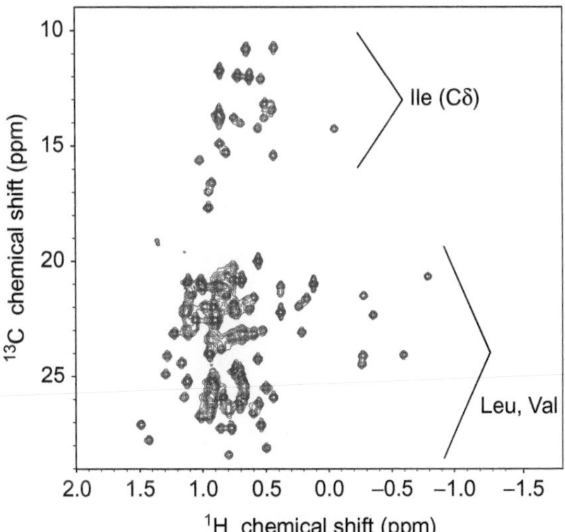

Figure 4 [^{1}H,^{13}C]-HMQC of Apo PKA-C acquired with a 220 μ*M* sample on an Avance III 900 MHz Bruker spectrometer. The spectrum was acquired with 2048 × 200 complex points and 16 transients at 27 °C for a total acquisition time of approximately 1½ h.

3.3 Materials

Potassium phosphate monobasic
Potassium chloride
Dithiothreitol
Magnesium chloride
Sodium azide
10 kDa Centrifugal concentrator
99+% Deuterium oxide
Long-tip Pipet, 13-1/4″
5 mm Shigemi tube

1. Dialyze purified PKA-C under NMR buffer (20 mM KH_2PO_4, 90 mM KCl, 10 mM DTT, 10 mM $MgCl_2$, 1 mM NaN_3 at pH 6.5) overnight at 4 °C. Initial concentration of PKA-C should <5 μM.
2. Concentrate PKA-C at 2400 × g using a 10 kDa cutoff centrifugal concentrator at 4 °C until the enzyme is concentrated to ~200 μM. The protein concentration can reliably be measured by absorbance measurement at 280 nm using an extinction coefficient of 52,060 1/Mcm.
3. Pipette 250 μL (for Agilent spectrometers) or 300 μL (for Bruker spectrometers) and add 12.5 or 15 μL of D_2O (~5%, v/v), respectively. Transfer the protein solution into a shigemi tube using a long-tip pipet. Place the plunger inside the tube until the plunger is evenly covered with no visible bubbles in the solution.
4. After spectrometer set up (spectrometer details are dependent on model and spectrometer) acquire an [^{1}H,^{13}C]-HMQC spectrum.

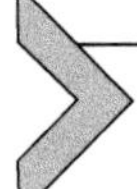

4. SEMIAUTOMATED METHYL-GROUP RESONANCE ASSIGNMENT STRATEGIES

For backbone studies, the assignment is carried out by walking through the peptide backbone through a series of triple resonance NMR experiments (Kay, Ikura, Tschudin, & Bax, 1990; Salzmann, Pervushin, Wider, Senn, & Wuthrich, 1998; Salzmann, Wider, Pervushin, Senn, & Wuthrich, 1999; Yang & Kay, 1999). For the assignment of methyl groups resonances, the classical approach requires TOCSY-based spectroscopy with the magnetization starting on the methyl groups and transferred *via* ^{13}C–^{13}C couplings to the protein backbone (Sattler, Schleucher, & Griesinger, 1999). This approach, however, is not applicable to methyl group side chains of large proteins, such as the 82-kDa malate synthase G with a correlation time of ~46 ns (Tugarinov & Kay, 2003b). In the latter case, the branched amino acids present a bifurcation of the magnetization pathways with a significant loss of transfer. Although several experiments have been tailored to correlate backbone Cα, Cβ, and C′ chemical shifts to methyl groups (Sheppard, Guo, & Tugarinov, 2009; Tugarinov & Kay, 2003a; Tugarinov, Venditti, & Marius Clore, 2014), the relatively fast transverse relaxation rates of these nuclei for larger proteins or conformational exchange phenomena (see the case of PKA-C; Srivastava et al., 2014) do not allow one to carry out sequential backbone assignments. A possible solution is to use a *divide and conquer* approach in which proteins are dissected into several smaller domains and the assignments are pieced together (Saio et al., 2014; Velyvis, Schachman, & Kay, 2009). However, there are many large monomeric proteins that are unamenable to this approach. An alternative strategy is to use site-specific mutations to exchange amino acids with methylated side chains for glycine or other amino acids. The latter method, however, is very cumbersome and time-consuming, particularly for very large systems. In response to this need, Matthews and coworkers proposed an automatic method that uses both NMR and X-ray structural data to back calculate methyl side chain assignments (Xu et al., 2009). This automated procedure was designed to rapidly assign methyl groups using full-length proteins and without mutagenesis (Xu et al., 2009). This approach requires a crystal structure as a starting point to calculate methyl group chemical shifts and synthetic NOEs. An algorithm is then used to rank the scores of the initial (seed) assignment against the experimental NMR data. The latest version of this method features an improved algorithm with no manual intervention in NOESY spectra peak picking.

Inspired by this work, we developed an approach that uses a similar philosophy, but utilizes a probabilistic method to tackle assignment problem. Specifically, we improved the efficiency of the phase space search using *fuzzy logic* coupled with Monte Carlo sampling. This approach, named FLAMEnGO (Fuzzy Logic Assignment of MEthyl GrOups), utilizes NOESY data as the primary input in concert with other data sets, such as methine-methyl TOCSY data, as well as paramagnetic relaxation enhancements (PREs). In our newer version of this software, FLAMEnGO 2.0 (Chao et al., 2014), we introduced a graphic interface that enables the interactive assignment of the methyl-group resonances with the option of including secondary constraints, such as mutagenesis or supplemental NMR data.

In the following synopsis, we describe the construction and usage of FLAMEnGO 2.0, as well as its application to the side-chain methyl groups of PKA-C.

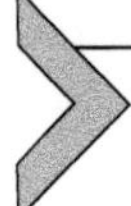

5. SEMIAUTOMATED ASSIGNMENT PROTOCOL USING FLAMEnGO 2.0

All of the available methyl auto-assignment algorithms share the same philosophy: searching for the best assignments by comparing experimental NOE connectivities and chemical shifts to back-calculated data from the X-ray coordinates (Chao et al., 2014; Chao, Shi, Masterson, & Veglia, 2012; Venditti, Fawzi, & Clore, 2011; Xu et al., 2009; Xu & Matthews, 2013). Starting from the available crystal structure or NMR structural bundle, FLAMEnGO 2.0 back calculates a 3D/4D NOESY spectrum from the internuclear methyl–methyl distances and compares synthetic data to an experimental NOESY spectrum (Fig. 5). The algorithm iteratively swaps the resonance assignments until the scoring function is maximized. This process is repeated iteratively for an array of different simulated NOE distances until no further improvement is found. Given the probabilistic nature of this algorithm, multiple runs are performed in parallel to provide a statistical weight to the resonance assignments. Although FLAMEnGO is designed to handle sparse NOE data sets, sufficient coverage of the NOE network and other restraints (i.e., methine-methyl TOCSY or methyl-HN COSY, PREs or site-directed mutagenesis) are required for a confident assignment. The outline of the procedure for using FLAMEnGO GUI is shown in Fig. 6.

Below, we detail the experimental data and data formatting, as well as brief instructions on usage of FLAMEnGO 2.0.

5.1 Materials

- 3D C,C,H ^{13}C-HMQC-NOESY-^{13}C-HMQC data set. (These pulse sequences are part of the TOPSPIN release for Bruker spectrometers, and are downloadable from the NMRFAM website http://pine.nmrfam.wisc.edu/download_pulseprogs.html for Agilent spectrometers.)
- 2D [^{1}H,^{13}C]-HMQC data set (required)
- Predicted chemical shift table (required, sources listed below)
- PRE data (optional)
- Methine-methyl TOCSY data (optional)

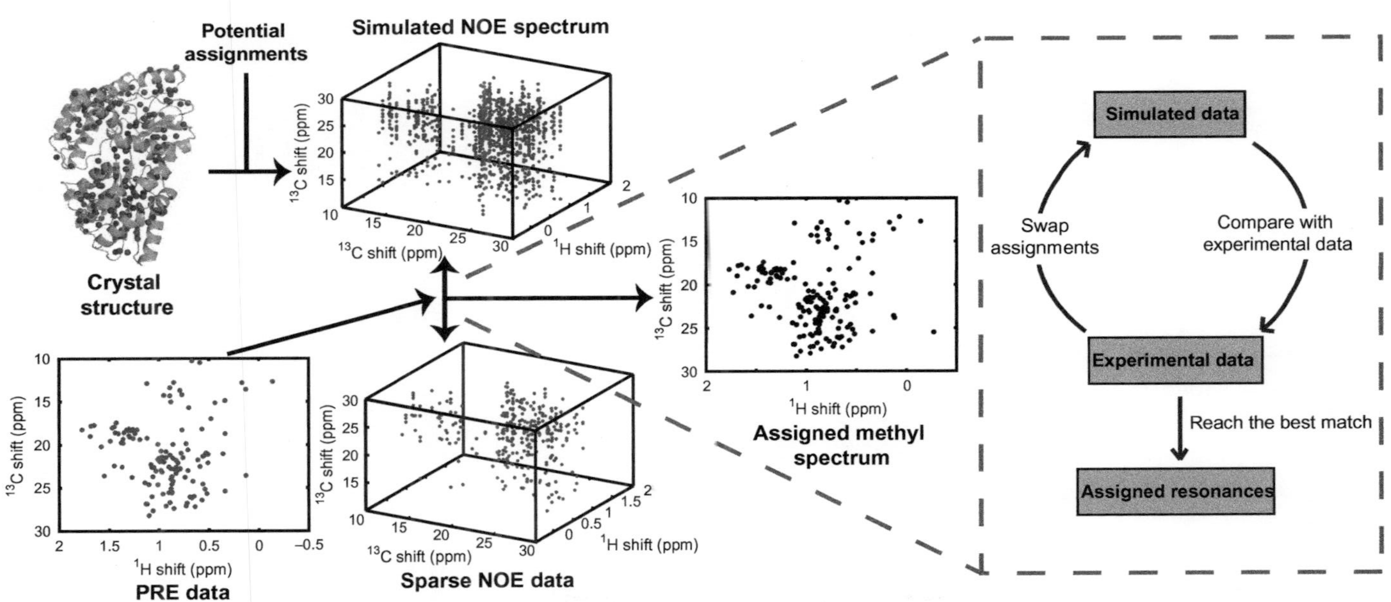

Figure 5 Overview of the FLAMEnGO algorithm. An input structure and the [^{1}H,^{13}C]-HMQC spectrum are used to simulate a NOE spectrum. This spectrum is compared with the experimental NOE data, the assignments swapped and the process is repeated until a best match is found. Other experimental restraints, such as PRE data, are used. *Adapted from Chao et al. (2012).* (See the color plate.)

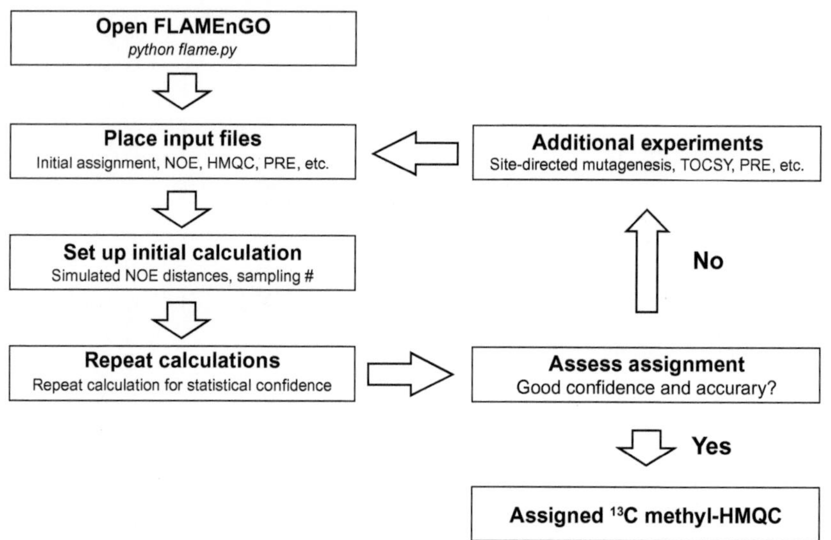

Figure 6 Outline for running FLAMEnGO GUI.

- Assigned amide chemical shifts (optional, if amide-methyl NOESY data are incorporated)
- PDB file for protein. Note that hydrogen atoms need to be added to the PDB. Molprobity (http://molprobity.biochem.duke.edu/) is a resource to perform this function.

5.2 System Requirements and Specifications

The python-based FLAMEnGO 2.0 program includes two main components: (1) the frontend GUI program flame.py, and (2) the backend engine FLAMEnGO.py. The former provides a convenient interface for configuring the setup and displaying the results, while the latter is the driver for a Monte Carlo search. The software is ready to run under Python 2.7, and requires wxPython and matplotlib modules for GUI and plotting. An integrated scientific python package, such as Enthought Canopy (https://www.enthought.com/products/canopy/) or Anaconda Python (https://store.continuum.io/cshop/anaconda/) can provide full functionality.

5.3 Input File Formats

FLAMEnGO 2.0 uses a Monte Carlo search algorithm to swap resonance assignments from given input data and computes the *fitness* of the resonance assignments based on a score function. The details of the scoring function are discussed in Chao et al. (2014, 2012).

5.3.1 Initial Random Assignment

The algorithm needs an *arbitrary* initial assignment to start the swapping process associated with the Monte Carlo algorithm. This *seed* assignment must be consistent with PDB numbering and included in all input data. The data are formatted in a series of columns by <assignment #> <amino acid type> <resonance atom ID> -> <assignment #> <amino acid type> <resonance atom ID>. Note that there is no change in residue, rather this indicates to the program which residues to assign. To constrain a specific resonance assignment (e.g., found using site-directed mutagenesis), the user should place an asterisk at the beginning of the line:

```
296   VAL HG2 -> 296 VAL HG2
* 301 ILE HD1 -> 301  ILE HD1
321   ILE HD1 -> 321  ILE HD1
323   VAL HG1 -> 323  VAL HG1
323   VAL HG2 -> 323  VAL HG2
325   ILE HD1 -> 325  ILE HD1
```

5.3.2 3D C,C,H ^{13}C-HMQC-NOESY-^{13}C-HMQC

The 3D C,C,H ^{13}C-HMQC-NOESY-^{13}C-HMQC data need to be formatted in a series of columns in the following order: <^{13}C (ω2) chemical shift> <^{1}H (ω3) chemical shift> <^{13}C (ω1) chemical shift> <^{13}C (ω2) linewidth> <^{1}H (ω3) linewidth> <^{13}C (ω1) linewidth>

```
18.32 1.28  27.41 0.4  0.025 0.4
22.76 0.88  19.53 0.4  0.025 0.4
22.76 0.88  15.69 0.4  0.025 0.4
22.76 0.88  21.55 0.4  0.025 0.4
22.76 0.88  18.66 0.4  0.025 0.4
20.97 0.78  17.77 0.4  0.025 0.4
25.05 0.68  23.33 0.4  0.025 0.4
```

5.3.3 [^{1}H,^{13}C]-HMQC/^{15}N Amide

The data is formatted in a series of columns by <assignment #> <amino acid type> <resonance atom ID> <chemical shift value>. Note that the assignment reported in this file at the beginning of the calculation is arbitrary for the [^{1}H,^{13}C]-HMQC. The amide chemical shifts are optional and only required if one is using amide to methyl NOE data. The amide chemical shifts must be assigned based on previous experiments.

```
 7 LEU HD1   0.34
 7 LEU HD2   0.56
 7 LEU CD1  25.39
 7 LEU CD2  24.56
20 LEU HD1   0.88
20 LEU HD2   0.64
20 LEU CD1  25.10
20 LEU CD2  27.22
```

5.3.4 Paramagnetic Relaxation Enhancement Data

Although the PRE data are listed as an option, we demonstrated that the inclusion of these constraints in the computation dramatically increases both the speed and accuracy of the resonance assignments (Chao et al., 2014, 2012). In particular, PRE data are crucial to solve the ambiguities caused by partial peak overlap (Fig. 7). PRE data are included as distances using the semiquantitative approach described by Battiste and Wagner (2000), in which the ratios of the peak intensities in the presence of the oxidized versus the reduced spin label are converted into three different categories: weakly affected (<20% of signal is lost through the PRE effect), moderately affected (20–80% of the signal is lost), or strongly effected (>80% of the signal is lost). These categories are flagged as W, M, and S, respectively, in the input file. If a resonance is unobserved regardless of the PRE effect, then a flag of 999 is used in the 4th column of the input file. The data are formatted in four columns, including <assignment #> <resonance atom ID> <amino acid type> <Flag (W, M, S, 999)>:

```
 89  HD1  LEU  999
 92  HD2  LEU  999
102  HD1  LEU  M
102  HD2  LEU  M
 71  HD1  ILE  S
 92  HD1  LEU  S
138  HD1  LEU  W
 13  HD1  LEU  W
```

5.3.5 Prediction of Methyl Group Chemical Shifts from the PDB File

Various algorithms are used to predict the side chain chemical shift values. Output from any commonly used program, such as CH3SHIFT (Sahakyan, Vranken, Cavalli, & Vendruscolo, 2011) (http://www-sidechain.ch.cam.ac.

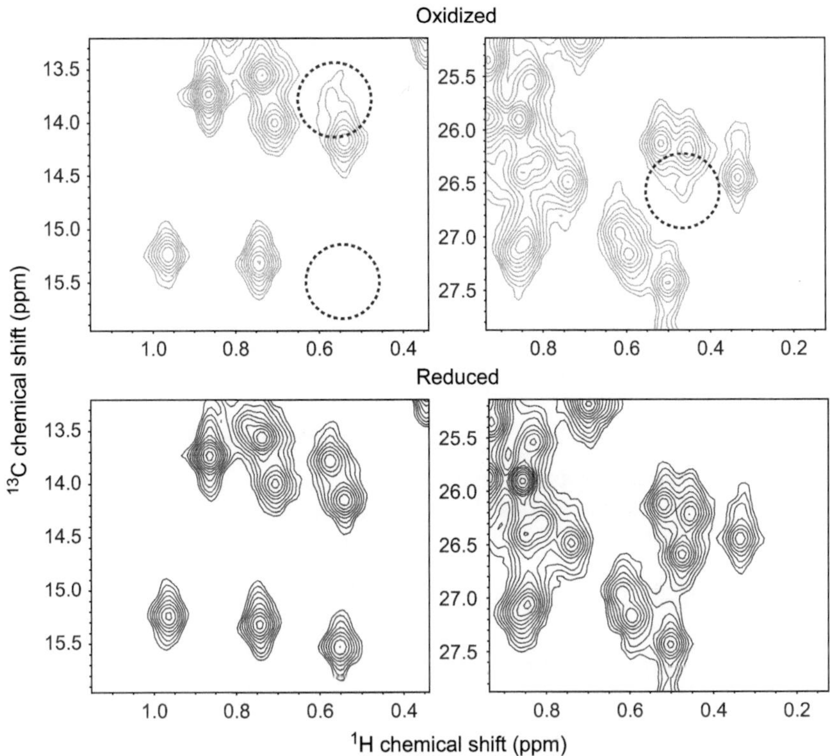

Figure 7 Spectra of oxidized (green) and reduced (red) PKA-C with a spin label on residue 244. Note the spectrum should be nearly identical before and after the reduction of the spin label. (See the color plate.)

uk/CH3Shift/), SHIFTX2 (Han, Liu, Ginzinger, & Wishart, 2011) (http://www.shiftx2.ca/), and PPM (Li & Bruschweiler, 2012) (http://spin.ccic.ohio-state.edu/index.php/ppm) can be implemented in FLAMEnGO 2.0. Since the output is not uniform between programs, data must be reformatted as follows: <residue #> <amino acid type> <resonance atom ID> <predicted chemical shift value> <error of prediction>

```
1 VAL CG2 19.992 1.128
1 VAL HG2 0.828 0.173
5 LEU CD1 25.747 1.412
5 LEU HD1 0.969 0.176
```

5.3.6 Residue Type

Amino acid-type assignments can be included as constraints in the search algorithm. The data format is identical to that of the initial assignment file.

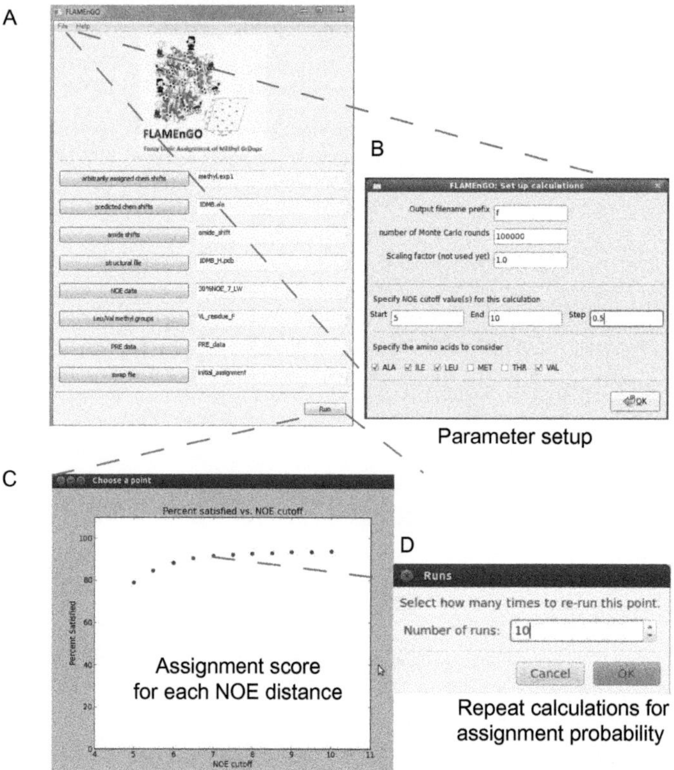

Figure 8 Graphical interface for FLAMEnGO. (A) Main menu of FLAMEnGO. Input files are directly placed inside the selection. (B) Parameter set up for the NOE distances, amino acid types you wish to assign, and the number of Monte Carlo steps. (C) Output from the run. Note that the program performs the calculation at each NOE distance. (D) Once the score has plateaued from increasing the NOE distance, select the point and repeat the calculation at this distance to provide a statistical assignment of each residue. *Adapted from Chao et al. (2014).*

5.4 Instructions to Run FLAMEnGO 2.0 GUI

1. Open FLAMEnGO

   "`python flame.py`"
2. Specify the location of each input file on the main window.
3. To set up FLAMEnGO go under **File** > **SetUp** (Fig. 8A). Set the NOE range between 5.0 and 10.0 with step size of 0.5. The sampling number is the number of Monte Carlo calculations run at each point and should be over 100 k (Fig. 8B).
4. Select the **Run** button. Please note that calculations take a considerable amount of time.

5. Once the runs are finished a plot of the score with respect to the NOE distance will appear (Fig. 8C). To minimize the uncertainty of the assignment, click on the point where the NOE value with the maximum score is first reached (Fig. 8D)
6. Select the number of calculations you wish to run (≥10). The final output will give the assignment and the statistical weight of each assignment.

5.5 Output Files

FLAMEnGO provides two types of output files. The first is the assignment file, which provides the calculated assignment at each given calculation at a specified NOE cutoff distance. Once the aggregate calculations are performed at a cutoff distance, the probability of each assignment is calculated. The probability of each assignment from the calculations is what is used to determine the *fitness* of the assignment.

5.5.1 Score and Assignment

The output file provides the user with the percent of the residues assigned, the assignment score, and the assignment from the single calculation. The assignment is provided as the following <initial assignment> -> <calculated assignment>. On the first line the first two columns provide the <final score> <% of residues assigned>.

```
(205.2936878260806,93.079554272662747,193,17.844744345905379)
##################################
158  LEU  HD1 -> 148  LEU  HD1
237  VAL  HG1 ->  90  VAL  HG2
321  ILE  HD1 -> 196  ILE  HD1
259  LEU  HD2 -> 191  LEU  HD2
197  LEU  HD1 ->  75  LEU  HD1
```

5.5.2 Summary

The summary file will provide the assignment, as well as the number of times the particular assignment was made across all the calculations. A confident assignment is defined in this case as one that is consistent through all calculations at least 80% of the time; a reliable statistic for this assessment requires ≥10 calculations. For each assignment the output is as follows: <initial assignment> -> <calculated assignment> <number of times the assignment was calculated>.

```
105  VAL  HG1 -> 137  VAL  HG1   10
254  LEU  HD1 ->  68  LEU  HD1   10
```

```
 75  LEU  HD2 -> 263  LEU  HD2   6
143  LEU  HD2 ->  45  LEU  HD2   7
 13  LEU  HD1 -> 259  LEU  HD1  10
241  VAL  HG2 -> 137  VAL  HG2  10
```

6. CONCLUSIONS AND PERSPECTIVES

Selective labeling of the methyl group side chains has extended the application of NMR methods to the analysis of both structure and dynamics of proteins up to 1 MDa. However, as the systems become larger, classical methods for resonance assignments fail. In this chapter, we reported a semi-empirical approach that enables the assignment of methyl-group resonances in large proteins. This structure-based approach requires the X-ray or the NMR ensemble of structures to generate a probability-based assignment. Inclusion of NOEs, PRE, site-specific mutagenesis data, methine-methyl TOCSY, and partial assignment from methyl-HN COSY data dramatically improves both the accuracy and the speed of the assignment procedure. This method promises to extend biomolecular NMR studies beyond the MDa limit.

ACKNOWLEDGMENTS

This work was supported in part by the NIH (GM100310 and GM72701 to G.V. and T32 AR007612 to J.K.). NMR experiments were carried out at the Minnesota NMR Center and FLAMEnGO 2.0 calculations at the Minnesota Supercomputing Institute.

REFERENCES

Ayala, I., Sounier, R., Use, N., Gans, P., & Boisbouvier, J. (2009). An efficient protocol for the complete incorporation of methyl-protonated alanine in perdeuterated protein. *Journal of Biomolecular NMR*, *43*(2), 111–119. http://dx.doi.org/10.1007/s10858-008-9294-7.

Battiste, J. L., & Wagner, G. (2000). Utilization of site-directed spin labeling and high-resolution heteronuclear nuclear magnetic resonance for global fold determination of large proteins with limited nuclear overhauser effect data. *Biochemistry*, *39*(18), 5355–5365.

Chao, F. A., Kim, J., Xia, Y., Milligan, M., Rowe, N., & Veglia, G. (2014). FLAMEnGO 2.0: An enhanced fuzzy logic algorithm for structure-based assignment of methyl group resonances. *Journal of Magnetic Resonance*, *245*, 17–23. http://dx.doi.org/10.1016/j.jmr.2014.04.012.

Chao, F. A., Shi, L., Masterson, L. R., & Veglia, G. (2012). FLAMEnGO: A fuzzy logic approach for methyl group assignment using NOESY and paramagnetic relaxation enhancement data. *Journal of Magnetic Resonance*, *214*(1), 103–110. http://dx.doi.org/10.1016/j.jmr.2011.10.008.

Crespi, H. L., & Katz, J. J. (1969). High resolution proton magnetic resonance studies of fully deuterated and isotope hybrid proteins. *Nature*, *224*(5219), 560–562.

Crespi, H. L., Rosenberg, R. M., & Katz, J. J. (1968). Proton magnetic resonance of proteins fully deuterated except for 1H-leucine side chains. *Science*, *161*, 795–796.

Gardner, K. H., & Kay, L. E. (1997). Production and Incorporation of15N,13C,2H (1H-δ1 Methyl) Isoleucine into proteins for multidimensional NMR studies. *Journal of the American Chemical Society*, *119*(32), 7599–7600. http://dx.doi.org/10.1021/ja9706514.

Gardner, K. H., Zhang, X., Gehring, K., & Kay, L. E. (1998). Solution NMR studies of a 42 KDa *Escherichia coli* maltose binding protein/β-cyclodextrin complex: Chemical shift assignments and analysis. *Journal of the American Chemical Society*, *120*(45), 11738–11748. http://dx.doi.org/10.1021/ja982019w.

Gelis, I., Bonvin, A. M., Keramisanou, D., Koukaki, M., Gouridis, G., Karamanou, S., et al. (2007). Structural basis for signal-sequence recognition by the translocase motor SecA as determined by NMR. *Cell*, *131*(4), 756–769. http://dx.doi.org/10.1016/j.cell.2007.09.039.

Goto, N. K., Gardner, K. H., Mueller, G. A., Willis, R. C., & Kay, L. E. (1999). A robust and cost-effective method for the production of Val, Leu, Ile (delta 1) methyl-protonated 15N-, 13C-, 2H-labeled proteins. *Journal of Biomolecular NMR*, *13*(4), 369–374. http://dx.doi.org/10.1023/A:1008393201236.

Han, B., Liu, Y., Ginzinger, S. W., & Wishart, D. S. (2011). SHIFTX2: Significantly improved protein chemical shift prediction. *Journal of Biomolecular NMR*, *50*(1), 43–57. http://dx.doi.org/10.1007/s10858-011-9478-4.

Hemmer, W., McGlone, M., & Taylor, S. S. (1997). Recombinant strategies for rapid purification of catalytic subunits of cAMP-dependent protein kinase. *Analytical Biochemistry*, *245*(2), 115–122. http://dx.doi.org/10.1006/abio.1996.9952.

Kainosho, M., Torizawa, T., Iwashita, Y., Terauchi, T., Mei Ono, A., & Guntert, P. (2006). Optimal isotope labelling for NMR protein structure determinations. *Nature*, *440*(7080), 52–57. http://dx.doi.org/10.1038/nature04525.

Kasinath, V., Valentine, K. G., & Wand, A. J. (2013). A 13C labeling strategy reveals a range of aromatic side chain motion in calmodulin. *Journal of the American Chemical Society*, *135*(26), 9560–9563. http://dx.doi.org/10.1021/ja4001129.

Kay, L. E., Ikura, M., Tschudin, R., & Bax, A. (1990). Three-dimensional triple-resonance NMR spectroscopy of isotopically enriched proteins. *Journal of Magnetic Resonance*, *89*(3), 496–514. http://dx.doi.org/10.1016/0022-2364(90)90333-5.

Li, D. W., & Bruschweiler, R. (2012). PPM: A side-chain and backbone chemical shift predictor for the assessment of protein conformational ensembles. *Journal of Biomolecular NMR*, *54*(3), 257–265. http://dx.doi.org/10.1007/s10858-012-9668-8.

Markley, J. L., Putter, I., & Jardetzky, O. (1968). High-resolution nuclear magnetic resonance spectra of selectively deuterated staphylococcal nuclease. *Science*, *161*(3847), 1249–1251. http://dx.doi.org/10.1126/science.161.3847.1249.

Meissner, A., & Sorensen, O. W. (1999). Optimization of three-dimensional TROSY-type HCCH NMR correlation of aromatic (1)H-(13)C groups in proteins. *Journal of Magnetic Resonance*, *139*(2), 447–450. http://dx.doi.org/10.1006/jmre.1999.1796.

Narayana, N., Cox, S., Shaltiel, S., Taylor, S. S., & Xuong, N. (1997). Crystal structure of a polyhistidine-tagged recombinant catalytic subunit of cAMP-dependent protein kinase complexed with the peptide inhibitor PKI(5-24) and adenosine. *Biochemistry*, *36*(15), 4438–4448. http://dx.doi.org/10.1021/bi961947.

Nietlispach, D., Clowes, R. T., Broadhurst, R. W., Ito, Y., Keeler, J., Kelly, M., et al. (1996). An approach to the structure determination of larger proteins using triple resonance NMR experiments in conjunction with random fractional deuteration. *Journal of the American Chemical Society*, *118*(2), 407–415. http://dx.doi.org/10.1021/ja952207b.

Pervushin, K., Riek, R., Wider, G., & Wüthrich, K. (1997). Attenuated T2 relaxation by mutual cancellation of dipole–dipole coupling and chemical shift anisotropy indicates an avenue to NMR structures of very large biological macromolecules in solution. *Proceedings of the National Academy of Sciences*, *94*(23), 12366–12371.

Putter, I., Barreto, A., Markley, J. L., & Jardetzky, O. (1969). Nuclear magnetic resonance studies of the structure and binding sites of enzymes. X. Preparation of selectively deuterated analogs of staphylococcal nuclease. *Proceedings of the National Academy of Sciences of the United States of America*, *64*(4), 1396–1403.

Religa, T. L., Ruschak, A. M., Rosenzweig, R., & Kay, L. E. (2011). Site-directed methyl group labeling as an NMR probe of structure and dynamics in supramolecular protein systems: Applications to the proteasome and to the ClpP protease. *Journal of the American Chemical Society*, *133*(23), 9063–9068. http://dx.doi.org/10.1021/ja202259a.

Religa, T. L., Sprangers, R., & Kay, L. E. (2010). Dynamic regulation of archaeal proteasome gate opening as studied by TROSY NMR. *Science*, *328*(5974), 98–102. http://dx.doi.org/10.1126/science.1184991.

Ruschak, A. M., Religa, T. L., Breuer, S., Witt, S., & Kay, L. E. (2010). The proteasome antechamber maintains substrates in an unfolded state. *Nature*, *467*(7317), 868–871. http://dx.doi.org/10.1038/nature09444.

Ruschak, A. M., Velyvis, A., & Kay, L. E. (2010). A simple strategy for (1)(3)C, (1)H labeling at the Ile-gamma2 methyl position in highly deuterated proteins. *Journal of Biomolecular NMR*, *48*(3), 129–135. http://dx.doi.org/10.1007/s10858-010-9449-1.

Sahakyan, A. B., Vranken, W. F., Cavalli, A., & Vendruscolo, M. (2011). Structure-based prediction of methyl chemical shifts in proteins. *Journal of Biomolecular NMR*, *50*(4), 331–346. http://dx.doi.org/10.1007/s10858-011-9524-2.

Saio, T., Guan, X., Rossi, P., Economou, A., & Kalodimos, C. G. (2014). Structural basis for protein antiaggregation activity of the trigger factor chaperone. *Science*, *344*(6184), 1250494. http://dx.doi.org/10.1126/science.1250494.

Salzmann, M., Pervushin, K., Wider, G., Senn, H., & Wuthrich, K. (1998). TROSY in triple-resonance experiments: New perspectives for sequential NMR assignment of large proteins. *Proceedings of the National Academy of Sciences of the United States of America*, *95*(23), 13585–13590. http://dx.doi.org/10.1073/pnas.95.23.13585.

Salzmann, M., Wider, G., Pervushin, K., Senn, H., & Wuthrich, K. (1999). TROSY-type triple-resonance experiments for sequential NMR assignments of large proteins. *Journal of the American Chemical Society*, *121*(4), 844–848. http://dx.doi.org/10.1021/Ja9834226.

Sattler, M., Schleucher, J., & Griesinger, C. (1999). Heteronuclear multidimensional NMR experiments for the structure determination of proteins in solutions employing pulsed field gradients. *Progress in Nuclear Magnetic Resonance Spectroscopy*, *34*, 93–158.

Sheppard, D., Guo, C., & Tugarinov, V. (2009). Methyl-detected 'out-and-back' NMR experiments for simultaneous assignments of Alabeta and Ilegamma2 methyl groups in large proteins. *Journal of Biomolecular NMR*, *43*(4), 229–238. http://dx.doi.org/10.1007/s10858-009-9305-3.

Shi, L., & Kay, L. E. (2014). Tracing an allosteric pathway regulating the activity of the HslV protease. *Proceedings of the National Academy of Sciences of the United States of America*, *111*(6), 2140–2145. http://dx.doi.org/10.1073/pnas.1318476111.

Sprangers, R., & Kay, L. E. (2007). Quantitative dynamics and binding studies of the 20S proteasome by NMR. *Nature*, *445*(7128), 618–622. http://dx.doi.org/10.1038/nature05512.

Srivastava, A. K., McDonald, L. R., Cembran, A., Kim, J., Masterson, L. R., McClendon, C. L., et al. (2014). Synchronous opening and closing motions are essential for cAMP-dependent protein kinase A signaling. *Structure*, *22*(12), 1735–1743. http://dx.doi.org/10.1016/j.str.2014.09.010.

Studier, F. W. (2005). Protein production by auto-induction in high density shaking cultures. *Protein Expression and Purification, 41*(1), 207–234.

Tugarinov, V., Hwang, P. M., Ollerenshaw, J. E., & Kay, L. E. (2003). Cross-correlated relaxation enhanced 1H[bond]13C NMR spectroscopy of methyl groups in very high molecular weight proteins and protein complexes. *Journal of the American Chemical Society, 125*(34), 10420–10428. http://dx.doi.org/10.1021/ja030153x.

Tugarinov, V., & Kay, L. E. (2003a). Ile, Leu, and Val methyl assignments of the 723-residue malate synthase G using a new labeling strategy and novel NMR methods. *Journal of the American Chemical Society, 125*(45), 13868–13878. http://dx.doi.org/10.1021/ja030345s.

Tugarinov, V., & Kay, L. E. (2003b). Side chain assignments of Ile delta 1 methyl groups in high molecular weight proteins: An application to a 46 ns tumbling molecule. *Journal of the American Chemical Society, 125*(19), 5701–5706. http://dx.doi.org/10.1021/ja021452.

Tugarinov, V., Venditti, V., & Marius Clore, G. (2014). A NMR experiment for simultaneous correlations of valine and leucine/isoleucine methyls with carbonyl chemical shifts in proteins. *Journal of Biomolecular NMR, 58*(1), 1–8. http://dx.doi.org/10.1007/s10858-013-9803-1.

Tzeng, S. R., & Kalodimos, C. G. (2012). Protein activity regulation by conformational entropy. *Nature, 488*(7410), 236–240. http://dx.doi.org/10.1038/nature11271.

Velyvis, A., Ruschak, A. M., & Kay, L. E. (2012). An economical method for production of (2)H, (13)CH3-threonine for solution NMR studies of large protein complexes: Application to the 670 kDa proteasome. *PLoS One*, 7(9), e43725. http://dx.doi.org/10.1371/journal.pone.0043725.

Velyvis, A., Schachman, H. K., & Kay, L. E. (2009). Assignment of Ile, Leu, and Val methyl correlations in supra-molecular systems: An application to aspartate transcarbamoylase. *Journal of the American Chemical Society, 131*(45), 16534–16543. http://dx.doi.org/10.1021/ja906978r.

Venditti, V., Fawzi, N. L., & Clore, G. M. (2011). Automated sequence- and stereo-specific assignment of methyl-labeled proteins by paramagnetic relaxation and methyl-methyl nuclear Overhauser enhancement spectroscopy. *Journal of Biomolecular NMR, 51*(3), 319–328. http://dx.doi.org/10.1007/s10858-011-9559-4.

Venters, R. A., Huang, C. C., Farmer, B. T., 2nd, Trolard, R., Spicer, L. D., & Fierke, C. A. (1995). High-level 2H/13C/15N labeling of proteins for NMR studies. *Journal of Biomolecular NMR, 5*(4), 339–344. http://dx.doi.org/10.1007/BF00182275.

Xu, Y., Liu, M., Simpson, P. J., Isaacson, R., Cota, E., Marchant, J., et al. (2009). Automated assignment in selectively methyl-labeled proteins. *Journal of the American Chemical Society, 131*(27), 9480–9481. http://dx.doi.org/10.1021/ja9020233.

Xu, Y., & Matthews, S. (2013). MAP-XSII: An improved program for the automatic assignment of methyl resonances in large proteins. *Journal of Biomolecular NMR, 55*(2), 179–187. http://dx.doi.org/10.1007/s10858-012-9700-z.

Yang, D. W., & Kay, L. E. (1999). TROSY triple-resonance four-dimensional NMR spectroscopy of a 46 ns tumbling protein. *Journal of the American Chemical Society, 121*(11), 2571–2575. http://dx.doi.org/10.1021/Ja984056t.

Yonemoto, W. M., McGlone, M. L., Slice, L. W., & Taylor, S. S. (1991). Prokaryotic expression of catalytic subunit of adenosine cyclic monophosphate-dependent protein kinase. *Methods in Enzymology, 200*, 581–596.

CHAPTER THREE

^{19}F-Site-Specific-Labeled Nucleotides for Nucleic Acid Structural Analysis by NMR

Lincoln G. Scott*, Mirko Hennig†,1

*Cassia LLC, San Diego, California, USA
†Nutrition Research Institute, University of North Carolina at Chapel Hill, Kannapolis, North Carolina, USA
1Corresponding author: e-mail address: mirkohennig@yahoo.com

Contents

Abstract

Naturally occurring RNA lacks fluorine-19 (^{19}F), thus, their specifically fluorinated counterparts are particularly well suited to noninvasively monitoring the dynamic conformational properties and ligand-binding interactions of the RNA. For nuclear magnetic resonance (NMR) spectroscopy, ^{19}F-NMR of fluorine-substituted RNA provides an attractive, site-specific probe for structure determination in solution. Advantages of ^{19}F include high NMR sensitivity (83% of ^{1}H), high natural abundance (100%), and the extreme sensitivity of ^{19}F to the chemical environment leading to a large range of chemical shifts. The preparation of base-substituted 2-fluoropurine and 5-fluoropyrimidine

Methods in Enzymology, Volume 566
ISSN 0076-6879
http://dx.doi.org/10.1016/bs.mie.2015.05.015

5′-triphosphates (2F-ATP/5F-CTP/5F-UTP) can be carried out using efficient enzymatic synthesis methods. Both pyrimidine analogs, 5-fluorouridine and 5-fluorocytidine, as well as, 2-fluoroadenosine are readily incorporated into RNA transcribed *in vitro* using T7 RNA polymerase.

1. INTRODUCTION

Fluorine-19 (^{19}F) nuclear magnetic resonance (NMR) investigations of RNA have been of interest for many years, due in large part to the favorable NMR properties (reviewed in Rastinejad, Evilia, & Lu, 1995). The ^{19}F chemical shift of a covalently bound fluorine atom is extraordinarily sensitive toward changes in the local microenvironment, which is primarily attributable to the anisotropic distribution of the electrons in the 2p orbitals. The van der Waals radius of a fluorine atom is 1.46 Å which is only slightly larger than a hydrogen (1.2 Å) (Alvarez, 2013) and smaller than a methyl group (2.0 Å) making it a promising candidate to substitute for either moiety without structural and functional alteration (Fig. 1).

We, and others, have demonstrated that the introduction of ^{19}F substitutions into the heterocyclic bases is nonperturbing and provides uniquely positioned, sensitive NMR reporter groups to monitor conformational properties of modified RNA (Hennig, Scott, Sperling, Bermel, & Williamson, 2007; Puffer et al., 2009; Scott, Geierstanger, Williamson, & Hennig, 2004). Additional advantages of using spin $I=½$ ^{19}F as an NMR probe are its high sensitivity (83% of the ^{1}H) combined with 100% natural abundance and a wide chemical shift distribution (~50-fold larger than that of ^{1}H) (Kitevski-LeBlanc & Prosser, 2012; Rastinejad et al., 1995).

Here, we describe a combination of RNA-labeling techniques as well as NMR methods and applications to a well-established model system, the human immunodeficiency virus (HIV) transactivation response element (TAR) (Fischer et al., 1996; Fischer, Gdaniec, Olejniczak, Bielecki, & Adamiak, 1999; Hennig et al., 2006, 2007; Huang, Varani, & Drobny,

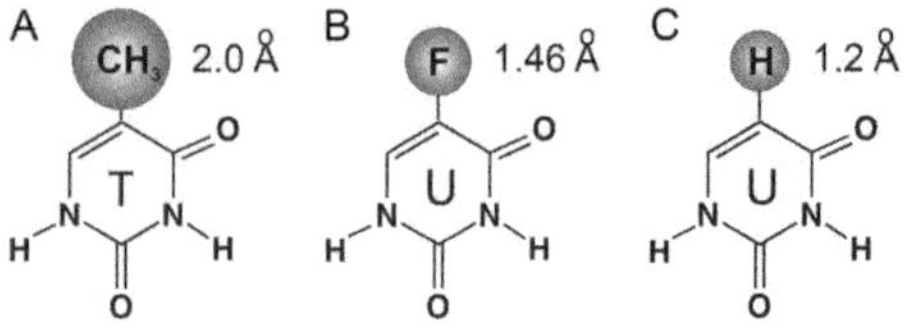

Figure 1 Van der Waals radii of methyl, fluorine, and hydrogen in (A) thymine, (B) 5-fluorouracil, and (C) uracil, respectively.

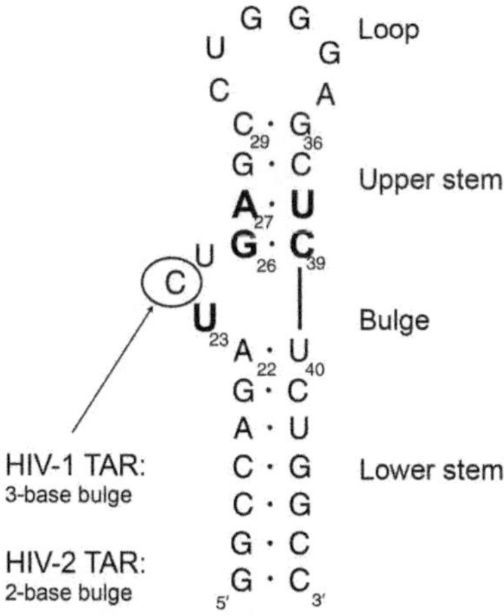

Figure 2 Sequence and secondary structure of the TAR RNA where bold typeface highlights nucleotides important for Tat recognition.

2010; Olejniczak et al., 2002; Olsen et al., 2005; Scott et al., 2004). The Tat (transcription activator)–TAR complex formation provides an essential gene regulatory function for HIV (Marciniak, Calnan, Frankel, & Sharp, 1990; Seelamgari et al., 2004). Peptides from the basic region of Tat, as well as argininamide, retain the specificity of TAR RNA binding although with greatly reduced affinity (Calnan, Biancalana, Hudson & Frankel, 1991). The nucleotides on the 30-nucleotide TAR RNA hairpin loop important for Tat binding are clustered around a three-nucleotide bulge (two nucleotides in HIV-2), as shown in Fig. 2.

2. GENERAL ENZYMATIC SYNTHESIS OF 2F-ATP, 5F-CTP, AND 5F-UTP

We established efficient and economical enzymatic syntheses for the fluorinated nucleotide analogues 5-fluorouridine-5′-triphosphate (5F-UTP), 5-fluorocytidine-5′-triphoshate (5F-CTP) (Hennig et al., 2006, 2007), and 2-fluoroadenosine-5′-triphosphate (2F-ATP) (Scott et al., 2004) and demonstrated that the nucleotides can be specifically incorporated into RNA using standard DNA template-directed T7 RNA polymerase methods (Fig. 3).

The strategy for the syntheses was carried out using an adaptation from Tolbert and Williamson (1996, 1997), using enzymes derived from the pentose phosphate and nucleotide salvage biosynthetic pathways (Table 1). The

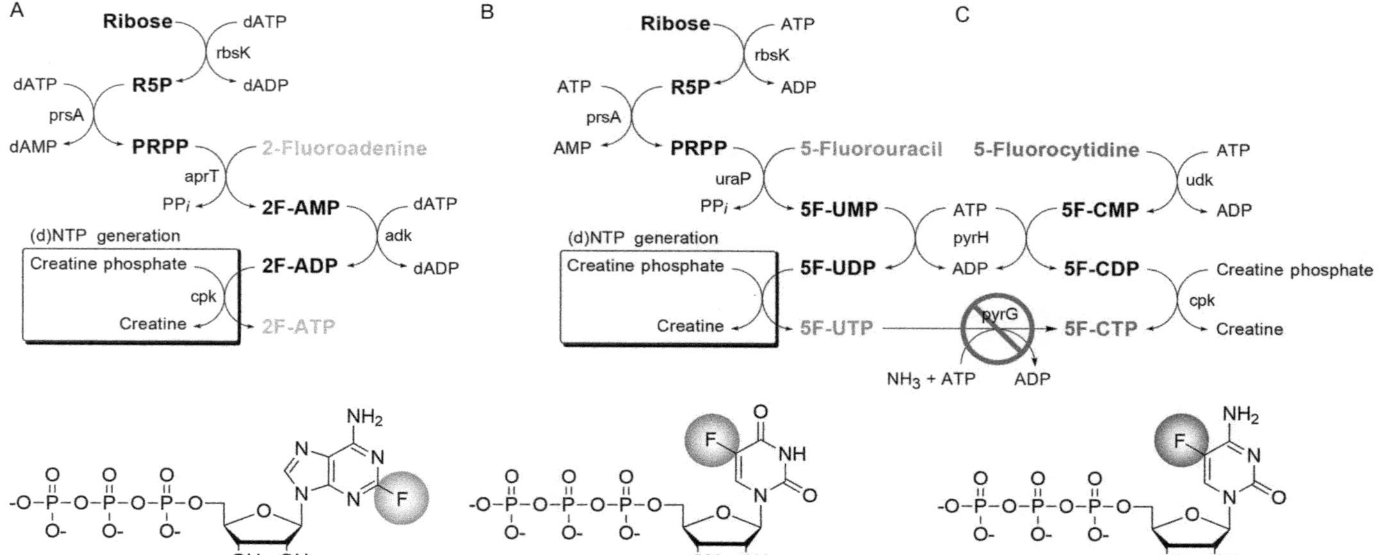

Figure 3 Enzymatic synthesis of fluorinated nucleotides. (A) Synthesis of 2-fluoroadenosine-5′-triphosphate (2F-ATP) from 2-fluoroadenine and ribose. (B) Synthesis of 5-fluorouridine-5′-triphosphate (5F-UTP) from 5-fluorouracil and ribose. (C) Synthesis of 5-fluorocytidine-5′-triphosphate (5F-CTP) from 5-fluorocytidine. Phosphorylation of 5-fluorocytidine by uridine kinase (udk) proceeds to produce 5F-CMP, which is then converted into 5F-CTP. In contrary, 5F-UTP proved not to be an acceptable substrate for the enzyme CTP synthetase (pyrG). (See the color plate.)

Table 1 Enzymes Used in the Synthesis of 5F-UTP, 5F-CTP, and 2F-ATP

Enzyme	Abbreviation	EC	Source	Vendor
Ribokinase	rbsK	2.7.1.15	*E. coli*	Recombinant
Phosphoribosylpyrophosphate synthetase	prsA	2.7.6.1	*E. coli*	Recombinant
Adenine phosphoribosyltransferase	aprT	2.4.2.7	*E. coli*	Recombinant
Uracil phosphoribosyltransferase	uraP	2.4.2.9	*E. coli*	Recombinant
Adenylate kinase	adk	2.7.4.3	*E. coli*	Recombinant
Uridine monophosphate kinase	pyrH	2.7.4.22	*E. coli*	Recombinant
Uridine kinase	udk	2.7.1.48	*E. coli*	Recombinant
Creatine phosphokinase	cpk	2.7.3.2	Rabbit muscle	Sigma-Aldrich

synthesis for 2F-ATP and 5F-UTP involves the conversion of ribose into 5-phospho-D-ribosyl-α-1-pyrophosphate, which is then efficiently converted with the appropriate phosphoribosyltransferase (aprT or uraP) and the nucleotide monophosphate kinase (adk or pyrH) to the corresponding nucleotide 5′-triphosphate (Hennig et al., 2007; Scott et al., 2004).

Because 5F-UTP is a potent inhibitor of CTP synthetase (pyrG), the synthesis of 5F-CTP was carried out from the corresponding, commercially available nucleoside 5-fluorocytidine through the tandem action of uridine kinase (udk), uridine monophosphate kinase (pyrH), and creatine phosphokinase (cpk). Our synthetic strategies routinely yield greater than 80% isolated products with chemical purities in excess of 95% (Hennig et al., 2007).

In addition to the high yields and purities, there are number of other advantages to our enzymatic syntheses approaches. First, the syntheses require five equivalents of phosphate, which are provided by regenerated catalytic dATP, which is later easily removed from the ribonucleotide product by phenylboronate affinity chromatography. Second, all the enzymes are routinely available commercially or by laboratory-scale bacterial expression in high yield and specific activity. Third, depending on available starting materials, the strategy is amiable to incorporation of ^{2}H, ^{13}C, and ^{15}N stable isotopes (Schultheisz, Szymczyna, Scott, & Williamson, 2008; Schultheisz, Szymczyna, Scott, & Williamson, 2011; Scott, Tolbert, & Williamson, 2000).

3. STABILITY OF MODIFIED RNA

3.1 Impact of Fluorine Substitutions on Base Pairing

The fluorine substitutions introduced above have a negative inductive effect effectively creating slightly polar (δ^+)C—F(δ^-) bonds. As expected, the considerable electronegativity of fluorine influences the pK values of neighboring groups. The presence of a fluorine atom at the C5-position of a pyrimidine has a pronounced acid-strengthening effect. The apparent N3 nitrogen pK_a for 5-fluorouridine and 5-fluorocytidine are 7.6 (Knobloch, Linert, & Sigel, 2005) and 2.3 (Oyelere & Strobel, 2000), respectively. The pK_a values are reduced by about −1.7 to −1.8 units with respect to their nonfluorinated analogs (Bevilacqua, Brown, Nakano, & Yajima, 2004). The near neutrality of the N3 pK_a of 5F-Ura substitutions leads to significant line broadening of imino protons in uniformly 5F-Ura-modified TAR RNA. Similarly, a fluorine substitution introduced into adenosine at the C2-position (2F-Ade) dramatically shifts the pK_a by more than −2.0 units (2F-Ade N1 p$Ka < 1.0$) (Suydam & Strobel, 2008). However, we have shown that the modified nucleotide analogues 5F-Cyt selectively base pair with guanine in 5-fluorocytidine-substituted RNA (Hennig et al., 2006, 2007), 5F-Ura selectively base pair with adenine in 5-fluorouridine-substituted RNA (Hennig et al., 2006, 2007), and 2F-Ade selectively base pair with uridine in 2-fluoroadenosine-substituted RNA in a nonperturbing way (Scott et al., 2004; Fig. 4).

The common alternative, specific labeling of RNA with ^{19}F at the C2′-position in the ribose can be problematic especially for regions adopting noncanonical conformations because of the conformational bias imposed by the fluorine substitution. 2′-Deoxy-2′-fluorinated nucleosides are

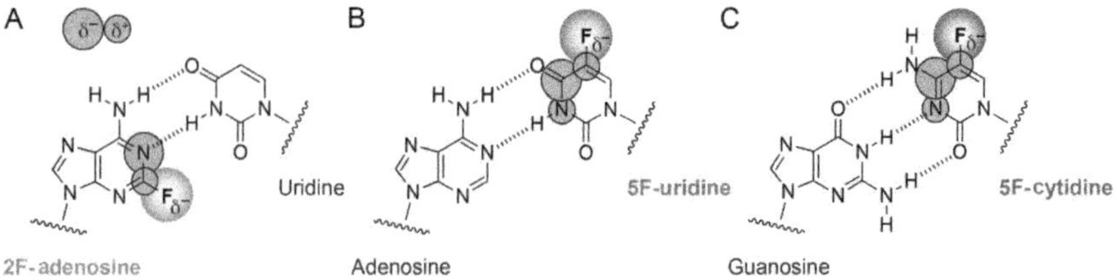

Figure 4 Schematic Watson–Crick base-pairing interactions involving fluorinated nucleobases: (A) A^{2F}-U, (B) A-U^{5F}, and (C) G-C^{5F}. Large and small circles symbolize δ^- and δ^+ to illustrate the negative inductive effect of an electron-withdrawing ^{19}F substituent.

effectively locked in the C3′-endo sugar pucker normally associated with A-form RNA geometry (Barchi, Jeong, Siddiqui, & Marquez, 1997; Deleavey & Damha, 2012; Kreutz, Kahlig, Konrat, & Micura, 2005; Reif et al., 1997; Thibaudeau, Plavec, & Chattopadhyaya, 1998). While attractive due to the increased sensitivity associated with three magnetically equivalent ^{19}F reporters (Fauster, Kreutz, & Micura, 2012), the 2′-trifluoromethylthio (2′-SCF_3) modification proposed by Micura and coworkers decreases the stability of regular A-form helices (Kosutic et al., 2014).

3.2 Thermal Stability of Fluoro-Modified RNA

Uniform 5F-Ura- and 5F-Cyt substitutions and incorporations into HIV-2 TAR RNA transcripts (in place of Ura and/or Cyt) do not notably affect the structures of modified RNA molecules in solution as demonstrated by thermodynamic stability measurements shown in Fig. 5. The melting profiles for all 5F-Ura- and/or 5F-Cyt-modified TAR RNA constructs were essentially the same as those measured for wild-type TAR (compiled in Table 2). The derived thermodynamic parameters are very similar, strongly suggesting that fluorine substitutions at the C5-position in pyrimidines do not change RNA stability. The 2F-Ade modifications (in place of Ade) slightly reduce the

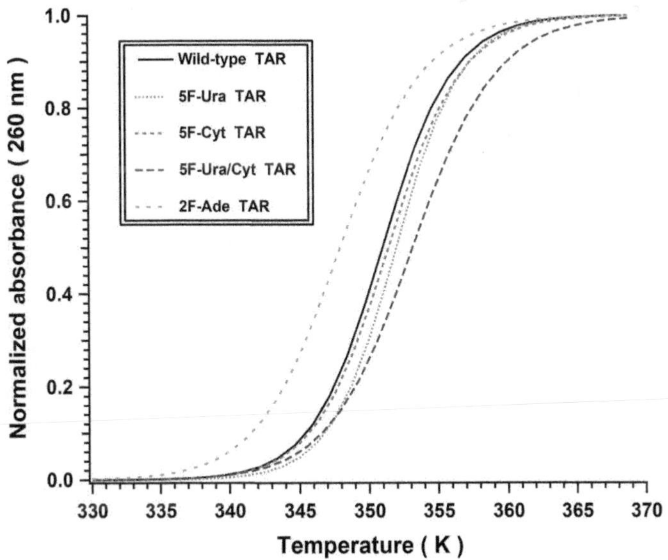

Figure 5 Thermal melting profiles obtained for wild-type (solid line), 2F-Ade- (dashed, wide gap line), 5F-Ura- (dotted line), 5F-Cyt- (dashed, narrow gap line), and 5F-Ura/Cyt-modified (long dashed, narrow gap line) HIV-2 TAR.

Table 2 Thermodynamic Parameters for Fluoro-Modified HIV-2 TAR RNA Obtained by UV Melting Profile Analysis

RNA	T_m (K)	$\Delta H_m°$ (kcal mol^{-1})	$\Delta S_m°$ (kcal mol^{-1} K^{-1})
wt-TAR	350.48 ± 0.69	−100.48 ± 1.44	−0.29 ± 0.01
5F-Ura-TAR	351.60 ± 0.58	−106.22 ± 1.38	−0.30 ± 0.01
5F-Cyt-TAR	351.19 ± 0.54	−99.33 ± 1.59	−0.28 ± 0.01
5F-Ura/Cyt-TAR	353.99 ± 0.96	−90.54 ± 7.90	−0.26 ± 0.02
2F-Ade-TAR	348.04 ± 0.20	−92.94 ± 0.74	−0.27 ± 0.01

thermodynamic stability of HIV-2 TAR as evident from a T_m value decreased by ca. 2.5 K (Table 2). Collectively, these findings represent important benchmarks and confirm the noninvasive character of the chosen nucleobase substitutions, which is a prerequisite for the interpretation of dynamic structures or binding properties obtained for such modified RNA targets.

4. DIAGNOSTICS OF ^{19}F-CHEMICAL SHIFTS

Trifluoroethanol (TFE) and trifluoroacetic acid (TFA) are frequently used as internal chemical shift referencing agents in ^{19}F NMR. However, we recommend (and apply throughout this work) indirect referencing using the published scaling factor of 0.940866982 (sphere ε-value for TFA) (Maurer & Kalbitzer, 1996). We have assigned the ^{19}F resonances for each 2-fluoroadenosine, 5-fluorouridine, and 5-fluorocytidine in modified TAR RNA samples. The ^{19}F nuclei of 2F-Ade-modified TAR are deshielded and shifted by ca. 115 ppm to lower fields with respect to ^{19}F nuclei located at the 5-position of the pyrimidine bases. The observed chemical shift dispersion in the 2F-Ade-, 5F-Ura-, and 5F-Cyt-substituted HIV-2 TAR RNAs is approximately 3 ppm reflecting the different microenvironments and stacking arrangements of involved fluorine nuclei (Fig. 6). It should be noted that this represents almost a fivefold increase in dispersion when compared to observed base H2 and H5 proton chemical shifts.

We further investigated the role of the bulged residues U23 and U25 by comparing ^{19}F chemical shifts obtained for 5F-Ura- and 5F-Cyt-substituted HIV-2 TAR with the bulge-deleted ΔTAR RNA. Elimination of the two-base bulge shifts the flanking U^{5F}40 by ca. 1.5 ppm to higher fields. In

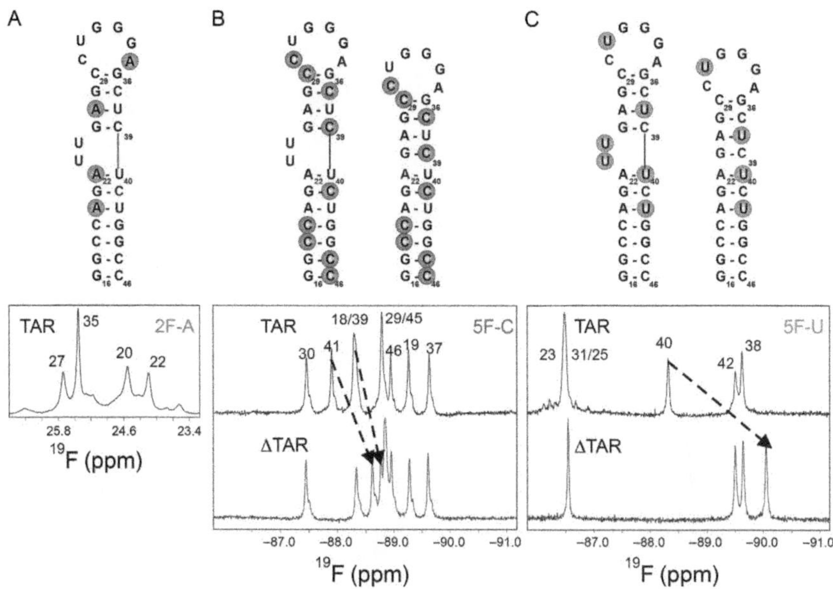

Figure 6 Sequence and secondary structural representation of HIV-2 TAR and bulge-deleted ΔTAR with (A) 2-fluoroadenosine substitutions, (B) 5-fluorocytidine substitutions, and (C) 5-fluorouridine substitutions. Circles highlight fluorine substitutions. ^{19}F one-dimensional spectrum of (A) 2F-Ade, (B) 5F-Cyt, and (C) 5F-Ura-substituted HIV-2 TAR and bulge-deleted ΔTAR, respectively. Assignments for ^{19}F resonances are indicated, and dashed lines connect the resonance frequencies of U40, C41, and C39 in the wild-type and deletion mutant RNA molecules.

analogy, the bulge-neighboring $C^{5F}39$ and 41 experience increased shielding and shift by approximately 0.5 and 1.0 ppm to higher fields, respectively.

Preparing full-length RNAs with homogenous 3′ ends is often a challenge (Milligan, Groebe, Witherell, & Uhlenbeck, 1987; Milligan & Uhlenbeck, 1989). We inverted the TAR G17-C45 base pair to a functionally tolerated wobble G17-U45 base pair, which provided a 5′-GUC-3′ sequence required for cleavage by an immediately adjacent hammerhead (HH) ribozyme in these constructs (Price, Ito, Oubridge, Avis, & Nagai, 1995; Wyszko et al., 1999).

The G·U wobble base pair has a comparable thermodynamic stability to that of Watson–Crick A-U and is nearly isomorphic (Varani & McClain, 2000). The ^{19}F chemical shifts serve as sensitive markers of RNA secondary structure, and significant shift differences are observed between Watson–Crick and the G·U wobble base pair. In agreement with Watson–Crick

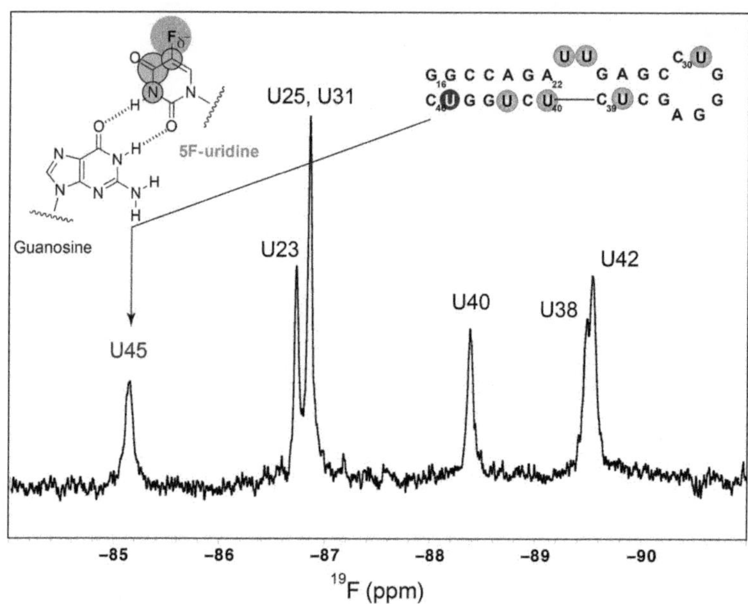

Figure 7 ^{19}F one-dimensional spectrum of 5F-Ura-substituted HIV-2 TAR featuring a hammerhead (HH) ribozyme sequence immediately preceded by a 5′-GU45C-3′ cleavage sequence. The left inset shows a schematic of a G·U^{5F} wobble base pair. Large and small circles symbolize δ^- and δ^+ illustrating the negative inductive effect of an electron-withdrawing ^{19}F substituent. The right inset shows sequence and secondary structural representation of the TAR-GU45C construct with 5-fluorouridine substitutions highlighted by circles. Assignments for ^{19}F resonances are given.

A-U^{5F} to G·U^{5F} wobble transitions observed in tRNA (Chu & Horowitz, 1989), a newly introduced U^{5F}45 is deshielded and shifted by approximately 4.5 ppm to lower fields compared with U^{5F}42 and U^{5F}38 that adopt Watson–Crick geometries. While the C4=O of uridine accepts a hydrogen bond from the N6-H amino group of adenosine in Watson–Crick A-U base-pairing interactions, it is the uridine's C2=O that acts as a hydrogen bond acceptor within G·U wobble base pairs. It should be noted that both 2F-Ade and 5F-Cyt modifications interfere with the HH-cleavage reaction acting in *cis* and are thus incompatible with this approach to obtaining homogenous 3′ ends of RNA transcripts (data not shown) (Fig. 7).

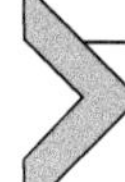

5. NMR-ASSIGNMENT STRATEGIES FOR FLUORO-MODIFIED RNA

Site-specific incorporation of 5F-Ura/Cyt and 2F-Ade modifications into RNA targets is incompatible with T7 RNA polymerase catalyzed *in vitro* transcription methods, thus necessitating an NMR assignment strategy to unambiguously assign the uniformly incorporated ^{19}F probes. However, the sequence-specific assignment of fluorine resonances in substituted RNA is straightforward and can be accomplished using an approach that combines ^{19}F,^{19}F-nuclear Overhauser effect spectroscopy (NOESY) and ^{1}H,^{1}H-NOESY, ^{19}F,^{1}H-heteronuclear NOESY (HOESY) (Rinaldi, 1983; Yu & Levy, 1984), and ^{1}H,^{19}F-heteronuclear multiple bond correlations (HMBC) (Hennig et al., 2006; Scott et al., 2004). In order to increase the ^{19}F spin density and to generate essentially one ^{19}F-"spy" probe per base pair, we synthesized a TAR RNA carrying both, 5F-Cyt- and 5F-Ura substitutions. The following ^{19}F,^{19}F-NOESY, ^{1}H,^{19}F-HOESY, and ^{1}H,^{19}F-HMBC experiments were recorded using a ca. 1.0 m*M* 5F-Ura/Cyt-labeled TAR RNA NMR sample in 500 μL of H_2O/D_2O (9:1) of NMR buffer (10 m*M* sodium phosphate buffer (pH 6.5), 150 m*M* NaCl).

5.1 General NMR Hardware Considerations

While (^{1}H-coupled) ^{19}F 1D- and 2D-homonuclear ^{19}F,^{19}F-NOESY experiments can be recorded without restriction (granted that the ^{1}H coil of the probe can be detuned to ^{19}F), all dual resonance (^{1}H/^{19}F) experiments require at least a dual resonance ^{1}H/^{19}F probe in combination with either two independent high-band channels or a (more commonly available) X-QNP switchbox allowing for transmitter output switching between ^{19}F and ^{1}H (within typically ≤20 μs; see, e.g., Bruker Topspin release pulse sequences *zgfhigqn* [or *zghfigqn*] for recording ^{1}H-decoupled ^{19}F- (or ^{19}F-decoupled ^{1}H-) 1D NMR experiments using a X-QNP switchbox). Other hardware requirements include band-selective ^{1}H- and ^{19}F-preamplifiers and/or radio frequency (rf) filter combinations of the type ^{19}F band-pass (BP)/^{1}H band-stop (BS) and ^{1}H BP/^{19}F BS that are inserted between probe and preamplifier.

All measurements shown below were carried out at 500 or 600 MHz, 298 K, on Bruker Avance spectrometers equipped with a 5-mm quadruple resonance (^{1}H/^{19}F-^{31}P/^{13}C-QXI) probe with shielded triple axis gradients along with a X-QNP switchbox and, e.g., a BLAXH 300/50 amplifier (featuring three rf-channels, two for ^{1}H/^{19}F, one for X frequencies) or alternatively with a 5-mm triple resonance (^{1}H-^{13}C/^{15}N-TXI) probe with the inner coil tuned to ^{19}F. Commonly achievable $\pi/2$-pulse widths using a ^{1}H/^{19}F-^{31}P/^{13}C-QXI probe and a BLAXH 300/50 amplifier operating at 600/564 MHz ^{1}H/^{19}F resonance frequencies are 9 μs (^{1}H, ca. 10–30 W) and 20 μs (^{19}F, ca. 30 W). Chemical shift anisotropy contributes significantly to transverse (T_2) relaxation of ^{19}F and, due to its square dependence on the external magnetic field, becomes the major source of ^{19}F line width at higher fields (>400 MHz) with adverse effects to sensitivity (Sykes & Hull, 1978).

5.2 ^{19}F,^{19}F-NOESY

While heteronuclear ^{19}F,^{1}H NOEs have been exploited on a regular basis to study fluorinated proteins and nucleic acids (Battiste & Newmark, 2006; Collier, Arnold, & Fisher, 1996), the usage of homonuclear 2D ^{19}F,^{19}F NOESY is not commonplace. Presumably, this can be attributed to the sparser nature of applications dealing with uniformly ^{19}F-labeled biomolecules. However, applications to small, fluorinated molecules have been described (Battiste, Jing, & Newmark, 2004). In order to observe transient ^{19}F–^{19}F NOEs interactions in 5F-pyrimidine-substituted RNA, mixing times (τ_{mix}) greater or equal to 400 ms should be employed (Hennig et al., 2007). Sequential dinucleotide steps, which involve fluoro-modified pyrimidines (5′-YpY-3′), are characterized by fluorine–fluorine distances of ca. 3.7 Å leading to observable ^{19}F-^{19}F NOESY correlations. A ^{19}F,^{19}F NOESY recorded on 5F-Ura/Cyt-modified TAR reveals that stacking interactions appear to be disrupted between the bulged 5F-uridines U^{5F}23 and U^{5F}25, as well as, C^{5F}30 and U^{5F}31 located in the apical loop. All other 5′-YpY-3′ steps generate NOE cross-peaks thereby greatly facilitating sequential assignments of 5F-Ura/Cyt-modified TAR (Fig. 8).

5.3 ^{19}F,^{1}H-HOESY

Heteronuclear ^{19}F,^{1}H NOEs are readily accessible owing to a high $\gamma(^{19}\text{F})$ and the 100% natural abundance of ^{19}F (Metzler, Leighton, & Lu, 1988). Importantly, like homonuclear ^{1}H,^{1}H NOEs, there is a value of the overall tumbling time at which the ^{19}F,^{1}H hetNOE is zero. The ^{19}F,^{1}H NOE assumes a value of 0.5 in the extreme narrowing limit, passes through zero before finally approaching its maximum value of -1 for slowly tumbling macromolecules. Whether one chooses to detect ^{19}F or ^{1}H in HOESY experiments depends on resolution and sensitivity criteria, satisfactory spectra can be obtained in either configuration (Hennig et al., 2007; Scott et al., 2004). Fluorine-19 detection has the advantage of negating the need for water suppression.

Typically, the ^{19}F,^{1}H dipole–dipole relaxation is dominated by the interaction between F5 and the proximal base H6 proton for 5-fluoropyrimidines

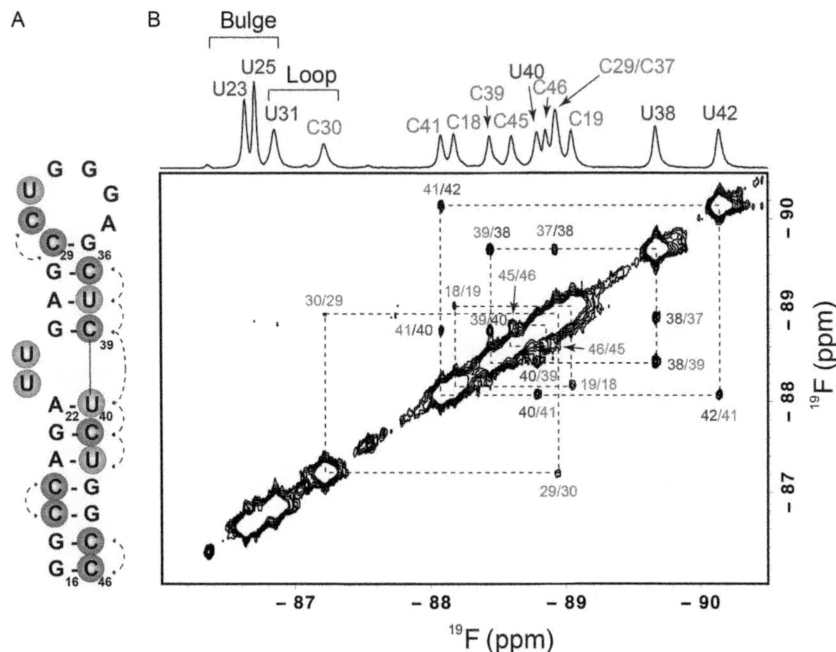

Figure 8 (A) Sequence and secondary structural representation of HIV-2 TAR with 5-fluorocytidine and 5 fluorouridine substitutions highlighted by circles. Dashed lines indicated observable cross-peaks between stacked 5F-Ura/Cyt-modified nucleobases. (B) Phase-sensitive ^{19}F,^{19}F-NOESY experiments (^{19}F resonance frequency of 470.5 MHz) showing fluorine connectivities between stacked bases in 5F-Ura/Cyt-substituted TAR. Assignments of individual homonuclear ^{19}F-NOESY cross-peaks are given. The NOESY mixing time τ_{mix} was 400 ms. (See the color plate.)

corresponding to a short internuclear distance of 2.58 Å. In H_2O, nucleotides carrying 5F-Cyt modifications show additional strong cross-peaks mediated by F5,H41 amino proton NOEs. However, spin diffusion between remotely dipolar-coupled protons efficiently spreads magnetization throughout the macromolecular RNA target at longer mixing times as shown in the conventional proton-detected ^{1}H,^{19}F-HOESY optimized for detecting exchangeable protons (Fig. 9; Scott et al., 2004).

In most cases, NMR assignments will be available for unmodified RNA targets before embarking on studies involving fluorinated base analogues. However, the transfer of previously obtained proton assignments is hampered by the pronounced proton (and carbon) chemical shift changes caused by fluorine substitutions. The influence of a fluorine substitution on the shielding of protons is expected to be strongest if these protons are in close

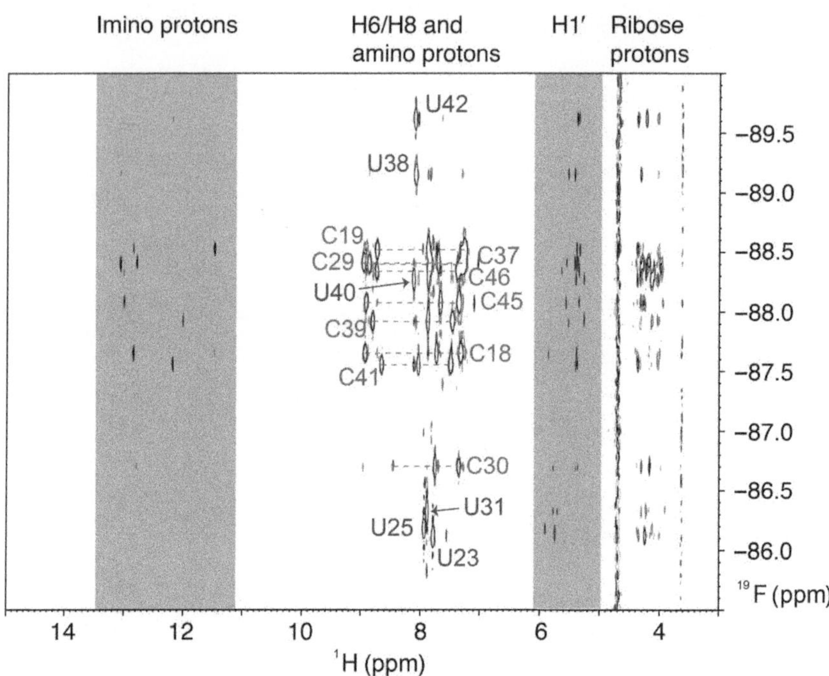

Figure 9 Proton-detected ^{1}H,^{19}F-HOESY (^{1}H resonance frequency of 600 MHz) of 5F-Ura/Cyt-modified TAR recorded in 90% H_2O/10% D_2O at 298 K showing cross-peaks between $^{19}F(\omega_1)$ nuclei and imino and anomeric (gray boxes) as well as aromatic, amino, and ribose ^{1}H protons (ω_2). While assignments of individual ^{1}H,^{19}F-HOESY cross-peaks are given for 5F-Ura F-H6 aromatic correlations, dashed lines connect assigned 5F-Cyt F-H42/H41 amino correlations. Efficient water suppression combined with minimal baseline distortion was achieved employing excitation sculpting (Hwang & Shaka, 1995). The HOESY mixing time τ_{mix} in the experiment was 400 ms.

proximity to the fluorine atom. Thus, exchangeable uracil imino protons in Watson–Crick-type A^{2F}-U base pairs are more shielded and experience pronounced upfield shifts ($\Delta\delta_{average}$(H3) ca. −2.6 ppm) compared to canonical A-U base pairs (Scott et al., 2004). Similarly, guanine imino protons in Watson–Crick-type G-C^{5F} base pairs shift upfield ($\Delta\delta_{average}$(H1) ca. −0.3 ppm) compared to canonical G-C base pairs. Cytosine amino protons in G-C^{5F} base pairs are deshielded and shifted by approximately 0.47 (H42) and 0.54 (H41) ppm, respectively, to lower fields compared with unmodified Watson–Crick geometries. The presence of an electron-withdrawing fluorine substituent in the 5-position causes pronounced line broadening, deshielding, and downfield shifts ($\Delta\delta_{average}$(H3) ca. 0.7 ppm) of uracil imino protons in A-U^{5F} base pairs with respect to regular A-U base pairs (Hennig et al., 2007). Our published proton chemical shift assignments reveal that all base H6 protons in U^{5F} and C^{5F}-modified nucleotides are deshielded ($\Delta\delta_{average}(H6) = 0.21 \pm 0.06$ ppm) as a consequence of the strongly electron-withdrawing nature of the fluorine substituent (Hennig et al., 2006). There is essentially no chemical shift effect for the anomeric H1′ protons ($\Delta\delta_{average}(H1') = -0.03 \pm 0.04$ ppm), thus sensitive H1′,F correlations as described below to transfer assignments in 5-fluoropyrimidine-substituted RNA are desirable.

5.4 ^{1}H,^{19}F-HMBC

We have shown previously that a proton-detected ^{1}H,^{19}F-HMBC experiment provides a valuable and sensitive alternative to ^{1}H,^{19}F-HOESY approaches for generating two-dimensional H6,F correlations that render overlaps in one-dimensional ^{19}F spectra. The gradient-selected, through-bond HMBC correlations (Cicero, Barbato, & Bazzo, 2001; Willker, Leibfritz, Kerssebaum, & Bermel, 1993) seem advantageous in the context of contemporary, inverse probe technology (such as the QXI) optimized for ^{1}H detection. As shown in Fig. 10, high-quality ^{1}H,^{19}F-HMBC

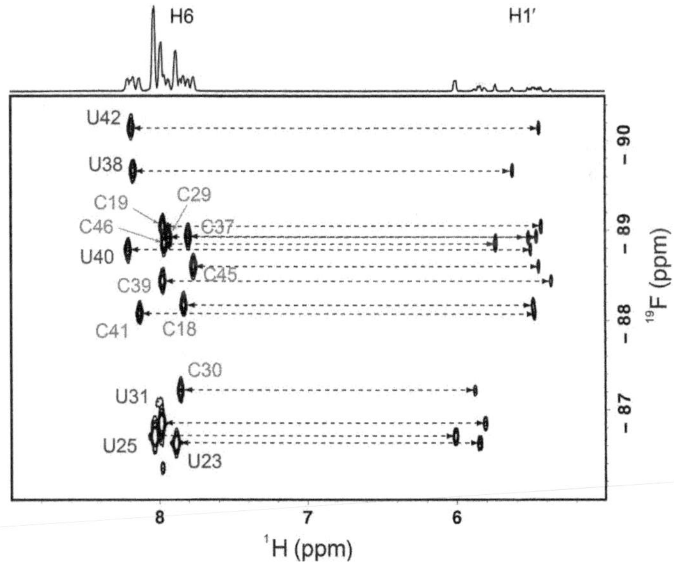

Figure 10 Proton-detected ^{1}H,^{19}F-HMBC (^{1}H resonance frequency of 500 MHz) of 5F-Ura/Cyt-modified TAR recorded in 90% H_2O/10% D_2O at 298 K. The delay for ^{1}H-^{19}F *J*-coupling evolution was set to 80 ms. Near absorption profiles in ω_1 were obtained prior to magnitude calculation in ω_2. Assignments of individual HMBC cross-peaks between ^{19}F nuclei and aromatic H6 protons are given. Dashed lines connect aromatic H6 and anomeric H1′ proton correlations mediated by ^3J(H6,F) and ^5J(H1′,F), respectively.

experiments tailored for the detection of base H6,F and anomeric H1′,F correlations (with the delay for ^{1}H-^{19}F long-range coupling evolution set to ~60–80 ms) can be obtained in about 2 h on a ca. 1.0 m*M* HIV-2 TAR RNA samples carrying both 5F-Cyt and 5F-Ura modifications. Due to the fact that anomeric H1′ proton chemical shift is essentially unaffected by fluorine substitutions in the nucleobases, the H1′ proton correlations provide valuable anchor points to transfer assignments from, e.g., an unmodified RNA sample.

6. COUPLING MEASUREMENTS IN FLUORO-MODIFIED RNA

6.1 Scalar *J*-Coupling Measurements

Early attempts to record ^{1}H,^{13}C-correlations in 5-fluoropyrimidine-substituted TAR lead to the observation of large chemical shift changes in the vicinity of 5F-Ura- and 5F-Cyt substitutions (pyrimidine C6 carbons are more shielded and shifted by ca. −15 ppm to higher fields) and unexpected splitting caused by scalar couplings involving the newly introduced ^{19}F (Fig. 11A). All altered chemical shifts and scalar couplings in 5-fluoro-modified uridine and cytidine are summarized in Table 3 and Fig. 11B and C. The large two-bond ^{2}J(C,F) couplings require ^{19}F-decoupling to obtain ^{1}H,^{13}C-correlations with optimal sensitivity but, on the other hand, can be advantageous in the context of residual dipolar coupling (RDC) measurements (Fig. 13).

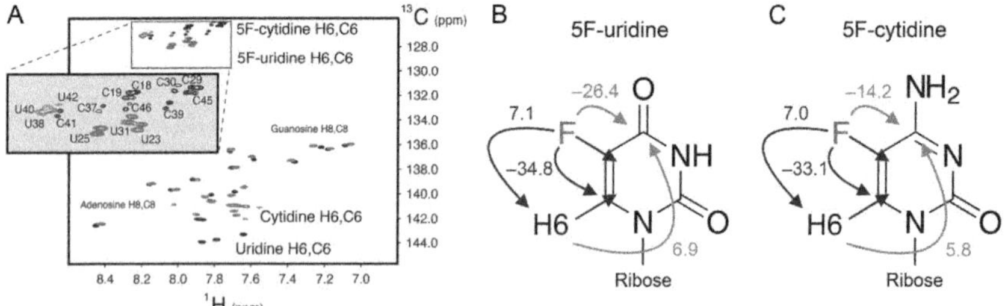

Figure 11 (A) ^{19}F-coupled ^{1}H,^{13}C HSQC experiments (^{1}H resonance frequency of 600 MHz) recorded for 5F-Ura- and 5F-Cyt-substituted TAR. Assignments of individual HSQC cross-peaks between aromatic H6 protons and C6 carbons (split in ω_1 by ^{2}J(C6,F) in 5-fluoropyrimidine-substituted TAR are given in the inset. Scalar *J*-couplings within (B) 5-fluorouridine and (C) 5-fluorocytidine obtained in d_4-methanol at 600 MHz and 293 K. (See the color plate.)

Table 3 ^{13}C- and ^{1}H Chemical Shifts and Scalar *J*-Couplings for 5F-Cytidine and 5F-Uridine, Respectively

Nuclei or Associated *J*-Coupling	5-Fluorocytidine	5-Fluorouridine
C2[a]	159.57 ppm	153.87 ppm
^{3}J(C2,H6)	6.3 Hz	7.8 Hz
^{3}J(C2,H1′)	1.4 Hz	1.2 Hz
C4	162.38 ppm	162.30 ppm
^{3}J(C4,H6)	5.8 Hz	6.9 Hz
^{2}J(C4,F5)	−14.2 Hz	−26.4 Hz
C5	141.21 ppm	144.62 ppm
^{1}J(C5,F5)	241.6 Hz	231.7 Hz
^{2}J(C5,H6)	−3.6 Hz	−4.5 Hz
C6	129.86 ppm	129.20 ppm
^{1}J(C6,H6)	182.3 Hz	182.3 Hz
^{2}J(C6,F5)	−33.1 Hz	−34.8 Hz
^{3}J(C6,H1′)	3.4 Hz	3.8 Hz
H6[b]	8.51 ppm	8.49 ppm
^{3}J(H6,F5)	7.0 Hz	7.1 Hz
H1′	5.99 ppm	6.06 ppm
^{3}J(H1′,H2′)	3.1 Hz	2.2 Hz
^{5}J(H1′,F5)	1.5 Hz	1.5 Hz

[a] ^{13}C chemical shifts are referenced indirectly using the ^{13}C/^{1}H (DSS) frequency ratio.
[b] ^{1}H chemical shifts are referenced according to external DSS.

We also measured long-range ^{5}J(H1′,F) scalar couplings in 5-fluoropyrimidine-substituted RNA that can be used to restrict the glycosidic torsion angle χ defining the orientation of the base with respect to the ribose sugar moiety. The observable nature of this five-bond interaction can be qualitatively explained on the basis of an extended W configuration formed by the intervening H1′—C1′—N1—C6—C5—F5 bonds (Fig. 12). A ^{1}H,^{19}F-spin state selective excitation (S^{3}E) (Meissner, Duus, & Sorensen, 1997a,b)-^{19}F-α,β-edited NOESY experiment simplifying the conventional exclusive correlation spectroscopy (E.COSY) (Griesinger, Sorensen, & Ernst, 1985, 1986, 1987)-type cross-peaks allowed

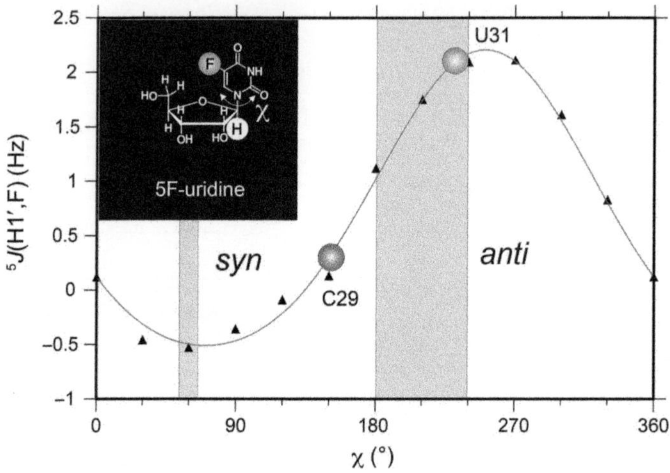

Figure 12 Theoretical Karplus relation ($y=0.37\cos^2(x-70.98°)-1.57\cos(x-70.98°)+0.62$) obtained for an isolated 5-fluorouridine adopting a C3′-endo sugar pucker. Experimental ^5J(H1′,F) couplings for $U^{5F}31$ and $C^{5F}29$ defining the upper and lower limit of measured values, respectively, are indicated. Regions associated with *syn* and *anti* orientations around the glycosidic χ torsion are highlighted using gray boxes. The inset shows 5F-uridine with the long-range coupled H1′ and F5 nuclei circled and the intervening glycosidic χ torsion shown as an arrow.

for the extraction of these scalar through bond interactions. *J*-couplings could be extracted from the inspection of intraresidual (H1′,H6) NOE cross-peaks in the resulting ^{19}F-α,β-edited NOESY subspectra (Hennig et al., 2006). Conceptually similar experiments for the measurement of scalar and dipolar *J*(H,F)-coupling interactions in RNA specifically labeled with ^{19}F at the 2′-ribose position were introduced by Luy and Marino (Luy, Barchi, & Marino, 2001; Luy & Marino, 2001). While theoretical Karplus relations obtained using density functional theory (DFT) calculations revealed a measurable dependence on the type of the nucleoside, the sugar pucker appeared to be negligible. Inevitably, all ^5J(H1′,F) scalar couplings measured for pyrimidines in 5F-Ura- or 5F-Cyt-modified TAR fall into ranges compatible with the preferred *anti* orientation of the χ glycosidic torsion.

6.2 RDC Measurements

The structure elucidation of elongated molecules such as nucleic acids is typically hampered by the lack of long-range NOEs. RDCs measured in aligned samples provide powerful orientational restraints for NMR structure

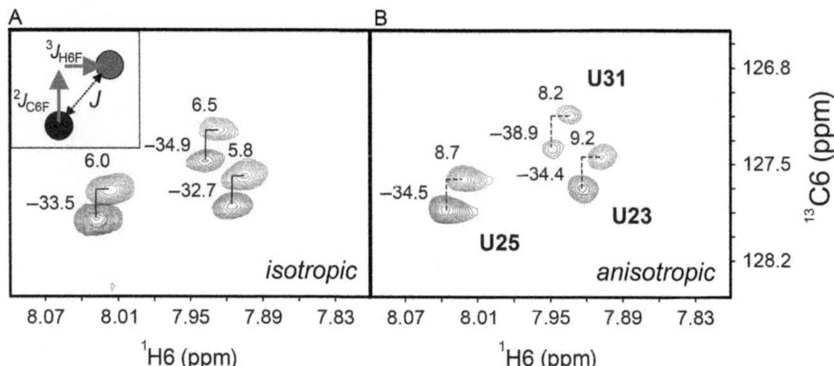

Figure 13 (S^{3}E)-^{19}F-α,β-edited ^{1}H6,^{13}C6-HSQC experiment for (A) 5F-Ura- and (B) 5F-Ura-substituted TAR partially aligned using ~25 mg mL^{-1} Pf1 phage. Individual subspectra shown are superimposed and correspond to the passive ^{19}F spins adopting α- and β-states, respectively. Assignments of U^{5F}23, U^{5F}25, and U^{5F}31 ^{1}H6,^{13}C6 cross-peaks are given and extracted *J* (*isotropic*) and *J*+*D* (*anisotropic*) couplings are reported. The inset shows a schematic superposition of the two subspectra (^{1}H6,^{13}C6 HSQC cross-peak) with the ^{19}F spin in either α- (red) and β-states (blue). The displacement vector *J* between the submultiplets consists of the ^2J(C6,F) coupling in ω_1 and the ^3J(H6,F) coupling in ω_2. (See the color plate.)

determination to some extent alleviating the absence of long-range distance restraints. RNA samples can be partially aligned by the addition of ~25 mg mL^{-1} Pf1 phage (Hansen, Hanson, & Pardi, 2000; Hansen, Mueller, & Pardi, 1998).

Due to the anisotropic distribution of electrons in the three 2p orbitals, scalar couplings involving ^{19}F are the largest known (Gakh, Gakh, & Gronenborn, 2000) and can be exploited in E.COSY experiments. In particular, the rather large two-bond ^2J(C6,F) scalar couplings observed in 5-fluoropyrimidines (Fig. 11) facilitate the extraction of otherwise inaccessible RDCs in an adapted S^{3}E-^{19}F-α,β-edited heteronuclear single quantum coherence (HSQC) experiment previously introduced to measure small couplings within nucleobases (Zidek, Wu, Feigon, & Sklenar, 2001). This permits the measurement of two different RDCs, ^2D(C6,F) and ^3D(H6,F) per 5-fluoroprimidine base for structure refinement of 5F-Ura- and/or 5F-Cyt-modified RNA at ^{13}C natural abundance (Fig. 13).

7. SOLVENT-INDUCED ISOTOPE SHIFTS

Solvent-induced isotope shift (SIIS) effects have been used to diagnose the accessibility of fluorine nuclei to solvent in fluoro-substituted nucleotide analogues and nucleic acids (Chen, Shafer, & Kuntz, 1997; Chu & Horowitz, 1991; Metzler & Lu, 1989). The assumption being made is that a solvent-exposed nucleotide will experience larger chemical shift changes when H_2O is replaced by D_2O. The fluorine signal from 5F-dU can serve as an "exposed" reference and is deshielded about 0.39 ppm when the solvent is changed from H_2O to D_2O (Metzler & Lu, 1989). On the other hand, buried-fluorinated nucleotides experience little-to-no SIIS. Considering solvent exposure in modified RNA is straightforward in practice by monitoring the linear dependence of ^{19}F chemical shift with replacement of H_2O with D_2O.

Figure 14A and B show series of 1D ^{19}F NMR spectra of the 5F-Cyt- and 5F-Ura-modified TAR as a function of H_2O/D_2O. After linear

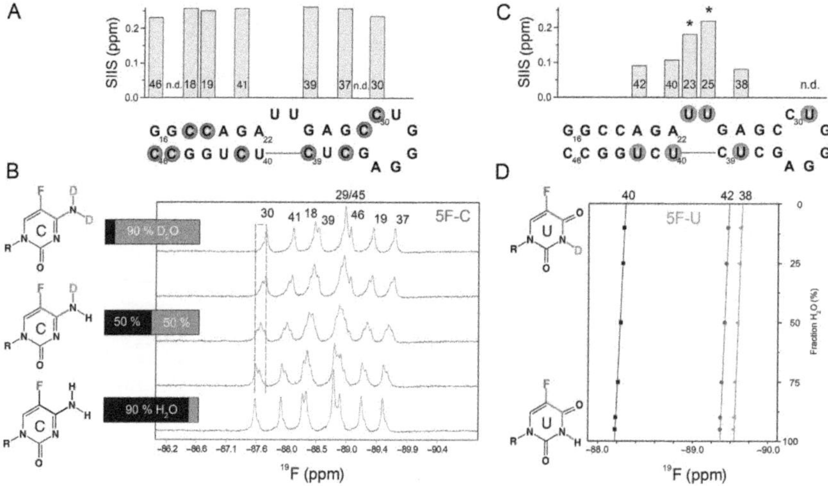

Figure 14 (A) SIIS of 5-fluorocytidine-substituted TAR. (B) ^{19}F NMR spectra of 5-fluorocytidine-substituted TAR obtained at 308 K (^{19}F resonance frequency of 564.6 MHz). After considering exchange with the bulk solvent, the insets indicate the most populated species of the exocyclic 5-fluorocytidine amino groups at 90% H_2O/10% D_2O ($N4H_2$), 50% H_2O/50% D_2O (N4HD), and 10% H_2O/90% D_2O ($N4D_2$), respectively. $C^{5F}29$ and $C^{5F}45$ overlap thus no SIIS was determined. (C) SIIS of 5-fluorouridine-substituted TAR. SIIS of 5F-Ura nucleotides marked with an asterisk were obtained from proton-detected ^{1}H,^{19}F-HMBC due to overlap in the 1D ^{19}F NMR spectra. No SIIS was determined for $U^{5F}31$ due to overlap and poor sensitivity. (D) Linear SIIS-H_2O/D_2O correlations obtained for $U^{5F}38$, 40, and 42 of 5-fluorouridine-substituted TAR obtained at 308 K (^{19}F resonance frequency of 564.6 MHz). After considering exchange with the bulk solvent, the insets indicate the most populated species of the imino 5-fluorouridine nitrogen at 90% H_2O/10% D_2O (N3H) and 10% H_2O/90% D_2O (N3D), respectively.

regression, the nucleotide-specific SIIS can be determined from the corresponding slopes. Invariably, upfield shifts are detected when H_2O was replaced by D_2O. The observed SIIS for 5F-Cyt-modified TAR is large and averages 0.25 ppm. However, as described previously (Puffer et al., 2009), the observed effect does not solely depend on solvent exposure of the ^{19}F nucleus. The observed SIIS appears to be dominated by two sequential, four-bond ^{2}H/^{1}H isotope effects stemming from solvent exchange of the exocyclic H41 and H42 amino protons in the modified nucleobases. Interestingly, the smallest chemical shift changes among the 5-fluorocytidines were observed for the terminal G16-C^{5F}46 (0.232 ppm) base pair and the unpaired C^{5F}30 (0.236 ppm) loop 5-fluorocytidine. Moreover, these nucleotides appear inversely correlated with intrinsic amino exchange rates. Detectable SIIS for 5F-Ura-modified TAR range from 0.220 to 0.082 ppm for the bulged and unpaired U^{5F}25 and the Watson–Crick paired U^{5F}38. Because multiple explanations continue to circulate to physically explain SIIS (Hansen, Dettman, & Sykes, 1985; Lau & Gerig, 1996; Muller, 1977), the possibility that an analogous four-bond ^{2}H/^{1}H isotope effect involving the N3 imino proton indirectly contributing to the SIIS of 5-fluorouridine is worthy of consideration.

8. LIGAND-INDUCED CHEMICAL SHIFT CHANGES

New methodologies to facilitate the screening of potential RNA drug targets are sorely needed. Fluorine-19 NMR can make important contributions to understand specificity determinants for the Tat-binding pocket of TAR at atomic resolution. Mutational studies have shown that U23 is the only essential nucleotide in the three-base HIV-1 TAR U23C24U25-bulge for Tat recognition. The two other bulged nucleotides can be freely substituted or even be replaced with nonnucleotides. The two-base U23U25-bulge HIV-2 TAR has similar Tat-binding affinity.

The binding of argininamide to fluorinated TAR RNA can be directly monitored by ^{19}F NMR. The ^{19}F chemical shifts of 2F-Ade (minor groove reporter) (Fig. 15) and 5F-Cyt (major groove reporter) (Fig. 16) and substitutions are shown as a function of total argininamide concentration. For both modified TAR constructs, chemical shift changes were observed in direct relation to increasing argininamide concentration. By identifying perturbations of assigned ^{19}F resonances, not only are ligands identified but also their respective binding sites are localized. Furthermore, for ligand–receptor (L–R) pairs in fast exchange on the NMR time scale ($K_D = 10\ \mu M$ or

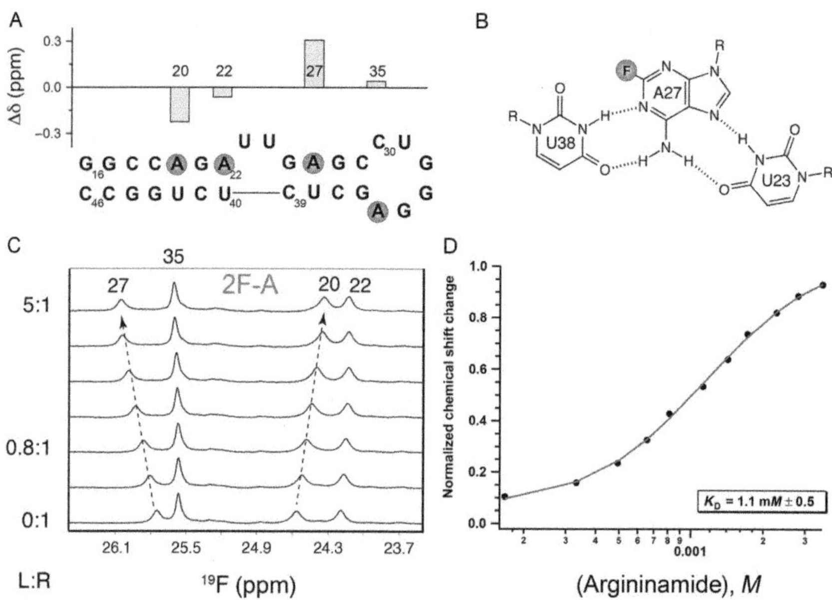

Figure 15 (A) Argininamide-induced chemical shift changes ($\Delta\delta = \delta$(2F-A TAR·Arg) $-$ δ(2F-A TAR) (ppm)) of 2-fluoroadenosine-substituted TAR. (B) Upon binding of argininamide (a Tat mimic), the TAR RNA undergoes a conformational change in the bulge region where the essential nucleotides, U38, $A^{2F}27$, and U23, form a base triple. (C) ^{19}F NMR spectra of 2-fluoroadenosine-substituted TAR obtained at 298 K (^{19}F resonance frequency of 564.6 MHz) as a function of ligand:receptor (L:R) ratio. (D) The K_D for the 2F-Ade TAR–argininamide interactions was established by fitting the ligand-induced chemical shift changes to the equation: $\Delta\delta = 0.5\ \Delta\delta_{max}\{1 + X + K_D/[R]_0 - [(1 + X + K_D/[R]_0)^2 - 4X]^{1/2}\}$ where X is the molar L:R ratio and $[R]_0$ is the total concentration of the receptor, 2F-Ade TAR.

greater) such as TAR–argininamide, affinities can be determined from an analysis of ligand-induced chemical shift changes ($\Delta\delta$).

Upon binding of argininamide (or Tat and Tat peptides), the TAR RNA undergoes a major conformational change in the bulge region. In the bound form, the essential nucleotides, U38, A27, and U23, form a base-triple (Fig. 15B), which results in an opening of the major groove for peptide recognition (Brodsky, Erlacher, & Williamson, 1998; Brodsky & Williamson, 1997; Hennig & Williamson, 2000; Puglisi, Tan, Calnan, Frankel, & Williamson, 1992). Consequently, the residue $A^{2F}27$ is deshielded and experiences the largest shift (0.31 ppm) among 2F-adenosine nucleotides in HIV-2 TAR to lower fields ($\Delta\delta = \delta$(2F-A TAR·Arg) $-$ δ(2F-A TAR) (ppm)) upon argininamide binding (Fig. 15A and C). Argininamide and TAR are in fast exchange on the NMR time scale; thus, collectively induced

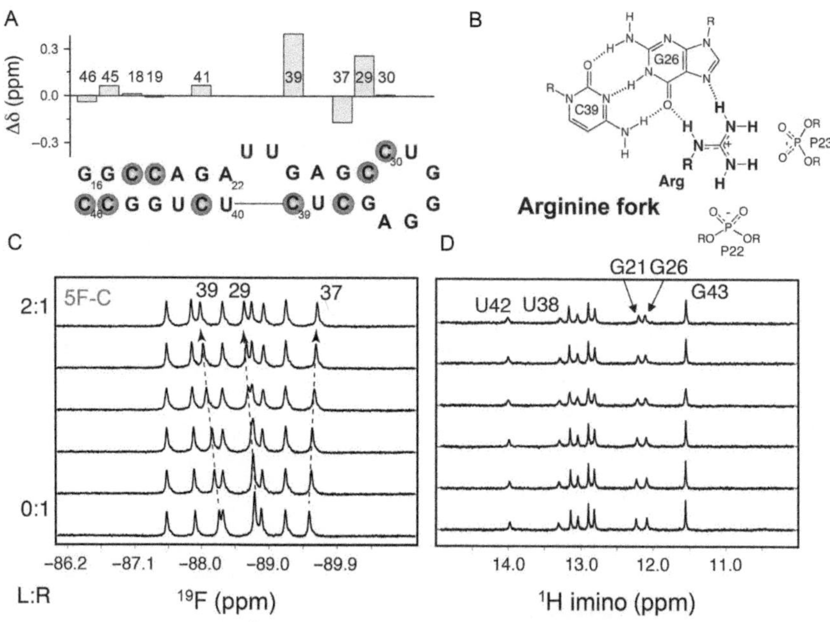

Figure 16 (A) Argininamide-induced chemical shift changes ($\Delta\delta = \delta$(5F-C TAR·Arg) − δ(5F-C TAR) (ppm)) of 5-fluorocytidine-substituted TAR. (B) Argininamide interacts specifically with the $C^{5F}39$-G26 base pair and two backbone phosphate groups form an arginine fork. (C) ^{19}F NMR spectra of 5-fluorocytidine-substituted TAR obtained at 298 K (^{19}F resonance frequency of 564.6 MHz) as a function of ligand:receptor (L:R) ratio. (D) ^{1}H imino NMR spectra of 5-fluorocytidine-substituted TAR obtained at 298 K (^{1}H resonance frequency of 600 MHz) as a function of ligand:receptor (L:R) ratio.

chemical shift changes of 2F-adenosines were fitted by nonlinear regression to yield a K_D of 1.1 ± 0.5 mM (Fig. 15D; Fielding, 2007), which is in excellent agreement with affinities obtained from fluorescence-based assays (Bradrick & Marino, 2004).

The guanidinium group of argininamide engages the bulge-flanking G26-C39, as well as, the phosphate oxygens of P22 and P23 in the major groove through hydrogen bonding interactions forming an arginine fork (Fig. 16B; Calnan, Tidor, Biancalana, Hudson & Frankel, 1991). As a result, $C^{5F}39$ is deshielded and significantly shifted by 0.4 ppm to lower fields ($\Delta\delta = \delta$(5F-C TAR·Arg) − δ(5F-C TAR) (ppm)) (Fig. 16A and C).

We and others (Fischer et al., 1999) also probed the TAR–argininamide interaction using an RNA receptor molecule carrying uniform 5-fluorouridine substitutions (data not shown). As expected, the dramatic change in chemical environment associated with $U^{5F}38$-A27-$U^{5F}23$ base-triple formation significantly shields $U^{5F}23$ with its ^{19}F resonance shifted upfield by −1.09 ppm. On the other hand, the $U^{5F}40$'s ^{19}F resonance frequency experiences a 0.33 ppm downfield shift, which is consistent with nearly coaxial stacking of upper and lower TAR stems in the argininamide-bound state.

Some remarkable features illustrating the value and complementarity of ^{19}F NMR-based screening should be emphasized: the chemical shift dispersion of ^{19}F 2-fluoroadenosine and 5-fluorocytidine resonances (~2.5–3.0 ppm) in the TAR equals the corresponding ^{1}H-imino dispersion (~3.0 ppm); the ^{19}F chemical shift change that $U^{5F}23$, $U^{5F}40$, $A^{2F}27$, $C^{5F}39$, and $C^{5F}37$ experience upon argininamide binding greatly exceeds the ^{1}H chemical shift change of the imino resonances (Fig. 16D). Finally, monitoring the ^{19}F chemical shift changes is not limited to lower temperatures, a necessity for the observation of exchangeable protons involved in noncanonical interactions.

9. CONCLUSION AND REMARKS

This chapter outlines methodology that can be applied to incorporate fluorinated nucleotide analogues 5F-UTP, 5F-CTP (Hennig et al., 2006, 2007), and 2F-ATP (Scott et al., 2004) into RNA. The ^{19}F chemical shifts serve as sensitive probes of RNA secondary structure. While ^{19}F nuclei shifted to the high-field edge of the spectrum are associated with canonical, Watson–Crick base-pairing interactions, the low-field edge of the ^{19}F 1D-spectrum signals irregular stacking interactions typically found in (single stranded) bulges or loops. The described 2F-Ade-, 5F-Ura-, and 5F-Cyt substitutions induce chemical shift changes to proximal proton and carbon nuclei thereby increasing the corresponding chemical shift dispersions. Scalar *J*-couplings are sensitive to the electronic environment of the coupled nuclear spins. The ^{19}F nucleus has lone pair electrons, which lead to large, observable long-range *J*-couplings such as $^{5}J(H1',F)$. Furthermore, large $^{2}J(C6,F)$ couplings can be utilized in E.COSY-type experiments to determine novel RDCs for RNA structure determination. The slightly lower sensitivity of ^{19}F with respect to ^{1}H, as well as, increased linewidth at higher magnetic fields are offset by the much larger chemical shift range and the extraordinary sensitivity of ^{19}F resonance frequencies to changes in their chemical environment induced by, e.g., solvents or ligands.

ACKNOWLEDGMENTS

We thank James R. Williamson (The Scripps Research Institute) for support and encouragement and Markéta Munzarová and Vladimír Sklenář (Masaryk University, Brno) for DFT calculations. We acknowledge the help of Wolfgang Bermel (Bruker BioSpin GmbH) with all aspects of ^{19}F NMR data collection and are grateful for many helpful discussions, suggestions, and comments. Support was provided by National Science Foundation (NSF) Grant MCB 0845512 and by National Center for Research Resources (NCRR), a component of the National Institutes of Health (NIH), Grant RR024442-01A1 to M.H. This content is solely the responsibility of the authors and does not necessarily represent the official views of the funding agencies.

REFERENCES

Alvarez, S. (2013). A cartography of the van der Waals territories. *Dalton Transactions*, *42*(24), 8617–8636. http://dx.doi.org/10.1039/c3dt50599e.

Barchi, J. J., Jr., Jeong, L. S., Siddiqui, M. A., & Marquez, V. E. (1997). Conformational analysis of the complete series of 2′ and 3′ monofluorinated dideoxyuridines. *Journal of Biochemical and Biophysical Methods*, *34*(1), 11–29.

Battiste, J. L., Jing, N., & Newmark, R. A. (2004). 2D ^{19}F/^{19}F NOESY for the assignment of NMR spectra of fluorochemicals. *Journal of Fluorine Chemistry*, *125*(9), 1331–1337. http://dx.doi.org/10.1016/j.jfluchem.2004.03.011.

Battiste, J., & Newmark, R. A. (2006). Applications of ^{19}F multidimensional NMR. *Progress in Nuclear Magnetic Resonance Spectroscopy*, *48*(1), 1–23. http://dx.doi.org/10.1016/j.pnmrs.2005.10.002.

Bevilacqua, P. C., Brown, T. S., Nakano, S., & Yajima, R. (2004). Catalytic roles for proton transfer and protonation in ribozymes. *Biopolymers*, *73*(1), 90–109. http://dx.doi.org/10.1002/bip.10519.

Bradrick, T. D., & Marino, J. P. (2004). Ligand-induced changes in 2-aminopurine fluorescence as a probe for small molecule binding to HIV-1 TAR RNA. *RNA*, *10*(9), 1459–1468. http://dx.doi.org/10.1261/rna.7620304.

Brodsky, A. S., Erlacher, H. A., & Williamson, J. R. (1998). NMR evidence for a base triple in the HIV-2 TAR C-G.C+ mutant-argininamide complex. *Nucleic Acids Research*, *26*(8), 1991–1995.

Brodsky, A. S., & Williamson, J. R. (1997). Solution structure of the HIV-2 TAR-argininamide complex. *Journal of Molecular Biology*, *267*(3), 624–639. http://dx.doi.org/10.1006/jmbi.1996.0879.

Calnan, B. J., Biancalana, S., Hudson, D., & Frankel, A. D. (1991). Analysis of arginine-rich peptides from the HIV Tat protein reveals unusual features of RNA-protein recognition. *Genes & Development*, *5*(2), 201–210.

Calnan, B. J., Tidor, B., Biancalana, S., Hudson, D., & Frankel, A. D. (1991). Arginine-mediated RNA recognition: The arginine fork. *Science*, *252*(5009), 1167–1171.

Chen, Q., Shafer, R. H., & Kuntz, I. D. (1997). Structure-based discovery of ligands targeted to the RNA double helix. *Biochemistry*, *36*(38), 11402–11407.

Chu, W. C., & Horowitz, J. (1989). F-19 NMR of 5-fluorouracil-substituted transfer-RNA transcribed in vitro—Resonance assignment of fluorouracil-guanine base-pairs. *Nucleic Acids Research*, *17*(18), 7241–7252.

Chu, W. C., & Horowitz, J. (1991). Recognition of *Escherichia coli* valine transfer RNA by its cognate synthetase: A fluorine-19 NMR study. *Biochemistry*, *30*(6), 1655–1663.

Cicero, D. O., Barbato, G., & Bazzo, R. (2001). Sensitivity enhancement of a two-dimensional experiment for the measurement of heteronuclear long-range coupling constants, by a new scheme of coherence selection by gradients. *Journal of Magnetic Resonance*, *148*(1), 209–213.

Collier, A. K., Arnold, J. R. P., & Fisher, J. (1996). NMR investigations of fluorine-labelled RNA: Application of two-dimensional heteronuclear experiments to study single-strand conformation. *Magnetic Resonance in Chemistry*, *34*(3), 191–196.

Deleavey, G. F., & Damha, M. J. (2012). Designing chemically modified oligonucleotides for targeted gene silencing. *Chemistry & Biology*, *19*(8), 937–954. http://dx.doi.org/10.1016/j.chembiol.2012.07.011.

Fauster, K., Kreutz, C., & Micura, R. (2012). 2′-SCF3 uridine—A powerful label for probing structure and function of RNA by ^{19}F NMR spectroscopy. *Angewandte Chemie International Edition in English*, *51*(52), 13080–13084. http://dx.doi.org/10.1002/anie.201207128.

Fielding, L. (2007). NMR methods for the determination of protein–ligand dissociation constants. *Progress in Nuclear Magnetic Resonance Spectroscopy*, *51*(4), 219–242. http://dx.doi.org/10.1016/j.pnmrs.2007.04.001.

Fischer, A., Gdaniec, Z., Biala, E., Łożyński, M., Milecki, J., & Adamiak, R. W. (1996). ^{19}F NMR of RNA. The structural and chemical aspects of 5-fluoro-cytidine and-uridine labelling of oligoribonucleotides. *Nucleosides and Nucleotides*, *15*(1–3), 477–488. http://dx.doi.org/10.1080/07328319608002399.

Fischer, A., Gdaniec, Z., Olejniczak, M., Bielecki, L., & Adamiak, R. W. (1999). Does 29-mer RNA hairpin of the HIV-1 TAR RNA sequence bind magnesium? ^{19}F-NMR and modelling studies. *Nucleic Acids Symposium Series*, *42*, 117–118.

Gakh, Y. G., Gakh, A. A., & Gronenborn, A. M. (2000). Fluorine as an NMR probe for structural studies of chemical and biological systems. *Magnetic Resonance in Chemistry*, *38*(7), 551–558.

Griesinger, C., Sorensen, O. W., & Ernst, R. R. (1985). Two-dimensional correlation of connected NMR transitions. *Journal of the American Chemical Society*, *107*(22), 6394–6396.

Griesinger, C., Sorensen, O. W., & Ernst, R. R. (1986). Correlation of connected transitions by two-dimensional NMR-spectroscopy. *Journal of Chemical Physics*, *85*(12), 6837–6852.

Griesinger, C., Sorensen, O. W., & Ernst, R. R. (1987). Practical aspects of the E-COSY technique—Measurement of scalar spin–spin coupling-constants in peptides. *Journal of Magnetic Resonance*, *75*(3), 474–492.

Hansen, P. E., Dettman, H. D., & Sykes, B. D. (1985). Solvent-induced deuterium isotope effects on ^{19}F chemical shifts of some substituted fluorobenzenes. Formation of inclusion complexes. *Journal of Magnetic Resonance (1969)*, *62*(3), 487–496. http://dx.doi.org/10.1016/0022-2364(85)90217-3.

Hansen, M. R., Hanson, P., & Pardi, A. (2000). Filamentous bacteriophage for aligning RNA, DNA, and proteins for measurement of nuclear magnetic resonance dipolar coupling interactions. *Methods in Enzymology*, *317*, 220–240.

Hansen, M. R., Mueller, L., & Pardi, A. (1998). Tunable alignment of macromolecules by filamentous phage yields dipolar coupling interactions. *Nature Structural Biology*, *5*(12), 1065–1074.

Hennig, M., Munzarova, M. L., Bermel, W., Scott, L. G., Sklenar, V., & Williamson, J. R. (2006). Measurement of long-range ^{1}H-^{19}F scalar coupling constants and their glycosidic torsion dependence in 5-fluoropyrimidine-substituted RNA. *Journal of the American Chemical Society*, *128*(17), 5851–5858.

Hennig, M., Scott, L. G., Sperling, E., Bermel, W., & Williamson, J. R. (2007). Synthesis of 5-fluoropyrimidine nucleotides as sensitive NMR probes of RNA structure. *Journal of the American Chemical Society*, *129*(48), 14911–14921.

Hennig, M., & Williamson, J. R. (2000). Detection of N-H…N hydrogen bonding in RNA via scalar couplings in the absence of observable imino proton resonances. *Nucleic Acids Research, 28*(7), 1585–1593.

Huang, W., Varani, G., & Drobny, G. P. (2010). ^{13}C/^{15}N-^{19}F intermolecular REDOR NMR study of the interaction of TAR RNA with Tat peptides. *Journal of the American Chemical Society, 132*(50), 17643–17645. http://dx.doi.org/10.1021/ja1051439.

Hwang, T. L., & Shaka, A. J. (1995). Water suppression that works—Excitation sculpting using arbitrary wave-forms and pulsed-field gradients. *Journal of Magnetic Resonance, Series A, 112*(2), 275–279.

Kitevski-LeBlanc, J. L., & Prosser, R. S. (2012). Current applications of ^{19}F NMR to studies of protein structure and dynamics. *Progress in Nuclear Magnetic Resonance Spectroscopy, 62*, 1–33. http://dx.doi.org/10.1016/j.pnmrs.2011.06.003.

Knobloch, B., Linert, W., & Sigel, H. (2005). Metal ion-binding properties of (N3)-deprotonated uridine, thymidine, and related pyrimidine nucleosides in aqueous solution. *Proceedings of the National Academy of Sciences of the United States of America, 102*(21), 7459–7464. http://dx.doi.org/10.1073/pnas.0501446102.

Kosutic, M., Jud, L., Da Veiga, C., Frener, M., Fauster, K., Kreutz, C., et al. (2014). Surprising base pairing and structural properties of 2′-trifluoromethylthio-modified ribonucleic acids. *Journal of the American Chemical Society, 136*(18), 6656–6663. http://dx.doi.org/10.1021/ja5005637.

Kreutz, C., Kahlig, H., Konrat, R., & Micura, R. (2005). Ribose 2′-F labeling: A simple tool for the characterization of RNA secondary structure equilibria by ^{19}F NMR spectroscopy. *Journal of the American Chemical Society, 127*(33), 11558–11559.

Lau, E. Y., & Gerig, J. T. (1996). Solvent effects on fluorine shielding in fluorobenzene. *Journal of the American Chemical Society, 118*(5), 1194–1200. http://dx.doi.org/10.1021/ja9526372.

Luy, B., Barchi, J. J., Jr., & Marino, J. P. (2001). S(3)E-E.COSY methods for the measurement of (19)F associated scalar and dipolar coupling constants. *Journal of Magnetic Resonance, 152*(1), 179–184.

Luy, B., & Marino, J. P. (2001). Measurement and application of ^{1}H-^{19}F dipolar couplings in the structure determination of 2′-fluorolabeled RNA. *Journal of Biomolecular NMR, 20*(1), 39–47.

Marciniak, R. A., Calnan, B. J., Frankel, A. D., & Sharp, P. A. (1990). HIV-1 Tat protein trans-activates transcription in vitro. *Cell, 63*(4), 791–802.

Maurer, T., & Kalbitzer, H. R. (1996). Indirect referencing of P-31 and F-19 NMR spectra. *Journal of Magnetic Resonance, Series B, 113*(2), 177–178.

Meissner, A., Duus, J. O., & Sorensen, O. W. (1997a). Integration of spin-state-selective excitation into 2D NMR correlation experiments with the heteronuclear ZQ/2Q pi rotations for 1JXH- resolved E.COSY-type measurements of heteronuclear coupling constants in proteins. *Journal of Biomolecular NMR, 10*(1), 89–94.

Meissner, A., Duus, J. O., & Sorensen, O. W. (1997b). Spin-state-selective excitation. Application for E.COSY-type measurement of J(HH) coupling constants. *Journal of Magnetic Resonance, 128*(1), 92–97.

Metzler, W. J., Leighton, P., & Lu, P. (1988). Two-dimensional heteronuclear NOE spectroscopy—Application to F-19-labeled oligodeoxyribonucleotides. *Journal of Magnetic Resonance, 76*(3), 534–539.

Metzler, W. J., & Lu, P. (1989). Lambda cro repressor complex with OR3 operator DNA. ^{19}F nuclear magnetic resonance observations. *Journal of Molecular Biology, 205*(1), 149–164.

Milligan, J. F., Groebe, D. R., Witherell, G. W., & Uhlenbeck, O. C. (1987). Oligoribonucleotide synthesis using T7 RNA polymerase and synthetic DNA templates. *Nucleic Acids Research, 15*(21), 8783–8798.

Milligan, J. F., & Uhlenbeck, O. C. (1989). Synthesis of small RNAs using T7 RNA polymerase. *Methods in Enzymology, 180*, 51–62.

Muller, N. (1977). Charge-transfer contributions to solvent shifts in fluorine magnetic resonance spectra. *Journal of Magnetic Resonance (1969), 25*(1), 111–121. http://dx.doi.org/10.1016/0022-2364(77)90124-X.

Olejniczak, M., Gdaniec, Z., Fischer, A., Grabarkiewicz, T., Bielecki, L., & Adamiak, R. W. (2002). The bulge region of HIV-1 TAR RNA binds metal ions in solution. *Nucleic Acids Research, 30*(19), 4241–4249.

Olsen, G. L., Edwards, T. E., Deka, P., Varani, G., Sigurdsson, S. T., & Drobny, G. P. (2005). Monitoring tat peptide binding to TAR RNA by solid-state ^{31}P-^{19}F REDOR NMR. *Nucleic Acids Research, 33*(11), 3447–3454.

Oyelere, A. K., & Strobel, S. A. (2000). Biochemical detection of cytidine protonation within RNA. *Journal of the American Chemical Society, 122*(42), 10259–10267. http://dx.doi.org/10.1021/ja001918t.

Price, S. R., Ito, N., Oubridge, C., Avis, J. M., & Nagai, K. (1995). Crystallization of RNA-protein complexes I. Methods for the large-scale preparation of RNA suitable for crystallographic studies. *Journal of Molecular Biology, 249*(2), 398–408.

Puffer, B., Kreutz, C., Rieder, U., Ebert, M. O., Konrat, R., & Micura, R. (2009). 5-Fluoro pyrimidines: Labels to probe DNA and RNA secondary structures by 1D ^{19}F NMR spectroscopy. *Nucleic Acids Research, 37*(22), 7728–7740. http://dx.doi.org/10.1093/nar/gkp862.

Puglisi, J. D., Tan, R., Calnan, B. J., Frankel, A. D., & Williamson, J. R. (1992). Conformation of the TAR RNA-arginine complex by NMR spectroscopy. *Science, 257*(5066), 76–80.

Rastinejad, F., Evilia, C., & Lu, P. (1995). Studies of nucleic acids and their protein interactions by ^{19}F NMR. *Methods in Enzymology, 261*, 560–575.

Reif, B., Wittmann, V., Schwalbe, H., Griesinger, C., Worner, K., JahnHofmann, K., et al. (1997). Structural comparison of oligoribonucleotides and their 2′-deoxy-2′-fluoro analogs by heteronuclear NMR spectroscopy. *Helvetica Chimica Acta, 80*(6), 1952–1971.

Rinaldi, P. L. (1983). Heteronuclear 2D-NOE spectroscopy. *Journal of the American Chemical Society, 105*(15), 5167–5168.

Schultheisz, H. L., Szymczyna, B. R., Scott, L. G., & Williamson, J. R. (2008). Pathway engineered enzymatic de novo purine nucleotide synthesis. *ACS Chemical Biology, 3*(8), 499–511. http://dx.doi.org/10.1021/cb800066p.

Schultheisz, H. L., Szymczyna, B. R., Scott, L. G., & Williamson, J. R. (2011). Enzymatic de novo pyrimidine nucleotide synthesis. *Journal of the American Chemical Society, 133*(2), 297–304. http://dx.doi.org/10.1021/ja1059685.

Scott, L. G., Geierstanger, B. H., Williamson, J. R., & Hennig, M. (2004). Enzymatic synthesis and F-19 NMR studies of 2-fluoroadenine-substituted RNA. *Journal of the American Chemical Society, 126*(38), 11776–11777.

Scott, L. G., Tolbert, T. J., & Williamson, J. R. (2000). Preparation of specifically ^{2}H- and ^{13}C-labeled ribonucleotides. *Methods in Enzymology, 317*, 18–38.

Seelamgari, A., Maddukuri, A., Berro, R., de la Fuente, C., Kehn, K., Deng, L. W., et al. (2004). Role of viral regulatory and accessory proteins in HIV-1 replication. *Frontiers in Bioscience, 9*, 2388–2413.

Suydam, I. T., & Strobel, S. A. (2008). Fluorine substituted adenosines as probes of nucleobase protonation in functional RNAs. *Journal of the American Chemical Society, 130*(41), 13639–13648. http://dx.doi.org/10.1021/ja803336y.

Sykes, B. D., & Hull, W. E. (1978). Fluorine nuclear magnetic resonance studies of proteins. *Methods in Enzymology, 49*, 270–295.

Thibaudeau, C., Plavec, J., & Chattopadhyaya, J. (1998). A new generalized Karplus-type equation relating vicinal proton-fluorine coupling constants to H-C-C-F torsion angles. *Journal of Organic Chemistry*, *63*(15), 4967–4984.

Tolbert, T. J., & Williamson, J. R. (1996). Preparation of specifically deuterated RNA for NMR studies using a combination of chemical and enzymatic synthesis. *Journal of the American Chemical Society*, *118*(34), 7929–7940.

Tolbert, T. J., & Williamson, J. R. (1997). Preparation of specifically deuterated and ^{13}C-labeled RNA for NMR studies using enzymatic synthesis. *Journal of the American Chemical Society*, *119*(50), 12100–12108.

Varani, G., & McClain, W. H. (2000). The G × U wobble base pair. A fundamental building block of RNA structure crucial to RNA function in diverse biological systems. *EMBO Reports*, *1*(1), 18–23. http://dx.doi.org/10.1093/embo-reports/kvd001.

Willker, W., Leibfritz, D., Kerssebaum, R., & Bermel, W. (1993). Gradient selection in inverse heteronuclear correlation spectroscopy. *Magnetic Resonance in Chemistry*, *31*(3), 287–292.

Wyszko, E., Fuerste, J. P., Barciszewska, M., Szymanski, M., Adamiak, R., Erdmann, V. A., et al. (1999). Preparation of HIV TAR RNA with RNA scissors. *Journal of Biochemistry*, *126*(2), 326–332.

Yu, C., & Levy, G. C. (1984). Two-dimensional heteronuclear NOE (HOESY) experiments: Investigation of dipolar interactions between heteronuclei and nearby protons. *Journal of the American Chemical Society*, *106*(22), 6533–6537.

Zidek, L., Wu, H., Feigon, J., & Sklenar, V. (2001). Measurement of small scalar and dipolar couplings in purine and pyrimidine bases. *Journal of Biomolecular NMR*, *21*(2), 153–160.

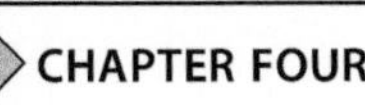

CHAPTER FOUR

Applying Thymine Isostere 2,4-Difluoro-5-Methylbenzene as a NMR Assignment Tool and Probe of Homopyrimidine/ Homopurine Tract Structural Dynamics

Robert G. Brinson[*,1], **Jennifer T. Miller**[†], **Jason D. Kahn**[‡], **Stuart F.J. Le Grice**[†], **John P. Marino**[*]

[*]Institute for Bioscience and Biotechnology Research, National Institute of Standards and Technology and the University of Maryland, Rockville, Maryland, USA
[†]Basic Research Laboratory, NCI-Frederick National Institutes of Health, Frederick, Maryland, USA
[‡]Department of Chemistry and Biochemistry, University of Maryland, College Park, Maryland, USA
[1]Corresponding author: e-mail address: brinson@ibbr.umd.edu

Contents

Methods in Enzymology, Volume 566
ISSN 0076-6879
http://dx.doi.org/10.1016/bs.mie.2015.05.009

Abstract

Proton assignment of nuclear magnetic resonance (NMR) spectra of homopyrimidine/homopurine tract oligonucleotides becomes extremely challenging with increasing helical length due to severe cross-peak overlap. As an alternative to the more standard practice of ^{15}N and ^{13}C labeling of oligonucleotides, here, we describe a method for assignment of highly redundant DNA sequences that uses single-site substitution of the thymine isostere 2,4-difluoro-5-methylbenzene (dF). The impact of this approach in facilitating the assignment of intractable spectra and analyzing oligonucleotide structure and dynamics is demonstrated using A-tract and TATA box DNA and two polypurine tract-containing RNA:DNA hybrids derived from HIV-1 and the *Saccharomyces cerevisiae* long-terminal repeat-containing retrotransposon Ty3. Only resonances proximal to the site of dF substitution exhibit sizable chemical shift changes, providing spectral dispersion while still allowing chemical shift mapping of resonances from unaffected residues distal to the site of modification directly back to the unmodified sequence. It is further illustrated that dF incorporation can subtly alter the conformation and dynamics of homopyrimidine/homopurine tract oligonucleotides, and how these NMR observations can be correlated, in the cases of the TATA box DNA, with modulation in the TATA box-binding protein interaction using an orthogonal gel assay.

1. INTRODUCTION

Fluorinated nucleoside analogs have been used as structural and chemical probes in several nuclear magnetic resonance (NMR) spectroscopy studies of nucleic acid structure and protein–nucleic acid interactions (Carlow, Short, & Wolfenden, 1996; Chu & Horowitz, 1991a,b; Hammann, Norman, & Lilley, 2001; Kwon, Jiang, Song, & Stivers, 2002; Liu, Chu, Liu, & Horowitz, 1997; Olejniczak et al., 2002; Rastinejad, Evilia, & Lu, 1995; Rastinejad & Lu, 1993). In these analogs, the fluorine (^{19}F) nuclei provide NMR signals that can serve as sensitive reporters of the structural and chemical environment around the site of substitution. In general, fluorine has several advantages as a NMR active nucleus (Rastinejad et al., 1995). Like the proton (^{1}H), fluorine (^{19}F) has a spin of ½ and is found in 100% natural abundance. Moreover, fluorine signals display a broader chemical shift range than protons (~700 ppm for ^{19}F vs. 15 ppm for ^{1}H), as well as greater sensitivity to local changes in chemical environment and conformational dynamics. Due to its high gyromagnetic ratio and 100% natural abundance, fluorine nuclei can be detected directly using sensitive, one-dimensional (1D) NMR methods much in the same way 1D ^{1}H spectra are acquired. In addition, two-dimensional (2D) heteronuclear Overhauser effect spectroscopy (^{1}H,^{19}F HOESY) experiments (Metzler, Leighton, & Lu, 1988), which are analogous to homonuclear Overhauser effect spectroscopy (^{1}H,^{1}H NOESY) experiments, can be readily measured to define the local (<5 Å radius) proton environment around a ^{19}F nucleus.

Solid-phase phosphoramidite chemistry has been the predominant approach used to synthesize oligonucleotides that are selectively substituted with fluorinated nucleoside analogs (Carlow et al., 1996; Chu and Horowitz, 1991a,b; Graber, Moroder, & Micura, 2008; Hammann et al., 2001; Kwon et al., 2002; Liu et al., 1997; Olejniczak et al., 2002; Puffer et al., 2009; Rastinejad et al., 1995; Rastinejad and Lu, 1993; Scott, Geierstanger, Williamson, & Hennig, 2004). Among the different fluorine analogues, selective incorporation of the fluorinated pyrimidines 5-fluorouracil (5-FU) and 5-fluorocytidine (5-FC) has been the most widely used to enable ^{19}F NMR studies of both RNA and DNA. When incorporated into an oligonucleotide strand that is then hybridized with its complement to form a helical duplex, these pyrimidine analogs place a ^{19}F NMR reporter nucleus in the major groove of the Watson–Crick helical structure, making them sensitive to detecting interactions and conformational changes associated with this groove. Alternatively, enzyme-based synthetic schemes have been developed, which generate ^{19}F-labeled ribonucleotide triphosphates that can be incorporated into RNA molecules using *in vitro* transcription with T7 RNA polymerase (Hennig, Scott, Sperling, Bermel, & Williamson, 2007; Scott et al., 2004). Using the approach, 2-fluoroadenosine-containing RNA oligonucleotides have been generated and shown by NMR to be relatively nonperturbing of helical structure and to provide a useful ^{19}F NMR reporter nucleus in the minor groove of the helical structure.

2,4-Difluoro-5-methylbenzene (dF), a nonpolar aromatic nonhydrogen bonding isostere of thymine, is another example of a fluorinated pyrimidine analog that has been used to probe the role of hydrogen bonding and steric effects in oligonucleotides (Kool & Sintim, 2006). When incorporated into an oligonucleotide strand that is then hybridized with its complement to form a helical duplex, dF places ^{19}F NMR reporter nuclei in both the major and minor grooves of the duplex structure (Fig. 1A), making it suitable for detecting interactions and conformational changes associated with both grooves. NMR studies have shown this analog is also generally

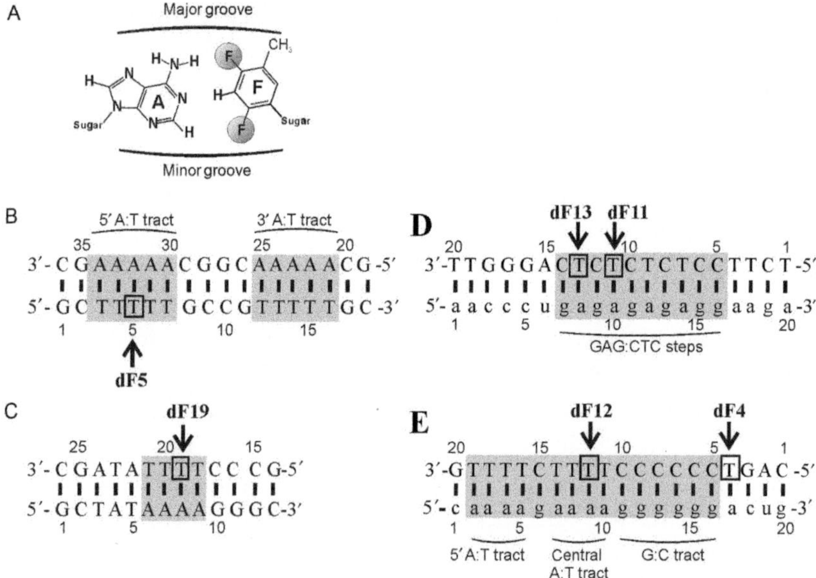

Figure 1 (A) dF structure and its positioning opposite an adenosine base, (B) A-tract and (C) TATA DNA duplex constructs, and the (D) Ty3 and (E) HIV-1 PPT-containing RNA/DNA constructs, respectively. DNA bases are in upper case and RNA bases in lower case. The numbering of bases is given from 5′ to 3′. The position of dF substitution is indicated by an arrow and a box. The r(GAG):d(CTC) steps in the Ty3 sequence, two (rA_4):(dT_4) tracts and the (rG_6):(dC_6) tract in the HIV-1 sequence, the two (dA_5):(dT_5) tracts in the A-tract sequence, and the (dA_4):(dT_4) tract in the TATA sequence are highlighted in gray.

nonperturbing to the helical geometry of a DNA duplex. Kool and coworkers showed that incorporating dF into an oligonucleotide that formed a 12-mer DNA duplex with its complement resulted in no significant observable changes in the structure of the duplex when compared to the unmodified control sample (Guckian, Krugh, & Kool, 1998). In this study, changes in the electronic environment adjacent to the site of substitution were found to cause an upfield shift of proximal imino protons between dF-substituted and control duplexes. Moreover, a slight destabilization of the duplex, observed by thermal melting, was observed and attributed to a lack of hydrogen bonding between dF and its base-pairing partner at the site of insertion. dF stacking within the helix, however, is stronger than that of thymine and therefore partially compensates for this lack of hydrogen bonding (Guckian et al., 1996).

Application of the dF labeling strategy for NMR studies is presented here as an tool for the analysis of homopyrimidine/homopurine-rich nucleic acid sequences: A-tract and TATA box DNA duplexes derived from *Arabidopsis thaliana* and two polypurine tract (PPT)-containing RNA:DNA hybrid duplexes derived from HIV-1 and the *Saccharomyces cerevisiae* long-terminal repeat (LTR)-retrotransposon Ty3. Using these constructs, we illustrate how dF insertion disperses overlapped NMR signals within highly redundant oligonucleotide sequences, thus facilitating proton resonance assignments of otherwise intractable spectra. We additionally show how NMR can be used to detect subtle perturbation in the conformation and dynamics of a duplex formed with a dF-substituted oligonucleotide, demonstrating how dF can also serve as a sensitive, *in situ* molecular probe (Rausch, Qu, Yi-Brunozzi, Kool, & Le Grice, 2003; Yi-Brunozzi et al., 2008). Lastly, it is shown how NMR observed changes in duplex conformation and dynamics can be further correlated, in the cases of the TATA box DNA, with modulation in the TATA box-binding protein interaction using an orthogonal gel-binding assay.

2. PREPARATION OF dF-SUBSTITUTED OLIGONUCLEOTIDES FOR NMR ANALYSIS

2.1 Design of Oligonucleotides

Strategic placement of dF within oligonucleotides can alleviate spectral crowding and facilitate assignment of ^{1}H resonances. To illustrate such placement, we chose sequences that contained varying degrees of redundancy, including A-tract and TATA box DNA duplexes derived from *A. thaliana* (Kim & Burley, 1994) and two PPT-containing RNA:DNA hybrid duplexes derived from HIV-1 and *S. cerevisiae* LTR-retrotransposon Ty3 (Fig. 1). For the A-tract, TATA box, and HIV PPT constructs, dF was inserted at the second or third position of a 4- or 5-nucleotide dT tract to disrupt redundancy. In a second HIV PPT construct, dF was inserted at position 4 to attempt to relieve, in part, the signal overlap arising from the rG:dC tract. For the Ty3 sequences, two separate constructs with dF inserted at positions 11 and 13 were prepared to break the redundancy of the r(GAG):d(CTC) runs.

2.2 Reagents

5′-Dimethoxytrityl-(2,4-difluorotoluyl nucleoside), 3′-[(2-cyanoethyl)-(*N*,*N*-diisopropyl)]-phosphoramidite (Glen Research Corp.).

1× TBE (tris–borate–EDTA) buffer: 89 m*M* Tris base, 89 m*M* boric acid, 2 m*M* ethylenediaminetetraacetic acid (EDTA), pH 8.3. All components were purchased from Sigma-Aldrich.

1× TBM buffer: 90 m*M* Tris base, 89 m*M* boric acid, 1 m*M* $MgCl_2$
NMR buffer: 80 m*M* NaCl, 10 m*M* Na_2HPO_4/NaH_2PO_4, pH 7.0, in 90% H_2O and 10% D_2O. All components were purchased from Sigma-Aldrich.

2.3 Synthesis of dF-Substituted Oligonucleotides

dF-substituted and canonical DNA and RNA oligonucleotides were synthesized at a 1 μmol scale using standard phosphoramidite chemistry on a PE Biosystems Expedite 8909 nucleic acid synthesizer. No changes were needed for the synthesis of dF-substituted oligonucleotides from the standard protocols supplied by the instrument manufacturer.

2.3.1 Purification of dF-Substituted Oligonucleotides

The individual strands were purified using preparative denaturing polyacrylamide gel electrophoresis (PAGE) using 20% acrylamide at 19:1 acrylamide: bisacrylamide. PAGE was performed at 30 W and approximately 50 °C in 1 × TBE buffer. The desired oligonucleotide product was visualized with ultraviolet (UV) irradiation and cut from the gel. The samples were then electroeluted from the gel strips at 200 V with ½× TBE buffer using an Elutrap device (Whatman Scientific). Fractions from the Elutrap were collected every hour for 6 h and analyzed via UV-vis spectroscopy to confirm the presence of oligonucleotide. All fractions containing sample were pooled.

2.3.2 Sample Conditioning for NMR Spectroscopy

Electroeluted nucleic acid was concentrated with a 3 kDa centrifugal filter and then buffer exchanged in the same filter sequentially with water (twice), 1 *M* NaCl (once), water (thrice), and the final NMR buffer (thrice) to a volume of 250 μl. Washing with 1 *M* NaCl is especially important with RNA to displace residual Tris buffer, which is found to bind to some RNA molecules. Each single-stranded oligonucleotide was microdialyzed against a 3 kDa membrane to assure full buffer exchange. Following concentration and buffer exchange, the single-stranded oligonucleotides were quantified spectrophotometrically at 260 nm, assuming a molar extinction coefficient equal to the sum of the ribo- and/or deoxyribonucleotides using the nearest-neighbor two-state model. DNA and RNA/DNA hybrid duplexes were formed by mixing equimolar amounts of the complementary strands, heating the solution to 90 °C for 3 min, and then slow cooling to room temperature. Homogeneous duplex formation was visualized and stoichiometry further confirmed using native PAGE (15% acrylamide at 19:1 acrylamide:

bisacrylamide in 1 × TBM buffer run at 40 W and 4 °C). Final duplex concentrations were typically 1.0 m*M* in 275 μl and 90% H_2O, 10% D_2O in NMR buffer. All samples were loaded into a 5-mm symmetrical, susceptibility-matched NMR microtube (Shigemi, Inc.) for NMR experiments.

2.3.3 Exchange of Samples into 99.96% D_2O

The following procedure was employed for NMR experiments requiring that the sample be in 99.96% D_2O. The plunger was removed from the Shigemi tube and set aside. The top of the tube was plugged with parafilm to prevent accidental expulsion of sample during the freeze-drying process. Crushed dry ice was used to freeze the upper portion of the tube above the sample solution. The tube was then removed from the dry ice and gently shaken to freeze the sample along its walls. This process was performed iteratively until the entire sample was frozen along the walls of the NMR tube. The RNA sample was subsequently lyophilized, reconstituted in 99.96% D_2O, frozen a second time using the same procedure, and lyophilized again. After the second lyophilization, the sample was once again reconstituted in 99.96% D_2O and used for NMR experiments.

3. NMR METHODS

3.1 Proton NMR Data Acquired in 90%:10% $H_2O:D_2O$

NMR experiments were collected on a 600 MHz Bruker AVANCE I NMR spectrometer equipped with a room temperature, triple resonance (i.e., ^1H, ^{13}C, and ^{15}N) TXI probe.

Initially, a standard 1D ^1H experiment was acquired using a water flip-back Watergate for water suppression. The imino ^1H region, 10.0–15.0 ppm, was inspected to assure that the duplex was properly formed. Duplex formation is ascertained from qualitatively visualizing peak shape (30 ± 10 Hz imino linewidth) and counting the appropriate number of peaks. Imino resonances from terminal base pairs are known to exchange rapidly with solvent and are usually not observed. For duplex oligonucleotides, most other imino resonances from the Watson–Crick pairs should be observed.

A 2D nuclear Overhauser effect spectroscopy (NOESY) experiment also employing a water flip-back Watergate for solvent suppression was then collected at typically two different mixing times (e.g., 100 and 300 ms). This standard experiment allows the user to map sequential correlations of imino resonances along the length of the sequences. This assignment process has been reviewed extensively (Flinders & Dieckmann, 2006).

3.2 Proton and Fluorine NMR Data Acquired in 99.96% D_2O

After exchange of the NMR sample into D_2O, another 2D NOESY, typically at a mixing time of 250 ms, and a double quantum filtered correlation spectroscopy (DQF-COSY) experiments were collected for analysis of non-exchangeable proton resonances from the base and sugars that comprise each strand of the duplex. This analysis has been reviewed extensively (Flinders and Dieckmann, 2006).

$^{19}F,^{1}H$-heteronuclear single-quantum coherence and $^{19}F,^{1}H$-heteronuclear Overhauser effect spectroscopy (HOESY, $\tau_{mix}=300$ ms) experiments were acquired using a dual resonance ($^{1}H,^{19}F$) HF probe. The HOESY experiment demonstrated proper helical stacking of the dF base based on the observed HF NOE patterns (Fig. 2). Specifically, the presence of cross-peaks to neighboring thymidines and cross-strand adenosines indicates that the dF-modified base is properly base stacked within the helix.

4. dF AS AN NMR ASSIGNMENT TOOL

The DNA duplex and RNA/DNA hybrids that instigated the development of this approach exhibit severe resonance overlap that arises from sequence redundancy, making complete spectral analysis a formidable challenge. As will be shown, dF can dramatically perturb the chemical shifts of proton resonances proximal to the site of substitution and thus significantly reduce spectral overlap. Since the electronic environment of distal resonances are unaffected by dF substitution, the chemical shifts of these resonances remain in the same position as the unmodified sequences. The reduction in spectral overlap, combined with the close matching of distal resonance signals between the modified and unmodified sequences, facilitates the assignment of duplexes formed by dF-substituted oligonucleotides as well as mapping of resonance assignments back to the unmodified sequence.

4.1 Illustration of the ^{1}H Imino Resonance Assignment Strategy Using dF Substitution

Representative 1D imino ^{1}H spectra, the "fingerprint" region of nucleic acid duplexes, are shown in Fig. 3. While the dispersion resulting from dF substitution of imino resonances from the TATA box and A-tract

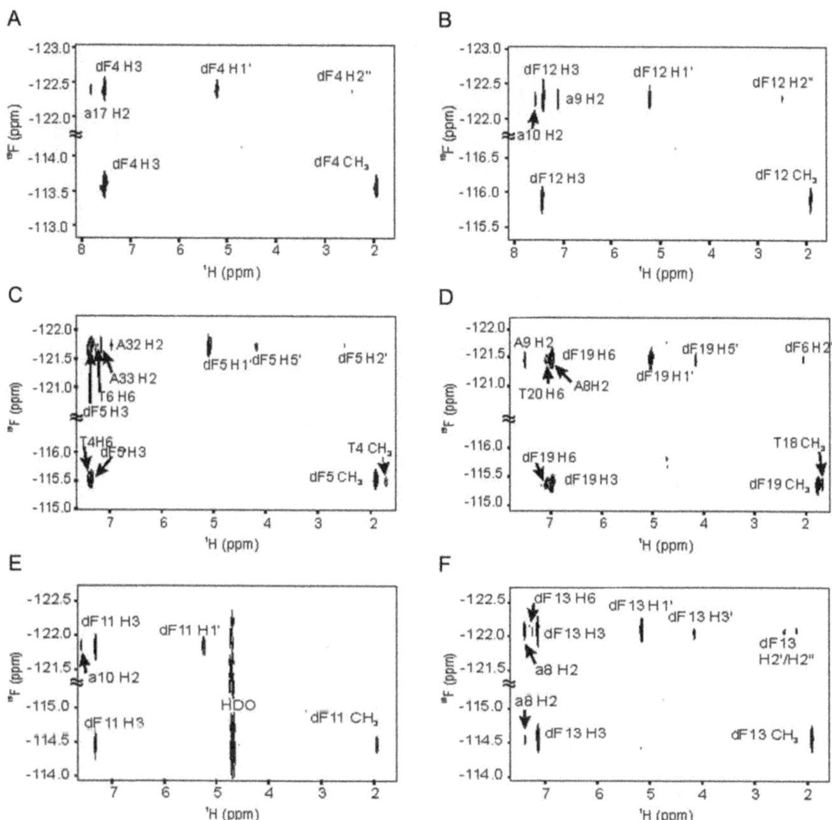

Figure 2 ^{1}H-^{19}F NOESY spectra of the (A) HIV dF4 duplex; (B) HIV dF12 duplex; (C) A-tract dF5 duplex; (D) TATA box dF19 duplex; (E) Ty3 dF11 duplex; (F) Ty3 dF13 duplex, all in 99.96% D_2O at 30 °C. The proton carrier frequency was set to 4.708 ppm and the fluorine carrier frequency to −118 ppm. Sweep widths were set to 6009 Hz for ^{1}H and 5646 Hz for ^{19}F, 128 complex points in t_1, 4 K complex points in t_2, and 480 scans per increment.

duplexes was negligible, the effect of dF substitution is clearly illustrated for the imino spectra of dF-modified Ty3 and HIV PPT duplexes. Four imino resonances of the Ty3 construct were unresolved in the wild-type sequence: T7, T9, T11, and T13. Both the Ty3 dF11 and dF13 sequences dispersed the PPT duplex signals, allowing assignment of resonances distal from the site of substitution. In a similar manner, the imino spectra of the HIV wild-type sequence exhibit resonance overlap in the thymine region (~13.5 ppm) and in the G:C tract (~12.3 ppm). Again, both the HIV dF12 and dF4 substitutions dispersed resonances and facilitated assignments of imino

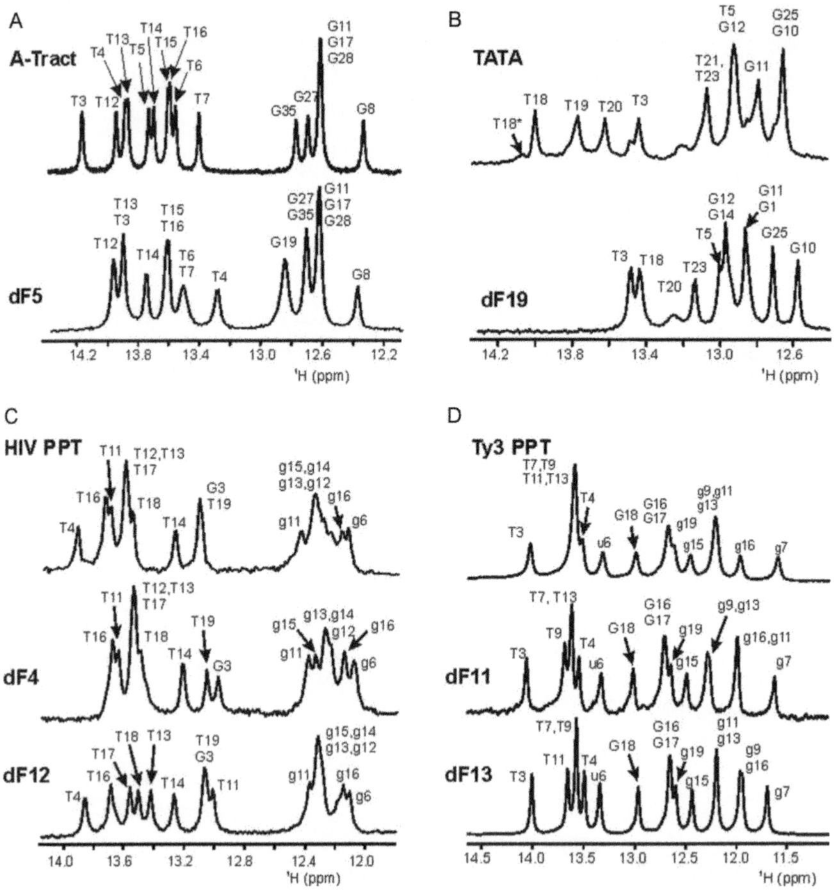

Figure 3 Comparison of the imino regions of the 1D water flip-back Watergate ^{1}H NMR spectra of (A) A-tract DNA; (B) TATA box DNA; (C) HIV PPT; and (D) Ty3 PPT duplexes, all at 10 °C. In all panels, the unmodified duplex is the top spectrum, and the position of dF substitution in the modified duplex is indicated in the bottom spectra. Imino proton assignments are indicated. RNA is lower case and DNA upper case. *Panels (C) and (D) are modified from published data Yi-Brunozzi et al. (2008).*

resonances associated with the T-tracts and G:C tract. For the HIV dF12 substitution, chemical shifts of the central rA:dT tract were significantly shifted to allow assignment of the thymine resonances distal from the central rA:dT tract, which remained in approximately the same position as found in the unmodified sample. Using a similar strategy, the HIV dF4 substitution yielded sufficient resolution in the G:C tract to facilitate assignment of all rG imino resonances.

4.2 Illustration of the ^{1}H Nonexchangeable Resonance Assignment Strategy Using dF Substitution

The process of assigning backbone resonances through inter/intramolecular NOEs, particularly aromatic (H6/H8/H2) and anomeric (H1′)/H5 correlations, can also be especially challenging with high sequence redundancy. The region of the 2D NOESY spectra containing these correlations is approximately from 6.8 to 8.3 ppm (H8/H6/H2) in 1D and 4.8 to 6.3 ppm (H1′/H5) in the second (Fig. 4). In order to obtain detailed information for the structure and dynamics of an oligonucleotide, as complete an assignment of these correlations in this region as is possible is desired. As was observed for the imino spectra, the utility of dF as a tool for NMR resonance assignment is clearly evident by its dispersion of overlapped signals in a 2D NOESY experiment on the dF5-substituted A-tract (Fig. 4A and B). This spectrum shows slight shifting of the H8–H1′ cross-peaks of residues A31–A35. The inset clearly shows subtle shifting of dF resonances proximal its site of substitution, facilitating assignment of the analogous cross-peaks in the distal A-tract associated with residues A21–A25.

The unmodified HIV PPT provides an even clearer example of a severely overlapped NOESY spectrum which is not possible to assign using standard proton-based approaches for resonance assignment. In addition, preparing RNA through *in vitro* transcription was problematic due to the run of six guanosines, so facile stable isotopic labeling with ^{13}C or ^{15}N was not practical. Through comparative analysis of the unmodified, dF12 and dF4 NOESY data, complete assignment could be made for all H8, H6, H2, H1′, and H5 resonances (Fig. 4C and D). In this case, the dF12 substitution resulted in dispersion of signals from the central A:T tract which facilitated assignments in the 5′ A:T tract, while dF4 substitution afforded resolution that facilitated assignments of resonances in the G:C tract.

4.3 Quantifying Chemical Shift Perturbations in the dF-Modified Constructs

4.3.1 Computation of Chemical Shift Differences

From the assignments of each unmodified and dF-substituted A-tract and TATA box DNA, HIV-1 and Ty3 RNA:DNA PPT hybrids (Yi-Brunozzi et al., 2008), a chemical shift difference histogram was computed to visualize the range in magnitude of the perturbations resulting from dF substitution for the anomeric to aromatic region (Fig. 5). The deviation of chemical shifts of all aromatic protons and H1′ were normalized for each

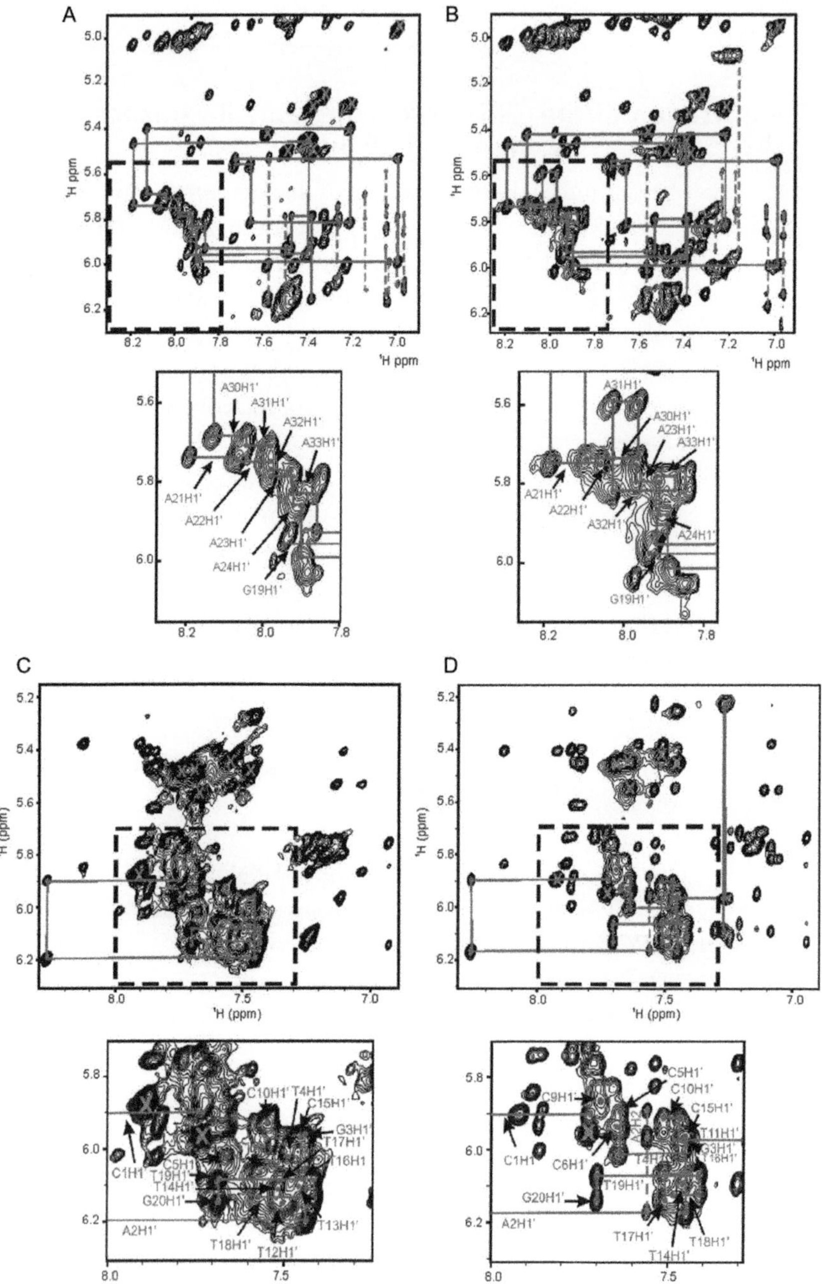

Figure 4 Expansion of the anomeric (H1′) to aromatic region of a NOESY (τ_{NOE} = 250 ms) experiment applied to a 1.0 m*M* sample of the (A) unmodified A-tract DNA duplex, (B) A-tract dF5 DNA duplex, (C) unmodified HIV, and (D) HIV dF12 duplex. The local dispersion of peaks due to dF substitution is emphasized by the expansions in the dotted boxes below each spectrum. The sequential assignment "NOE walk" for residues G19 through C36 is shown in red lines for the A-tract and C1 though G20 for the HIV spectra. For clarity, assignment labels were removed from the full spectrum but retained in the expansions. Dashed green lines are used to annotate correlations of adenosine H2 protons with 5′ sequential and 3′ cross-strand anomeric (H1′) protons. An "X" in cyan indicates a cytosine H5–H6 correlation. (See the color plate.)

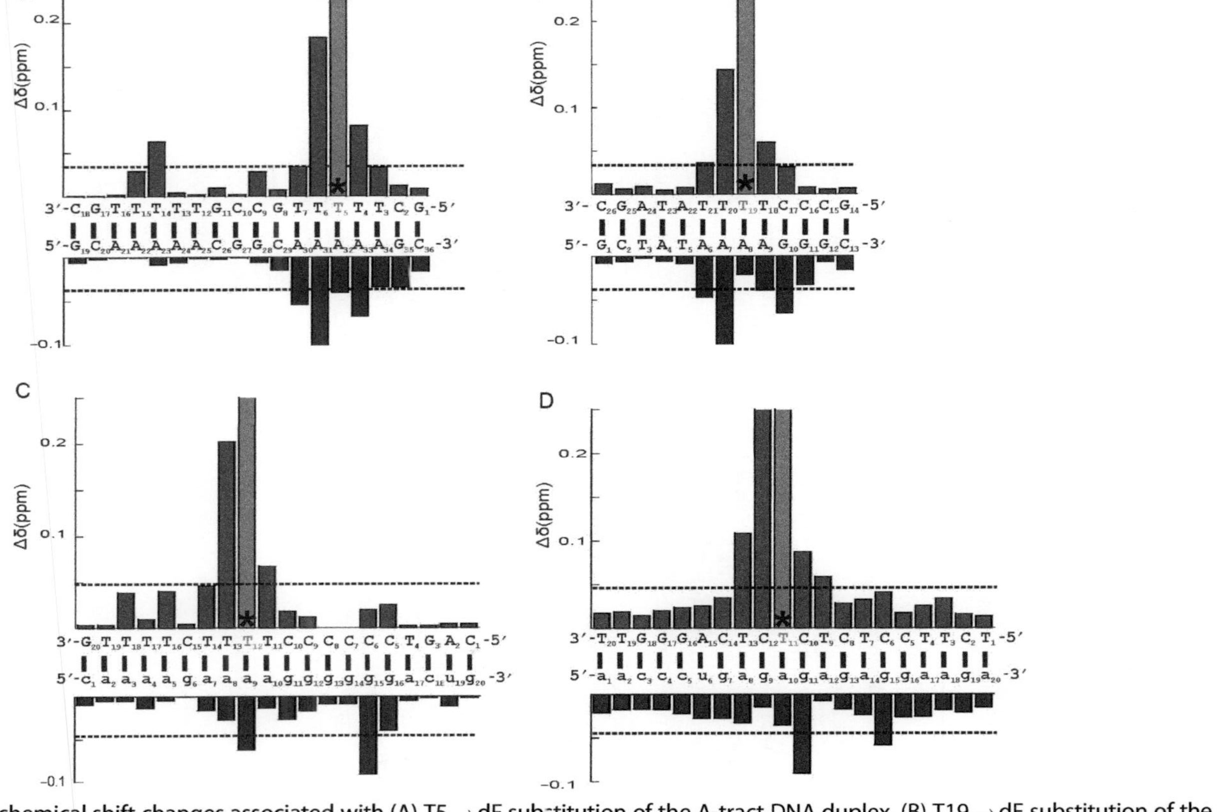

Figure 5 Proton chemical shift changes associated with (A) T5 → dF substitution of the A-tract DNA duplex, (B) T19 → dF substitution of the TATA box DNA duplex, (C) T12 → dF substitution of the HIV-1 PPT, and (D) T11 → dF substitution of the Ty3 PPT. DNA and RNA sequences of each hybrid are shown in upper case and lower case letters, respectively. Positions of dT → dF substitution are highlighted in green and an asterisk "*." Each bar represents the average difference in normalized ppm values obtained for equivalent nucleotides within dF-substituted and unmodified duplexes. All aromatic and H1′ resonances were included in the average. The dotted gray line designates the threshold of a significant chemical shift difference. Chemical shift changes (ppm) for dF-substituted DNA strands are shown above the sequence schematics (red bars), while unmodified strands are depicted below (blue bars). (See the color plate.)

nucleotide, and each nucleotide in a dF-substituted duplex was compared with the corresponding residue in the unmodified duplex. A similar plot could be created for any other region (or for a combined plot of all or other resonances).

4.3.2 Visualization of Chemical Shift Differences for All Constructs

Introduction of dF clearly results in all cases in a significant local alteration of chemical shift for the dF base protons, while proximal bases were perturbed by approximately 0.2 ppm or less. These bases included T4, T6, and A30-33 for the A-tract DNA duplex; A6, A7, G10, T18, and T20 for the TATA DNA duplex; T11, T13, a9, and g15 for HIV; and T9, C10, C12, T13, g11, and g15 for Ty3. Most other resonances of nucleotides distal to the site of substitution were shifted in the spectrum from the unmodified positions by an insignificant average of approximately 0.05 ppm or less. These histograms confirm and quantify what is observed visually in the spectra; that is, perturbations resulting from dF substitution are primarily local and assignment of signals from nucleotides distal from the site of isostere substitution can readily be mapped onto the unmodified sequence.

5. STRUCTURAL AND DYNAMIC PERTURBATIONS OBSERVED BY dF SUBSTITUTION

While these examples highlight the utility of single-site dF substitutions as a means of alleviating spectral overlap for the purpose of assigning resonances, it was additionally observed that dF modifications in these sequences can function as a subtle structural probe of conformational dynamics. In these instances, a subtle change in structure afforded by the dF substitution appears to select for a sampling of different conformational states with different exchange rates that can be detected on the NMR time scale.

5.1 Altered Sugar Pucker Conformations in the PPT dF Series

The unmodified Ty3 PPT construct was observed to have predominantly C3′-endo sugar puckers on the RNA strand, which is the conformation found in an A-form RNA helix (Yi-Brunozzi et al., 2008). In a C3′-endo sugar pucker, the torsional angle between the H1′ and H2′ protons is such that the three bond coupling constant ($^3J_{HH}$) is less than 1 Hz, and no cross-peak is observed between the H1′ and H2′ protons in a DQF-COSY spectrum. However, if the conformational equilibrium of the sugar shifts to a mixture of C3′- and C2′-endo puckers, the $^3J_{HH}$ coupling constant between

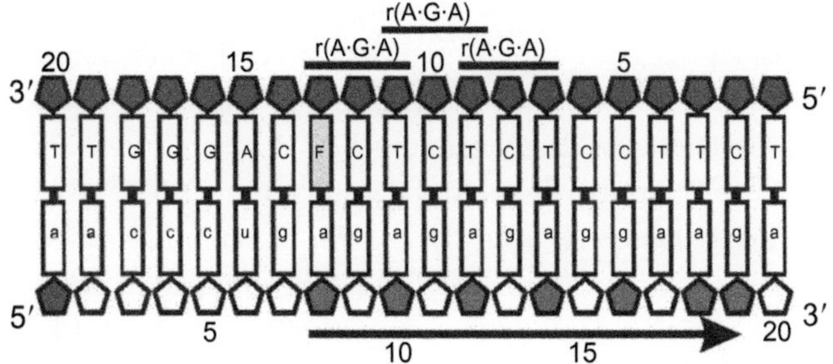

Figure 6 A schematic representation of the Ty3 dF13 PPT hybrid sequence. dF13 substitution is in yellow. Wild-type ribose sugars in mixed C3′/C2′-endo conformation are shown in red. The symmetric 5′-r(A·G·A)-3′ steps are labeled. Highlighted collectively in green are ribose sugars for which a C3′-endo to C3′/C2′-endo sugar pucker switch is observed when the PPT hybrids, and arrow indicates the propagation of the sugar pucker switch to the 3′ end of the sequence. *Figure was modified from published data Yi-Brunozzi et al. (2008).* (See the color plate.)

H1′ and H2′ becomes measureable, and a cross-peak in the DQF-COSY becomes visible. For the Ty3 dF13 and dF11 samples, the dF substitution on the DNA strand resulted in a shift to a mixed sugar pucker conformation on the RNA strand that propagated to every other base down the length of the PPT to residue g16 (Fig. 6). This observed propagation of altered sugar puckering along the RNA strand was attributed to stacking of repeated 5′-r(A·G·A)-3′ symmetric steps in the TY3 sequence (Yi-Brunozzi et al., 2008). To confirm this was a phenomenon specific to the helical stacking in the Ty3 PPT sequence, a C10T substitution was made and shown to also produce a similar altering of sugar pucker that propagated down the RNA strand of the PPT sequence.

In contrast to the TY3 PPT, the HIV PPT dF construct series showed negligible long-range effects on PPT conformation and dynamics. This suggested that this RNA/DNA hybrid PPT may not exhibit the same structural coupling due to an absence of repeated symmetric base stacking step that serve in the Ty3 system to propagate structural perturbation down the length of the PPT sequence.

5.2 Changes in the Dynamic Equilibrium of the TATA Box DNA Duplex

Changes in the spectral heterogeneity and line widths of the TATA DNA samples further demonstrates that dF can introduce subtle changes in the conformational dynamics of a duplex. The unmodified TATA box duplex, in general, displayed greater conformational heterogeneity (e.g., additional imino proton signals from minor species) than the dF19-substituted duplex. Indeed, the heterogeneity observed in the TATA box duplex 1D imino proton spectra was previously observed for imino resonance T18 but not discussed (Fig. 3B; Chen & Russu, 2004). An effect of dF substitution on the conformational dynamics of the TATA box duplexes is further illustrated in the 2D DQF-COSY. In this experiment, unassigned H6–H5 resonance cross-peaks from cytosines from minor conformation(s) are clearly evident for the unmodified sequence and absent in the dF19-substituted duplex (Fig. 7, boxed region). Analogous cross-peaks can also be observed in the 2D NOESY spectrum of the unmodified duplex (data not shown). In total, the NMR data suggest that dF substitution biases the conformational equilibrium of the TATA box DNA duplex to a subset of conformational states sampled by the unmodified duplex. As the TATA DNA duplex is well known to be malleable and unwound upon binding to the TATA-binding protein (TBP) (Davis, Majee, & Kahn, 1999; Juo et al., 1996; Kim and Burley, 1994), we surmised that the NMR observed effect of dF substitution on the TATA box DNA duplex might have an impact on the formation of

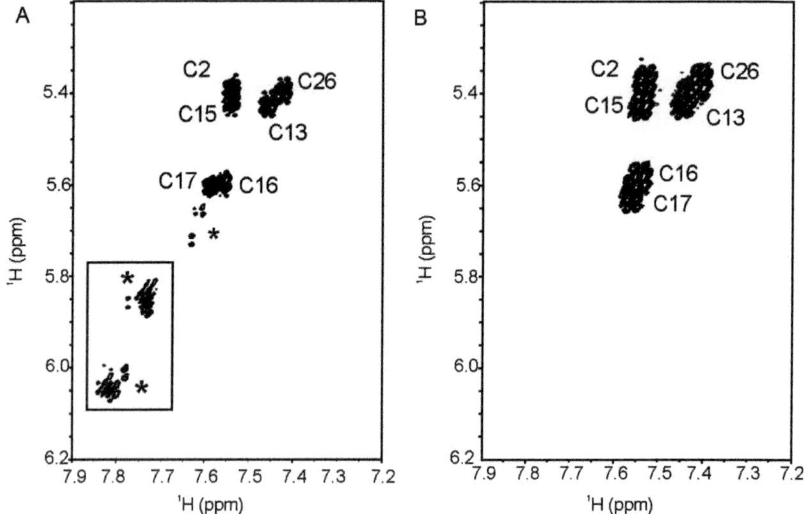

Figure 7 Expansion of the H6–H5 region of a DQF-COSY of the (A) unmodified and (B) dF19-containing TATA duplexes collected in 99.96% D_2O at 30 °C. An asterisk indicates an unassigned peak. The boxed region highlights cross-peaks of interest that are further discussed in the text. For clarity, positive and negative contours are black.

the TBP–TATA complex. The next section (5.3) describes gel-based methods that were used to explore the differential binding of the two TATA substrates to TBP.

5.3 An Orthogonal Gel Assay Correlating dF NMR Results from TATA Box DNA with TATA Box Protein Binding

The dF-induced changes in the DNA conformational ensembles, as observed in the NMR experiments can be predictors of altered or impaired function. To test these predictions, an orthogonal functional assay can be performed using the modified and unmodified sequences to make a direct comparison. For the TATA box DNA duplex example, an electrophoretic mobility shift assay (EMSA) for binding to the TATA box-binding protein (TBP) is described that was used to compare the behavior of the unmodified and dF19-modified TATA box DNA duplexes.

5.3.1 TBP-Binding Assay

Unmodified and dF-substituted DNA oligonucleotides were 5′-end labeled using T4 polynucleotide kinase and $\gamma - {}^{32}$P-ATP, purified by denaturing PAGE, and eluted overnight in 0.3 *M* sodium acetate, pH 5.2, 1.0 m*M* EDTA at 37 °C. DNA was recovered by ethanol precipitation and reconstituted in water. Labeled oligonucleotides were mixed with equimolar amounts of their unlabeled complements at 0.4 μ*M* final concentration in 10 m*M* Tris, pH 7.4, 25 m*M* KCl and annealed in a thermal cycler by heating to 85 °C for 3 min followed by slow cooling to 4 °C at 0.1 °C/s. TBP was purified essentially as previously reported (Petri, Hsieh, & Brenowitz, 1995), and the active fraction, determined by EMSA at saturating DNA, was 13%. TBP was serially diluted in 25 m*M* HEPES (pH 8.0), 0.3 *M* KCl, 20% glycerol, 1 m*M* EDTA, and 1 m*M* DTT. Final active TBP concentrations in the binding reactions were 180, 91.8, 46.9, 23.5, 5.9, 2.9, 1.5, and 0.73 n*M*. Binding reactions contained 4 n*M* substrate, 50 m*M* HEPES (pH 8.0), 200 m*M* potassium glutamate, 1 m*M* $CaCl_2$, 2.5 m*M* DTT, 10 m*M* MgOAc, 100 μg/ml gelatin, 50 μg/ml BSA, 0.05% NP-40 (Igepal), and 10% glycerol. Negative controls contained TBP dilution buffer. Reactions were assembled at 4 °C, then placed on ice, and incubated a further 5 min. One-tenth volume of 50% glycerol containing 0.05% (w/v) bromophenol blue and xylene cyanol was added, and samples were loaded onto a 15% acrylamide gel (75:1 acrylamide:bis; prerun at 4 °C for 15 min at 230 V) containing 0.045 *M* Tris base, 0.045 *M* boric acid (pH 8.3), 1 m*M* Na_2EDTA, 2 m*M* DTT, 3 m*M* $MgCl_2$, and 1% glycerol. Electrophoresis was performed at 4 °C at 350 V for 3 h. The gel was dried before exposure to a phosphor screen. Screens were scanned with a Typhoon Trio Plus variable mode imager and bands quantified by ImageQuant Total Lab software (GE Healthcare).

5.3.2 Binding of TBP to Unmodified and dF-Substituted TATA Box DNA Duplexes

As shown in Fig. 8A, the dF-substituted duplex supported significantly more shifted product than the unmodified sequence overall TBP concentrations tested. At the highest protein:DNA concentration, 100% of the dF-substituted TATA duplex was bound, compared to only 18% with unmodified DNA. The apparent increased affinity and lifetime of protein–DNA complex formation for the dF19-substituted TATA duplex suggests an increased helical malleability with respect to the unmodified sequence that likely results from changes in the stacking geometry and stability at

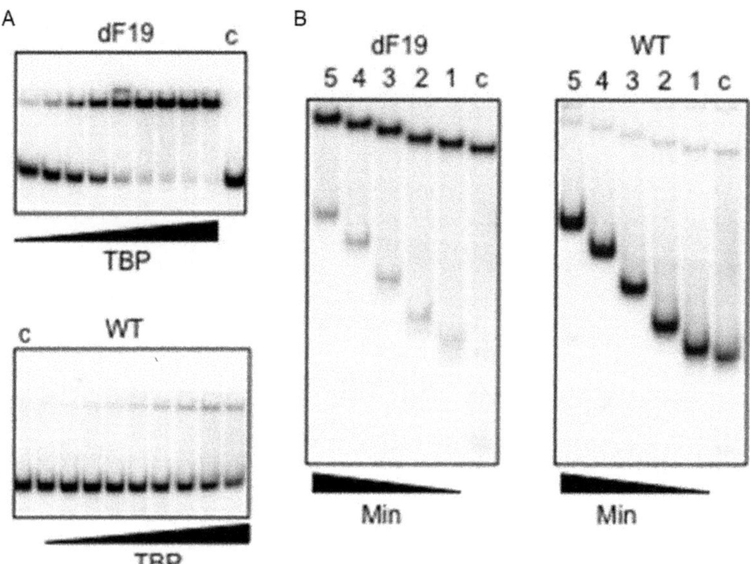

Figure 8 TBP binding to unmodified and dF-substituted TATA box DNA duplexes. (A) Nondenaturing polyacrylamide gels showing unmodified and dF19-substituted duplexes (4 n*M*) binding to TBP, varying in twofold increments (left to right) from 0.73 to 180 n*M*, Lanes c: no protein. (B) Competition assays of TBP complexes with unmodified and dF-substituted TATA box DNA duplex. Lanes from right to left for each substrate: Lanes c, no competitor; Lane 1, 10 min; Lane 2, 30 min; Lane 3, 60 min; Lane 4, 90 min; Lane 5, 120 min. Both the disappearance of bound DNA and appearance of free DNA are noticeably slower for the dF-substituted substrate.

the site of dF substitution. In contrast, since the A-T pairs in the TATA box DNA complex remain hydrogen bonded, the lack of H-bonding for dF seems unlikely to stabilize the protein–DNA complex.

5.3.3 Relative Thermodynamics and Kinetics of TBP:DNA Complexes

If the enhanced affinity reflects preorganization of the dF-substituted TATA duplex, this would create a lower-barrier pathway for the normally slow formation of the TBP–DNA complex (Hoopes, LeBlanc, & Hawley, 1992). dF substitution might then be expected to increase the on-rate, but this is difficult to measure directly, and affinity changes are more frequently reflected in the off-rate, as suggested by the initial binding assay results. To further address the basis for the enhanced thermodynamic stability of the TBP complex with dF19-substituted TATA duplexes, one can instead carry out a competition experiment that measures the stability of the complex in solution as a function of time in the presence of excess competitor DNA. In each experiment, the labeled double-stranded DNA substrate (4 n*M* final) was incubated with TBP (92 n*M* active concentration) for 15 min on ice to form the complex prior to addition of a 50-fold excess of cognate cold competitor (Fig. 8B). Binding reactions were performed under the same buffer conditions and analyzed via PAGE as described above. One sample was loaded on each gel (lane "c") before adding competitor. After addition of competitor, aliquots were removed at 10, 30, 60, 90, and 120 min. One-tenth volume of 50% glycerol containing 0.05% (w/v) bromophenol blue and xylene cyanol was added, and samples were loaded onto the running gel. The gel was dried, exposed to a phosphor screen, and analyzed as above. Based on a qualitative comparison of the rates of appearance of free DNA and disappearance of bound DNA for the two TBP–DNA complexes, the off-rate of the complex formed with the dF-modified duplex was found to be significantly slower than the off-rate of the unmodified duplex. However, the difference in solution off-rates alone does not seem to explain the dramatic difference in gel stability. The molecular determinants of stability in the nonequilibrium EMSA are not well understood, but presumably they include responses to ionic strength, mechanical interactions with the gel matrix, and differential effects on on-rates versus off-rates. If the on-rate for the TBP–dF oligonucleotide is fast, reassociation in the gel matrix cage may be the main source of stabilization.

The observed slower rate of dissociation, the implied increased rate of association, and the enhanced affinity observed between the dF-modified construct and the TBP supports the idea that the DNA may be preorganized to bind TBP and that affinity depends on the energetic cost of the DNA deformation needed for TBP complex formation, in accord with earlier work (Arndt, Wobbe, Ricupero-Hovasse, Struhl, & Winston, 1994; Davis et al., 1999; Strahs, Barash, Qian, & Schlick, 2003). The dF substitution also appears to affect less well-understood mechanisms that control kinetic stability in the EMSA.

6. CONCLUSIONS

In this chapter, single-site dF substitutions in A-tact and TATA box DNA and two RNA/DNA PPT-containing hybrid duplexes were used to demonstrate the utility of this base analog as a tool for spectral assignment and as a probe of structure/dynamics in highly redundant, homopurine-rich DNA sequences. dF substitution offers an alternative and cost-effective means of assigning spectra that are otherwise intractable. In all cases, insertion within pyrimidine tracts results in local perturbation of resonances of adjacent base pairs due to the proximity of the electronegative ^{19}F but leaves the chemical shift of all other nucleotides relatively unaffected. dF substitution significantly facilitated assignment of proton resonances, particularly in the imino, aromatic, and H1′ regions of the spectrum. In addition, no significant change in the relaxation behavior of resonances associated with residues more than one base pair from the site of dF insertion nor deterioration of spectral quality was observed, as is often seen with other labeling (e.g., ^{13}C) approaches. Taken together, data from the DNA duplexes demonstrate that dF can be used as a tool to facilitate proton resonance assignments and could be easily extended to other pyrimidine/purine-rich sequences found in regulatory DNAs of other biological systems.

In addition to its use as a tool for assignment of ^{1}H NMR signals, incorporation of dF into an oligonucleotide may be exploited as a subtle structural probe. dF substitutions of the DNA strand of the Ty3 PPT constructs induced long-range sugar pucker switches in the RNA complement, suggesting an inherent PPT flexibility that may induce recognition by the cognate reverse transcriptase. We additionally used an orthogonal assay of TBP–TATA box DNA binding to demonstrate that the altered conformational dynamics observed by NMR had a measurable impact on the kinetics and thermodynamics of the protein–DNA interaction. Indeed, it has become apparent that while dF introduces minimal changes in ground-state structure, it does decrease the overall thermal stability of the modified duplex (Kool and Sintim, 2006). In addition, in sequences like TATA or HIV-1, which adopt alternative low energy structures, the lack of H-bonding or changes in the stacking interactions at the site of dF incorporation appears to lower the energetic barriers to interconversion between structures or to "relax" the structure to one dominant conformation.

In conclusion, our results from applications to purine-rich sequence elements show that single-site substitution of dF: (1) facilitates assignment of otherwise intractable NMR spectra of highly redundant, A:T tract nucleic acid duplexes and (2) serves as a probe of subtle differences in the structure and dynamics of nucleic acid duplexes.

ACKNOWLEDGMENTS

This work was supported in part by the NIH (GM 59107) to J.P.M. S.F.J.L.G. is supported by the intramural research program of the National Cancer Institute, National Institutes of Health. NMR instrumentation at IBBR was purchased in part with support from the W.M. Keck Foundation, the NIH/NCRR and NIST. The authors additionally acknowledge support of the Hollings Marine Laboratory NMR facility.

Note: Certain commercial equipment, instruments, and materials are identified in this paper in order to specify the experimental procedure. Such identification does not imply recommendation or endorsement by the National Institute of Standards and Technology, nor does it imply that the material or equipment identified is necessarily the best available for the purpose.

REFERENCES

Arndt, K. M., Wobbe, C. R., Ricupero-Hovasse, S., Struhl, K., & Winston, F. (1994). Equivalent mutations in the two repeats of yeast TATA-binding protein confer distinct TATA recognition specificities. *Molecular and Cellular Biology*, *14*, 3719–3728.

Carlow, D. C., Short, S. A., & Wolfenden, R. (1996). Role of glutamate-104 in generating a transition state analogue inhibitor at the active site of cytidine deaminase. *Biochemistry*, *35*, 948–954.

Chen, C., & Russu, I. M. (2004). Sequence-dependence of the energetics of opening of at basepairs in DNA. *Biophysical Journal*, *87*, 2545–2551.

Chu, W. C., & Horowitz, J. (1991a). Fluorine-19 NMR studies of the thermal unfolding of 5-fluorouracil-substituted Escherichia coli valine transfer RNA. *FEBS Letters*, *295*, 159–162.

Chu, W. C., & Horowitz, J. (1991b). Recognition of Escherichia coli valine transfer RNA by its cognate synthetase: A fluorine-19 NMR study. *Biochemistry*, *30*, 1655–1663.

Davis, N. A., Majee, S. S., & Kahn, J. D. (1999). TATA box DNA deformation with and without the TATA box-binding protein. *Journal of Molecular Biology*, *291*, 249–265.

Flinders, J., & Dieckmann, T. (2006). NMR spectroscopy of ribonucleic acids. *Progress in Nuclear Magnetic Resonance Spectroscopy*, *48*, 137–159.

Graber, D., Moroder, H., & Micura, R. (2008). ^{19}F NMR spectroscopy for the analysis of RNA secondary structure populations. *Journal of the American Chemical Society*, *130*, 17230–17231.

Guckian, K. M., Krugh, T. R., & Kool, E. T. (1998). Solution structure of a DNA duplex containing a replicable difluorotoluene-adenine pair. *Nature Structural Biology*, *5*, 954–959.

Guckian, K. M., Schweitzer, B. A., Ren, R. X., Sheils, C. J., Paris, P. L., Tahmassebi, D. C., et al. (1996). Experimental measurement of aromatic stacking affinities in the context of duplex DNA. *Journal of the American Chemical Society*, *118*, 8182–8183.

Hammann, C., Norman, D. G., & Lilley, D. M. (2001). Dissection of the ion-induced folding of the hammerhead ribozyme using ^{19}F NMR. *Proceedings of the National Academy of Sciences of the United States of America, 98*, 5503–5508.

Hennig, M., Scott, L. G., Sperling, E., Bermel, W., & Williamson, J. R. (2007). Synthesis of 5-fluoropyrimidine nucleotides as sensitive NMR probes of RNA structure. *Journal of the American Chemical Society, 129*, 14911–14921.

Hoopes, B. C., LeBlanc, J. F., & Hawley, D. K. (1992). Kinetic analysis of yeast TFIID-TATA box complex formation suggests a multi-step pathway. *The Journal of Biological Chemistry, 267*, 11539–11547.

Juo, Z. S., Chiu, T. K., Leiberman, P. M., Baikalov, I., Berk, A. J., & Dickerson, R. E. (1996). How proteins recognize the TATA box. *Journal of Molecular Biology, 261*, 239–254.

Kim, J. L., & Burley, S. K. (1994). 1.9 A resolution refined structure of TBP recognizing the minor groove of TATAAAAG. *Nature Structural Biology, 1*, 638–653.

Kool, E. T., & Sintim, H. O. (2006). The difluorotoluene debate—A decade later. *Chemical Communications, 35*, 3665–3675.

Kwon, K., Jiang, Y. L., Song, F., & Stivers, J. T. (2002). ^{19}F NMR studies of vaccinia type IB topoisomerase. Conformational dynamics of the bound DNA substrate. *The Journal of Biological Chemistry, 277*, 353–358.

Liu, M., Chu, W. C., Liu, J. C., & Horowitz, J. (1997). Role of acceptor stem conformation in tRNAVal recognition by its cognate synthetase. *Nucleic Acids Research, 25*, 4883–4890.

Metzler, W. J., Leighton, P., & Lu, P. (1988). Two-dimensional heteronuclear NOE spectroscopy; application to ^{19}F-labeled oligodeoxyribonucleotides. *Journal of Magnetic Resonance, 76*, 534–539.

Olejniczak, M., Gdaniec, Z., Fischer, A., Grabarkiewicz, T., Bielecki, L., & Adamiak, R. W. (2002). The bulge region of HIV-1 TAR RNA binds metal ions in solution. *Nucleic Acids Research, 30*, 4241–4249.

Petri, V., Hsieh, M., & Brenowitz, M. (1995). Thermodynamic and kinetic characterization of the binding of the TATA binding protein to the adenovirus E4 promoter. *Biochemistry, 34*, 9977–9984.

Puffer, B., Kreutz, C., Rieder, U., Ebert, M. O., Konrat, R., & Micura, R. (2009). 5-Fluoro pyrimidines: Labels to probe DNA and RNA secondary structures by 1D ^{19}F NMR spectroscopy. *Nucleic Acids Research, 37*, 7728–7740.

Rastinejad, F., Evilia, C., & Lu, P. (1995). Studies of nucleic acids and their protein interactions by ^{19}F NMR. *Methods in Enzymology, 261*, 560–575.

Rastinejad, F., & Lu, P. (1993). Bacteriophage T7 RNA polymerase. ^{19}F-nuclear magnetic resonance observations at 5-fluorouracil-substituted promoter DNA and RNA transcript. *Journal of Molecular Biology, 232*, 105–122.

Rausch, J. W., Qu, J., Yi-Brunozzi, H. Y., Kool, E. T., & Le Grice, S. F. (2003). Hydrolysis of RNA/DNA hybrids containing nonpolar pyrimidine isosteres defines regions essential for HIV type 1 polypurine tract selection. *Proceedings of the National Academy of Sciences of the United States of America, 100*, 11279–11284.

Scott, L. G., Geierstanger, B. H., Williamson, J. R., & Hennig, M. (2004). Enzymatic synthesis and ^{19}F NMR studies of 2-fluoroadenine-substituted RNA. *Journal of the American Chemical Society, 126*, 11776–11777.

Strahs, D., Barash, D., Qian, X., & Schlick, T. (2003). Sequence-dependent solution structure and motions of 13 TATA/TBP (TATA-box binding protein) complexes. *Biopolymers, 69*, 216–243.

Yi-Brunozzi, H. Y., Brinson, R. G., Brabazon, D. M., Lener, D., Le Grice, S. F., & Marino, J. P. (2008). High-resolution NMR analysis of the conformations of native and base analog substituted retroviral and LTR-retrotransposon PPT primers. *Chemistry & Biology, 15*, 254–262.

SECTION II

Scattering Techniques

CHAPTER FIVE

Biomolecular Deuteration for Neutron Structural Biology and Dynamics

Michael Haertlein*, Martine Moulin*, Juliette M. Devos*, Valerie Laux*, Orla Dunne*,†, V. Trevor Forsyth*,‡,1

*Life Sciences Group, Institut Laue-Langevin, Grenoble Cedex 9, France
†Department of Structural and Molecular Biology, University College London, London, United Kingdom
‡Macromolecular Structure Group, Keele University, Staffordshire, United Kingdom
[1]Corresponding author: e-mail address: tforsyth@ill.fr

Contents

Abstract

Neutron scattering studies provide important information in structural biology that is not accessible using other approaches. The uniqueness of the technique, and its complementarity with X-ray scattering, is greatest when full use is made of deuterium labeling. The ability to produce tailor-made deuterium-labeled biological macromolecules allows neutron studies involving solution scattering, crystallography, reflection, and dynamics to be optimized in a manner that has major impact on the scope, quality, and throughput of work in these areas. Deuteration facilities have now been developed

Methods in Enzymology, Volume 566
ISSN 0076-6879
http://dx.doi.org/10.1016/bs.mie.2015.11.001

at many neutron centres throughout the world; these are having a crucial effect on neutron studies in the life sciences and on biologically related studies in soft matter. This chapter describes methods that have been developed for the efficient production of deuterium-labeled samples for a wide range of neutron scattering applications. Examples are given that illustrate the use of these samples for each of the main techniques. Perspectives for biological deuterium labeling are discussed in relation to developments at current facilities and those that are planned in the future.

1. THE MOTIVATION FOR MACROMOLECULAR DEUTERATION APPROACHES

Neutron scattering studies offer unique insights to structural biology. This results mainly from the fact that the scattering power for neutrons of hydrogen is comparable to the other atoms typically found in biological macromolecules, contrasting strongly with the situation for X-rays where hydrogen atoms scatter very weakly. This is of crucial importance given that about half of the atoms in biological molecules are hydrogen and that they are often highly significant to biological structure and function. Table 1 shows a comparison of the X-ray and neutron scattering lengths for atoms commonly found in biological macromolecules. X-ray scattering lengths increase monotonically as a function of atomic number, reflecting the fact

Table 1 X-Ray and Neutron Coherent Scattering Lengths of Atoms Commonly Found in Biological Macromolecules

Atom	Atomic Number	Scattering Length for Neutrons (10^{-13} cm)	Scattering Length for X-Rays (10^{-13} cm)
Hydrogen	1	−3.74	2.8
Deuterium	1	6.67	2.8
Carbon	6	6.65	16.9
Nitrogen	7	9.40	19.7
Oxygen	8	5.80	22.5
Phosphorous	15	5.10	42.3
Sulfur	16	2.85	45.0
Iron	26	9.51	73.0

Note the large difference in the neutron scattering lengths of hydrogen and its isotope deuterium.

that they are scattered by the atomic electrons. In the case of neutron scattering, where the scattering event is associated with the interaction between the neutron and the atomic nucleus, the variation of scattering length over the periodic table does not follow any such trend and may vary considerably from isotope to isotope. This means, for example, that hydrogen atoms can be imaged more easily in neutron crystallographic studies than in corresponding X-ray work. In addition, the ability to exchange hydrogen by its heavier (and more strongly scattering) deuterium (^2H or D) isotope offers a completely unique scope to neutron scattering studies of the structure and dynamics of biological systems in essentially all relevant sample habits.

A further aspect of neutron scattering that is important to this discussion is the fact that the two spin states of the hydrogen nucleus give rise to a large incoherent scattering cross section. For structural studies (diffraction, small-angle scattering, reflection), the useful signal derives from the coherent scattering; incoherent scattering from hydrogen is typically treated as a background that restricts the accuracy with which the coherent observations can be measured. This issue is particularly significant for high-angle diffraction work (e.g., crystallography and fiber diffraction where the higher resolution data tend to be weaker and more difficult to extract) and has serious implications for sample size and data quality. Hydrogen incoherent scattering can however be exploited in neutron studies of the dynamics of biological systems; the large incoherent signal from hydrogen can be used to measure the mean square displacement of the hydrogen atoms in solvent and protein components of a macromolecule.

Hydrogen and deuterium have close but clearly distinguishable physical properties. Light and heavy water (H_2O and D_2O) have slightly different melting points (0 and 3.8 °C, respectively), boiling points (100 and 101.4 °C, respectively), and densities (1 and 1.108 g cm^{-3}, respectively). There are also subtle but measurable differences between the parameters associated with bonding involving H and D atoms. Despite these differences, the behaviour of the two isotopes in the context of solvent interactions and covalent incorporation into macromolecules is such that isomorphism is usually reliable to the extent that the differences in neutron scattering properties can be heavily exploited.

Microorganisms such as *Escherichia coli* can cope, after adaptation, with both solvent (D_2O) and macromolecular isotope effects—both of which impact on growth. Deuterium effects on the central carbon metabolism have been studied by Hochuli, Szyperski, and Wüthrich (2000), who note an increase of the anaplerotic supply of the tricarboxylic acid (TCA) cycle. For animals and higher plants, high concentrations of D_2O are toxic. Weak physicochemical differences, mainly related to protein stability, have been observed between hydrogenated and perdeuterated proteins (Brockwell et al., 2001). However, there have been numerous X-ray crystallographic studies comparing perdeuterated and hydrogenated proteins that find little difference between the structures (e.g., Artero, Haertlein, McSweeney, & Timmins, 2005; Hazemann et al., 2005). Fisher and Helliwell (2008) have surveyed structural changes relating to deuterium substitution in small molecules and concluded that the main effects are those associated with the differing bond geometries involved. These effects are rather subtle and unlikely to be of concern given the resolutions typically obtained in neutron crystallographic or small-angle neutron scattering (SANS) studies of proteins. However, it is important to note that biochemical and biophysical characterization of deuterated samples should be carried out whenever possible and compared with those of the corresponding hydrogenated molecules. Differences in aggregation properties between hydrogenated and deuterated proteins are sometimes noted (e.g., White et al., 2015)—an effect that may well occur more frequently for soft proteins.

The production of deuterated biomolecules in the quantities needed for neutron scattering studies is technically challenging and intrinsically expensive. It is very difficult for individual users to prepare samples of this type on a "one-off" basis and the number of users who are able to operate in this way is limited. To remove these limitations, centralized facilities for the production of deuterated macromolecules were proposed (Forsyth, Myles, Timmins, & Haertlein, 2001) to provide the necessary continuity of expertise, dedicated laboratory facilities, and economies of scale. Protein deuteration requires large quantities of heavy water, deuterated carbon sources, and other reagents. The operation of a centralized laboratory by experienced staff that have a wide range of skills in microbiology, biochemistry, molecular biology and bioprocessing is essential and, allows the efficient production of tailor-made deuterated biomolecules as well as the full exploitation of deuterated biomass (including hydrolysis to small biomolecules) and the recycling of D_2O from used growth media.

The purpose of this chapter is to review the scope of deuterium labeling for neutron scattering investigations in structural biology and dynamics. It summarizes the methods that have been used thus far, and discusses those that will be needed in the future given likely developments in biotechnology and those at current and future neutron beam sources.

1.1 Biological Small-Angle Neutron Scattering

At low resolution, the differing scattering lengths of hydrogen and deuterium can be used in contrast variation studies where the visibility of particular parts of a molecular system can be changed by adjusting the D_2O/H_2O ratio within the solvent (Jacrot, Chauvin, & Witz, 1977).

The impact of contrast variation for low resolution neutron work (both in solutions and in crystals) is enormous; no comparable capability exists for analogous small-angle X-ray scattering (SAXS) approaches. The large variation of contrast available over the range from 0% D_2O (100% H_2O) to 100% D_2O (0% H_2O) is summarized in Fig. 1. In its simplest form, contrast variation can be performed by varying the D_2O/H_2O solvent composition to match out hydrogenated protein (~40% D_2O), nucleic acid (~70% D_2O), or lipid (~5–20% D_2O) components of a system (Fig. 1)—exploiting the natural contrast that exists in hydrogenated complexes. However, for multi-component protein systems, where specific parts of a structure may be of high relevance to biological function, no natural contrast exists, and deuterium labeling of the constituent parts is necessary to allow the different

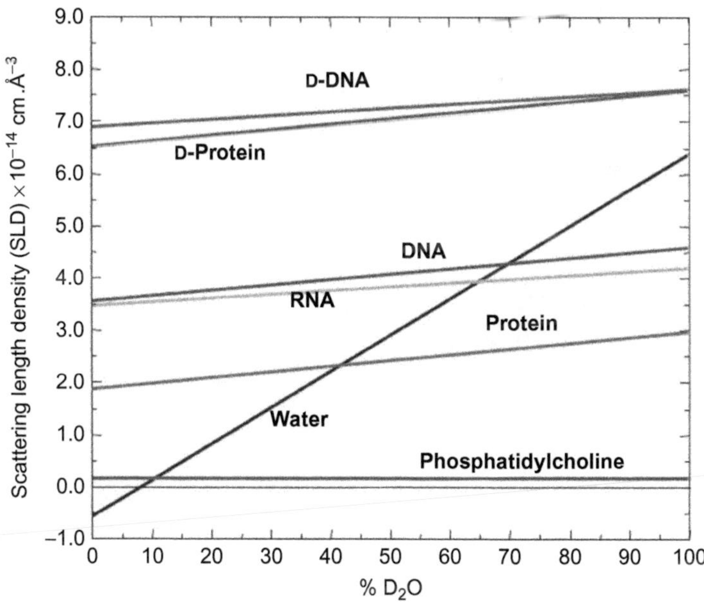

Figure 1 Scattering length densities (SLDs) plotted as a function of D_2O/H_2O solvent composition for various types of biological macromolecule. The points at which these lines intersect that for the isotopically mixed water defines the match points at which that biological component becomes invisible in terms of its coherent scattering. (See the color plate.)

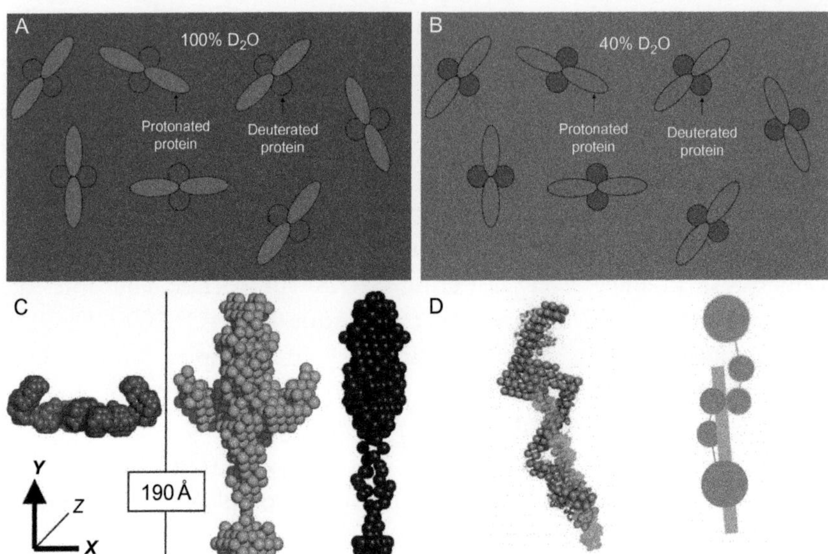

Figure 2 Graphical representation of the use of solvent contrast variation and some examples. (A) Schematic representation of the matchout of the deuterated protein in a complex in 100% D_2O solvent, (B) analogous representation of matchout of hydrogenous parts of the same complex in 40% D_2O, (C) restriction modification system studied by contrast variation in conjunction with selective deuteration (Callow, Sukhodub, Taylor, & Kneale, 2007), and (D) contrast variation study of a mariner transposase system using deuterated protein in complex with hydrogenated DNA (Cuypers, Trubitsyna, et al., 2013). (See the color plate.)

components to be modeled (Fig. 2). For example, it is possible to investigate and model the shape of a single labeled protein subunit within a multisubunit protein complex or to measure the distance between two labeled subunits within a complex. Specific aspects of such sample preparation will be discussed in more detail in Section 2 of this chapter.

1.2 Medium and High Resolution: Protein Crystallography and Fiber Diffraction

At medium to high resolution, the ability to determine and distinguish H/D or H_2O/D_2O positions in single crystals provides unique information and insight to catalysis, protein–ligand interactions, and protein–solvent interfaces (Blakeley, 2009; Blakeley et al., 2010; Casadei et al., 2014; Fisher et al., 2014; Kovalevsky et al., 2010, 2011). At atomic and subatomic resolution, neutron crystallography is able to provide remarkable additional detail in visualizing structured solvent networks and in the identification of rare ionic species. For the vast majority of cases, this sort of detail cannot be determined by X-ray analysis alone. The importance of this type of information is highlighted by the development of neutron diffractometers at steady state and spallation neutron sources throughout the world (Teixeira et al., 2008), and by the high demand for these instruments. In fiber diffraction, the advantages of using neutron diffraction for the investigation of solvent structure and hydrogen bonding around molecules such as DNA (Forsyth et al., 1989; Langan et al., 1992; Shotton et al., 1997), cellulose (Nishiyama, Langan, Wada, & Forsyth, 2010; Sawada et al., 2013; Thomas et al., 2013), chitin (Sawada et al., 2012), amyloid (Tiggelaar et al., 2011), and synthetic polymers (Gardner, English, & Forsyth, 2004) are very clear—no comparable information has been obtained from X-ray diffraction, and there is consistent use of this approach in the study of partially ordered systems.

A major challenge confronting medium- and high-resolution neutron diffraction work is the requirement for large sample size. This problem is exacerbated by incoherent scattering from hydrogen in the solvent and within the macromolecule itself. Additionally, the negative scattering length of hydrogen tends to cause cancellation effects in Fourier maps if the resolution available is not relatively high.

It is relatively straightforward to replace H_2O solvent by D_2O, and in doing so to exchange the labile hydrogen atoms attached of the N–H and O–H groups in a biological molecule; this results in a large decrease in hydrogen incoherent scattering, significantly eases sample volume requirements and augments data quality. The effect of H_2O/D_2O solvent replacement on the observed diffraction data is illustrated in Fig. 3, which shows high-angle neutron fiber diffraction data recorded from a sample of DNA in the A conformation using ILL's D19 diffractometer. The sample was initially hydrated by humidification in H_2O using a system that has been described by Shotton et al. (1998). The sequence shows the variation of the A-DNA diffraction pattern as a function of time following the replacement of the H_2O by D_2O in the humidification system. The change in the level of the hydrogen incoherent scattering from the solvent is very clear, as is the visibility of the coherent diffraction features. The central part of the diffraction pattern shows contrast variation effects of the type described in Section 1.1. While solvent exchange has a very significant effect on the effective signal-to-noise ratio for the coherent data, there still remains a very large incoherent scattering contribution from hydrogen that is covalently bound to carbon atoms in the structure. This places major limitations on

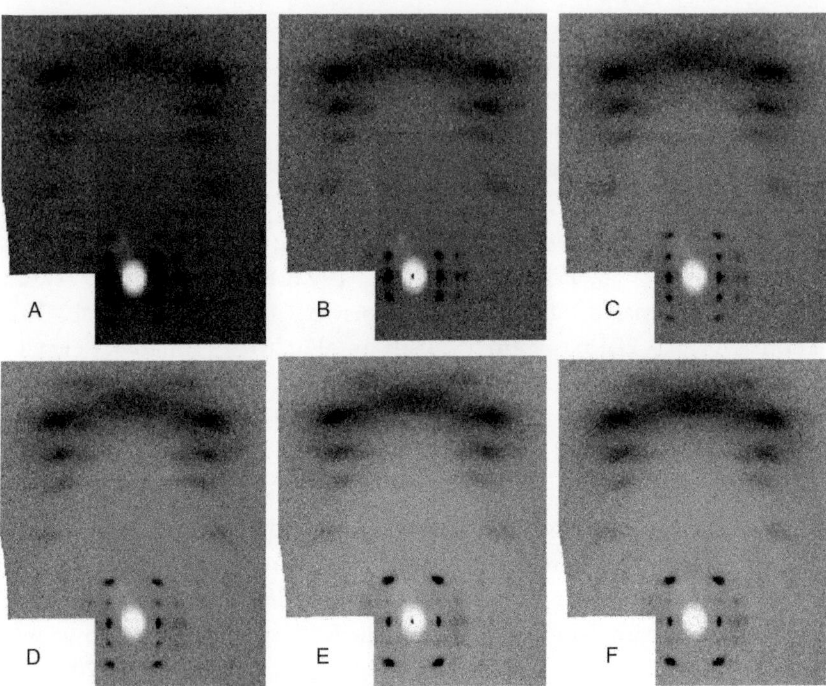

Figure 3 Neutron high-angle fiber diffraction patterns recorded from a sample of A-DNA using the D19 diffractometer at the ILL. The sequence from (A) to (F) shows how the diffraction pattern varies as the hydration of the sample (via the humidity in the sample environment) is changed from H_2O to D_2O. The change in the level of the hydrogen incoherent scattering is very clear as is the improvement in the visibility of the coherent diffraction peaks.

sample size, data collection times, data interpretation, and indeed on the quality of the final analysis. As a rough guide (neglecting the effect of unit cell size), it has been estimated that for a neutron diffraction study of a typical macromolecule, one can expect a gain of a factor of $\sim$10 in the sample volume required to produce a high quality neutron diffraction dataset through sample perdeuteration, as illustrated in the study of perdeuterated A-DNA by Shotton et al. (1997).

In neutron crystallography, numerous examples of the benefits of using perdeuterated protein have been described. Figures 4 and 5 show neutron scattering density maps calculated for perdeuterated transthyretin (TTR) (Haupt et al., 2011, 2014) and perdeuterated rubredoxin, respectively (Cuypers, Mason, et al., 2013). Other good examples include neutron crystallographic studies of aldose reductase (Hazemann, Dauvergne, et al., 2005) and type-III antifreeze protein (Howard, Blakeley, et al., 2011).

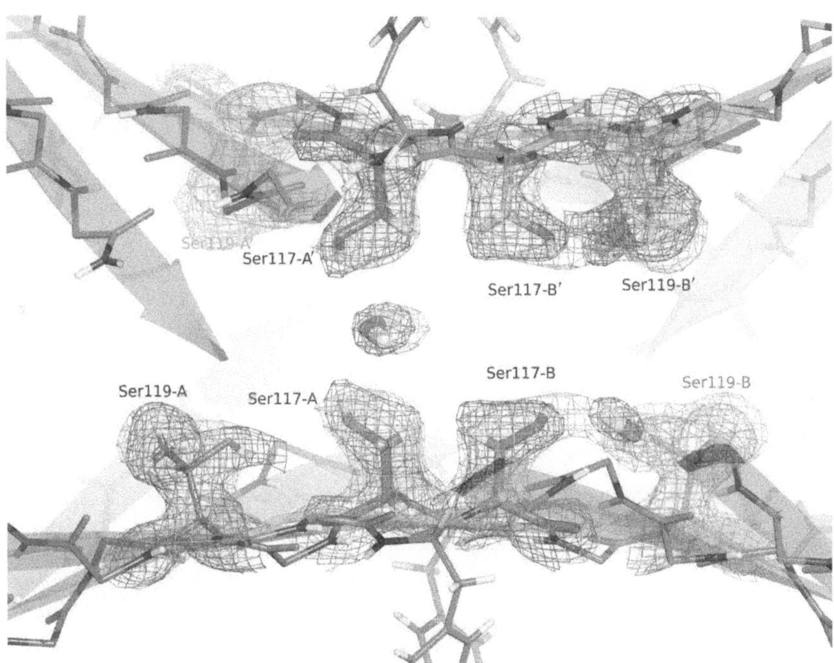

Figure 4 View of the central part of the bivalent binding pocket of tetrameric human transthyretin (TTR). The neutron study used perdeuterated TTR, with data recorded from both LADI-III and D19 diffractometers at the ILL. The neutron scattering density maps show a key water molecule trapped between hydroxyl groups of serine residues in adjacent monomers. This imparts an intrinsic asymmetry to the TTR tetramer in a region that is very important for the binding of thyroxine and of ligands that have been designed to prevent tetramer dissociation and subsequent amyloid formation. *Picture from Haupt et al. (2014) with permission of the International Union of Crystallography.* (See the color plate.)

1.3 Neutron Reflection

Neutron reflection (NR) approaches, can provide valuable information on the structure of interfaces and on the interaction of model membranes with proteins or peptides. The arguments concerning the deployment of deuteration in NR studies are in concept similar to those for SANS (see Section 1.1). While there is a good natural contrast between lipid fatty acid tails and protein molecules that can be exploited using contrast variation, one of the difficulties is that the lipid head groups have a significantly different scattering length density (SLD) and may be difficult to distinguish from protein components. The use of deuterated protein in NR studies of this type alleviates this problem by providing a markedly increased contrast

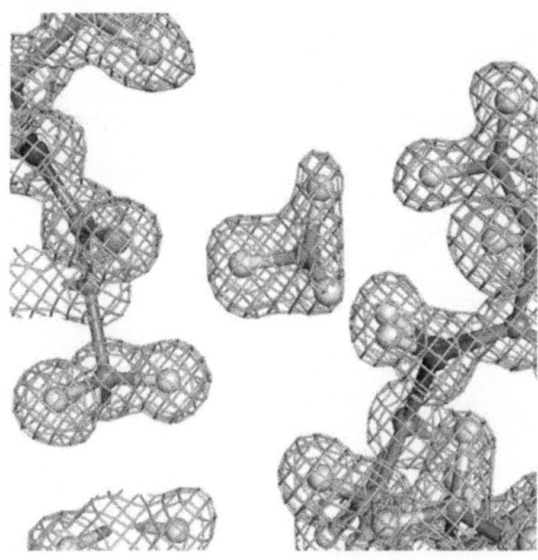

Figure 5 High-resolution studies (current limit 0.88 Å) of both oxidized and reduced forms of perdeuterated *Pyrococcus furiosus* rubredoxin have been carried out on the D19 diffractometer at the ILL and show unparalleled detail. A tautomeric shift is noted following the change from oxidized to reduced forms (Cuypers, Mason, et al., 2013). Highly ordered water networks are observed that provide definitive evidence of hydronium ions that may play a role in charge transport in the redox protein. (See the color plate.)

between membrane and protein components of a system (e.g., Cowsill et al., 2014; Grage et al., 2011; Hellstrand et al., 2013).

1.4 Dynamics

Neutrons scattering is a powerful probe for the study of molecular dynamics in proteins (Gabel et al., 2002; Zaccai, 2011). The main impact of macromolecular deuteration for the study of dynamics relates to the fact that the deuterium-labeled part of a multicomponent system contributes only minimally to the incoherent scattering. This can be exploited by making samples in which specific parts of an otherwise deuterated system are hydrogenated, providing a strong incoherent scattering signal from a particular part of the system while the contribution from the remainder is effectively masked. Three different types of experiment make use of this approach: (i) those in which the protein part of a sample is perdeuterated and the hydration water is H_2O. This allows the dynamics of hydration water to be studied and is a powerful complement to the standard approach in which protein dynamics is studied using natural abundance protein with D_2O solvent; (ii) samples in which deuterated protein is reverse labeled to study the dynamics of specific natural abundance amino acids, for example, tryptophan around the active site of bacteriorhodopsin (bR) (Réat et al., 1998) or amino acids that are either hydrophobic in character (e.g., leucine, isoleucine, valine), and hence located in the core of the protein, or hydrophilic (e.g., lysine) and located close to the surface Wood et al. (2013). This labeling approach allows the dynamics of the interior of a protein to be distinguished from the dynamics of the exterior, as illustrated in Fig. 6; and (iii) samples in which the effects of molecular crowding on protein dynamics may be studied using deuterated biomolecular crowding agents/molecules.

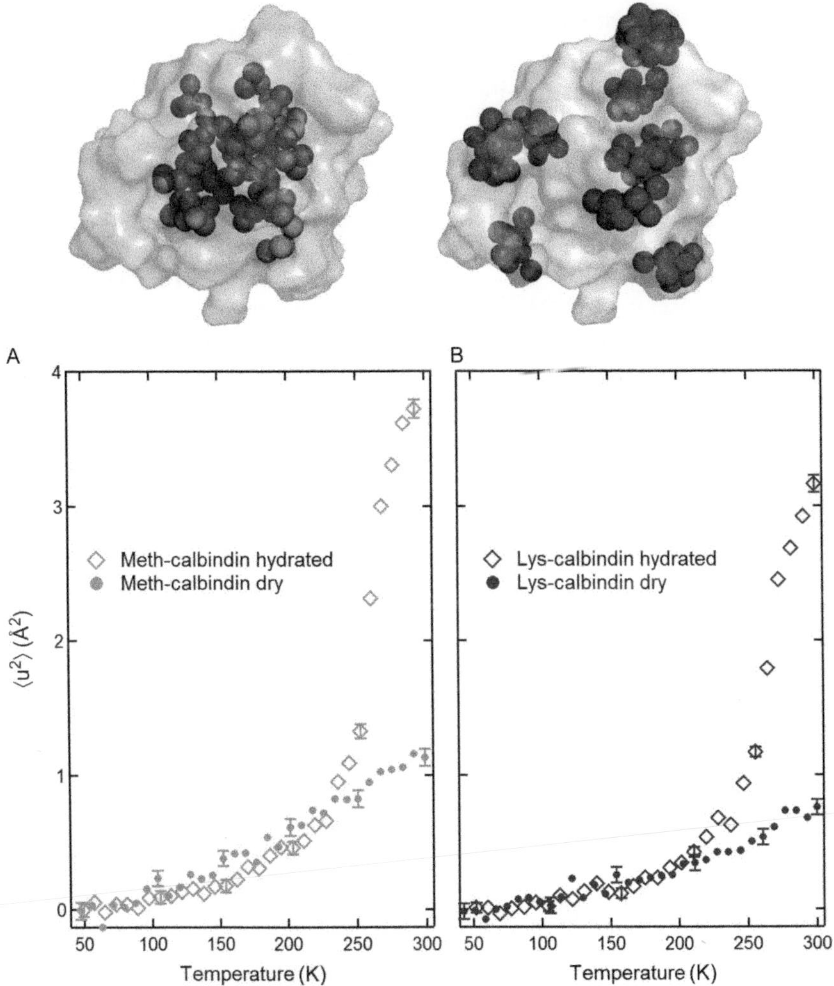

Figure 6 Coupling of protein and hydration water nanosecond dynamics as measured from the interior and exterior of calbindin by neutron elastic incoherent scattering (see Wood et al., 2013). (See the color plate.)

2. CONCEPTS OF *IN VIVO* DEUTERATION

In principle, there are two different routes for the production of deuterated proteins: chemical synthesis and biosynthesis (*in vivo/in vitro*). However, in practice, the applicability of chemical synthesis is restricted to relatively short peptides. As the length of the peptide increases, preparative scale synthesis becomes difficult and *in vivo* techniques are frequently used for the production of the longer peptides. In general, the scope and relevance of *in vivo* protein production depends critically on the organism used. Gram-negative bacteria such as *E. coli* are routinely used for protein production but are often limited by the lack of co- and post-translational modification effects such as disulfide bond formation, glycosylation, and phosphorylation. These can lead to limited solubility of the recombinant protein and the formation of aggregated protein within inclusion bodies. While smaller proteins can in certain situations be refolded, the refolding step is not always totally effective and this can be especially difficult if performed in deuterated buffer. Furthermore, the fact that *E. coli* has a double-membraned envelope has the consequence that proteins are not usually secreted directly into the growth medium and purification generally has to occur from the crude extract of lysed cells.

Many of these problems can be avoided through the use of eukaryotic expression systems in yeast, insect, and mammalian cells. The advantage of these systems arises from the fact that post-translational modifications can occur, and proteins can be secreted directly into the growth medium if needed. Hence, one of the most important issues for correct folding—the formation of disulfide bonds within the proteins—can occur during secretion. If for structural studies post-translational modifications such as glycosylation are important, the recombinant protein should, if possible, be expressed in cell types that closely match the native context.

If *in vivo* protein production is not feasible (for example, as a result of toxicity issues of the protein involved), cell-free (*in vitro*) expression approaches can be envisaged. These systems are "open" (i.e., non-cellular) and allow the addition of chaperones and disulfide bridge formation factors.

For macromolecular deuteration, the choice of an appropriate expression system is essential. The ideal system is one in which the host organism can be easily adapted to growth in deuterated minimal media containing simply heavy water, a deuterated carbon source, and mineral salts. The early work of Katz and Crespi (1966) demonstrated that most microorganisms can be adapted to grow in perdeuterated media (for a review, see Carlstedt, Crespi, Blake, & Katz, 1970). Several types of algae and bacteria have been used for production of perdeuterated metabolites and various media based on perdeuterated biomass are commercially available. In general terms, bacterial and yeast systems can be readily adapted to high levels of deuteration, whereas insect and mammalian cell systems are more problematic to label at a high level due to the higher toxicity of deuterium to the cells and the need for deuterated precursors (e.g., essential amino acids that cannot be synthesized in these cell types) in the medium used. In principle, cell-free systems offer great potential for deuterium labeling given the potential for amino acid-specific labeling. However, low yields and the requirement for labeled precursors has restricted the use of this approach.

2.1 Expression of Deuterated Macromolecules in the Prokaryote *E. coli*

2.1.1 Adaptation to Deuterium-Containing Media

The capacity of *E. coli* cells to adapt to growth in D_2O media is largely strain dependent. The most frequently used *E. coli* host strain for recombinant protein expression (strain BL21) readily adapts to D_2O. Other widely and successfully used strains for the production of deuterated proteins are MRE600 and its derivative, rifampicin-resistant MRE600Rif (Venters et al., 1995). However, in general, the change to deuterated media requires an adaptation procedure that typically occurs over several days. Paliy, Bloor, Gilbert, and Barber (2003) have described a multi-stage adaptation protocol whereby repeated plating and colony selection results in deuterium-tolerant cell cultures. Three *E. coli* strains, JM109, MRE600, and MRE600Rif, have been adapted to growth on deuterated succinate medium in this way. The adapted strains showed good and consistent growth rates and were capable of being transformed with plasmids. Expression of heterologous proteins in these strains is consistently high. These authors suggest that *E. coli* cells are inherently capable of growth on deuterated media, but that nonspecific mutations enhance deuterium tolerance. Thus plating and colony selection leads to highly deuterium-tolerant strains. A general scheme for the adaptation process to deuterated minimal medium used in the ILL's Life Sciences Group is shown in Fig. 7.

Deuterated cultures of *E. coli* can also be grown in "rich" deuterated media that in principle supplies all the cells' needs and minimizes the biosynthesis effort. Such cultures tend to grow more rapidly and it is often

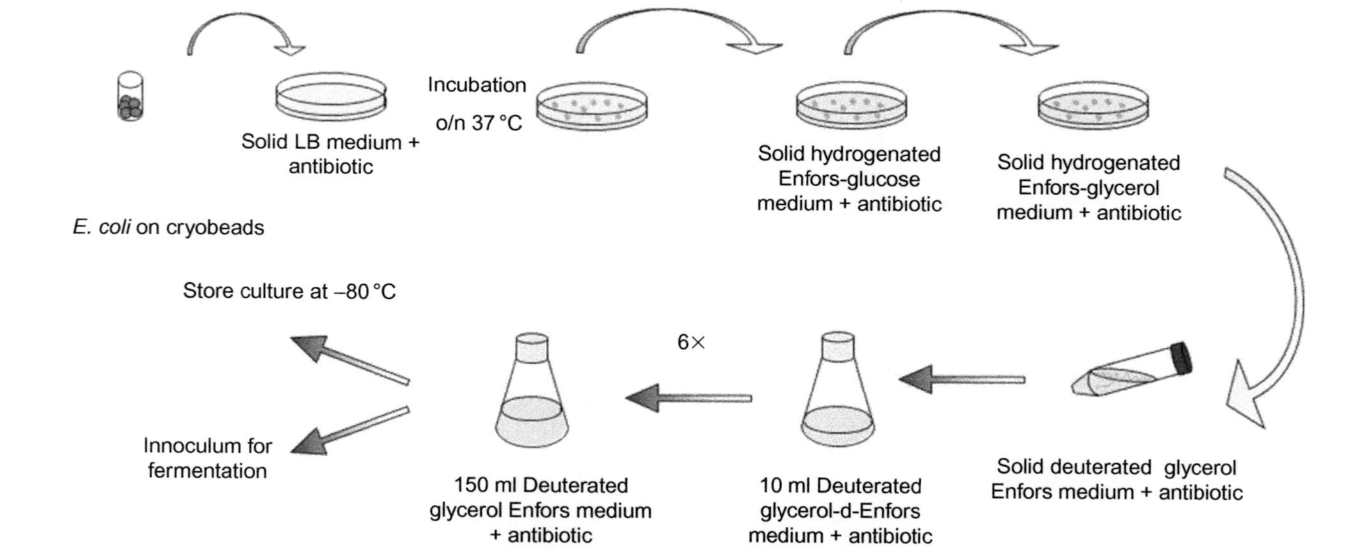

Figure 7 Graphical representation of the adaptation of *E. coli* cells to deuterated minimal medium. Recombinant *E. coli* cells stored in cryobeads are first grown on Luria Bertani (LB) agar plates, then on agar plates containing hydrogenated minimal medium with glucose as carbon source. In the following step, plates containing hydrogenated minimal medium with glycerol as carbon source are used, followed by incubation in a tube containing deuterated minimal medium glycerol agar. Cells grown on this agar are transferred to a flask containing 10 ml of deuterated glycerol minimal medium. The cells are then diluted and regrown in fresh deuterated medium. This process is repeated several times until cells grow at a reasonable growth rate. A 150 ml culture is then prepared as preculture for fermenter high cell density culture.

argued by suppliers of such media that the reduced stress on the cells obviates the need for adaptation procedures. However, growth in deuterated rich media is typically slower than that in nondeuterated media, and induction conditions need to be adapted.

2.1.2 Choice of Media and Culture Conditions

The ease of cultivation of *E. coli* depends greatly on the choice of the deuterated medium. Rich media based on hydrolysate from deuterated algae or from chemolithoautotrophic bacteria are available but the cost of these can be prohibitive. The most economic method to obtain deuterated proteins is through bacterial expression from cells grown on minimal media containing a deuterated carbon source such as deuterated acetate, succinate, or d_8-glycerol. Different minimal media can be used for deuteration. The most frequently used is M9 medium (6 g Na_2HPO_4, 3 g NaH_2PO_4, 0.5 g NaCl, 0.5 g NH_4Cl). For high cell density cultures (HCDC) different media are used (see below). This method has proved successful in the deuteration of fully deuterated cell components (Vanatalu et al., 1993) and several recombinant proteins including membrane proteins were produced in large quantity. It requires a prototrophic *E. coli* host strain such as BL21 (DE3) and is especially attractive when used in combination with high cell density cultivation. If a prototrophic *E. coli* host strain cannot be used, the minimal medium has to be supplemented with "rich media," or with the relevant deuterated amino acid.

As noted above, growth rates and protein expression may be increased through the use of "rich media" that contain more complex substrates rather than simple carbon sources and salts. "Fully rich" media containing mixtures of amino acids, peptides, nucleic acids, and complex carbohydrates can decrease doubling times and increase total protein expression. The use of rich medium may therefore be an alternative to deuterated minimal medium if (i) fermentation equipment for high cell density cultivation is not available, (ii) the required quantities of labeled protein are relatively low, and (iii) the expression system being used is highly productive.

Various types of rich media are available commercially. Some are derived from bacterial hydrolysate and contain primarily amino acids, some low-molecular-weight oligopeptides, and almost no carbohydrates. The bacterial strain used for the hydrolysate production is the chemolithoautotrophic strain *Ralstonia eutropha* grown on inorganic isotopically labeled substrates (Heumann, 2004). These media have been optimized to ensure reproducible cell density and high yields of heterologously expressed proteins. They are available for ^{2}H, ^{13}C, and ^{15}N forms as well as for $^{2}H/^{15}N$, $^{13}C/^{15}N$ double, and $^{2}H/^{13}C/^{15}N$ triple labeled forms. Similar growth media based on labeled algal extracts are also available for labeling of proteins in bacteria.

2.1.3 Large-Scale Preparation of Fully Deuterated Cell Components

Vanatalu et al. (1993) have developed a protocol for the isolation of deuterated ribosomes. Here, the *E. coli* strain MRE600rif was cultured in minimal media in a fed-batch mode using deuterated acetate and succinate in 100% D_2O. The cultivation was carried out in substrate-limiting conditions whereby the growth rate was kept at a low level and the accumulation of substrate or by-products in the medium was avoided. Cells were harvested in kilogram quantities, and the extracted ribosomes showed a normal complement of proteins, containing intact rRNA, and were fully active.

2.1.4 Large-Scale Preparation of Fully Deuterated DNA

Shotton et al. (1997) have used perdeuterated chromosomal DNA isolated from MRE600 *E. coli* cells grown in 95% D_2O using deuterated succinate as a carbon source. The ability to isotopically replace hydrogen in the DNA sample by deuterium allowed the large coherent scattering length of deuterium to be utilized in imaging the water distribution around various conformations of polymeric DNA, and provided a very significant advantage over previous work of this type that had been based on the use of unlabeled DNA of both natural and synthetic origin (Forsyth et al., 1989; Langan et al., 1992). The interest in this type of neutron study derived from time-resolved X-ray fiber diffraction studies of structural polymorphism in the DNA double helix (Mahendrasingam et al., 1986; Forsyth et al., 1986; Fuller, Forsyth, & Mahendrasingam, 2004). Zhu et al. (2010) have used perdeuterated plasmid DNA to study the effects of crowding on the conformation of underwound DNA strands using a zero averaged contrast approach.

2.1.5 In Vivo Deuteration of Transfer RNA

Fully deuterated tRNAs can be prepared from cells grown in deuterated minimal medium, but the tRNA content in these conditions is low due to regulation of tRNA biosynthesis in response to the slow growth of cells. Jünemann et al. (1996) have reported large-scale preparation of deuterated tRNA species using overexpression constructs; this improves the yield of specific deuterated tRNAs by a factor of ~2–10, depending on the tRNA and growth conditions tested. Purification yields nearly homogenous deuterated tRNAs with a chargeability of 1400–1500 pmol amino acid/A260 unit. The deuterated tRNAs were of excellent biological activity.

2.1.6 Expression of Deuterated Recombinant Proteins

The Gram-negative bacterium *E. coli* remains the most versatile host for the production of heterologous proteins (Baneys, 1999). The production of a functional protein is intimately related to the cellular machinery of the organism producing the protein. *E. coli* is the "factory" of choice for the expression of many proteins because its genome has been fully mapped and the organism is easy to handle. Currently, the most widely used approach involves the cloning of the target gene under the control of bacteriophage T7 transcription and subsequent expression of the protein in a host strain that can supply T7 RNA polymerase (BL21 (DE3)). Inducible expression can be achieved by using an isopropyl β-D-1-thiogalactopyranoside (IPTG)-inducible *lacUV5* promoter to regulate expression of the chromosomally encoded T7 RNA polymerase (Studier, Rosenberg, Dunn, & Dubendorf, 1990). For the expression of toxic gene products where undesirable basal expression has to be avoided, an *E. coli* strain in which the chromosomally encoded T7 RNA polymerase is under control of the araBAD promoter (BL21-AI, Invitrogen) should be used. Formation of disulfide bonds in *E. coli* cytoplasm is in some cases possible in a trxB thioredoxin reductase mutant where the periplasmic enzyme DsbC, which catalyzes the disulfide isomerization, is engineered to be expressed in the cytoplasm (Bessette, Aslund, Beckwith, & Georgiou, 1999).

2.1.6.1 A General Protocol for Perdeuteration of Recombinant Proteins Using High Cell Density Cultures, Minimal Medium, and a Bioreactor

Production of many recombinant proteins can be carried out in HCDC. Most cultivation strategies are based on a batch phase followed by a fed-batch phase. A well-known phenomenon is the accumulation of acetate as a by-product. This is due to the cell specific growth rate transcending a critical value following an excess of the main carbon source (e.g., glycerol). This metabolic overflow leads to inhibition effects and low product yields.

A general protocol has been developed in the Deuteration Laboratory within ILL's Life Sciences Group and has been successfully applied to large-scale expression of several bacterial, yeast, mammalian, and human deuterated proteins (Achterhold et al., 2011; Appolaire et al., 2014; Artero et al., 2005; Blakeley et al., 2008; Brouette et al., 2013; Chenal et al., 2009; Clarke et al., 2010; Compton et al., 2014; Cuypers, Mason, et al., 2013; Cuypers, Trubitsyna, et al., 2013; Di Costanzo, Moulin, Haertlein, Meilleur, & Christianson, 2007; Edlich-Muth et al., 2015; Fichou et al., 2015; Gallat, Brogan, et al., 2012; Gallat, Laganowsky, et al., 2012; Grossmann et al., 2008; Haupt et al., 2014, 2011; Hazemann et al., 2005; Hellstrand et al., 2013; Howard et al., 2011; Jasnin, Moulin, Haertlein, Zaccai, & Tehei, 2008a, 2008b; Jasnin, Tehei, Moulin, Haertlein, & Zaccai, 2008; Kennaway et al., 2012; Laulumaa et al., 2015; Meilleur, Dauvergne, Schlichting, & Myles, 2005; Niemann et al., 2008; Paciaroni et al., 2009, 2008, 2012; Petit-Haertlein et al., 2009, 2010; Rochel et al., 2011; Schirò et al., 2015; Schneck et al., 2013; Seyedarabi et al., 2010; Taylor, Swiderska, Artero, Callow, & Kneale, 2012; Vijayakrishnan et al., 2011, 2010; Wood, Froelich, et al., 2008; Wood et al., 2013; Wood, Grudinin, et al., 2008; Wu et al., 2009; Zhu et al., 2010). The deuterated proteins include cytochrome P450cam, gamma E-crystallin, human aldose reductase, human arginase, *Listeria monocytogenes* Invasion Protein InlB, maltose-binding protein (MBP), yeast TIM9, anti-freeze protein, human ApoI, diphtheria toxin T domain, myoglobin, human pyruvate dehydrogenate complex, human TTR, pectate lyase, calbindin, restriction endonulease EcoR124, α-synuclein, rubredoxin, oligomeric aminopeptidase TET, the intrinsically disordered protein tau, and the *E. coli* membrane protein MscL. The deuterated proteins have been produced on 100 mg L^{-1} to 1 g L^{-1} scales.

The expression strain used is BL21 (DE3). This prototrophic strain grows in minimal media without any amino acid supplement. In addition, this strain can be easily adapted to growth in D_2O minimal media. For high cell density growth in deuterated medium, a kanamycin resistance marker is used. Ampicillin selection is very widely used in molecular biology, but is not suitable in this context since plasmid stability may be affected by leakage of beta-lactamase into the medium. If the expression construct carries an ampicillin marker, the gene/cDNA should be recloned into a kanamycin resistant pET variant. Freshly transformed cells are plated on Luria Bertani (LB) solid medium. A colony is then selected and plated on solid hydrogenated minimal medium with hydrogenated glycerol as carbon source. After overnight growth, cells are plated on fully deuterated solid minimal medium with deuterated d_8-glycerol as carbon source. Liquid deuterated minimal medium is then inoculated with the adapted cells. When the cells reach the late exponential phase, a new culture is inoculated (1:10 dilution) and this is repeated until the cells attain a growth rate similar to that of a culture in hydrogenated minimal medium. The composition of the minimal medium used is the following: 6.86 g L^{-1} $(NH_4)_2SO_4$, 1.56 g L^{-1} KH_2PO_4, 6.48 g L^{-1} $Na_2HPO_4 \cdot 2H_2O$, 0.49 g L^{-1} $(NH_4)_2HC_6H_5O_7$ (diammonium hydrogen citrate), 0.25 g L^{-1} $MgSO_4 \cdot 7H_2O$, with 1.0 ml L^{-1} of trace metal stock solution (0.5 g L^{-1} $CaCl_2 \cdot 2H_2O$, 16.7 g L^{-1} $FeCl_3 \cdot 6H_2O$, 0.18 g L^{-1} $ZnSO_4 \cdot 7H_2O$, 0.16 g L^{-1} $CuSO_4 \cdot 5H_2O$, 0.15 g L^{-1} $MnSO_4 \cdot 4H_2O$, 0.18 g L^{-1} $CoCl_2 \cdot 6H_2O$, 20.1 g L^{-1} EDTA), 5 g L^{-1} glycerol. The medium is supplemented with 40 mg L^{-1} kanamycin to select for the recombinant plasmid used. For preparation of a fully deuterated medium, hydrated mineral salts are dried and labile H atoms exchanged for deuterium by dissolving in a minimal volume of D_2O and then freeze drying. Perdeuterated d_8-glycerol is used as a carbon source.

High cell density cultures (HCDC) are performed to maximize the protein production. A Labfors 2.3 L computer-controlled fermenter (Infors, France) as shown in Fig. 8A is used, using IRIS software (http://www.infors-ht.com). Deuterated minimal medium (1–1.5 L) is inoculated with the adapted cells (150 ml preculture). During batch and fed-batch phases, the pD is maintained at 6.4 by the addition of NaOD. The gas flow rate of sterile filtered air is $\geq$0.5 L min^{-1}. Stirring is adjusted to ensure a dissolved oxygen tension of 30%. The initial glycerol supply is consumed during the batch phase. Cells are then fed continuously with fresh feeding solution containing 12% d_8-glycerol in an exponential way (fed-batch phase). The glycerol has to be supplied in a limiting way to ensure that the cell specific growth rate does not reach its maximal value.

Figure 8B shows the variation of the key parameters during a typical computer-controlled HCDC run for a deuterated *E. coli* expression system. Optimal cell density for induction and induction time are largely dependent on the expression construct used (leakiness and toxicity of expression play an important roles). These parameters can be obtained from pilot expression experiments in deuterated flask cultures which indicate the optimal growth phase for induction and for the stability of expression of the gene product. Cells are subsequently harvested by centrifugation, washed with a suitable buffer, and the cell paste stored at −80 °C until purification.

Purification of perdeuterated recombinant proteins is usually carried out in hydrogenated buffers using similar protocols to those used for hydrogenated proteins. The level of deuterium incorporation is determined by mass spectrometry (MS) (examples shown in Fig. 9). ESI–MS spectra are recorded for both hydrogenated and perdeuterated proteins purified in hydrogenated buffers. The molecular weight difference between the deuterated and hydrogenated analogs allows the level of deuteration to be quantified—this

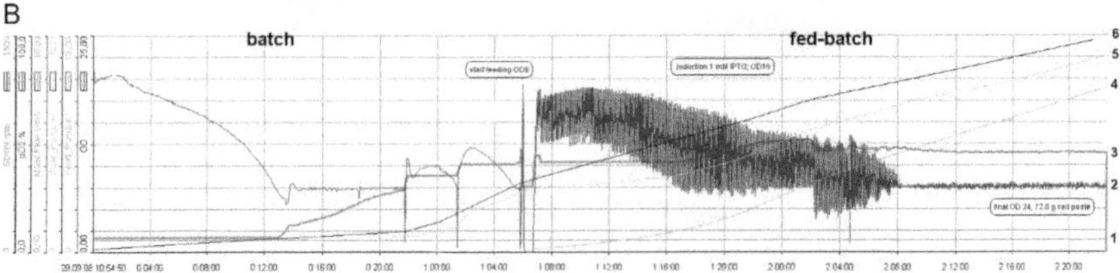

Key: 1 (pink) - airflow; 2 (blue) - pO2; 3 (red) - stirring speed; 4 (orange) - feeding; 5 (turquoise) - NaOD; 6 (black) - OD

Figure 8 (A) Fermentation equipment used for high cell density cultures and (B) tracking of fermentation parameters for a typical deuterated HCDC run, showing stirrer speed, dissolved oxygen concentration, air flow volumes, as well as feed and base additions. Following inoculation, cellular growth (oxygen consumption) is monitored by a decrease in dissolved oxygen concentration (pO_2, in blue) at constant stirrer speed. When the set-point for pO_2 of 30% is reached, the stirrer speed (rpm in red) increases to compensate for the increased oxygen demand of the growing cells. After about 30 h in the batch phase at an OD_{600} of about 4–5, the deuterated glycerol (5 g L^{-1}) is consumed by the growing cells, as demonstrated by the sudden increase in pO_2. d_8-glycerol feeding initiates the fed-batch phase. Cells are typically induced with 1 m*M* IPTG at an OD_{600} of 15 and growth stopped when cells reach an OD_{600} of ~20. (See the color plate.)

A
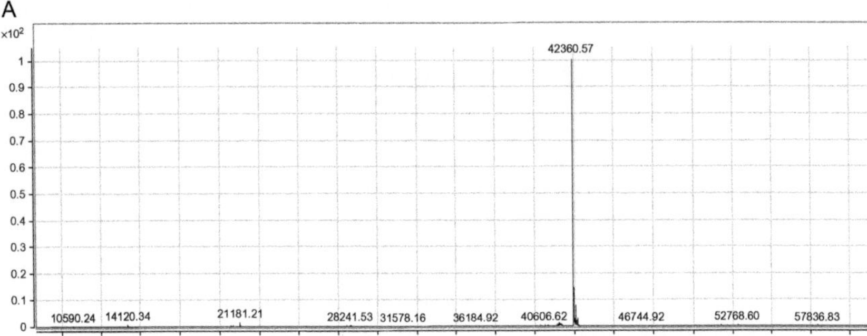

B
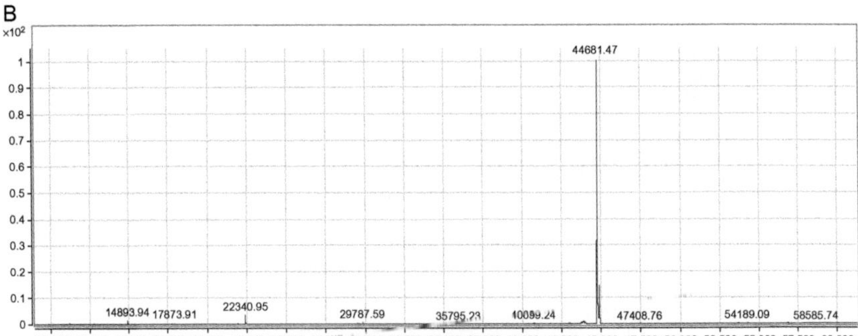

C
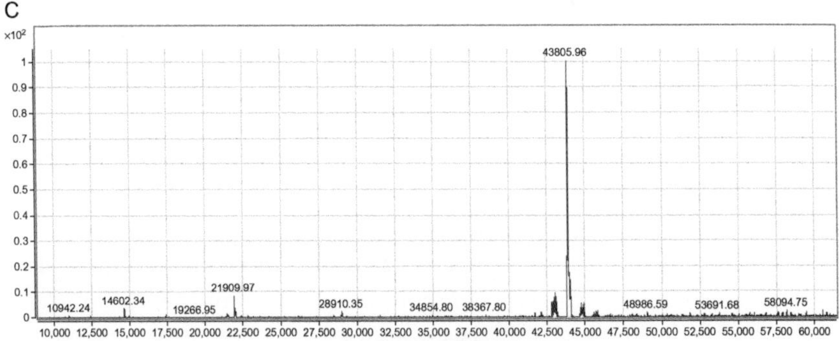

Figure 9 Mass spectrometry analysis (LC-TOF) for three differently labeled samples of maltose-binding protein (MBP): (A) hydrogenated (unlabeled) MBP, (B) perdeuterated MBP, and (C) "matchout labeled" MBP. Mass spectrometry measurements were carried out in hydrogenated buffers; the mass differences between the various samples arise from the non-exchangeable deuterium atoms incorporated into the samples. (See the color plate.)

is usually very close to 100%. After purification, the deuterated protein is normally dialyzed against a D_2O-based buffer to replace exchangeable hydrogens by deuterium on both macromolecule and solvent.

2.1.6.2 Matchout Protein Deuteration in High Cell Density Cultures

The production of matchout labeled proteins for SANS applications requires a different protocol in which hydrogenated glycerol is used as a carbon source with a lower D_2O content (85%). This approach optimizes the versatility of samples for SANS studies. Additionally, the protocol is highly cost efficient given that it allows the use of recycled D_2O (e.g., from previous perdeuterated cultures) and avoids the use of expensive deuterated carbon sources. Details of this labeling approach have been described by Dunne et al. (submitted) for both *E. coli* and *P. pastoris* expression systems.

2.1.6.3 Alternative Protocols for Deuteration of Recombinant Proteins

A similar protocol has been used by Shu, Ramakrishnan, and Schoenborn (2000) to express fully deuterated myoglobin in minimal medium but without the use of a fermenter, and by Brockwell et al. (2001) to express perdeuterated glutathione *S*-transferase in d_4-sodium succinate medium. Tuominen et al. (2004) have reported the production in high yield of perdeuterated yeast inorganic pyrophosphatase (96% deuterated) using a rich medium. Perdeuterated staphylococcal nuclease has also been produced by Gamble, Clauser, and Kossiakoff (1994), with a mean perdeuteration level of 96% using a medium containing deuterated amino acids.

2.1.6.4 Specific Labeling Approaches for the Study of Membrane Proteins by SANS

Membrane protein systems are notoriously difficult to crystallize and there has been considerable interest in SANS studies in which deuterium labeling can be used in conjunction with contrast variation to eliminate the inhomogeneous contributions to SANS data from the lipids, detergents, and surfactant components that are typically used to solubilize these proteins. Breyton et al. (2013) have reviewed a number of these approaches using deuterated proteins.

An alternative approach is the use of membrane-containing nanodiscs as carriers of specific target membrane proteins (Denisov, Grinkova, Lazarides, & Sligar, 2004). This has been exploited in nuclear magnetic resonance (NMR) (Hagn, Etzkorn, Raschle, & Wagner, 2013) and in SAXS (Skar-Gislinge & Arleth, 2011) studies and involves the construction of discs

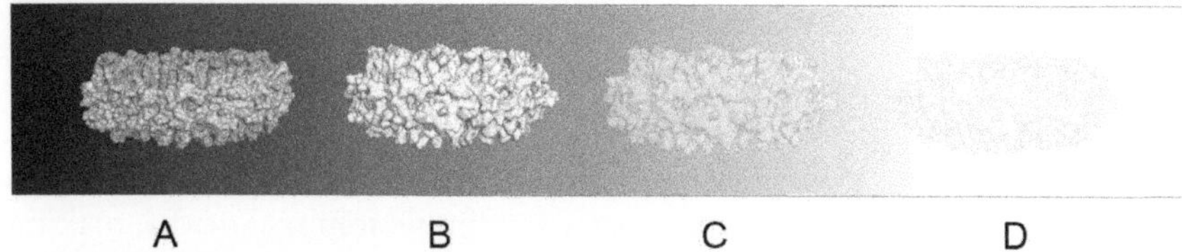

Figure 10 Schematic representation of the visibility by SANS of the "stealth" nanodisc in buffer solution with a gradually increasing levels of solvent D_2O. (A) Hydrogenated nanodisc consisting of a phospholipid bilayer (green) and membrane scaffold protein (MSP) (light blue). (B) Stealth nanodisc comprised of deuterated lipid and deuterated MSP at 60% D_2O. (C) Stealth nanodisc at 80% D_2O. (D) Stealth nanodisc at 100% D_2O. *Picture reproduced from Maric et al. (2014) with permission of the International Union of Crystallography.* (See the color plate.)

in which the protein of interest is embedded in a lipid bilayer that is contained (and defined in extent) by a belt protein. Such an approach is of particular interest for SANS work since there exists the possibility to deuterium label lipid and belt protein components in such a way that contrast variation can be used to study the membrane protein without coherent scattering contributions from the carrier itself. This may allow the construction of a low resolution image of the membrane protein in a lipid environment. Protocols for the specific recombinant deuteration in *E. coli* of the protein component (membrane scaffold protein, e.g., MSP1D1), the lipid component (e.g., phosphatidyl-choline) selectively deuterated in its head-group, glycerol, and fatty acid parts have been developed by Maric et al. (2014, 2015). These "stealth" carrier discs are invisible in 100% D_2O and may be used as a general platform for low resolution structural studies of membrane proteins using SANS. Figure 10 illustrates the variation in the visibility of a "stealth" nanodisc in buffers having various D_2O concentrations. In Fig. 10D, the nanodisc is almost perfectly matched out and it would be at this sort of contrast that an unlabeled protein could be modeled in the absence of scattering contribution from the carrier.

2.2 Expression of Deuterated Macromolecules in the Eukaryote *P. pastoris*

As noted in Section 2, *E. coli* is a prokaryote and lacks intracellular organelles, such as the endoplasmic reticulum and the Golgi apparatus that are present in eukaryotes, and which are responsible for post-translational modifications of the proteins being produced. Some eukaryotic proteins can be produced in *E. coli* but in a non-functional, unfinished form, since glycosylation or post-translational modifications do not occur. Therefore, researchers have turned to eukaryotic yeast and mammalian expression systems for protein production. Heterologous expression in *P. pastoris* has many of the advantages of eukaryotic expression, including proper folding and disulfide bond formation, glycosylation, and secretion. Contrary to other eukaryotic systems, protein production from *P. pastoris* occurs in simple minimal media, making this system attractive for production of labeled proteins for neutron scattering and NMR analysis. *P. pastoris* is therefore the expression system of choice for proteins that cannot be refolded from inclusion bodies in *E. coli* or that require post-translational modifications for proper folding or function. The yield of expressed proteins from *P. pastoris* depends critically on growth conditions, and attainment of high cell densities by fermentation has been shown to improve protein yields by 10–100-fold. Simple low cost production of perdeuterated biomass using *P. pastoris* grown on methanol in deuterated media may be of interest and has been described by Haon, Auge, Tropis, Milon, and Lindley (1993).

2.2.1 The P. pastoris Expression System

Compared to other eukaryotic expression systems, *Pichia* offers many advantages because it does not have the endotoxin problem associated with Gram negative bacteria nor the viral contamination problem of proteins produced in animal cell cultures (for a review, see Cregg & Higgins, 1995). Furthermore, *P. pastoris* can utilize methanol as a carbon source in the absence of glucose. The *P. pastoris* expression system uses the methanol-induced alcohol oxidase (AOX1) promoter, which controls the gene that codes for the expression of alcohol oxidase, the enzyme which catalyzes the first step in the metabolism of methanol. This promoter has been characterized and incorporated into a series of *P. pastoris* expression vectors. Since the proteins produced in *P. pastoris* are typically folded correctly and secreted into the medium, the fermentation of genetically engineered *P. pastoris* provides an excellent alternative to *E. coli* expression systems.

2.2.2 Perdeuterated Lipids

Perdeuterated lipids are extremely useful in studying biological membranes (or model membranes) as well as membrane proteins by neutron scattering methods and by neutron reflectivity in particular. Chemical synthesis of unsaturated perdeuterated lipids is challenging and rarely used. However, de Ghellinck et al. (2014) have demonstrated that perdeuterated phospholipids and sterols can be extracted from *P. pastoris* grown in deuterated minimal medium where the phospholipid and ergosterol homeostasis is maintained, but the fatty acid unsaturation level is modified by comparison with hydrogenated cultures. Significant production of perdeuterated unsaturated lipids is feasible when *P. pastoris* is grown at lower temperature.

The multilamellar organization of fully deuterated lipid extracts of *P. pastoris* membranes has been analyzed by neutron diffraction (Gerelli et al., 2014). At high relative humidity, hydrogenated and deuterated lipids are similar in their multilamellar organization, whereas at low relative humidity, hydrogenated lipids are characterized by a larger single lamellar structure than observed for the deuterated samples. Perdeuterated lipids have been used to characterize structural changes in the membrane of *P. pastoris* induced by the antifungal Amphotericin B (de Ghellinck et al., 2015).

2.2.3 Perdeuteration of Proteins in P. pastoris Using Minimal Growth Medium

The ILL Deuteration Laboratory is developing protocols for deuteration of proteins in *P. pastoris* using minimal media. Perdeuterated recombinant human serum albumin (HSA) and hen egg lysozyme have been obtained in suitable quantities. Biophysical studies of HSA in hydrogenated and perdeuterated forms are under way (White et al., 2015). Protocols for the random-fractional deuteration (*matchout* labeling) for proteins to study protein–protein complexes will be described elsewhere (Dunne et al., submitted).

2.3 Alternative Expression Systems

2.3.1 Cyanobacterial Expression Systems

Cyanobacterial expression systems could become attractive alternatives for cost-effective deuteration of recombinant proteins since CO_2 can be used as carbon source and expensive organic carbon sources such as deuterated glucose or glycerol are not needed. The feasibility of such an approach has already been demonstrated for ^{13}C and ^{15}N labeling of *E. coli* gyrase B (Desplancq et al., 2001) using the photoautotrophic cyanobacterium *Anabaena* species PCC 7120. Perdeuteration of a recombinant protein using this approach has never been reported, probably because the cyanobacterial expression systems are more difficult to use than the standard *E. coli* systems. However, native perdeuterated C-phycocyanin has been isolated from blue-green algae cultured in D_2O and deuterated up to 99%. Figure 11 shows the culture of the cyanobacterium *Anabaena* in a photobioreactor system used in ILL's Deuteration Laboratory.

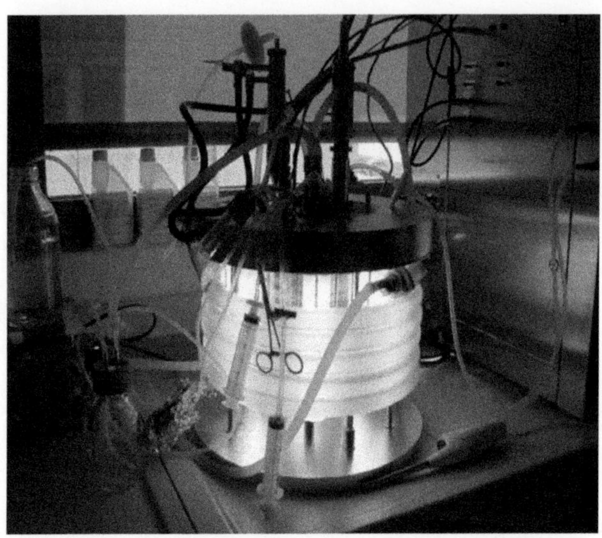

Figure 11 Photobioreactor system (Biostat PBR2S, Sartorius) being used for the production of biomass from the cyanobacterium *Anabaena*. This system allows control of temperature, pH (pD), recirculation rate, and light intensity. (See the color plate.)

2.3.2 Deuteration in Halobacteria

Halobacterium can be adapted to growth in deuterated media and is an ideal host for amino acid-specific labeling since it possesses multiple auxotrophies for amino acids. The bR within the purple membrane (PM) of *Halobacterium* has been studied extensively by NMR and neutron scattering methods using isotope-labeled samples. For solution NMR studies, Patzelt et al. (1997) have used labeled bR in detergent micelles: samples with ^{2}H or ^{13}C labels in the chromophore and/or the protein were prepared using labeled peptone from the green algae *Scenedesmus obliquus* grown in D_2O. In the field of solid-state NMR, Varga et al. (2007) have studied the effect of deuteration on the ^{13}C linewidths of U-^{13}C, ^{15}N, ^{2}H crystalline bR using different deuterium levels.

Wood, Grudinin, et al. (2008) have produced three specifically labeled PM samples where each was deuterated with various types of amino acid (isoleucines, leucines, and tyrosines) specifically hydrogenated, allowing the study of the dynamics of specific amino acid types in bR within PM of *Halobacterium salinarum*. Similar studies have been carried out previously on deuterated PM reverse labeled with H-retinal, H-tryptophan, and H-methionine (Réat et al., 1998).

2.3.3 Baculovirus and Mammalian Cell Expression Systems

Media for insect cell cultures used for Baculovirus expression and media for mammalian expression systems are not commercially available for deuterium labeling. These expression systems have not been used so far for sample preparation for neutron scattering experiments. Although it is possible to obtain deuterated proteins from mammalian or insect cell expression at a low deuterium level, the prohibitive costs for the deuterated growth media components (individual deuterated amino acids) make these expression systems unattractive.

2.3.4 Cell-Free Synthesis

The *in vitro* synthesis of proteins is performed in a cell-free system that contains all necessary components for the transcription and translation. These components are prepared from a cell lysate. The gene carrying the information of the expressed protein and the amino acids required for synthesis of the labeled protein are added to the lysate and the labeled protein is purified from the lysate. Several cell-free systems, based on *E. coli* S30, HeLa, rabbit reticulocyte, and wheat germ extracts are available, but the *E. coli* S30 and wheat germ extracts seem to be most suitable for preparative scale synthesis.

In vitro synthesis of proteins is an alternative to *in vivo* expression when the protein of interest tends to be toxic, precipitates in cells as inclusion bodies, or is sensitive to proteolytic digestion in cells. In addition, the advantage of cell-free synthesis for isotope labeling is that a single amino acid species can be labeled in an otherwise unlabeled protein or that the protein can be partially labeled with a free choice of amino acid species.

Low yields of recombinant protein expression, the difficulties of preparation of cell-free extracts and the costs of commercially available preparative *in vitro* synthesis kits have limited the use of this method. Nevertheless, Kigawa et al. (1999) have used modified *E. coli* S30 cell extracts successfully for stable-isotope labeling of a protein with $^{13}C/^{15}N$ labeled amino acids for NMR spectroscopy. The use of cell-free systems for labeling applications in NMR work has become more popular and has been expanded to the labeling of membrane proteins (Sobhanifar et al., 2010). Cell-free expression systems allow the direct production of membrane proteins in membrane-mimetic environments that may be suitable for liquid-state NMR measurements.

The feasibility of perdeuteration for neutron crystallography using wheat germ extracts has been demonstrated by Masuda et al. (2005) for dihydrofolate reductase (DHFR) using a deuterated Chlorella hydrolysate missing asparagine, glutamine, cysteine, and tryptophan amino acids that were supplied separately as "H" amino acids. Mass analysis of the non-deuterated DHFR and deuterated DHFR indicated that the cell-free system can yield 100% deuteration provided that the reactions are carried out using fully deuterated amino acids.

Etezady-Esfarjani, Hiller, Villalba, and Wüthrich (2007) describe cell-free protein synthesis protocols for uniformly deuterated proteins using deuterated *E. coli* cell-extract (D-S30), and demonstrated that the 800-kDa chaperonin GroEL synthesized with the D-S30 cell-free system had a uniform deuteration level of about 95% and assembled into its biologically active oligomeric form. However, the yields of protein obtained are too low for most neutron scattering applications. This type of system could be used for amino acid-specific deuteration on a 10 mg level (e.g., for SANS experiments.)

3. LEVELS OF SPECIFICITY IN DEUTERIUM LABELING

Regardless of the organism used for protein expression, the strategy for deuteration depends critically on the neutron application and the level and type of labeling required. As noted in Section 1, neutron crystallography and high-angle fiber diffraction typically benefit from the highest possible level of deuteration, whereas SANS and reflection studies exploit contrast variation using samples that are matchout labeled (Section 1.1). More advanced options that allow specific regions of a protein to be highlighted include amino acid-selective labeling and segmental labeling. Figure 12 summarizes schematically the various labeling regimes that can be exploited in neutron scattering work.

3.1 Perdeuteration

Perdeuteration (Figure 12B), the complete exchange of hydrogen atoms by deuterium, can be achieved by growing suitable microorganisms (*E. coli*, *P. pastoris*, cyanobacteria, algae) after an adaptation process in fully deuterated media. For heterotrophic organisms, a fully deuterated carbon source is required. As noted earlier, perdeuteration is extremely useful in neutron crystallography, where the elimination of hydrogen incoherent scattering leads to major gains in terms of sample volume requirements. In addition, the replacement of hydrogen atoms (negative scattering length) by deuterium (positive scattering length) leads to all-positive density maps without cancelation effects that can obscure interpretation if resolution is limited.

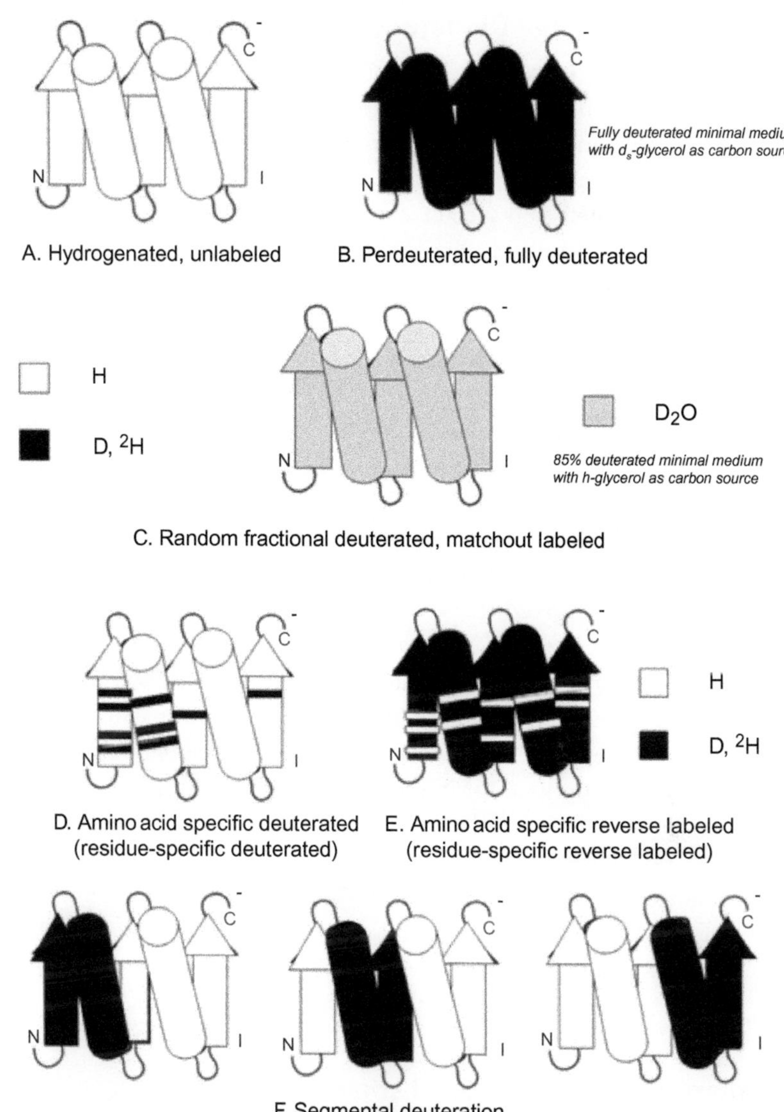

Figure 12 Schematic drawing summarizing different deuterium-labeling regimes.

Perdeuteration has also been used extensively for the study of macromolecular and solvent water dynamics using elastic incoherent neutron scattering (EINS) (Fichou et al., 2015; Jasnin, Moulin, et al., 2008a, 2008b; Jasnin, Tehei, et al., 2008; Paciaroni et al., 2009, 2008; Schirò et al., 2015; Wood, Froelich, et al., 2008; Wood et al., 2013; Wood, Grudinin, et al., 2008).

3.2 Random Fractional Deuteration (Partial Deuteration)

Random deuteration of recombinant proteins (Figure 12C) in *E. coli* is widely used for protein structure determination by NMR and sample preparations for SANS using contrast variation to study multisubunit complexes.

As described earlier in this chapter, selective labeling is required in SANS work to distinguish between specific components of a system in which the SLDs are similar (e.g., a multisubunit protein system). This technique was first used to study the 30S and 50S subunits of the ribosome (Capel et al., 1987; May, Nowotny, Nowotny, Voss, & Nierhaus, 1991) where the distances between the component protein subunits were found by triangulation. Examples in which labeling of selected protein components have allowed the characterization of protein–protein complexes are given by Callow et al. (2007) and Niemann et al. (2008). Since the SLD of perdeuterated protein is higher than that of pure D_2O, 100% deuterated protein cannot be matched out. Dunne et al. (submitted) have described protocols for the production of "matchout labeled" proteins in HCDC. This approach is highly cost effective given that the deuteration is carried out without a labeled carbon source using 85% D_2O.

Perkins (1981), and Leiting, Marsilio, and O'Connell (1998) have studied systematically the influence of the D_2O content in the growth medium, in the presence and absence of deuterated glucose, on the deuteration level of recombinant proteins used for NMR applications. With D_2O as the sole deuterium source 0–86% of the chemically nonexchangeable hydrogen atoms could be deuterated. Higher levels of deuteration required perdeuterated glucose in combination with D_2O. As an example, recombinant peptide deformylase from *Bacillus subtilis* was overexpressed, deuterated to various degrees, purified, and analyzed by MS and NMR.

3.3 Amino Acid Type-Specific Labeling

In neutron scattering experiments, residue-specific deuteration can be used to highlight regions or domains in a protein structure (Figure 12D). This type of labeling has been used in 2D NMR spectroscopy as a means of providing additional information on structure and dynamics (e.g., Gettins, Backovic, & Peterson, 2004). The labeling itself can be carried out either by adding a deuterated amino acid to a culture growing in an otherwise hydrogenated minimal medium, or by adding a hydrogenated amino acid to a culture growing in an otherwise deuterated minimal medium (reverse labeling).

Two concerns arise when the label is to be restricted to specific types of amino acids. Firstly, the labeled amino acid added to the culture may be diluted in the final sample by endogenous amino acid biosynthesis. Secondly, the added labeled amino acid may be scrambled to other types of residue by specific metabolic conversion. Both concerns can be mitigated to some degree by supplementing the medium with a high concentration of all remaining amino acids in unlabeled form (Ramesh et al., 1994) since amino acid biosynthesis is regulated by feedback inhibition. The degree of isotope scrambling depends largely on the amino acid used for residue-specific labeling. Gettins et al. (2004) have reported that by supplementing culture media with all amino acids, there is little leakage of their ^{15}N label from alanine into other residues, even if an auxotrophic strain of *E. coli* is not used. These authors also found little scrambling of leucine, whereas valine gave significant leakage into alanine. In contrast, Hoffman and Spicer (1991) reported have that preparation of proteins containing deuterated alanine methyl groups is problematic since alanine exchanges continuously with pyruvate via a non-repressible pathway. A growth medium containing deuterated alanine, no other amino acids, and growth limiting amounts of glucose resulted in the deuteration of several other amino acids in addition to alanine. Ayala, Sounier, Usé, Gans, and Boisbouvier (2009) developed an efficient protocol for the complete incorporation of methyl-protonated alanine in perdeuterated protein based on coaddition of perdeuterated forms of α-keto isovalerate-d_7, succinate-d_4, and L-isoleucine-d_{10} with labeled L-alanine.

The use of auxotrophic strains is highly recommended if a high occupancy of a labeled residue is needed, especially if feedback inhibition of the biosynthesis of this specific amino acid is not sufficient. An alternative approach has been suggested by Lee, Androphy, and Baleja (1995) that avoids the need for auxotrophic strains. Rifampicin (which inhibits *E. coli* RNA polymerase) is added to a culture after induction of the gene product to be labeled, thereby lowering endogenous metabolic activity, leaving the transcription of the recombinant gene under the control of the encoded plasmid T7 promoter. Using this method, these authors reported high ^{15}N phenylalanine labeling without scrambling. Significant direct histidine incorporation without isotope scrambling has also been described for the labeling of cytochrome *c* peroxidase (Savenkova, Satterlee, Erman, Siems, & Helms, 2001).

Highly efficient labeling results have also been reported for several aromatic amino acids where the amino acid precursor shikimic acid was added to the growth media of auxotrophic mutant strains containing a plasmid coding for shikimic acid transporter protein (Rajesh et al., 2003). Genetic tools for the construction of mutant strains for selective labeling of proteins are available (Waugh, 1996). For selective labeling of specific residues in *E. coli*, high cell density fermenter cultures can be used (Studts & Fox, 1999).

For neutron scattering applications, Laux et al. (2008) have described protocols for the production of MBP that has been selectively labeled with either deuterated tryptophan or deuterated methionine (single labeling), or both (double labeling). The selective labeling was carried out in *E. coli* high cell density cultures using auxotrophic mutants in minimal medium containing the required deuterated precursors. Samples were prepared of singly labeled MBP with the tryptophan residues deuterated, singly labeled MBP with methionine residues deuterated, and doubly labeled MBP with both tryptophan and methionine residues deuterated. Figure 13 shows a MS analysis of unlabeled and tryptophan labeled MBP using this method.

The conditional methionine auxotroph *E. coli* B834 has also be used for CD_3-methionine labeling (Sreenath et al., 2005). For labeling with cysteine, a similar protocol could be applied using the auxotroph BL21(DE3)cys (Strub et al., 2003).

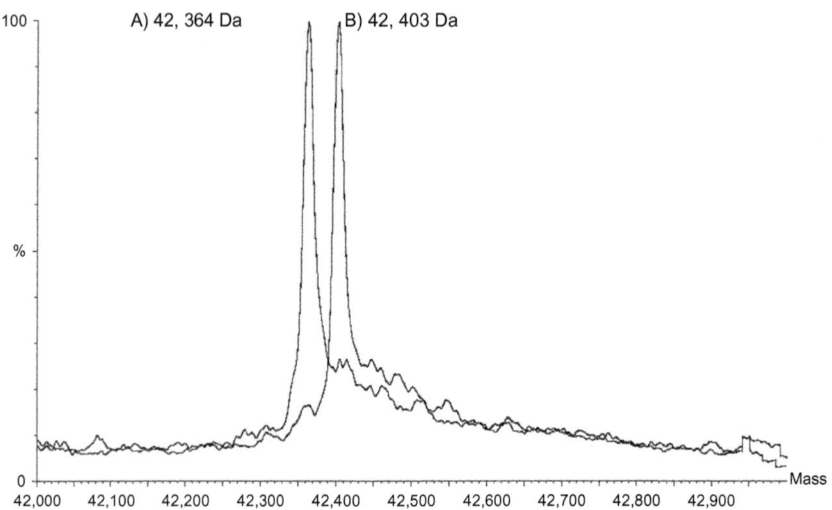

Figure 13 Mass spectrometry (electrospray Q-TOF) analysis of (A) unlabeled (42,364 Da) and (B) indole-d_5 tryptophan labeled (42,403 Da) maltose-binding protein (MBP). The mass difference of 39 Da is very close to the 40 Da expected between 8 d_5-indole tryptophan residues and 8 tryptophan residues.

Amino acid-specific protonation of lysine residues (with all other residues are in their deuterated forms) has been used to study the insertion of the diphtheria T domain into a model membrane (Chenal et al., 2009). Furthermore, antifreeze protein from ocean pout has been used as a model protein to develop protocols for amino acid-specific hydrogen reverse labeling (Figure 12E) of methyl groups in leucine and valine residues using *E. coli* high density cell cultures supplemented with the amino acid precursor α-keto isovalerate (Petit-Haertlein et al., 2010). Amino acids-specific reverse labeling has been used to study the dynamics of the model protein calbindin using labeling of an otherwise deuterated protein in one case with hydrogenated lysine (to monitor the dynamics at the surface of the protein) and in another with hydrogenated amino acid precursors to label the methyl groups of valine and isoleucine (to access the dynamics of core part of the protein) (Wood et al., 2013).

If expression in *E. coli* is not feasible, baculovirus-infected insect cells can be used for amino acid type-selective isotope labeling of proteins (Strauss et al., 2003). These authors report high incorporation rates for essential ^{15}N labeled amino acids (Phe, Tyr, and Val) and lower incorporation for the essential amino acid glycine. A similar incorporation level would be expected for deuterated amino acids. Cell-free systems of the type described in Section 2 are of course very attractive for amino acid-specific labeling that avoid isotope scrambling (Kigawa et al., 1999; Ludlam et al., 1995).

3.4 Segmental Isotope Labeling

Segmental labeling (Figure 12F) may allow any part of a given protein to be selectively deuterated at will, and as such it is particularly interesting for neutron scattering studies of biological structure and dynamics. A number of methods have been developed for solution state NMR work (reviewed by Skrisovska, Schubert, & Allain, 2009), including native chemical ligation, expressed protein ligation, protein trans-splicing, and sortase-catalyzed ligation (Refaei et al., 2011). To our knowledge, this technology has not been used so far for deuterium labeling of protein segments for neutron scattering studies, although the approach is currently being developed in our laboratories.

3.5 Deuterated Peptides

Production of highly deuterated peptides exceeding 10 mg L^{-1} of culture can be achieved by *in vivo* expression using a fusion protein involving the immunoglobulin-binding domain of streptococcal protein G (Koenig, Rogowski, & Louis, 2003). In a more classical way, deuterated or specifically labeled peptides can be synthesized using solid-phase methods (Bringezu et al., 2007; Chan & White, 2000; Liu, Crocker, Siminovitch, & Smith, 2003). FMOC protection chemistry can be used to achieve correct peptide bonding during synthesis. Some deuterated amino acid FMOC derivatives are available commercially. Approaches whereby deuterated amino acids are overexpressed in bacteria are under development in our group, where the focus is to allow the synthesis of tailor-made perdeuterated or selectively deuterated peptides. Such a capability will have a profound effect on the type of biological systems that can be studied by neutron scattering. For example, the ability to produce suitably labeled peptides would be of great value in studies of interactions involving antibacterial peptides and of amyloid-forming peptides.

4. CONCLUSION AND PERSPECTIVE

Given that the potential for deuterium labeling in neutron scattering studies of biomolecular structure and dynamics has been known for many decades, it is perhaps surprising that its exploitation has only been systematically grasped in the relatively recent past. The creation of a Deuteration Laboratory within the Life Sciences Group at the ILL has had a decisive impact on the scope and throughput of neutron work in the biological sciences, particularly for work involving SANS, neutron crystallography, and dynamics. The ILL user community in this area has grown correspondingly and there has been major diversification in the range of experiments beyond those possible without deuterium labeling. Similar initiatives for macromolecular deuteration have occurred at facilities throughout the world such as Oak Ridge National Laboratory (ORNL) in the USA, National Institute of Standards and Technology (NIST) in the USA, and ANSTO in Australia. At ISIS in the UK, facilities are being built-up from a well-developed base of expertise in the production of deuterated surfactants and lipids (Li, Thomas, & Penfold, 2014; Penfold et al., 2014; Penfold, Thomas, Li, Petkov, et al., 2015; Penfold, Thomas, Li, Xu, et al., 2015).

Structural biology has evolved rapidly over the last few decades. The emergence of synchrotron radiation and associated technologies (e.g., crystal cryoprotection, detector quality and speed) has resulted in a trend toward rapid structure determination and high-throughput methods—to some extent at the expense of classical structure–function work. This trend, in combination with the huge amount of genomic and bioinformatics information, has led to an explosion of information in the field of structural genomics. Since then there has been increasing recognition of the fact that biological systems have to be studied in a far more integrated way, using methods that span resolution boundaries from atomic to cellular levels, and employing a wide range of techniques for sample preparation as well as for biophysical and structural characterization. In many ways, this represents something of a return toward more structure–function-based approaches, combined with a growing awareness of the information gap that exists between molecular and cellular levels of organization. The Grenoble Partnership for Structural Biology (PSB) (http://www.psb-grenoble.eu/) has been developed very much in this spirit, bringing together the ILL, the ESRF (European Synchrotron Radiation Facility), the EMBL (European Molecular Biology Laboratory), and the IBS (Institute of Structural Biology) so that the extensive capabilities of each institute can be used in support of integrated and interdisciplinary approaches for structural biology. ILL's Life Sciences Group and its Deuteration Laboratory are located within the PSB and benefit from the shared availability of over 20 technical platforms in addition to the neutron and synchrotron beam sources. These include electron microscopy (negative stain and cryo), high field NMR, high-throughput crystallization robotics, mass spectrometry, analytical ultracentrifugation, cellular imaging, platforms for cloning, protein expression and purification, as well as an extensively equipped biophysical platform (DLS, MALLS, MST, ITC, SPR, etc.). Access to the Deuteration Laboratory occurs by rapid electronic peer review (http://www.ill.fr/lsg/deuteration) and the associated neutron scattering work may be carried out at any member state facility.

Future developments in macromolecular deuterium labeling will be driven in parallel with instrumental developments at existing sources (ILL, ISIS, NIST, ORNL, ANSTO) and future facilities such as the European Spallation Source (ESS) being built in Sweden. Neutron structural biology is being given high priority at these institutes and it can be expected that future instruments will have improved fluxes, optics, and detector systems. At the ILL, for example, there are order of magnitude improvements possible for existing instruments and sample preparation developments to maximize the impact of such investments are a continuous priority. For crystallography, this means that experiments will be faster and that the demand for perdeuterated samples will increase. For SANS, the combination of optimized deuterated samples and better instrumentation is likely to allow improved access to (for example) time-resolved measurements of molecular interactions. In the case of work on dynamics, improved instruments will allow the effective concentrations of the hydrogenous (i.e., signal producing) part of the otherwise deuterated sample to be reduced. This means that segmental labeling (for example) may in the future allow greater specificity in the analysis of protein dynamics.

The facility/instrumentation developments currently being planned will undoubtedly require the broadest possible scope for the provision of high quality deuterated samples. Considerable work is required for the production of deuterated macromolecules using expression systems that are currently challenging (e.g., insect and mammalian cells, cell-free synthesis). Such approaches will open up completely new ways in which—for example—suppressor tRNA technologies are used *in vivo* or *in vitro* to specifically deuterate given residues within a protein. This type of labeling has been successfully applied by Yabuki et al. (1998) in NMR work where milligram quantities of residue specifically isotope-labeled protein were produced using a cell-free system involving amber suppression. An alternative *in vivo* approach could be used in a deuterated culture of evolved aminoacyl-tRNA synthetases (Kwon, Wang, & Tirrell, 2006) that charge a suppressor tRNA with a specifically modified hydrogenated amino acid. In principle, this could lead to the production of designer samples in which defined sections of residues are selectively labeled—with potential applications in SANS and dynamics. Such applications will need significantly improved neutron beam fluxes, and are likely to be heavily exploitable in the future for priority research areas in health and biotechnology.

ACKNOWLEDGMENTS

The UK Engineering and Physical Sciences Research Council (EPSRC) is thanked for support under grants EP/C015452/1 and GR/R99393/01 that allowed the creation of the Deuteration Laboratory at the Institut Laue-Langevin, and also for grant GR/R47950/01 that allowed associated diffractometer developments at the ILL. The EU is thanked for its support of methodological developments that occurred under contracts RII3-CT-2003505925 and NMP4-CT-2006-033256. The authors acknowledge all current and previous members of ILL's Life Sciences Group. We thank Phil Callow for his invaluable contributions to SANS-related development work at ILL/Keele and for the provision of images used in Figure 2. Ed. Mitchell of ESRF's Business Development Office (BDO) is thanked for his active involvement in many projects combined X-ray and neutron applications, particularly those linking with industry. Keele University is thanked for studentship support, and for its contribution in the form of a joint appointment lecturer position between Keele and the ILL. All members of the Grenoble Partnership for Structural Biology (PSB) are thanked for the collaborative spirit that has resulted in such a remarkable interdisciplinary working environment.

REFERENCES

Achterhold, K., Ostermann, A., Moulin, M., Haertlein, M., Unruh, T., Fritz, G., et al. (2011). Dynamical properties of the hydration shell of fully deuterated myoglobin. *Physical Review E, 84*, 041930.

Appolaire, A., Girard, E., Colombo, M., Durá, M. A., Moulin, M., Haertlein, M., et al. (2014). Small-angle neutron scattering reveals the assembly mode and oligomeric architecture of TET, a large, dodecameric aminopeptidase. *Acta Crystallographica D, 70*(Pt. 11), 2983–2993.

Artero, J.-B., Hartlein, M., McSweeney, S., & Timmins, P. (2005). A comparison of refined X-ray structures of hydrogenated and perdeuterated rat gammaE-crystallin in H_2O and D_2O. *Acta Crystallographica. Section D, Biological Crystallography, 61*(Pt. 11), 1541–1549.

Ayala, I., Sounier, R., Usé, N., Gans, P., & Boisbouvier, J. (2009). An efficient protocol for the complete incorporation of methyl-protonated alanine in perdeuterated protein. *Journal of Biomolecular NMR, 43*(2), 111–119.

Baneys, F. (1999). Recombinant protein expression in *E. coli*. *Current Opinion in Biotechnology, 10*, 411–421.

Bessette, P. H., Aslund, F., Beckwith, J., & Georgiou, G. (1999). Efficient folding of proteins with multiple disulfide bonds in the *Escherichia coli* cytoplasm. *Proceedings of the National Academy of Sciences of the United States of America, 96*(24), 13703–13708.

Blakeley, M. P. (2009). Neutron macromolecular crystallography. *Crystallography Reviews, 15*, 157–218.

Blakeley, M. P., Ruiz, F., Cachau, R., Hazemann, I., Meilleur, F., Mitschler, A., et al. (2008). Quantum model of catalysis based on a mobile proton revealed by subatomic X-ray and neutron diffraction studies of h-aldose reductase. *Proceedings of the National Academy of Sciences of the United States of America, 105*(6), 1844–1848.

Blakeley, M. P., Teixeira, S. C., Petit-Haertlein, I., Hazemann, I., Mitschler, A., Haertlein, M., et al. (2010). Neutron macromolecular crystallography with LADI-III. *Acta Crystallographica D, 66*, 1198–1205.

Breyton, C., Gabel, F., Lethier, M., Flayhan, A., Durand, G., Jault, J. M., et al. (2013). Small angle neutron scattering for the study of solubilised membrane proteins. *The European Physical Journal E, Soft Matter, 36*(7), 13071–13076.

Bringezu, F., Wen, S., Dante, S., Hauss, T., Majerowicz, M., & Waring, A. (2007). The insertion of the antimicrobial peptide dicynthaurin monomer in model membranes: Thermodynamics and structural characterization. *Biochemistry, 46*(19), 5678–5686.

Brockwell, D., Yu, L., Cooper, S., McCleland, S., Cooper, A., Attwood, D., et al. (2001). Physicochemical consequences of the perdeuteriation of glutathione S-transferase from S. japonicum. *Protein Science, 10*(3), 572–580.

Brouette, N., Fragneto, G., Cousin, F., Moulin, M., Haertlein, M., & Sferrazza, M. (2013). A neutron reflection study of adsorbed deuterated myoglobin layers on hydrophobic surfaces. *Journal of Colloid and Interface Science, 390*(1), 114–120.

Callow, P., Sukhodub, A., Taylor, J., & Kneale, G. (2007). Shape and subunit organisation of the DNA methyltransferase M.Ahdl. *Journal of Molecular Biology, 69*(1), 177–185.

Capel, M. S., Engelman, D. M., Freeborn, B. R., Kjeldgaard, M., Langer, J. A., Ramakrishnan, V., et al. (1987). A complete mapping of the proteins in the small ribosomal subunit of *Escherichia coli*. *Science, 238*, 1403–1406.

Carlstedt, B. C., Crespi, H. L., Blake, M. I., & Katz, J. J. (1970). Biosynthesis of deuterated benzylpenicillins. I. Solvent deuterium oxide participation. *Journal of Pharmaceutical Sciences, 59*(10), 1456–1460.

Casadei, C., Gumiero, A., Metcalfe, C. L., Murphy, E. J., Basran, J., Concilio, M. G., et al. (2014). Heme enzymes. Neutron cryo-crystallography captures the protonation state of ferryl heme in a peroxidase. *Science, 345*, 193–197.

Chan, W. C., & White, P. D. (Eds.), (2000). *FMOC solid phase peptide synthesis: A practical approach*. New York, NY: Oxford University Press.

Chenal, A., Prongidi-Fix, L., Perier, A., Aisenbrey, C., Vernier, G., Lambotte, S., et al. (2009). Deciphering membrane insertion of the diphtheria toxin T domain by specular neutron reflectometry and solid-state NMR spectroscopy. *Journal of Molecular Biology, 391*(5), 872–883.

Clarke, M. J., Artero, J.-B., Moulin, M., Callow, P., Carver, J. A., Griffiths, P. C., et al. (2010). Investigation of gamma E-crystallin target protein binding to bovine lens alpha-crystallin by small-angle neutron scattering. *Biochimica et Biophysica Acta, 1800*(3), 392–397.

Compton, E. L., Page, K., Findlay, H. E., Haertlein, M., Moulin, M., Zachariae, U., et al. (2014). Conserved structure and domain organization among bacterial Slc26 transporters. *The Biochemical Journal, 463*(2), 297–307.

Cowsill, B. J., Zhao, X., Waigh, T. A., Eapen, S., Davies, R., Laux, V., et al. (2014). Interfacial structure of immobilized antibodies and perdeuterated HSA in model pregnancy tests measured with neutron reflectivity. *Langmuir, 30*(20), 5880–5887.

Cregg, J. M., & Higgins, D. R. (1995). Production of foreign proteins in the yeast Pichia pastoris. *Canadian Journal of Botany, 73*(Suppl. 1), S891–S897.

Cuypers, M. G., Mason, S. A., Blakeley, M. P., Mitchell, E. P., Haertlein, M., & Forsyth, V. T. (2013). Near-atomic resolution neutron crystallography on perdeuterated Pyrococcus furiosus rubredoxin: Implication of hydronium ions and protonation state equilibria in redox changes. *Angewandte Chemie, 52*(3), 1022–1025.

Cuypers, M. G., Trubitsyna, M., Callow, P., Forsyth, V. T., & Richardson, J. M. (2013). Solution conformations of early intermediates in Mos1 transposition. *Nucleic Acids Research, 41*(3), 2020–2033.

de Ghellinck, A., Fragneto, G., Laux, V., Haertlein, M., Jouhet, J., Michele Sferrazza, M., et al. (2015). Lipid polyunsaturation determines the extent of membrane structural changes induced by Amphotericin B in Pichia pastoris yeast, *BBA. Biomembranes, 1848*, 2317–2325.

de Ghellinck, A., Schaller, H., Laux, V., Haertlein, M., Sferrazza, M., Maréchal, E., et al. (2014). Production and analysis of perdeuterated lipids from *Pichia pastoris* cells. *PLoS One, 9*(4), e92999.

Denisov, I. G., Grinkova, Y. V., Lazarides, A. A., & Sligar, S. G. (2004). Directed self-assembly of monodisperse phospholipid bilayer nanodiscs with controlled size. *Journal of the American Chemical Society, 126*(11), 3477–3487.

Desplancq, D., Kieffer, B., Schmidt, K., Posten, C., Forster, A., Oudet, P., et al. (2001). Cost-effective and uniform ^{13}C- and ^{15}N-labeling of the 24-kDa N-terminal domain of the Escherichia coli gyrase B by overexpression in the photoautotrophic cyanobacterium Anabaena sp. PCC 7120. *Protein Expression and Purification, 23*(1), 207–217.

Di Costanzo, L., Moulin, M., Haertlein, M., Meilleur, F., & Christianson, D. W. (2007). Expression, purification, assay, and crystal structure of perdeuterated human arginase. *Archives of Biochemistry and Biophysics, 465*(1), 82–89.

Dunne, O., Moulin, M., Callow, P., Martel, A., Weidenhaupt, M., Perkins, S. M., et al. Matchout deuterium labelling of proteins using prokaryotic and eukaryotic expression systems. *European Biophysics Journal*. (submitted).

Edlich-Muth, C., Artero, J.-B., Callow, P., Przewloka, M. R., Watson, A. A., Zhang, W., et al. (2015). The pentaplasmic nucleoplasmin fold is present in Drosophila FKBP39 and a large number of chromatin-related proteins. *Journal of Molecular Biology, 427*(10), 1949–1963.

Etezady-Esfarjani, T., Hiller, S., Villalba, C., & Wüthrich, K. (2007). Cell-free protein synthesis of perdeuterated proteins for NMR studies. *Journal of Biomolecular NMR, 39*(3), 229–238.

Fichou, Y., Schirò, G., Gallat, F. X., Laguri, C., Moulin, M., Combet, J., et al. (2015). Hydration water mobility is enhanced around tau amyloid fibers. *Proceedings of the National Academy of Sciences of the United States of America*, *112*(20), 6365–6370.

Fisher, S. J., Blakeley, M. P., Howard, E. I., Petit-Haertlein, I., Haertlein, M., Mitschler, A., et al. (2014). Perdeuteration: Improved visualization of solvent structure in neutron macromolecular crystallography. *Acta Crystallographica D*, *70*, 3266–3272.

Fisher, S. J., & Helliwell, J. R. (2008). An investigation into structural changes due to deuteration. *Acta Crystallographica*, *A64*(Pt. 3), 359–367.

Forsyth, V. T., Greenall, R. J., Hussain, R., Mahendrasingam, A., Nave, C., Pigram, W. J., et al. (1986). X-ray high-angle fibre diffraction studies of nucleic acid structure using the Daresbury synchrotron radiation source. *Biochemical Society Transactions*, *14*, 553–557.

Forsyth, V. T., Mahendrasingam, A., Pigram, W. J., Greenall, R. J., Bellamy, K. A., Fuller, W., et al. (1989). Neutron fibre diffraction studies of DNA hydration. *International Journal of Biological Macromolecules*, *11*, 236–240.

Forsyth, V. T., Myles, D., Timmins, P. A., & Haertlein, M. (2001). A laboratory to support the deuteration of biological macromolecules. *Report to the ILL millennium programme* pp. (32–34).

Fuller, W., Forsyth, V. T., & Mahendrasingam, A. (2004). Water–DNA interactions by X-ray and neutron fibre diffraction. *Philosophical Transactions of the Royal Society*, *359*(1448), 1237–1248.

Gabel, F., Bicout, D., Lehnert, U., Tehei, M., Weik, M., & Zaccai, G. (2002). Protein dynamics studied by neutron scattering. *Quarterly Reviews of Biophysics*, *35*(4), 327–367.

Gallat, F. X., Brogan, A. P., Fichou, Y., McGrath, N., Moulin, M., Härtlein, M., et al. (2012). A polymer surfactant corona dynamically replaces water in solvent-free protein liquids and ensures macromolecular flexibility and activity. *Journal of the American Chemical Society*, *134*(32), 13168–13171.

Gallat, F. X., Laganowsky, A., Wood, K., Gabel, F., van Eijck, L., Wuttke, J., et al. (2012). Dynamical coupling of intrinsically disordered proteins and their hydration water: Comparison with folded soluble and membrane proteins. *Biophysical Journal*, *103*(1), 129–136.

Gamble, T. R., Clauser, K. R., & Kossiakoff, A. A. (1994). The production and X-ray structure determination of perdeuterated Staphylococcal nuclease. *Biophysical Chemistry*, *53*(1–2), 15–25.

Gardner, K. H., English, A., & Forsyth, V. T. (2004). New insights into the structure of poly(p-phenylene terephthalamide) from neutron fiber diffraction studies. *Macromolecules*, *37*, 9654–9656.

Gerelli, Y., de Ghellinck, A., Jouhet, J., Laux, V., Haertlein, M., & Fragneto, G. (2014). Multi-lamellar organization of fully deuterated lipid extracts of yeast membranes. *Acta Crystallographica D*, *70*, 3167–3176.

Gettins, P. G., Backovic, M., & Peterson, F. C. (2004). Use of NMR to study serpin function. *Methods*, *32*(2), 120–129.

Grage, S. L., Keleshian, A. M., Turdzeladze, T., Battle, A. R., Tay, W. C., May, R. P., et al. (2011). Bilayer-mediated clustering and functional interaction of MscL channels. *Biophysical Journal*, *100*(5), 1252–1260.

Grossmann, J. G., Callaghan, A. J., Marcaida, M. J., Luisi, B. F., Alcock, F. H., Tokatlidis, K., et al. (2008). Complementing structural information of modular proteins with small angle neutron scattering and contrast variation. *European Biophysics Journal*, *37*, 719.

Hagn, F., Etzkorn, M., Raschle, T., & Wagner, G. (2013). Optimized phospholipid bilayer nanodiscs facilitate high-resolution structure determination of membrane proteins. *Journal of the American Chemical Society*, *135*(5), 1919–1925.

Haon, S., Auge, S., Tropis, M., Milon, A., & Lindley, N. D. (1993). Low cost production of perdeuterated biomass using methylothrophic yeasts. *Journal of Labelled Compounds and Radiopharmaceuticals*, *XXXIII*(11), 1053–1063.

Haupt, M., Blakeley, M. P., Fisher, S. J., Mason, S. A., Cooper, J. B., Mitchell, E. P., et al. (2014). Binding site asymmetry in human transthyretin: Insights from a joint neutron and X-ray crystallographic analysis using perdeuterated protein. *IUCrJ Biology & Medicine*, *1*, 429–438.

Haupt, M., Blakeley, M. P., Teixeira, S. C. M., Mason, S. A., Mitchell, E. P., Cooper, J. B., et al. (2011). Preliminary neutron crystallographic study of human transthyretin. *Acta Crystallographica*, *F67*, 1428–1431.

Hazemann, I., Dauvergne, M.-T., Blakeley, M. P., Meilleur, F., Haertlein, M., Van Dorsselaer, A., et al. (2005). High-resolution neutron protein crystallography with radically small crystal volumes: Application of perdeuteration to human aldose reductase. *Acta Crystallographica D*, *61*, 1413–1417.

Hellstrand, E., Grey, M., Ainalem, M. L., Ankner, J., Forsyth, V. T., Fragneto, G., et al. (2013). Adsorption of α-synuclein to supported lipid bilayers: Positioning and role of electrostatics. *ACS Chemical Neuroscience*, *4*(10), 1339–1351.

Heumann, H. (2004). Method for labeling biopolymers using isotopes. US patent 6783751 B2.

Hochuli, M., Szyperski, T., & Wüthrich, K. (2000). Deuterium isotope effects on the central carbon metabolism of Escherichia coli cells grown on a D_2O-containing minimal medium. *Journal of Biomolecular NMR*, *17*(1), 33–42.

Hoffman, D. W., & Spicer, L. D. (1991). Isotopic labeling of specific amino acids as an aid to NMR spectrum assignment of the methionine repressor protein. In J. J. Villafranca (Ed.), *Techniques in protein chemistry: II* (pp. 409–416). San Diego, CA: Academic Press.

Howard, E. I., Blakeley, M. P., Haertlein, M., Petit-Haertlein, I., Mitschler, A., Fisher, S. J., et al. (2011). Neutron structure of type-III antifreeze protein allows the reconstruction of AFP-ice interface. *Journal of Molecular Recognition*, *24*(4), 724–732.

Jacrot, B., Chauvin, C., & Witz, J. (1977). Comparative neutron small-angle scattering study of small spherical RNA viruses. *Nature*, *266*(5601), 417–421.

Jasnin, M., Moulin, M., Haertlein, M., Zaccai, G., & Tehei, M. (2008b). *In vivo* measurement of internal and global macromolecular motions in E. coli. *Biophysical Journal*, *95*(2), 857–864.

Jasnin, M., Moulin, M., Haertlein, M., Zaccai, G., & Tehei, M. (2008a). Down to atomic scale intracellular water dynamics. *EMBO Reports*, *9*(6), 545–547.

Jasnin, M., Tehei, M., Moulin, M., Haertlein, M., & Zaccai, G. (2008). Solvent isotope effect on macromolecular dynamics in E. coli. *European Biophysics Journal*, *37*(5), 613–617.

Jünemann, R., Wadzack, J., Triana-Alonso, F. J., Bittner, J. U., Caillet, J., Meinnel, T., et al. (1996). In vivo deuteration of transfer RNAs: Overexpression and large-scale purification of deuterated specific tRNAs. *Nucleic Acids Research*, *24*(5), 907–913.

Katz, J. J., & Crespi, H. L. (1966). Deuterated organisms: Cultivation and uses. *Science*, *151*, 1187–1194.

Kennaway, C. K., Taylor, J. E., Song, C. F., Potrzebowski, W., Nicholson, W., White, J. H., et al. (2012). Structure and operation of the DNA-translocating Type I DNA restriction enzymes. *Genes & Development*, *26*, 92D–104D.

Kigawa, T., Yabukia, T., Yoshida, Y., Tsutsuic, M., Itod, Y., Shibatad, T., et al. (1999). Cell-free production and stable-isotope labeling of milligram quantities of proteins. *FEBS Letters*, *442*, 15–19.

Koenig, B. W., Rogowski, M., & Louis, J. M. (2003). A rapid method to attain isotope labeled small soluble peptides for NMR studies. *Journal of Biomolecular NMR*, *26*(3), 193–202.

Kovalevsky, A. Y., Hanson, L., Fisher, S. Z., Mustyakimov, M., Mason, S. A., Forsyth, V. T., et al. (2010). Metal ion roles and the movement of hydrogen during reaction catalyzed by D-xylose isomerase: a joint X-ray and neutron diffraction study. *Structure*, *18*(6), 688–699.

Kovalevsky, A. Y., Hanson, B. L., Mason, S. A., Yoshida, T., Fisher, S. Z., Mustyakimov, M., et al. (2011). Identification of the elusive hydronium ion exchanging roles with a proton in an enzyme at lower pH values. *Angewandte Chemie International Edition in English*, *50*(33), 7520–7523.

Kwon, I., Wang, P., & Tirrell, D. A. (2006). Design of a bacterial host for site-specific incorporation of p-bromophenylalanine into recombinant proteins. *Journal of the American Chemical Society*, *128*(36), 11778–11783.

Langan, P., Forsyth, V. T., Mahendrasingam, A., Pigram, W. J., Fuller, W., & Mason, S. A. (1992). A high-angle neutron fibre diffraction study of the hydration of the A conformation of the DNA double helix. *Journal of Biomolecular Structure and Dynamics*, *10*(3), 489–503.

Laulumaa, S., Blakeley, M. P., Raasakka, A., Moulin, M., Härtlein, M., & Petri Kursula, P. (2015). Production, crystallization, and neutron diffraction of fully deuterated human myelin peripheral membrane protein P2. *Acta Crystallographica F*, *71*, 1391–1395.

Laux, V., Callow, P., Svergun, D. I., Timmins, P. A., Forsyth, V. T., & Haertlein, M. (2008). Selective deuteration of tryptophan and methionine residues in maltose binding protein: A model system for neutron scattering. *European Biophysics Journal*, *37*(6), 815–822.

Lee, K. M., Androphy, E. J., & Baleja, J. D. (1995). A novel method for selective isotope labeling of bacterially expressed proteins. *Journal of Biomolecular NMR*, *5*(1), 93–96.

Leiting, B., Marsilio, F., & O'Connell, J. F. (1998). Predictable deuteration of recombinant protein in Escherichia coli. *Analytical Biochemistry*, *265*, 351–355.

Li, P. X., Thomas, R. K., & Penfold, J. (2014). Limitations in the use of surface tension and the Gibbs equation to determine surface excesses of cationic surfactants. *Langmuir*, *30*(23), 6739–6747.

Liu, W., Crocker, E., Siminovitch, D. J., & Smith, S. O. (2003). Role of side-chain conformational entropy in transmembrane helix dimerization of glycophorin A. *Biophysical Journal*, *84*(2 Pt. 1), 1263–1271.

Ludlam, C. F., Sonar, S., Lee, C. P., Coleman, M., Herzfeld, J., RajBhandary, U. L., et al. (1995). Site-directed isotope labeling and ATR-FTIR difference spectroscopy of bacteriorhodopsin: The peptide carbonyl group of Tyr 185 is structurally active during the bR→N transition. *Biochemistry*, *34*(1), 2–6.

Mahendrasingam, A., Forsyth, V. T., Hussain, R., Greenall, R. J., Pigram, W. J., & Fuller, W. (1986). Time-resolved X-ray diffraction studies of the B to D structural transition in the DNA double helix. *Science*, *233*, 195–197.

Maric, S., Skar-Gislinge, N., Midtgaard, S., Thygesen, M. B., Schiller, J., Frielinghaus, H., et al. (2014). Stealth carriers for low-resolution structure determination of membrane proteins in solution. *Acta Crystallographica D*, *70*(Pt. 2), 317–328.

Maric, S., Thygesen, M. B., Schiller, J., Marek, M., Moulin, M., Haertlein, M., et al. (2015). Biosynthetic preparation of selectively deuterated phosphatidylcholine in genetically modified Escherichia coli. *Applied Microbiology and Biotechnology*, *99*(1), 241–254.

Masuda, H., Murthy, M. S. R. C., Suto, K., Takenaka, Y., Kumar, P. K. R., & Mizuno, H. (2005). Expression and purification of deuterated proteins for neutron crystallography, Chapter 1, 47–60. In N. Nimura, H. Mizuno, J. R. Helliwell, & E. Westhof (Eds.), *Hydrogen- and hydration-sensitive structural biology. Proceedings of Symposium on Development of New Structural Biology including Hydrogen and Hydration, Mito, Japan (2000).*

May, R. P., Nowotny, V., Nowotny, P., Voss, H., & Nierhaus, K. H. (1991). Inter-protein distances within the large subunit from Escherichia coli ribosomes. *EMBO Journal*, *11*, 373–378.

Meilleur, F., Dauvergne, M. T., Schlichting, I., & Myles, D. (2005). Production and X-ray crystallographic analysis of fully deuterated cytochrome P450cam. *Acta Crystallographica. Section D, Biological Crystallography*, *61*(Pt. 5), 539–544.

Niemann, H. H., Petoukhov, M. V., Härtlein, M., Moulin, M., Gherardi, E., Timmins, P., et al. (2008). X-ray and neutron small-angle scattering analysis of the complex formed by the met receptor and the Listeria monocytogenes invasion protein InlB. *Journal of Molecular Biology*, *377*(2), 489–500.

Nishiyama, Y., Langan, P., Wada, M., & Forsyth, V. T. (2010). Looking at hydrogen bonds in cellulose. *Acta Crystallographica D*, *66*, 1172–1177.

Paciaroni, A., Cornicchi, E., Marconi, M., Orecchini, A., Petrillo, C., Haertlein, M., et al. (2009). Coupled relaxations at the protein–water interface in the picosecond timescale. *Journal of the Royal Society, Interface*, *6*, S635–S640.

Paciaroni, A., Orecchini, A., Cornicchi, E., Marconi, M., Petrillo, C., Haertlein, M., et al. (2008). Dynamical signature of protein hydration water reveals two populations of amorphous ice like behavior. *Physical Review Letters*, *101*(14), 148104.

Paciaroni, A., Orecchini, A., Haertlein, M., Moulin, M., Conti Nibali, V., De Francesco, A., et al. (2012). Vibrational collective dynamics of dry proteins in the terahertz region. *The Journal of Physical Chemistry B*, *116*(12), 3861–3865.

Paliy, D., Bloor, D. B., Gilbert, P., & Barber, J. (2003). Improved methods of cultivation and production of deuteriated proteins from *E. coli* strains grown on fully deuteriated minimal medium. *Journal of Applied Microbiology*, *94*(4), 580–586.

Patzelt, H., Ulrich, A. S., Egbringhoff, H., Düx, P., Ashurst, J., Simon, B., et al. (1997). Towards structural investigations on isotope labelled native bacteriorhodopsin in detergent micelles by solution-state NMR spectroscopy. *Journal of Biomolecular NMR*, *10*, 95–106.

Penfold, J., Thomas, R. K., Li, P., Petkov, J. T., Tucker, I., Cox, A. R., et al. (2014). Impact of the degree of ethoxylation of the ethoxylated polysorbate nonionic surfactant on the surface self-assembly of hydrophobin-ethoxylated polysorbate surfactant mixtures. *Langmuir*, *30*(32), 9741–9751.

Penfold, J., Thomas, R. K., Li, P. X., Petkov, J. T., Tucker, I., Webster, J. R. P., et al. (2015a). Adsorption at air–water and oil–water interfaces and self-assembly in aqueous solution of ethoxylated polysorbate nonionic surfactants. *Langmuir*, *31*(10), 3003–3011.

Penfold, J., Thomas, R. K., Li, P., Xu, H., Tucker, I. M., Petkov, J. T., et al. (2015b). Multivalent-counterion-induced surfactant multilayer formation at hydrophobic and hydrophilic solid–solution interfaces. *Langmuir*, *31*(24), 6773–6781.

Perkins, S. J. (1981). Estimation of deuteration levels in whole cells and cellular proteins by 1H NMR spectroscopy and neutron scattering. *Biochemical Journal*, *199*, 163–170.

Petit-Haertlein, I., Blakeley, M. P., Howard, E., Hazemann, I., Mitschler, A., Haertlein, M., et al. (2009). Perdeuteration, purification, crystallization and preliminary neutron diffraction of an ocean pout type III antifreeze protein. *Acta Crystallographica F*, *65*(4), 406–409.

Petit-Haertlein, I., Blakeley, M. P., Howard, E., Hazemann, I., Mitschler, A., Podjarny, A., et al. (2010). Incorporation of methyl-protonated valine and leucine residues into deuterated ocean pout Type III anti-freeze protein: Expression, crystallization and preliminary neutron diffraction studies. *Acta Crystallographica F*, *66*(6), 665–669.

Rajesh, S., Nietlispach, D., Nakayama, H., Takio, K., Laue, E. D., Shibata, T., et al. (2003). A novel method for the biosynthesis of deuterated proteins with selective protonation at the aromatic rings of Phe, Tyr and Trp. *Biomolecular NMR*, *27*(1), 81–86.

Ramesh, V., Frederick, R. O., Syed, S. E., Gibson, C. F., Yang, J. C., & Roberts, G. C. (1994). The interactions of *Escherichia coli* trp repressor with tryptophan and with an operator oligonucleotide. *European Journal of Biochemistry*, *225*, 601–608.

Réat, V., Patzelt, H., Ferrand, M., Pfister, C., Oesterhelt, D., & Zaccai, G. (1998). Dynamics of different functional parts of bacteriorhodopsin: H-^{2}H labeling and neutron scattering. *Proceedings of the National Academy of Sciences of the United States of America*, *95*(9), 4970–4975.

Refaei, M. A., Combs, A., Kojetin, D. J., Cavanagh, J., Caperelli, C., Rance, M., et al. (2011). Observing selected domains in multi-domain proteins via sortase-mediated ligation and NMR spectroscopy. *Journal of Biomolecular NMR*, *49*(1), 3–7.

Rochel, N., Ciesielski, F., Godet, J., Moman, E., Roessle, M., Peluso-Iltis, C., et al. (2011). Common architecture of nuclear receptor heterodimers on DNA direct repeat elements with different spacings. *Nature Structural and Molecular Biology*, *18*(5), 564–570.

Savenkova, M. I., Satterlee, J. D., Erman, J. E., Siems, W. F., & Helms, G. L. (2001). Expression, purification, characterization, and NMR studies of highly deuterated recombinant cytochrome *c* peroxidase. *Biochemistry*, *40*(40), 12123–12131.

Sawada, D., Nishiyama, Y., Langan, P., Forsyth, V. T., Kimura, S., & Wada, M. (2012). Water in crystalline fibers of dihydrate beta-chitin results in unexpected absence of intramolecular hydrogen bonding. *PLoS One*, 7(6), e39376.

Sawada, D., Nishiyama, Y., Petridis, L., Parthasarathi, R., Gnanakaran, S., Forsyth, V. T., et al. (2013). Structure and dynamics of a complex of cellulose with EDA: Insights into the action of amines on cellulose. *Cellulose*, *20*, 1563–1571.

Schirò, G., Fichou, Y., Gallat, F.-X., Wood, K., Gabel, F., Moulin, M., et al. (2015). Translational diffusion of hydration water correlates with functional motions in folded and intrinsically disordered proteins. *Nature Communications*, *16*(6), 6490.

Schneck, E., Schollier, A., Halperin, A., Moulin, M., Härtlein, M., Sferrazza, M., et al. (2013). Neutron reflectometry elucidates density profiles of deuterated proteins adsorbed onto surfaces displaying poly(ethylene glycol) brushes: Evidence for primary adsorption. *Langmuir*, *29*(46), 14178–14187.

Seyedarabi, A., To, T.-T., Ali, S., Hussain, S., Fries, M., Madsen, R., et al. (2010). Structural insights into substrate specificity and the anti beta-elimination mechanism of pectate lyase. *Biochemistry*, *49*(3), 539–546.

Shotton, M. W., Pope, L. H., Forsyth, V. T., Denny, R. C., Archer, J., Langan, P., et al. (1998). New developments in instrumentation for fibre diffraction experiments. *Journal of Applied Crystallography*, *31*(56), 758–776.

Shotton, M. W., Pope, L. H., Forsyth, V. T., Langan, P., Denny, R. C., Giesen, U., et al. (1997). A high-angle neutron fibre diffraction study of the hydration of deuterated A-DNA. *Biophysical Chemistry*, *69*(1), 85–96.

Shu, F., Ramakrishnan, V., & Schoenborn, B. P. (2000). Enhanced visibility of hydrogen atoms by neutron crystallography on fully deuterated myoglobin. *Proceedings of the National Academy of Sciences of the United States of America*, *97*(8), 3872–3877.

Skar-Gislinge, N., & Arleth, L. (2011). Small-angle scattering from phospholipid nanodiscs: Derivation and refinement of a molecular constrained analytical model form factor. *Physical Chemistry Chemical Physics*, *13*(8), 3161–3170.

Skrisovska, L., Schubert, M., & Allain, F. H. (2009). Recent advances in segmental isotope labeling of proteins: NMR applications to large proteins and glycoproteins. *Journal of Biomolecular NMR*, *46*(1), 51–65.

Sobhanifar, S., Reckel, S., Junge, F., Schwarz, D., Kai, L., Karbyshev, M., et al. (2010). Cell-free expression and stable isotope labelling strategies for membrane proteins. *Journal of Biomolecular NMR*, *46*(1), 33–43.

Sreenath, H. K., Bingman, C. A., Buchan, B. W., Seder, K. D., Burns, B. T., Geetha, H. V., et al. (2005). Protocols for production of selenomethionine-labeled proteins in 2-L polyethylene terephthalate bottles using auto-induction medium. *Protein Expression and Purification*, *40*(2), 256–267.

Strauss, A., Bitsch, F., Cutting, B., Fendrich, G., Graff, P., Liebetanz, J., et al. (2003). Amino-acid-type selective isotope labeling of proteins expressed in Baculovirus-infected insect cells useful for NMR studies. *Journal of Biomolecular NMR*, *26*(4), 367–372.

Strub, M. P., Hoh, F., Sanchez, J. F., Strub, J. M., Bock, A., Aumelas, A., et al. (2003). Selenomethionine and selenocysteine double labeling strategy for crystallographic phasing. *Structure*, *11*(11), 1359–1367.

Studier, F. W., Rosenberg, A. H., Dunn, J. J., & Dubendorf, J. W. (1990). Use of T7 RNA polymerase to direct expression of cloned genes. *Methods in Enzymology*, *185*, 60–89.

Studts, J. M., & Fox, B. G. (1999). Application of fed-batch fermentation to the preparation of isotopically labelled or selenomethionyl-labelled proteins. *Protein Expression and Purification*, *16*(1), 109–119.

Taylor, J. E., Swiderska, A., Artero, J.-B., Callow, P., & Kneale, G. (2012). Structural and functional analysis of the symmetrical type I restriction endonuclease R.EcoR124I$_{NT}$. *PLoS One*, 7(4), e35263.

Teixeira, S., Ankner, J. F., Bellissant-Funel, M. C., Bewley, R., Blakeley, M. P., Coates, L., et al. (2008). New sources and instruments for neutrons in biology. *Chemical Physics*, *345*(2–3), 133–151.

Thomas, L., Forsyth, V. T., Sturcova, A., Kennedy, C. J., May, R. P., Altaner, C. M., et al. (2013). Structure of cellulose microfibrils in primary cell-walls from collenchymas. *Plant Physiology*, *161*, 465–476.

Tiggelaar, S. M., Mossou, E., Callow, P., Callow, S., Teixeira, S. C., Mitchell, E. P., et al. (2011). Neutron fibre diffraction studies of amyloid using H_2O/D_2O isotopic replacement. *Acta Crystallographica F*, *67*(3), 332–335.

Tuominen, V. U., Myles, D., Dauvergne, M.-T., Lahti, R., Heikinheimo, P., & Goldman, A. (2004). Production and preliminary analysis of perdeuterated yeast inorganic pyrophosphatase crystals suitable for neutron diffraction. *Acta Crystallographica*, *D60*(3), 606–609.

Vanatalu, K., Paalme, T., Vilu, R., Burkhardt, N., Junemann, R., May, R., et al. (1993). Large-scale preparation of fully deuterated cell components: Ribosomes from Escherichia coli with high biological activity. *European Journal of Biochemistry*, *216*(1), 315–321.

Varga, K., Aslimovska, L., Parrot, I., Dauvergne, M. T., Haertlein, M., Forsyth, V. T., et al. (2007). NMR crystallography: The effect of deuteration on high resolution ^{13}C solid state NMR spectra of a 7-TM protein. *Biochimica et Biophysica Acta*, *1768*(12), 3029–3035.

Venters, R. A., Huang, C. C., Farmer, B. T., Trolard, R., Spicer, L. D., & Fierke, C. A. (1995). High-level $^{2}H/^{13}C/^{15}N$ labeling of proteins for NMR studies. *Journal of Biomolecular NMR*, *5*(4), 339–344.

Vijayakrishnan, S., Callow, P., Nutley, M. A., McGow, D., Gilbert, D., Kropholler, P., et al. (2011). Mammalian pyruvate dehydrogenase complex E2/E3BP core organization: A new variable composition model. *Biochemical Journal*, *437*, 565–574.

Vijayakrishnan, S., Kelly, S. M., Gilbert, R. J. C., Callow, P., Bhella, D., Forsyth, V. T., et al. (2010). Solution structure and characterisation of the human pyruvate dehydrogenase complex core assembly. *Journal of Molecular Biology*, *399*, 71–93.

Waugh, D. (1996). Genetic tools for selective labeling of proteins with alpha-^{15}N-amino acids. *Journal of Biomolecular NMR*, *8*, 184–192.

White, J., Heß, D., Raynes, J., Laux, V., Haertlein, M., Forsyth, V. T., et al. (2015). The aggregation of native human serum albumin. *European Biophysics Journal*, *44*(5), 367–371.

Wood, K., Froelich, A., Paciaroni, A., Moulin, M., Härtlein, M., Zaccai, G., et al. (2008). Coincidence of dynamical transitions in a soluble protein and its hydration water: Direct measurements by neutron scattering and MD simulations. *Journal of the American Chemical Society*, *130*(14), 4586–4587.

Wood, K., Gallat, F. X., Otten, R., van Heel, A. J., Lethier, M., van Eijck, L., et al. (2013). Protein surface and core dynamics show concerted hydration-dependent activation. *Angewandte Chemie*, *52*(2), 665–668.

Wood, K., Grudinin, S., Kessler, B., Weik, M., Johnson, M., Kneller, G. R., et al. (2008). Dynamical heterogeneity of specific amino acids in bacteriorhodopsin. *Journal of Molecular Biology*, *380*(3), 581–591.

Wu, Z., Gogonea, V., Lee, X., Wagner, M. A., Li, X. M., Huang, Y., et al. (2009). The double super helix model of high density lipoprotein. *The Journal of Biological Chemistry*, *284*(52), 36605–36619.

Yabuki, T., Kigawa, T., Dohmae, N., Takio, K., Terada, T., Ito, Y., et al. (1998). Dual amino acid-selective and site-directed stable-isotope labeling of the human c-Ha-Ras protein by cell-free synthesis. *Journal of Biomolecular NMR*, *11*(3), 295–306.

Zaccai, G. (2011). Neutron scattering experiments for protein dynamics. *Journal of Non-Crystalline Solids*, *357*, 615–621.

Zhu, X., Ng, S. Y., Gupta, A. N., Feng, Y. P., Ho, B., Lapp, A., et al. (2010). Effect of crowding on the conformation of interwound DNA strands from neutron scattering measurements and Monte Carlo simulations. *Physical Review E*, *81*, 061905.

CHAPTER SIX

Deuterium Labeling Together with Contrast Variation Small-Angle Neutron Scattering Suggests How Skp Captures and Releases Unfolded Outer Membrane Proteins

Nathan R. Zaccai*[,1], Clifford W. Sandlin*, James T. Hoopes[†], Joseph E. Curtis[‡], Patrick J. Fleming*, Karen G. Fleming*, Susan Krueger[‡,1]

*T.C. Jenkins Department of Biophysics, Johns Hopkins University, Baltimore, Maryland, USA
[†]Institute for Bioscience and Biotechnology Research, University of Maryland, Rockville, Maryland, USA
[‡]National Institute of Standards and Technology, Gaithersburg, Maryland, USA
[1]Corresponding authors: e-mail address: nrz20@cam.ac.uk; susan.krueger@nist.gov

Contents

Methods in Enzymology, Volume 566
ISSN 0076-6879
http://dx.doi.org/10.1016/bs.mie.2015.06.041

Abstract

In Gram-negative bacteria, the chaperone protein Skp forms specific and stable complexes with membrane proteins while they are transported across the periplasm to the outer membrane. The jellyfish-like architecture of Skp is similar to the eukaryotic and archaeal prefoldins and the mitochondrial Tim chaperones, that is the α-helical "tentacles" extend from a β-strand "body" to create an internal cavity. Contrast variation small-angle neutron scattering (SANS) experiments on Skp alone in solution and bound in two different complexes to unfolded outer membrane proteins (uOMPs), OmpA and OmpW, demonstrate that the helical tentacles of Skp bind their substrate in a clamp-like mechanism in a conformation similar to that previously observed in the apo crystal structure of Skp. Deuteration of the uOMP component combined with contrast variation analysis allowed the shapes of Skp and uOMP as well as the location of uOMP with respect to Skp to be determined in both complexes. This represents unique information that could not be obtained without deuterium labeling of the uOMPs. The data yield the first direct structural evidence that the α-helical Skp tentacles move closer together on binding its substrate and that the structure of Skp is different when binding different uOMPs. This work presents, by example, a tutorial on performing SANS experiments using both deuterium labeling and contrast variation, including SANS theory, sample preparation, data collection, sample quality validation, data analysis, and structure modeling.

1. INTRODUCTION

In Gram-negative bacteria, unfolded outer membrane beta-barrel proteins (uOMPs) interact with periplasmic chaperones in order to be trafficked to the outer membrane. Two of these chaperones, Skp and SurA, form stable complexes with uOMPs as they travel across the aqueous periplasm. These chaperones exhibit holdase activity because they protect their substrates from aggregation, as well as from cleavage by the periplasmic protease DegP (Wu et al., 2011).

Even though Skp and SurA appear to have redundant functions, their relative activities are dependent on bacterial species and whether or not the bacteria are under stress (Denoncin, Schwalm, Vertommen, Silhavy, & Collet, 2012; Rhodius, Suh, Nonaka, West, & Gross, 2006; Sklar, Wu, Kahne, & Silhavy, 2007). Although not a lethal mutation in *Escherichia coli*, loss of Skp results in a relative decline in uOMP concentration in the bacterial outer membrane (Chen & Henning, 1996).

Skp binds a wide range of bacterial uOMPs with nanomolar affinity (Moon, Zaccai, Fleming, Gessmann, & Fleming, 2013; Qu, Mayer, Behrens, Holst, & Kleinschmidt, 2007). At least 19 different *E. coli* uOMPs interact with Skp (Jarchow, Lück, Görg, & Skerra, 2008). These substrates vary in sequence composition and in size (20–150 kDa). In bacteria, Skp–uOMP complexes form in the periplasm near the bacterial inner membrane (Schäfer, Beck, & Müller, 1999). The Skp–uOMP complex is then transported to the outer membrane where the uOMP folding is thought to be mediated by the Bam complex (Webb, Heinz, & Lithgow, 2012). *In vivo*, the Skp-presented bound uOMP could directly interact with the Bam complex, or uOMP could first be delivered by Skp to SurA, which then transports it to the Bam complex (Ieva, Tian, Peterson, & Bernstein, 2011; Schwalm, Mahoney, Soltes, & Silhavy, 2013; Sklar et al., 2007). Skp may also transfer the uOMP directly to the bacterial outer membrane as *in vitro* experiments demonstrate that Skp-bound uOMPs retain the ability to fold into lipid bilayers containing phosphatidylcholine and phosphatidylglycerol (Bulieris, Behrens, Holst, & Kleinschmidt, 2003; McMorran, Bartlett, Huysmans, Radford, & Brockwell, 2013).

The holdase activity of Skp is not limited to specific membrane proteins (Jarchow et al., 2008). Skp is able to form a complex with the periplasmic (PP) domain of the auto transporter EspP (Ieva et al., 2011). It can also inhibit the aggregation and assist the folding of a number of soluble proteins. Notable examples include single-chain antibodies (Entzminger, Chang, Myhre, McCallum, & Maynard, 2012) and lysozyme (Walton & Sousa, 2004).

Skp forms a stable trimer (50 kDa) in solution as determined by gel filtration as well as cross-linking methods (Schlapschy et al., 2004) and as supported by crystal structures (Korndörfer, Dommel, & Skerra, 2004; Walton & Sousa, 2004). Three α-helical "tentacles" extend out from the β-strand Skp "body" to create a cavity sufficiently large to surround a

25-kDa substrate. This domain architecture, termed jellyfish-like fold, had been previously described for other holdases, including the archaeal and eukaryotic prefoldins and the mitochondrial small Tim proteins (Stirling, Bakhoum, Feigl, & Leroux, 2006). Like Skp, these holdases have the common property of protecting their substrates from aggregation. However, no sequence identity to Skp was present, and unlike the trimeric Skp, these proteins are heterohexamers, with six α-helical tentacles each.

X-ray and subsequent NMR data could not directly provide an explanation for the ability of Skp to bind a wide range of uOMPs of different sizes. It was hypothesized that an uOMP transmembrane (TM) domain is bound within an adaptable cavity formed by the α-helical tentacles of Skp. NMR analysis of Skp–OmpA and Skp–OmpX complexes indicated that the uOMP TM region contains little secondary structure while in complex with Skp (Burmann & Hiller, 2012; Burmann, Wang, & Hiller, 2013; Callon, Burmann, & Hiller, 2014; Walton, Sandoval, Fowler, Pardi, & Sousa, 2009). However, the ability of Skp to bind diverse substrates begs the question as to whether changes occur in the tertiary structure of Skp upon binding a client substrate.

In this study, a series of contrast variation small-angle neutron scattering (SANS) experiments were performed on Skp alone in solution and on two different Skp–uOMP complexes: Skp–OmpA and Skp–OmpW, in order to clarify the structural basis of uOMP presentation by Skp. OmpA has a role in *E. coli* cell morphology and stability. Folded OmpA (35 kDa) has a C-terminal PP domain that folds independently from its TM domain (Danoff & Fleming, 2011; Walton et al., 2009). OmpW (21 kDa) is an integral membrane protein that is required for resistance to phagocytosis (Wu et al., 2013). The atomic resolution structures of the TM domains of both these proteins revealed an eight-stranded beta-barrel when folded into membranes (Hong, Patel, Tamm, & van den Berg, 2006; Pautsch & Schulz, 1998). The folding properties of these proteins have been extensively investigated (Burgess, Dao, Stanley, & Fleming, 2008; Moon et al., 2013). Selective labeling by deuteration of OmpW and OmpA in the Skp–uOMP complexes allowed the determination of the individual structures of uOMP and Skp as well as their relationship to each other in complexes.

The SANS data indicate that Skp interacts with the uOMP TM domain in a manner analogous to the binding mechanism of prefoldins binding their unfolded substrates (Martín-Benito et al., 2002, 2007). The bulk of the unfolded TM domain of both OmpW and OmpA is within the Skp cavity but some parts must be on the exterior of Skp near the tips of the tentacles to be in agreement with the SANS data. The PP domain of OmpA can assume a number of different positions outside the Skp cavity and still be consistent with the data. These results yield the first direct structural evidence that the α-helical Skp tentacles move closer together on binding its substrate and that the structure of the individual helices of the tentacles is different when binding different uOMPs. Importantly, the SANS data represent unique information that could not be obtained without deuterium labeling of the uOMPs. Analysis of the SANS data provides experimental support for a simple clamp-like mechanism used by jellyfish-like chaperones (Stirling et al., 2006). It also allows postulation of a mechanism of Skp binding and delivery of uOMPs to the bacterial outer membrane.

2. SANS FROM BIOLOGICAL MOLECULES IN SOLUTION

SANS is able to provide the size, molecular mass, and shape of a macromolecular complex in solution on length scales between approximately 10 Å to about 1000 Å (Jacques & Trewhella, 2010; Jacrot, 1976; Moore, 1982; Zaccai, 2012). While analytical ultracentrifugation (AUC) can be used to ascertain the Skp–uOMP stoichiometry and small-angle X-ray scattering (SAXS) can yield the overall size and shape of the Skp–uOMP complex, analysis of contrast variation SANS data can uniquely retrieve the internal structure and organization of the complex. Recent reviews (Heller, 2010; Whitten & Trewhella, 2009) as well as classic papers (Engelman & Moore, 1975; Ibel & Stuhrmann, 1975; Jacrot, 1976) describe the contrast variation technique in detail.

2.1 Contrast and Scattering Intensity

Because neutrons interact with atomic nuclei, the strength of the scattering interaction is not dependent on the atomic number, Z, of the atom, as is the case for X-rays. Therefore, the light elements such as H, C, N, and O can scatter neutrons just as strongly as heavier elements. Also, neutrons are sensitive to different isotopes of an element, such as hydrogen and deuterium. Since SANS does not provide information on atomic length scales, the strength of the scattering interaction can be described in terms of a uniform scattering length density of the entire molecule, ρ, within the molecular volume, V (in cm^3 or $Å^3$). For biological molecules in solution, the strength of the scattering is further defined as the difference in the scattering length densities of the molecule and the solvent within the same molecular volume.

This difference in scattering length densities is known as the contrast, and is written as

$$\Delta\rho = \rho - \rho_s, \tag{1}$$

where ρ_s is the scattering length density of the solvent. When the scattering length densities of the molecule and solvent are the same, $\Delta\rho = 0$. This is called the contrast match point. The scattering length densities and contrast are usually expressed in units of cm^{-2} or cm $\mathrm{Å}^{-3}$, but can be found stated in units of $\mathrm{Å}^{-2}$.

The measured SANS intensity from a solution of monodisperse, randomly oriented biological macromolecules can be written in terms of the contrast as

$$I(q) = n(\Delta\rho)^2 V^2 \left\langle \left|F(\vec{q})\right|^2 \right\rangle, \tag{2}$$

where $\left|F(\vec{q})\right|^2$ is the scattering form factor, which depends on the shape of the molecule and n is the number density of molecules (in cm^{-3}). The brackets represent an averaging over all orientations of the molecule. The magnitude of the scattering vector $\vec{q}$ can be written as

$$q = \frac{4\pi \sin(\theta)}{\lambda}, \tag{3}$$

where 2θ is the scattering angle (in degrees), measured from the axis of the incoming neutron beam, and λ is the neutron wavelength. The wavelength is usually expressed in nm or Å, such that q is stated in units of nm^{-1} or $\mathrm{Å}^{-1}$. It can be seen from Eq. (2) that the scattering intensity at the contrast match point, $\Delta\rho = 0$, is zero.

2.2 Radius of Gyration and Forward Scattering Intensity

The radius of gyration, R_g, and the forward scattering intensity, $I(0)$, which is the scattering intensity at $q = 0$, are two important model-independent parameters that are obtained from SANS data. R_g provides information about the size of the molecule. A shape must be assumed for the molecule to relate R_g to the molecular dimensions. For example, a solid sphere has a radius, r, which is equal to $1.3R_g$.

By definition, the scattering form factor in Eq. (2) is equal to 1.0 at $q = 0$. Thus, $I(0)$ is related to the number density, n, as

$$I(0) = n(\Delta\rho)^2 V^2. \tag{4}$$

The number density can be written in terms of the concentration of the molecule, c (g cm^{-3}), as

$$n = \frac{cN_A}{M_w}, \tag{5}$$

where M_w is the molecular weight of the molecule (in Da, where 1 Da = 1 g mol^{-1}) and N_A is Avogadro's number. Equations (4) and (5) can be used to relate $I(0)$ to the M_w of the molecule, if the SANS data are on an absolute scale, usually in units of cm^{-1}.

The Guinier approximation (Guinier & Fournet, 1955),

$$I(q) = I(0)\exp\left(-q^2 \frac{R_g^2}{3}\right), \tag{6}$$

can be used on the low-q portions of the data to obtain values for R_g and $I(0)$. This "low-q" analysis is valid only in the region where $qR_g \lesssim 1.3$. R_g and $I(0)$ are found by plotting the natural log of Eq. (6) such that

$$\ln(I(q)) = \ln(I(0)) - q^2 \frac{R_g^2}{3}. \tag{7}$$

A linear fit of $\ln(I(q))$ versus q^2 (Eq. 7) to the low-q portion of the data allows the determination of R_g from the slope and $I(0)$ from the intercept. If there are aggregates of the molecule in the solution, Eq. (7) will not be linear. Rather it will have some curvature and the fit to a straight line will be poor. The effects can be subtle or very obvious depending on the severity of the aggregation (Jacques & Trewhella, 2010). If aggregation is present, whether subtle or severe, the R_g value found from Eq. (7) no longer represents that of a single monomer in a monodisperse solution. Rather, it is influenced by the larger aggregates present in the solution.

Another method to obtain R_g, which makes use of all of the data rather than a limited data set at small q values, is to use the distance distribution function, $P(r)$ versus r, to determine R_g and $I(0)$ (Glatter & Kratky, 1982). This function represents the probability distribution of distances, r, between all pairs of atoms in the molecule. The result is a smooth histogram-like plot that peaks at the most probable distance in the molecule. Thus, the shape of the $P(r)$ versus r curve depends strongly on the shape of the molecule.

$P(r)$ is typically obtained from the SANS data using an indirect Fourier transformation method (Glatter, 1977; Moore, 1982; Semenyuk & Svergun, 1991) using the relation

$$I(q) = 4\pi V \int_0^{D_{\max}} P(r) \frac{\sin(qr)}{qr} \mathrm{d}r. \tag{8}$$

This analysis requires a stipulation by the user of a maximum dimension, $D_{\max}$, beyond which $P(r) = 0$. Typically, several values of $D_{\max}$ are explored in order to find the range over which the $P(r)$ function does not change as a function of $D_{\max}$. The $P(r)$ function is also sensitive to aggregation. The more severe the aggregation, the more difficult it is to determine $D_{\max}$ (Jacques & Trewhella, 2010).

2.3 Contrast Variation

The scattering length density of water is shown as a function of % D_2O in the solvent in Fig. 1, along with scattering length densities of some typical

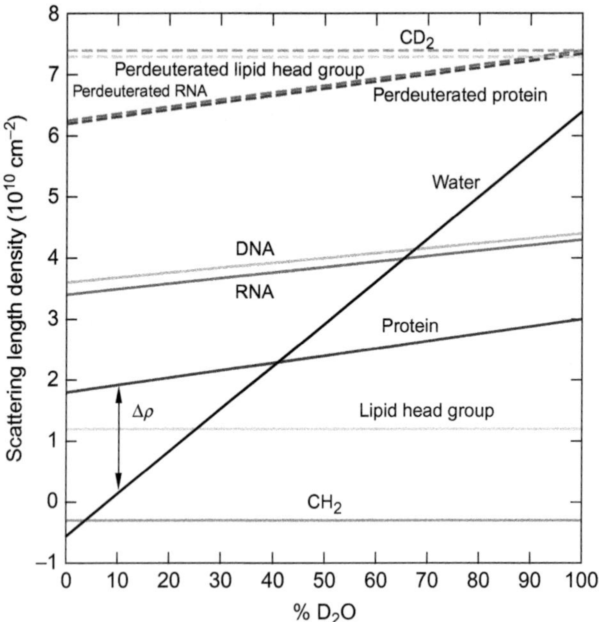

Figure 1 Plot of the neutron scattering length density versus % D_2O in the solvent for water compared to those for protein, RNA, DNA, and the components of lipids (lipid head group and CH_2), along with their perdeuterated counterparts. (See the color plate.)

biological molecules and compounds. The lines are straight for CH_2 and CD_2 because there is no exchange of deuterium for hydrogen as the % D_2O increases in the solvent. However, for proteins and nucleic acids, labile hydrogen atoms, i.e., those bound to nitrogen and oxygen, will exchange with deuterium in the solvent, so their scattering length densities vary with increasing % D_2O. A vertical line representing the contrast, $\Delta\rho$, between protein and water is shown on the graph for 10% D_2O in the solvent. Note that the protein and water lines cross at 40% D_2O, which is the contrast match point for most typical proteins. The match points for DNA, RNA, lipid head groups, and CH_2 can be found in the same manner. Note that the scattering length densities for perdeuterated molecules do not cross the water line. Thus, the match points for these molecules are greater than 100% D_2O and cannot be reached in practice.

In cases where a molecule is actually a complex consisting of components that have different scattering length densities, the contrast is different for each of the components. If the ratio of H_2O to D_2O in the solvent is varied, the contrast of each component will change as a function of the concentration of D_2O in the solvent. Thus, contrast match points exist for each of the two components as well as the entire complex. By varying the amount of D_2O in the solvent, one component can be essentially transparent at its match point while the other is still visible. It is this feature of SANS that makes the method so powerful for selective measurement of individual components within a complex. From Fig. 1, it is clear that proteins and nucleic acids have different contrast match points. The protein match point is around 40% D_2O, meaning that only the DNA or RNA is visible at this contrast. The DNA and RNA match points are around 65% D_2O such that only the protein is visible under these conditions. Therefore, complexes consisting of proteins and nucleic acids are ideal candidates for contrast variation experiments.

For a complex consisting of two proteins, replacement of the nonexchangeable hydrogen atoms, i.e., those bound to carbon, with deuterium in one of the components is required in order for the two match points to be different. This is typically accomplished by expressing one of the protein components using bacterium grown in deuterium-enriched media. Figure 1 shows that the match point of perdeuterated proteins, in which all nonexchangeable hydrogen atoms have been replaced by deuterium, is above 100% D_2O. Therefore, partially deuterated proteins are generally used for contrast variation experiments so that the match point of the deuterated component is somewhere between 60% D_2O and 100% D_2O. The exact match point of a deuterated component is dependent on the amount of deuteration achieved. The contrast variation experiment can be used to verify this parameter, especially if a reliable determination cannot be made by other methods such as mass spectrometry.

2.4 Contrast Match Point Analysis of the SANS Contrast Variation Data

For a two-component complex in which the components have different contrasts, the contrast match points for the complex are determined from $I(0)$. Expanding on Eq. (4),

$$I(0) = n(\Delta\rho)^2 V^2 = n(f_1 \Delta\rho_1 V_1 + f_2 \Delta\rho_2 V_2)^2. \tag{9}$$

The number density, n, is defined as in Eq. (5), but it is now in terms of the concentration and M_w of the entire complex. $\Delta\rho$ now refers to the mean contrast of the entire complex and V is the volume of the complex. The expression on the right is now written in terms of the two components, where f_1 and f_2 are the mass fractions of the 1st and 2nd components in the complex, $\Delta\rho_1$ and $\Delta\rho_2$ are the scattering contrasts of the 1st and 2nd components, and V_1 and V_2 are the volumes of the 1st and 2nd components. Thus, Eq. (9) can be combined with Eq. (5) to relate $I(0)$ to the M_w of the complex if the SANS data are on an absolute scale. This is a good quality assurance test on the data. If the complex has the expected M_w at all contrasts, then it is likely that the stoichiometry of the two components is correct at all contrasts. It is important to recognize that care should be taken to obtain the concentrations of the complexes as accurately as possible at each contrast because uncertainty in this value is a major source of error on the calculation of the M_w.

Because $I(0)$ is proportional to $\Delta\rho^2$ (Eqs. 4 and 9), which is in turn dependent on the fraction of D_2O in the solvent, f_{D_2O}, and n is proportional to c (Eq. 5), the contrast match point of the complex can be determined from the x-intercept of a linear fit to $\sqrt{I(0)/c}$ versus f_{D_2O}. The match point of the complex can also be calculated from the sequences of the components (Sarachan, Curtis, & Krueger, 2013; Whitten, Cai, & Trewhella, 2008). Using both approaches to determine the match point provides another quality assurance test on the data in that the calculated and experimentally determined match points should agree with each other.

2.5 Separation of the Radii of Gyration in a Two-Component Complex

The R_g values obtained at each contrast are related by the relationship (Ibel & Stuhrmann, 1975),

$$R_g^2 = R_m^2 + \frac{\alpha}{\Delta\rho} - \frac{\beta}{\Delta\rho^2}, \tag{10}$$

where R_m is the R_g value of the equivalent complex with a homogeneous scattering length density, α is the second moment of the density fluctuations, and β is the first moment of the density fluctuations. For two-component systems with different scattering length densities, the term α relates to the distribution of scattering length densities relative to the center of mass (CM) of the complex, and the term β provides the separation of the scattering CM of the two components (Moore, 1982). A Stuhrmann plot (Ibel & Stuhrmann, 1975) of R_g^2 versus $\Delta\rho^{-1}$ (Eq. 10) is used to determine R_m, α, and β. If the plot is linear, then $\beta = 0$ and the CM of the two components are concentric. In this case, the sign of the slope of the line, α, is an indication of whether the component with the higher scattering length density is on the interior or exterior of the complex. In practice, it is not always easy to distinguish between a linear and a parabolic Stuhrmann plot, especially if α is close to zero and if R_g values are not available close to the individual match points of the two components.

Similar information can be obtained from the parallel axis theorem

$$R_g^2 = f_1 R_1^2 + f_2 R_2^2 + f_1 f_2 D_{CM}^2, \tag{11}$$

where R_1 and R_2 are the radii of gyration of the components and D_{CM} is the distance between the scattering CM (Engelman & Moore, 1975). Here,

$$f_i = \frac{\Delta\rho_i V_i}{\Delta\rho V}, \tag{12}$$

where $\Delta\rho_i$ and V_i refer to the individual components and $\Delta\rho$ and V refer to the complex. The parallel axis theorem provides the radii of gyration of the components directly, whereas they are calculated from the definitions of α and β when the Stuhrmann analysis is used (Whitten et al., 2008).

2.6 Separating the Scattering Intensities in a Two-Component Complex

The scattering intensity from a two-component system with different scattering length densities can be approximated by (Whitten et al., 2008)

$$I(q) = \Delta\rho_1{}^2 I_1(q) + \Delta\rho_1 \Delta\rho_2 I_{12}(q) + \Delta\rho_2{}^2 I_2(q), \tag{13}$$

where $I_1(q)$ and $I_2(q)$ are the scattering intensities of components 1 and 2, respectively, and $I_{12}(q)$ is the scattering intensity due to the interference between the two components that occurs because they have different scattering length densities. $I_1(q)$ and $I_2(q)$ are related to the shapes of the two components and $I_{12}(q)$ is related to their spatial distribution. For a given set of measured contrast variation intensities, $I(q)$, and known values for the contrasts, $\Delta\rho_1$ and $\Delta\rho_2$, the three unknowns, $I_1(q)$, $I_2(q)$, and $I_{12}(q)$, are found by solving Eq. (13) simultaneously at all contrasts.

This analysis can be a useful tool to model each component separately and then put them in the proper position with respect to each other in the complex. Data must be obtained at a minimum of three contrasts to solve for the three unknowns in Eq. (13). However, in practice, successful studies have employed at least five contrast points and high-quality data were obtained at the match points of the individual components. This method worked particularly well in a recent study of a kinase, KinA, in complex with an inhibitor, Sda (Whitten et al., 2007).

2.7 Structure Modeling

Both SAXS and SANS are being used for structural determination of large protein complexes and for proteins containing flexible regions in solution. Many options are available for modeling multimeric protein complexes using a combination of rigid body and atomistic approaches, as described in recent reviews (Putnam, Hammel, Hura, & Tainer, 2007; Rambo & Tainer, 2010). The SASSIE software suite (Curtis, Raghunandan, Nanda, & Krueger, 2012) is one tool that is available to assist in the atomistic and rigid body modeling of the structures of biological molecules for comparison to SAXS and SANS data. SASSIE allows users of these techniques access to molecular dynamics (MD), Monte Carlo, docking, and rigid body modeling methods to assist in structure modeling and assessment of how well the models fit the data. Constraints can be incorporated from other techniques such as NMR and AUC. For example, SASSIE has been used for the structure modeling of intrinsically disordered monomeric proteins

(Curtis et al., 2012), large protein complexes (Krueger, Shin, Curtis, Rubinson, & Kelman, 2014; Krueger, Shin, Raghunandan, Curtis, & Kelman, 2011), and single-stranded nucleic acids (Peng, Curtis, Fang, & Woodson, 2014). It has also been applied to the study of monoclonal antibodies using free-energy analysis (Clark et al., 2013). A Web version is available (SASSIE-web:Beta, n.d.) for ease of access and to handle the intensive computational requirements of the structural modeling and data analysis.

For a two-component complex, SANS and contrast variation experiments provide the added structural information from the individual components as constraints for modeling the entire complex. If obtainable, the scattering intensities of the separate components (Eq. 13) can be helpful for the modeling of the individual components and for construction of the model structure for the entire complex (Whitten et al., 2008). However, the R_g and CM distance constraints found by the Stuhrmann (Eq. 10) and parallel axis theorem (Eqs. 11 and 12) analyses add unique information that can be used in the modeling process even in the absence of the component scattering intensities. Often, structural information for one or both of the components alone in solution is used as a starting point for their structures in the complex.

Structure models that fit the SANS data at all contrasts take full advantage of the information content of the contrast variation data set. Whether or not models are constructed from the scattering intensities of the separate components, the model structures should always be judged against the entire contrast variation data set. Working model structures are tested against the data by calculating SANS curves from the model structures and comparing them to the measured SANS curves at all contrasts. The model structures are then revised as necessary to obtain the best global fit to the entire SANS contrast variation data set.

3. MATERIALS AND METHODS[1]

3.1 uOMP and Skp Expression and Purification

The uOMPs were cloned and expressed to inclusion bodies as previously described, except that no detergent was included in the inclusion body washes (Burgess et al., 2008). Deuteration of the uOMPs was accomplished

[1] Certain commercial equipment, instruments, or materials are identified in this chapter to foster understanding. Such identification does not imply recommendation or endorsement by the National Institute of Standards and Technology, nor does it imply that the materials or equipment identified are necessarily the best available for the purpose.

at the NIST/University of Maryland Biomolecular Labeling Laboratory (BL^2). To produce the deuterated uOMPs, the *OMP* genes were recloned into the kanamycin-resistant pet28 vector. Expression was performed with HMS cells in deuterated M9 media, containing 60% D_2O. After $OD_{600} \approx 0.6$ was reached, expression was induced with 1 mmol L^{-1} (m*M*) IPTG and the cells were allowed to grow overnight at room temperature. Deuterated inclusion bodies were prepared with the same buffers as for the hydrogenated inclusion bodies. Inclusion bodies were subsequently stored at −20 °C.

The inclusion bodies were dissolved in 20 m*M* Tris (pH 8), 8.0 *M* urea, and, after centrifugation at 18,000 rpm (1 rpm = 1/60 Hz) in a Beckman J2-MI with a type 21 rotor for 1 h, the clarified supernatant was stored at −80 °C until use. The uOMP concentrations were determined using their calculated extinction coefficient. The respective coefficients used were 52,955 M^{-1} cm^{-1} for OmpA and 39,420 M^{-1} cm^{-1} for OmpW.

The expression and purification of Skp has been previously described (Moon et al., 2013). Briefly, the *E. coli skp* gene (Gene ID 944861) was amplified by using the following primers: AGGAGATATACCATGGCT GACAAAAT and GTGATGG-TGATGTTTAACCTGTTTCA, and was inserted by ligation-independent cloning into the pOPINE expression vector (Berrow et al., 2007). The resultant plasmid (NZ100) was then transformed into BL21 (DE3) pLysS *E. coli* (Novagen) for expression of Skp with a C-terminal six-histidine tag. Typically, 2XYT media supplemented with 1% glucose was inoculated from a frozen cell stock and grown overnight at 37 °C. After 1/50 dilution into fresh media, Skp expression was induced 3 h later with 1 m*M* IPTG and continued at room temperature for 20 h. Cells were harvested by centrifugation and stored at −20 °C for future use. The frozen cell pellet was resuspended in buffer A [50 m*M* Tris, pH 8, 500 m*M* NaCl, 10% (v/v) glycerol, and 20 m*M* imidazole] with one complete EDTA-free protease inhibitor tablet (Roche) per 50 mL buffer and DNase I (Sigma). Cells were lysed by sonication and cell debris removed by centrifugation at 19,000 rpm in a size 21 rotor in a Beckman J2-MI centrifuge. The supernatant was applied to a pre-equilibrated Nickel Sepharose High-Performance column (GE Healthcare) and washed multiple times with buffer A (without glycerol). Then, recombinant protein was eluted by addition of a mixture of 50% (v/v) buffer A and 50% (v/v) buffer B (50 m*M* Tris, pH 8, 500 m*M* NaCl, and 500 m*M* imidazole). Pooled fractions were subjected to gel filtration on a Superdex 200 10/300 GL column (GE Healthcare) and equilibrated in GF buffer (20 m*M* Tris, pH 8, and 200 m*M* NaCl) for SANS measurements. When required, the SANS buffer was also prepared with D_2O instead of H_2O. Because the 1 *M* Tris (pH 8) stock did not contain deuterium, this buffer contained 98% (v/v) D_2O and 2% (v/v) H_2O (98% D_2O buffer). The Skp trimer concentration was determined using a calculated extinction coefficient ($\varepsilon_{280} = 4470\ M^{-1}\ \text{cm}^{-1}$). Nominal Skp concentrations employed were 5.3 mg mL^{-1} in H_2O (0 % D_2O) buffer and 3.6 mg mL^{-1} in 98% D_2O buffer.

3.2 Formation of Skp–uOMP Complex

The concentration of Skp and uOMP were adjusted so that the uOMP in urea was ≈10 times more concentrated than Skp trimer in GF buffer. The Skp–uOMP complex was then assembled by dropwise addition of the uOMP solution until the first signs of precipitation were observed. The final urea concentration was estimated to be less than 1 *M*. At each addition of uOMP, the solution was checked for aggregation. After 30 min incubation at room temperature, the complex sample was repeatedly diluted twofold with GF buffer every 10 min, until the final urea concentration was under 0.1 *M*. Any aggregates were removed with a 0.22-μm filter and the protein was concentrated to between 2 and 5 mg mL^{-1} with an Ultra-15 Ultracel 30k Centrifugal Filter (Millipore UFC903024) prior to being purified on a Superdex 200 10/300 GL column (GE Healthcare) equilibrated in GF buffer. Fractions containing the Skp–uOMP complex were identified from their OD230/OD280 ratio, and the expected stoichiometry was confirmed by SDS-PAGE. The complex was assembled in hydrogenated GF buffer before being also purified by size-exclusion chromatography into the deuterated GF buffer.

Prior to data collection, the Skp–uOMP complex was concentrated. Final D_2O concentrations were obtained by diluting with the appropriate D_2O and H_2O GF buffers. The resultant protein concentration was estimated from calculated extinction coefficients. The protein complexes were stored at 4 °C and used within 24 h after gel filtration.

3.3 SANS Data Collection

SANS measurements were performed on the NG3 30-m SANS instrument (Glinka et al., 1998) at the NIST Center for Neutron Research (NCNR) in Gaithersburg, Maryland. The neutron wavelength, λ, was 6 Å, with a wavelength spread, $\Delta\lambda/\lambda$, of 0.15. Scattered neutrons were detected with a 64 cm × 64 cm two-dimensional, position-sensitive detector with 128 × 128 pixels at a resolution of 0.5 cm pixel^{-1}. Data reduction was accomplished using Igor Pro software (WaveMetrics, Lake Oswego, OR) with SANS macros developed at the NCNR (Kline, 2006). Raw counts were normalized to a common monitor count and then corrected for empty cell counts, ambient room background counts, and nonuniform detector response. Data were placed on an absolute scale by normalizing the scattering intensity to the incident beam flux for each individual pixel. Finally, the data were radially averaged to produce the scattering intensity $I(q)$ to plot as $I(q)$ versus q curves. Sample-to-detector distances of 5.0 m and 1.5 m were used in order to cover the range of $0.01\ \text{Å}^{-1} \leq q \leq 0.4\ \text{Å}^{-1}$. The $I(q)$ versus q scattering data obtained at the two instrument configurations were merged using the NCNR SANS reduction software (Kline, 2006). The $I(q)$ versus q scattering data for the buffer was then subtracted from the data for the corresponding sample. The q-range covered by the data after buffer subtraction was dependent on the H_2O/D_2O ratio in the buffer.

3.4 Match Point and Contrast Variation Data Analysis

Determination of $I(0)$ and R_g were performed using the Guinier approximation (Eqs. 6 and 7). The GNOM program (Semenyuk & Svergun, 1991) was used to determine the distance distribution function (Eq. 8) to further confirm the R_g and $I(0)$ values.

Theoretical $\Delta\rho_1$ and $\Delta\rho_2$ values were calculated for the Skp and uOMP components and theoretical $\sqrt{I(0)/c}$ values were calculated for the complexes as a function of f_{D_2O} from the protein sequences using the Contrast Calculator (Sarachan et al., 2013) module of the SASSIE (Curtis et al., 2012) software assuming different percentages of deuteration for the OmpW and OmpA. The $\sqrt{I(0)/c}$ values were compared to those obtained from the Guinier analysis, allowing for experimental verification of the complex match point and the amount of deuteration of the uOMP component in the measured samples. The match points of the individual components were then calculated from linear fits to $\Delta\rho_1$ versus f_{D_2O} and $\Delta\rho_2$ versus f_{D_2O}.

Both the Stuhrmann and parallel axis theorem analyses were performed using the program MULCh (Whitten et al., 2008). The contrasts and volumes were determined from the protein and buffer compositions and the radii of gyration and CM distances were determined using the Guinier R_g and $I(0)$ values obtained from the data.

3.5 Molecular Dynamics Simulation of Skp-OmpA Complexes

The Skp homotrimer was modeled from the Skp crystal structure (PDB ID: 1SG2; Korndörfer et al., 2004). The missing residues were included based on threefold symmetry considerations. The tentacles of the Skp homotrimer were then splayed out to specific separations using biased MD with NAMD (Phillips et al., 2005). A collective variable consisting of the distance between two groups defined as the α-carbon atoms of residues 35–80 in adjacent Skp monomers was defined. Then a harmonic potential with force constant 3.0 kcal mol^{-1} Å^{-2} was applied to the centers of the collective variable groups to bias the distance between adjacent tentacles to defined values. Simulations were done *in vacuo* using the CHARMM-22 force field parameters (MacKerell et al., 1998). Stable distances were achieved in 10,000 steps of 1 fs. An ensemble of Skp homotrimer structures with separation distances of 10–119 Å at the tips of the helices and corresponding R_g values of 28.5–39.4 Å was generated.

Because the PP domain of OmpA folds as an independent unit (Danoff & Fleming, 2011) and is folded when it is in complex with Skp (Walton et al., 2009), all-atom model structures of OmpA were constructed with an unfolded, but collapsed TM domain and natively folded PP domain as follows: a homology model of the *E. coli* PP domain (residues 187–316) was generated using Swiss-Model (Schwede, Kopp, Guex, & Peitsch, 2003). An extended structure ($\theta = -78$, $\varphi = 149$) containing the TM domain sequence of *E. coli* OmpA (residues 1–186) was built using PyMOL (The PyMOL Molecular Graphics System, version 1.7.4 Schrödinger, LLC, n. d.). This extended structure was partially collapsed using a torsion angle Monte Carlo procedure developed at Johns Hopkins University (REDUX, n.d.), resulting in an ensemble of structures with R_g values of 34–36 Å. These partially collapsed TM domain structures were manually connected to the above homology model of the PP domain. The combined TM–PP domain models were solvated with TIP3 water, neutralized with 0.2 *M* NaCl and subjected to standard MD equilibration using NAMD (Phillips et al., 2005) with the CHARMM-22 force field (MacKerell et al., 1998). Successive rounds of equilibration with decreasing R_g potentials were carried out using the collective variables module in NAMD. Two of the resulting structures, one with an R_g of 42 Å and the other with 39 Å, were chosen for manual docking in PyMOL to the Skp models described above. Inspection of the docked structures suggested that the Skp model with $R_g = 34$ Å and tip separation of 46 Å fit the OmpA models best. Finally, two Skp–OmpA complex models with different orientations of the OmpA docked to Skp were solvated with TIP3 water, neutralized with 0.2 *M* NaCl and subjected to equilibration with standard MD using NAMD and the CHARMM-22 force field. The equilibrated complex model M1 contains

Skp with $R_g = 31.2$ Å (S1) and OmpA with $R_g \approx 43$ Å and complex model M2 contains Skp with $R_g = 32.5$ Å (S2) and OmpA with $R_g \approx 30$ Å.

3.6 Modeling of the Skp–uOMP Complexes

Even though SANS is a low-resolution structural method, high-resolution structures derived from X-ray crystallography and NMR (along with MD and geometrical models) can be positioned to fit the SANS data in order to produce quasi-atomic structural models. Although the crystal structure of Skp is available, the bound uOMP TM domains remained unfolded when bound to Skp (Burmann & Hiller, 2012; Walton et al., 2009). There is also no available data that links a specific region of Skp to a specific region of uOMP (Callon et al., 2014). There is moreover intrinsic interplay of different regions of the uOMP on binding Skp. For example, in the case of PhoE, its N-terminal region is important for Skp binding, but its influence is strongly modulated by the OMP's C-terminal region (Harms et al., 2001). Therefore, OmpW and the TM portion of OmpA were modeled by a prolate ellipsoid of uniform scattering length density.

The ellipsoid is the simplest model, i.e., a single geometric shape, that can be used to describe the uOMP TM domains, and it is adequate for the fitting of the SANS data, especially in the Guinier region. However, it should be noted that other options such as bead models, rigid body models consisting of multiple geometric shapes, and all-atom models can be more useful under certain circumstances. Such models are generally less symmetric than single geometric shapes and may better represent the overall shape of the biological molecule, especially at the shorter length scales on the order of 10 Å. This is especially important for fitting SAXS data, which usually have better signal-to-noise than SANS data at q values beyond the Guinier range (Putnam et al., 2007; Rambo & Tainer, 2010). All-atom models provide the additional advantage of allowing modern MD methods to be employed to generate biologically relevant structures for comparison to SANS and SAXS data. For this reason, an all-atom model of Skp–OmpA was also tested against the SANS data, as is discussed later.

The prolate ellipsoid used for the TM domain of the uOMPs in both complexes was found by testing models of Skp and OmpW that best fit the Skp–OmpW contrast variation data. Three models were tested in which the axes were adjusted to create ellipsoids with R_g values of 21, 24, and 27 Å, (e21, e24, and e27, respectively) which match the range of possible R_g values found for bound OmpW, and volumes that essentially filled the Skp cavity, as suggested by both the $I(q)$ and $P(r)$ curves as a function of contrast. The ellipsoids were paired with Skp S1 and S2, from the M1 and M2 Skp–uOMP models described above and oriented such that the long axis was parallel to that of Skp. The CM distance between Skp and OmpW was varied for each model to correspond to a range calculated based on the R_g of Skp and the R_m value of the complex determined by the Stuhrmann and parallel axis theorem analyses. To reduce the R_g value of Skp to better match that obtained from the Skp–OmpW data, the S2 Skp structure was minimized for 1000 steps, subjected to a 10 ps MD simulation *in vacuo,* and then minimized again for 1000 steps using NAMD (Phillips et al., 2005), resulting in Skp model S3.

To determine how well each model represented the experimental SANS data, SANS curves were calculated from each of the model structures using the programs SIMUL and SCAT (Hansen, 1990) as well as Xtal2sas (Heidorn & Trewhella, 1988; Krueger et al., 1998). These curves were then compared to the Skp–OmpW SANS data to evaluate which Skp–OmpW model best fit the data. SIMUL was used to build the OmpW ellipsoid and to populate it randomly with points of uniform scattering contrast, $\Delta\rho$, corresponding to that of the deuterated OmpW and the D_2O composition of the buffer. Xtal2sas was used to read the atomic coordinates of the Skp structure (in PDB format) and to populate the structure randomly with points of scattering contrast that matched that of the amino acid residue in which the point was located. Care was taken to insure that the two volumes (Skp and OmpW) were filled with the same number density of points. SCAT was then used to calculate the scattering intensity of the complex, first by calculating $P(r)$ as a function of r, using the contrast values associated with each scattering point, and then by performing a Fourier transform of $P(r)$ to obtain $I(q)$ as a function of q.

The calculated SANS curves were compared to the data at each contrast and goodness-of-fit was determined using the χ^2 equation

$$\chi^2 = \frac{1}{(N-1)} \sum\nolimits_q \frac{\left(I_{\text{exp}}(q) - I_{\text{calc}}(q)\right)^2}{\sigma_{\text{exp}}(q)^2}, \tag{14}$$

where $I_{\text{exp}}(q)$ is the experimentally determined SANS intensity curve, $I_{\text{calc}}(q)$ is the calculated intensity curve from the model structure and $\sigma_{\text{exp}}(q)$ is the q-dependent error of the $I_{\text{exp}}(q)$ values. The sum was taken over $N=60$ data points. The goodness-of-fit to the entire contrast variation data set was determined by the average χ^2 value

$$\chi^2(\text{avg}) = \frac{1}{N}\sum_i \chi_i^2, \tag{15}$$

where N is the number of contrast variation $I(q)$ versus q scattering curves and χ_i^2 is the χ^2 value for the ith scattering curve.

Once reasonable models for the Skp–OmpW complex were found, the Skp–OmpA complex was modeled using an ellipsoid of the same dimension for the TM portion of OmpA as found for OmpW. Guided by the structure of a closely related protein (PDB ID: 1R1M; Grizot & Buchanan, 2004), the PP domain of OmpA was built using Swiss-Model (Schwede et al., 2003) with ≈10 disordered residues of OmpA added at the N-terminus using the PSFGEN module of NAMD (Phillips et al., 2005) to act as a "tether." Significantly, this and other OmpA-like structures (PDB ID: 2K1S; Ramelot et al., unpublished) and (PDB ID: 2L26; Li et al., 2012) have similar three-dimensional folds and theoretical R_g values.

Several models were tested in which the position of the TM domain of OmpA and the location at which the PP domain of OmpA was "tethered" approximately satisfied the R_g and CM distance parameters obtained from the Stuhrmann and parallel axis theorem analyses. The complex Monte Carlo module in SASSIE (Curtis et al., 2012) was used to allow the disordered region of the PP domain of OmpA to vary to generate ensembles of structures from the starting Skp–OmpA structures for comparison to SANS data. Accepted (nonoverlapping) configurations were generated by sampling backbone dihedral angles using CHARMM-22 all-atom protein force field parameters (MacKerell et al., 1998). The new configuration was checked for overlap of basis atoms, which were chosen as α-carbon atoms in this case. If the overlap distance between basis atoms was ≥3 Å, the new structure was accepted. Every 20th accepted structure was selected for further analysis in order to eliminate correlated dihedral angle moves.

A Skp–OmpA model with an all-atom representation of the TM domain of OmpA was also tested against the SANS data. Torsion angle MD (Chen, Im, & Brooks, 2005) was performed on the TM domain of OmpA (residues 1 to 171) from the M2 Skp–uOMP model described above in order to increase the R_g to match the value obtained for OmpW from the Skp–OmpW contrast variation data. This region was oriented such that the long axis was parallel to that of S3 Skp, with the C-terminal region near the top of the Skp trimer. The PP domain of OmpA, as well as the disordered residues between the TM and PP domains and at the C-terminus, were then added using the PSFGEN module in NAMD (Phillips et al., 2005) to form the complete OmpA molecule (residues 1–326).

The structure of the entire S3-OmpA complex was minimized using 1000 steps, subjected to a 10 ps MD simulation *in vacuo* to insure its stability, and then minimized again using 1000 steps, using NAMD (Phillips et al., 2005). The Complex Monte Carlo module in SASSIE (Curtis et al., 2012) was used to explore possible conformations of the PP domain of OmpA as described above.

SANS curves were calculated for the ensembles of Skp–OmpA model structures as a function of contrast as described above for the Skp–OmpW models. In this case, Xtal2sas was also used to populate the PP domain of the OmpA structure (or both the PP and TM domains of OmpA in the all-atom model) with points of scattering contrast that matched that of the amino acid residue in which the point was located. Care was taken to insure that the volumes (Skp, TM-OmpA, and PP-OmpA) were filled with the same number density of points. Goodness-of-fit to the SANS data was determined using the same χ^2 relations as above.

Examination of a plot of χ^2 versus R_g at each individual contrast, or χ^2(avg) versus R_g(avg) for the entire contrast variation data set, provides an idea of how well the individual structures generated from each starting structure fit the data as well as which starting structure produces the overall best fit to the data. The best- (lowest χ^2(avg)) and worst-fit (highest χ^2(avg)) model SANS curves were noted for each case, along with the average model SANS curve from the entire ensemble of accepted structures. These curves were plotted along with the experimental SANS data to aid in the visualization of the quality of the fits to the data. Surface density plots representing the total configuration space examined by all of the accepted structures were generated and compared to that representing the best-fit family of structures in each case. The best-fit family of structures was chosen based on those giving the lowest $\approx$10% of the χ^2(avg) values for each series. This cutoff was chosen arbitrarily based on the shape of the χ^2(avg) versus R_g(avg) curves and used to analyze the results further.

4. RESULTS

4.1 SANS Data

The SANS data for Skp alone in 0% and 98% D_2O buffers are shown in Fig. 2. Guinier analysis of the data resulted in R_g values of 34 ± 2 Å in 0% D_2O buffer and 32.9 ± 0.3 Å in 98% D_2O buffer. The Guinier-derived $I(0)$ values of 0.125 ± 0.005 cm^{-1} in 0% D_2O buffer and 0.160 ± 0.002 cm^{-1} in 98% D_2O buffer agree well with the calculated values of 0.14 and 0.15 cm^{-1}, respectively, using Eqs. (4) and (5) along with the M_w calculated from the

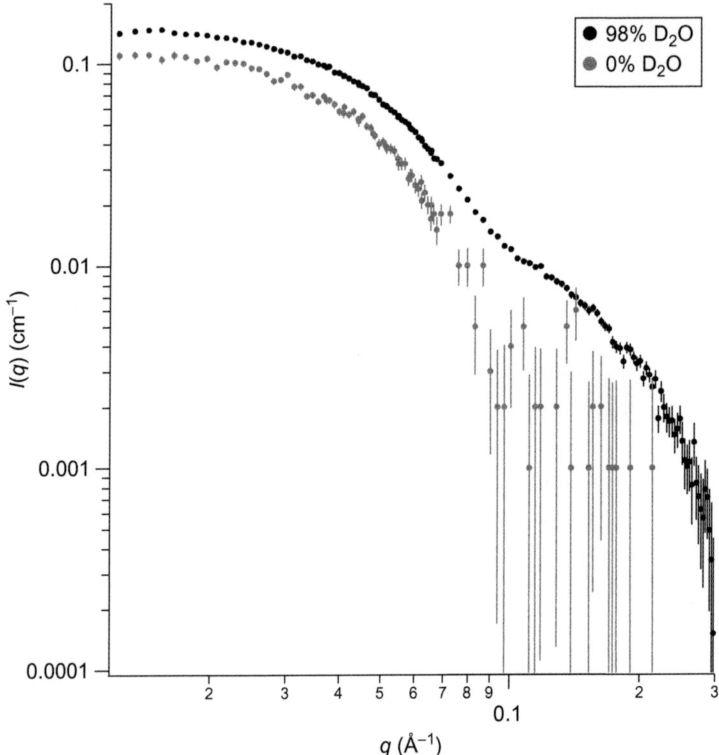

Figure 2 SANS data on an absolute scale for Skp in D_2O and H_2O buffers. Error bars represent the standard error of the mean with respect to the number of pixels used in the data averaging.

Skp sequence. This verifies that Skp is a trimer in solution, in agreement with the crystal structures (Korndörfer et al., 2004; Walton & Sousa, 2004).

Four SANS data sets were collected on both the Skp–OmpW and Skp–OmpA complexes. The Skp–OmpW data were collected in buffers with 0%, 30%, 80%, and 98% D_2O while the Skp–OmpA data were collected in buffers with 0%, 15%, 30%, and 98% D_2O, as shown in Fig. 3, along with their corresponding distance distribution functions, $P(r)$ versus r. The R_g and $I(0)$ values obtained from both Guinier and $P(r)$ analyses are listed in Tables 1 and 2 for Skp–OmpW and Skp–OmpA, respectively.

4.2 Match Point Analysis and Quantification of Omp Deuteration

$\sqrt{I(0)/c}$ versus f_{D_2O} plots using the Guinier-derived values in Tables 1 and 2 are shown for the Skp–OmpW and Skp–OmpA complexes in Fig. 4, along

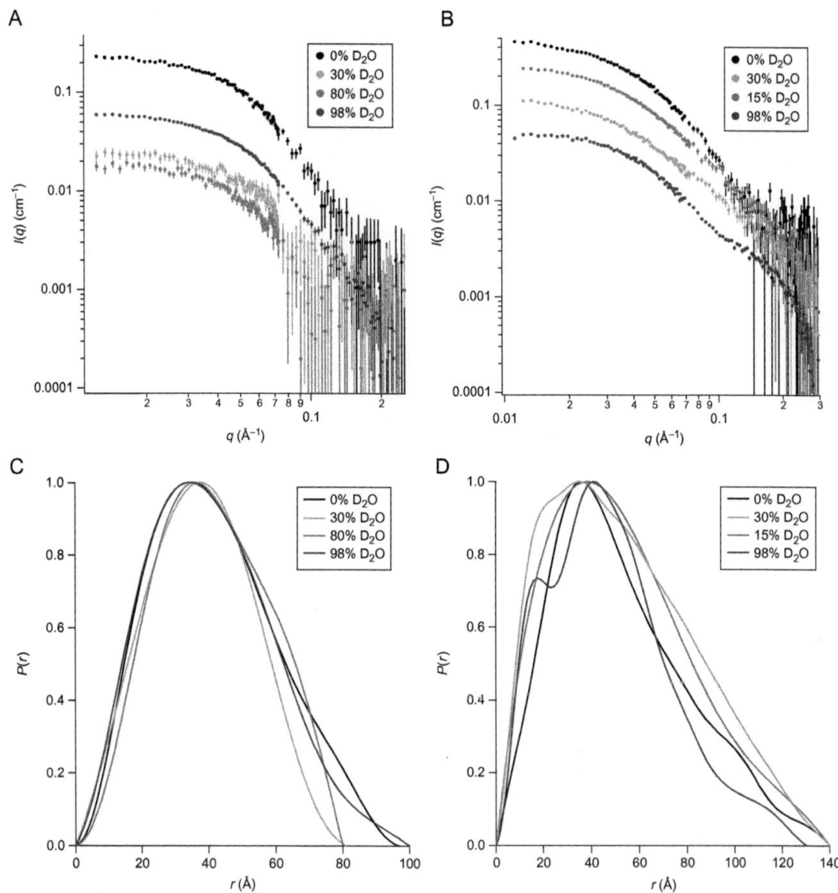

Figure 3 SANS data on an absolute scale for (A) Skp–OmpW and (B) Skp–OmpA. Error bars represent the standard error of the mean with respect to the number of pixels used in the data averaging. Distance distribution functions for (C) Skp–OmpW and (D) Skp–OmpA. The peak values are scaled to 1.0 so that the differences in shape can be easily observed.

with the calculated curves assuming different percentages of deuteration for the uOMP component (30%, 40%, and 50% deuteration for OmpW and 40%, 50%, and 60% deuteration for OmpA). The match point for the Skp–OmpW complex was found to be 51% $D_2O \pm 2\%$ D_2O, while that for the Skp–OmpA complex was found to be 57% $D_2O \pm 2\%$ D_2O.

It can be seen from Fig. 4A that the $\sqrt{I(0)/c}$ values for Skp–OmpW as a function of f_{D_2O} fall closer to the curve for 30% deuteration of OmpW. This value seems low given that OmpW was grown in a medium with a starting

Table 1 Skp–OmpW Parameters

	0% D_2O	30% D_2O	80% D_2O	98% D_2O
Concentration[a] (mg mL^{-1})	3.45	3.8	1.14	1.25
Guinier R_g (Å)	32.7 ± 0.2	31 ± 1	31.5 ± 0.4	31.6 ± 0.2
Guinier $I(0)$ (cm^{-1})	0.223 ± 0.001	0.027 ± 0.001	0.020 ± 0.001	0.063 ± 0.001
R_g from $P(r)$ (Å)	32.9 ± 0.1	29.0 ± 0.2	31.32 ± 0.03	31.3 ± 0.2
$I(0)$ from $P(r)$ (cm^{-1})	0.240 ± 0.001	0.025 ± 0.001	0.020 ± 0.001	0.063 ± 0.001
D_{max} (Å)	105 ± 5	83.5 ± 1.5	89 ± 4	93.5 ± 1.5

[a]Errors on the concentration values are about 5%.
Errors on R_g, $I(0)$, and D_{max} represent the standard error of the mean for three different Guinier or $P(r)$ fits.

Table 2 Skp–OmpA Parameters

	0% D_2O	15% D_2O	30% D_2O	98% D_2O
Concentration[a] (mg mL^{-1})	5.6	5.4	5.3	1.31
Guinier R_g (Å)	39.7 ± 0.1	39.8 ± 0.3	39.1 ± 0.3	36.3 ± 0.7
Guinier $I(0)$ (cm^{-1})	0.462 ± 0.002	0.261 ± 0.004	0.119 ± 0.001	0.056 ± 0.001
R_g from $P(r)$ (Å)	41.1 ± 0.4	42.0 ± 0.8	41.8 ± 0.3	38.1 ± 0.1
$I(0)$ from $P(r)$ (cm^{-1})	0.475 ± 0.003	0.263 ± 0.002	0.120 ± 0.001	0.056 ± 0.001
D_{max} (Å)	135 ± 5	137 ± 7	133 ± 3	127 ± 3

[a]Errors on the concentration values are about 5%.
Errors on R_g, $I(0)$, and D_{max} represent the standard error of the mean for three different Guinier or $P(r)$ fits.

value of 60% D_2O. The data point for 0% D_2O buffer does match that for 50% deuterated OmpW. On the other hand, the $\sqrt{I(0)/c}$ values for Skp–OmpA as a function of f_{D_2O} fall mainly on the curve for 50% deuterated OmpA, which seems more reasonable, given the starting value of 60% D_2O in the growth medium. The data point for 98% D_2O agrees better with the 40% deuterated OmpA curve. It is possible that the concentration estimated by extinction coefficients is higher than the actual concentration of the complex during the SANS measurement, resulting in a $\sqrt{I(0)/c}$ value that is low for this data point. Given the results shown in Fig. 4B, the amount of

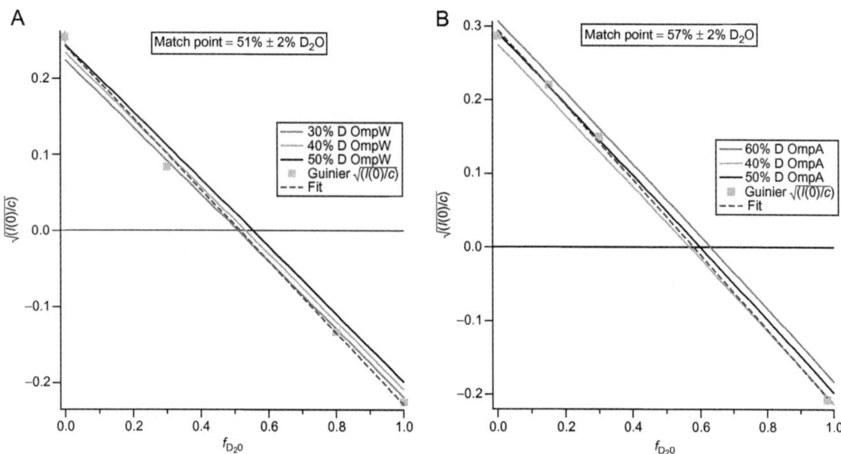

Figure 4 Experimental and calculated $\sqrt{I(0)/c}$ versus f_{D_2O} for (A) Skp–OmpW and (B) Skp–OmpA. Error bars on the experimental values represent the standard error of the mean based on three different Guinier fits.

deuteration for OmpA was estimated at 50% ± 5%. There is no compelling reason why the amount of deuteration should be different in OmpW and OmpA, as they were prepared under the same conditions. Because using either 30% or 50% for the amount of deuteration for OmpW did not change the subsequent Stuhrmann and parallel axis theorem analyses results significantly, 50% ± 5% deuteration was also estimated for OmpW. Given this value for both OmpW and OmpA, the match points of the Skp–OmpW and Skp–OmpA complexes were calculated to be 55% D_2O ± 2% D_2O and 60% D_2O ± 2% D_2O, respectively. Additionally, the match points of the components were determined to be 44% D_2O for Skp and 80% D_2O ± 7% D_2O for OmpW and OmpA.

The results in Fig. 4A and B indicate that the Skp–uOMP complexes were present in all samples analyzed and were formed by three Skp monomers bound to a single uOMP. This stoichiometric ratio is consistent with previous binding data (Moon et al., 2013; Qu et al., 2007). It is important to emphasize that no urea was present, and without Skp, the uOMPs would have precipitated over the course of the SANS data collection.

4.3 Stuhrmann and Parallel Axis Theorem Analyses

Both Stuhrmann and parallel axis theorem analyses were performed on the contrast variation SANS data and the results are listed in Tables 3 and 4,

Table 3 Stuhrmann Analysis for Skp–uOMP Complexes

	Skp–OmpW	Skp–OmpA
α	64 ± 55	242 ± 34
β	41 ± 162	461 ± 66
R_m (Å)	32.3 ± 0.5	39.3 ± 0.2
χ^2	4.75	0.4
R_g (Skp) (Å)	31.5 ± 0.5	33.5 ± 0.8
R_g (Omp) (Å)	32 ± 6	36.1 ± 0.8
D_{CM} (Å)	13 ± 26	38 ± 3

50% deuteration was assumed for OmpW and OmpA. χ^2 is from the fit to Eq. (10).

Table 4 Parallel Axis Theorem Analysis for Skp–uOMP Complexes

	Skp–OmpW	Skp–OmpA
R_g (Skp) (Å)	31.3 ± 0.6	32.7 ± 0.9
R_g (Omp) (Å)	32 ± 6	35.7 ± 0.9
D_{CM} (Å)	14 ± 28	40 ± 3

50% deuteration was assumed for OmpW and OmpA.

respectively. 50% deuteration was assumed for both OmpW and OmpA, with the $\Delta\rho$ values calculated accordingly for each contrast.

R_g changes very little as a function of contrast for the Skp–OmpW complex. Therefore, $R_g{}^2$ versus $\Delta\rho^{-1}$ is nearly flat and the fit to Eq. (10) results in large uncertainties for both α and β. A similar argument can be made for fits to Eq. (11). As a result, these analyses only allow determination of R_g for the Skp component of the complex, while R_g for the OmpW component and the CM distance, D_{CM}, have large uncertainties. Importantly, R_m, the radius of gyration of an equivalent complex of uniform scattering length density, is also well determined from the Stuhrmann analysis (Eq. 10). Thus, three possible R_g (OmpW) versus D_{CM} curves were calculated from Eq. (10) using the well-determined values obtained for R_g (Skp) and R_m (along with the values obtained by adding and subtracting their corresponding errors) and the results are shown in Fig. 5. This information was used to constrain the size of OmpW and the CM distance between Skp and OmpW in the Skp–OmpW model structures as described below.

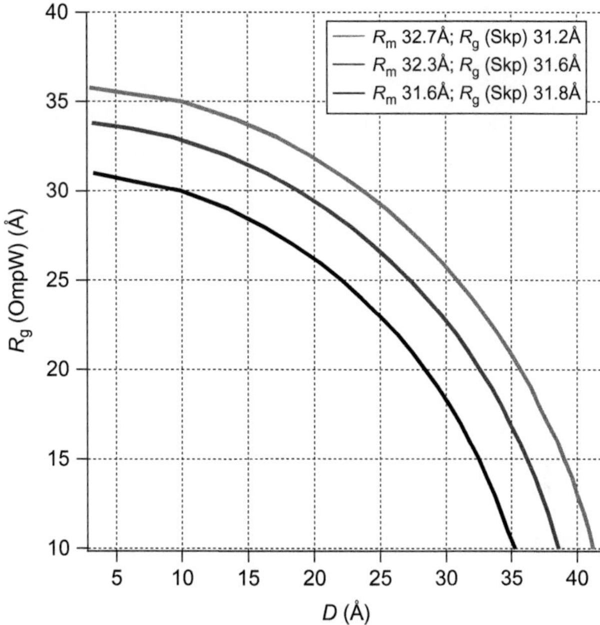

Figure 5 Calculation of possible values for R_g(OmpW) and CM distance, D_{CM}, based on R_m and R_g(Skp) obtained from the Stuhrmann analysis. Top curve: R_m = 32.7Å; Middle curve: R_m = 32.3Å; Bottom curve: R_m = 31.5Å.

Conversely, the change in R_g as a function of contrast is more pronounced for the Skp–OmpA complex. Thus, R_g for both Skp and OmpA, as well as D_{CM}, were well determined in this case and these values were used directly to build the starting model structures for the Skp–OmpA complex.

4.4 Skp Structure in Solution

Figure 6A shows the starting model for Skp based on the X-ray crystal structure as described in Section 3. The calculated R_g based on the atomic coordinates is 30.6 Å. This is about 3 Å smaller than the Guinier-derived R_g. The model structure obtained after the R_g-constrained MD simulation, shown in Fig. 6B, has a calculated R_g of 33.5 Å, in good agreement with the Guinier-derived R_g value. Importantly, the Skp β-strand "body" through which oligomerization occurs did not alter in conformation between the two models. This region merely served as a "hinge" point for the opening of the Skp helical tentacles. This structural pivot was also identified from NMR data collected from the OmpX–Skp complex (Burmann et al., 2013).

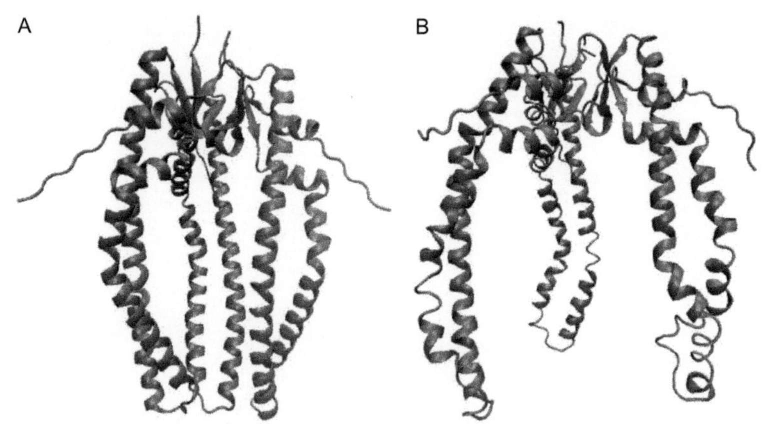

Figure 6 All-atom model structures representing Skp (A) in the crystal ($R_g \approx 30$ Å) and (B) in solution ($R_g \approx 33$ Å).

4.5 Skp–OmpW Structure in Solution

The fact that R_g does not change appreciably with contrast for the Skp–OmpW complex means that the R_g values of the two components are close to each other. Thus, a number of possible values for R_g (OmpW) and CM distance, D_{CM}, in Fig. 5 can be ruled out, given the R_g (Skp) and R_m values from the Stuhrmann analysis in Table 3. Several model structures were tested with Skp structures S1, S2, and S3, with R_g values of 31.2, 32.5, and 31.6 Å, respectively, OmpW represented by e21, e24, and e27 ellipsoids with R_g values of 21, 24, and 27 Å, respectively, and D_{CM} in the range of 25–30 Å. Representative model structures tested are pictured in Fig. 7 and are listed in Table 5, along with the R_g values of the individual components and the CM distance between them.

Table 1 shows that R_g is largest for the Skp–OmpW complex in 0% D_2O buffer and it is about 1 Å smaller in 98% D_2O. A comparison of these results with those in Table 6, which lists R_g as a function of contrast for the model structures, shows that only model S1e27 satisfies this requirement. However, R_g in 80% D_2O is slightly low in this model. This means that the R_g value for Skp is slightly too small, because Skp dominates the scattering at this contrast. Furthermore, the S1 Skp structure is too symmetric, as can be seen in Fig. 8A, which shows the model SANS curves for S1, S2, and S3 Skp. The curve for S1 Skp (shown in the S1e27 complex in Fig. 7) shows a pronounced peak at $q \approx 0.15$ Å^{-1} that is not observed in the data, which is due to the threefold symmetry of the three tentacles. Even the S2 Skp model (shown in the S2e21 and S2e27 complexes in Fig. 7) results in a SANS curve

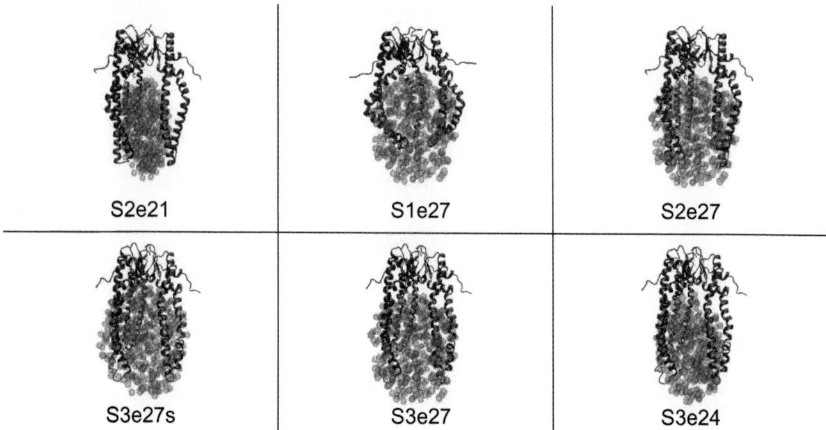

Figure 7 Skp–OmpW model structures tested against the SANS data as described in Section 3. The ball representation of the ellipsoids is for clarity. The ellipsoids were represented by nonoverlapping points for SANS curve calculations, as described in Section 3. (See the color plate.)

Table 5 Skp–OmpW Model Structures

Model	R_g (Skp) (Å)	R_g (OmpW) (Å)	R_g (Complex) (Å)	D_{CM} (Å)
S2e21	32.5	21.2	32.7	30.5
S1e27	31.2	27.6	32.1	28.6
S2e27	32.5	27.6	32.1	28.6
S3e27s	31.6	27.6	31.9	25.0
S3e27	31.6	27.6	32.1	29.2
S3e24	31.6	23.9	32.0	30.9

Values are from the coordinates only. Contrast was not taken into account.

that shows some structure at this q value, which was reduced substantially by performing a 10 ps MD simulation *in vacuo*, as described in Section 3, to create the less symmetric S3 Skp (shown in the remaining complexes in Fig. 7). In addition, the model SANS curve for an ellipsoid with $R_g = 27$ Å features a strong peak at $q \approx 0.18$ Å^{-1}, since ellipsoids are highly symmetric structures. The peak is less pronounced for ellipsoids of smaller R_g, as shown in Fig. 8B.

Given these results, model structures with S1 Skp and e27 ellipsoids do not represent the data well. Model S2e21, consisting of S2 Skp and the

Table 6 Skp–OmpW Model Structure R_g as a Function of Contrast

Model	0% D_2O R_g (Å)	30% D_2O R_g (Å)	80% D_2O R_g (Å)	98% D_2O R_g (Å)
S2e21	31.7	27.8	32.4	32.8
S1e27	32.3	30.9	30.4	31.6
S2e27	33.1	30.1	32.4	33.3
S3e27s	32.4	31.2	31.9	32.4
S3e27	33.3	31.5	31.9	33.0
S3e24	32.4	29.8	31.9	32.6
Guinier	32.7 ± 0.2	31 ± 1	31.5 ± 0.4	31.6 ± 0.2
P(*r*)	32.9 ± 0.1	29.0 ± 0.2	31.32 ± 0.03	31.3 ± 0.2

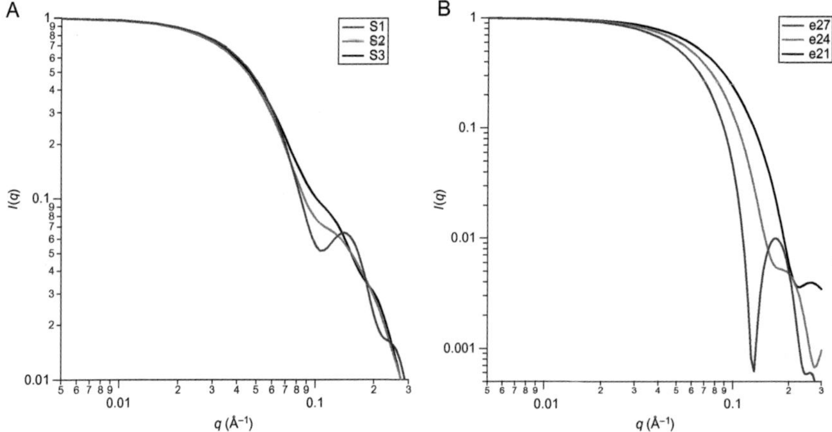

Figure 8 Model SANS curves from (A) the Skp component represented by the S1, S2, and S3 all-atom structures and (B) the OmpW component represented by the e21, e24, and e27 ellipsoids. *I*(0) is arbitrarily scaled to 1.0.

smaller e21 ellipsoid with $R_g = 21$ Å, also is not a good representation of the data; the R_g is larger in 98% D_2O than in 0% D_2O and R_g in 30% D_2O is too small, indicating that the e21 ellipsoid is too small, because OmpW dominates the scattering at this contrast.

On the other hand, a reasonable working model of the structure of Skp–OmpW in solution was obtained using S3 Skp and the e24 ellipsoid with $R_g = 24$ Å, with the two components positioned such that $D_{CM} \approx 30$ Å to satisfy the blue (dark gray in the print version) middle curve in Fig. 5 in

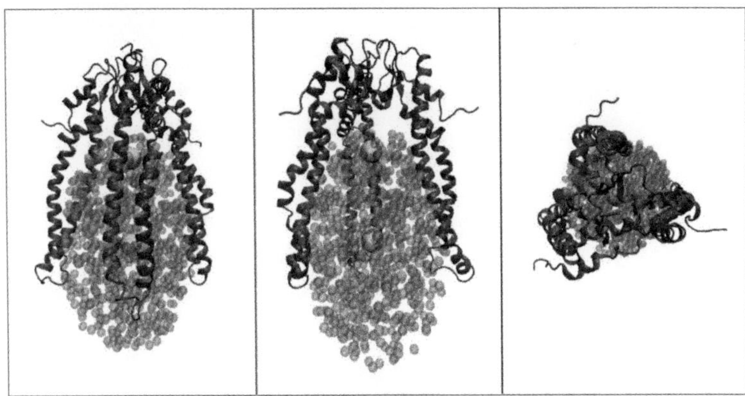

Figure 9 Several views of Skp–OmpW model S3e24, which was the best-fit model for this complex. The larger green balls are at the CM locations of S3 Skp and the e24 ellipsoid. The ball representation of the ellipsoids is for clarity. The ellipsoids were represented by nonoverlapping points for SANS curve calculations, as described in Section 3. (See the color plate.)

which R_g(skp) = 31.6 Å and R_m = 32.3 Å. While R_g in 98% D_2O is not smaller than that in 0% D_2O, the two values are equal and the discrepancy may be due to the CM distance, D_{CM}, being slightly too large. Several views of S3e24 are shown in Fig. 9, which also shows the location of the CM for both the Skp and OmpW components.

Model SANS curves from S3e24 are shown along with the data in Fig. 10. They agree well with the data for all contrasts below $q = 0.1$ Å^{-1}, beyond which the data for 30% D_2O and 80% D_2O become too noisy for comparison. The model SANS curves for 0% D_2O and 30% D_2O show a peak at $q \approx 0.18$ Å^{-1} that is not seen in the data due to the symmetry of the ellipsoid model for OmpW.

4.6 Skp–OmpA Structure in Solution

Starting model structures for Skp–OmpA were constructed based on the working model for Skp–OmpW above, incorporating the e24 ellipsoid to represent the TM domain of OmpA. The ellipsoid was positioned in the same location as for the Skp–OmpW model and also higher in the Skp cavity for comparison. The PP domain of OmpA was tethered near the center of the Skp cavity in slightly different positions as described in Section 3 such that the starting values of R_g (OmpA), including both the TM and PP domains, and D_{CM} approximately agreed with the values obtained in Tables 3 and 4. The starting values of these parameters did not need to

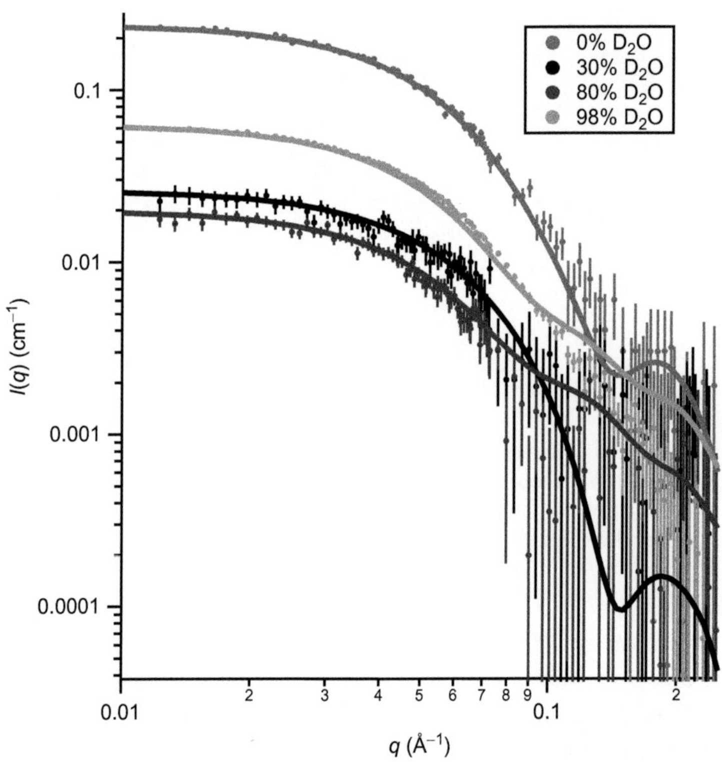

Figure 10 Skp–OmpW SANS data on an absolute scale (points) along with model SANS curves (solid lines) from S3e24. Error bars represent the standard error of the mean with respect to the number of pixels used in the data averaging.

be exact because the PP domain would be repositioned during the SASSIE runs to find the best-fit structures to the data.

Representative starting models for Skp–OmpA are shown in Fig. 11, including an all-atom model, in which the residues at the N-terminus of the PP domain are located high in the Skp cavity. The starting R_g values are shown as a function of contrast in Table 7. Model S3e24PP1 was not pursued further because the R_g values at all contrasts are too small and the disordered residues that tether the PP domain are shorter than in the other models. Thus, structures with larger R_g values were unlikely to be generated during the SASSIE runs. Short SASSIE runs producing a few hundred structures were performed on the rest of the starting model structures. Models S3e24PP2 and S3e24PP3 were found to be subsets of S3e24PP4 and S3e24PP5, respectively. Thus, longer SASSIE runs were only performed on models S3e24PP4 and S3e24PP5.

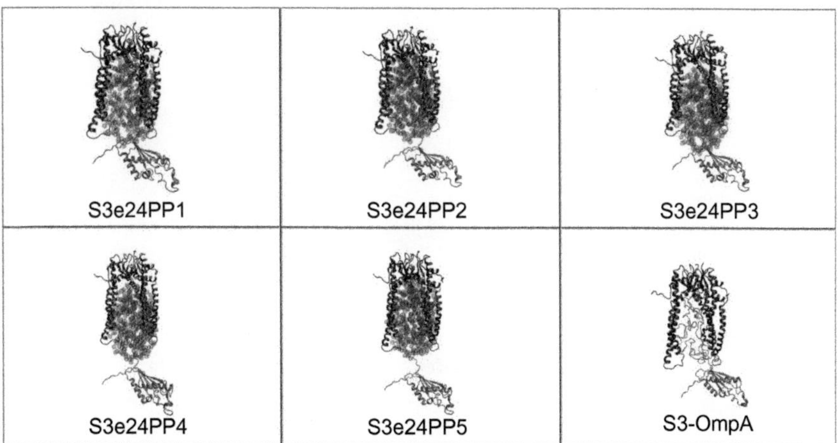

Figure 11 Skp–OmpA starting structure models for SASSIE runs as described in Section 3. The ball representation of the ellipsoids is for clarity. The ellipsoids were represented by nonoverlapping points for SANS curve calculations, as described in Section 3. (See the color plate.)

Table 7 Skp–OmpA Starting Model Structure R_g as a Function of Contrast

Model	**0% D_2O R_g (Å)**	**30% D_2O R_g (Å)**	**98% D_2O R_g (Å)**
S3e24PP1	37.4	35.7	35.9
S3e24PP2	40.3	38.9	37.2
S3e24PP3	40.1	37.6	37.7
S3e24PP4	42.1	40.3	38.3
S3e24PP5	42.3	41.4	37.9
S3-OmpA	39.1	37.5	36.3
Guinier	39.7 ± 0.1	39.1 ± 0.3	36.3 ± 0.7
$P(r)$	41.1 ± 0.4	41.8 ± 0.3	38.1 ± 0.1

SASSIE runs performed on models S3e24PP4, S3e24PP5, and the all-atom model, S3-OmpA, produced 1650 accepted structures for S3e24PP4 and S3e24PP5 and 2144 accepted structures for S3-OmpA. χ^2 versus R_g plots are shown as a function of contrast for models S3e24PP4 and S3e24PP5 in Fig. 12. The larger χ^2 values observed for the model SANS curves in 98% D_2O indicate that the S3 model for Skp does not describe the data as well for Skp–OmpA as it does for Skp–OmpW. Although the match

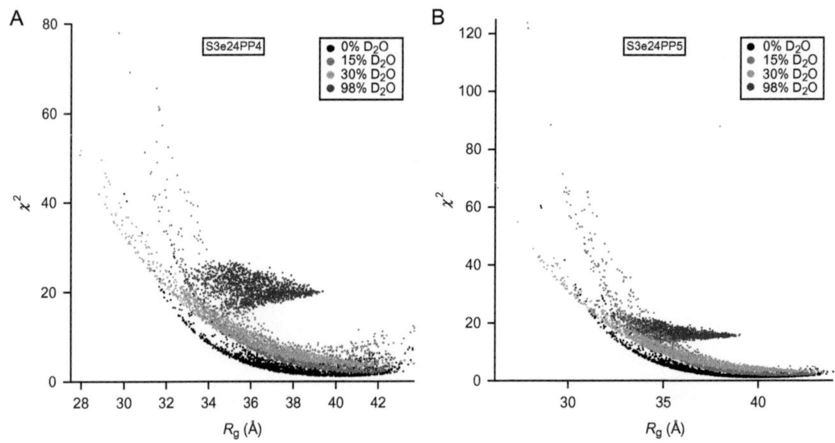

Figure 12 χ^2 versus R_g plots as a function of contrast from SASSIE runs exploring the conformation space of the OmpA PP domain for models (A) S3e24PP4 and (B) S3e24PP5.

point of the uOMPs is around 80% D_2O, the scattering from this component is still relatively weak in 98% D_2O compared to that of the Skp component. Thus, the differences in the 98% D_2O data can be ascribed mainly to the structure of Skp. The 98% D_2O data for Skp–OmpA (Fig. 3B) shows a distinct peak at $q \approx 0.18$ Å^{-1}, whereas the data at the same contrast for Skp–OmpW (Fig. 3A) lacks this peak. Thus, the structure of Skp bound to OmpA is more symmetric with respect to the location and shape of the three tentacles. This leads to the conclusion that a model more like S1 Skp, but with a larger R_g value more like S3 Skp, would result in a better fit of the Skp–OmpA models to the 98% D_2O data.

The χ^2(avg) versus R_g(avg) plots for both the S3e24PP4 and S3e24PP5 model structures are shown in Fig. 13. Better χ^2(avg) values are obtained for R_g(avg) > 34 Å for model S3e24PP5, although there are not as many structures with R_g(avg) > 38 Å, represented by the rectangular region, which denotes the lowest ≈10% of χ^2(avg) values. The χ^2(avg) values in this best-fit region are lower for S3e24PP5, suggesting that the TM domain of OmpA sits higher in the Skp cavity in the Skp–OmpA complex than OmpW in the Skp–OmpW complex.

χ^2 versus R_g plots are shown as a function of contrast for the all-atom Skp–OmpA model structure in Fig. 14. A comparison to Fig. 12 reveals the same high χ^2 values in 98% D_2O, as expected because the S3 model for Skp was used in both cases. On the other hand, the χ^2 values overall

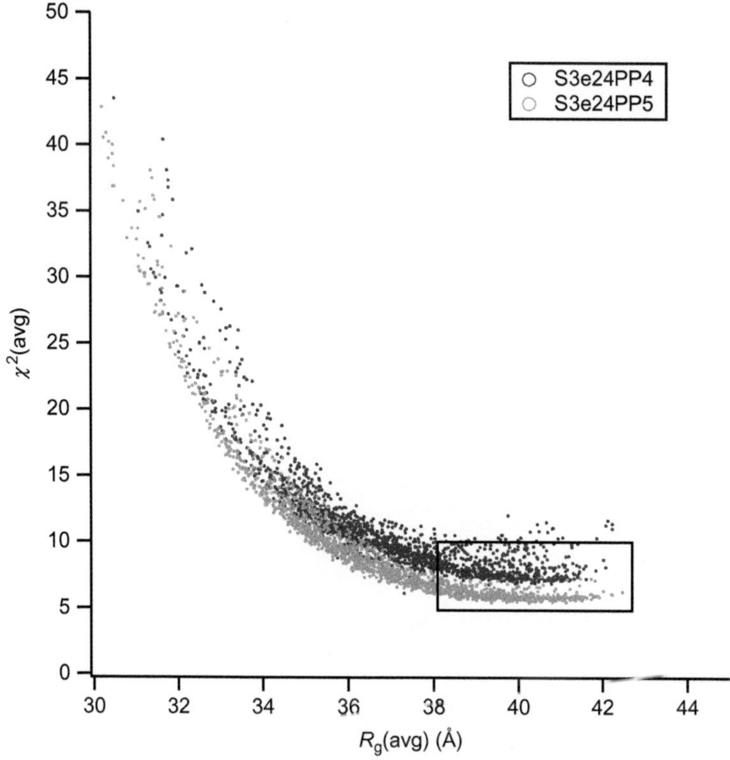

Figure 13 χ^2(avg) versus R_g(avg) plots from the plots in Fig. 12 for models S3e24PP4 (dark gray in the print version) and S3e24PP5 (light gray in the print version).

are smaller at all contrasts compared to the S3e24PP5 model. The same is true for the χ^2(avg) versus R_g(avg) plot for Skp–OmpA in Fig. 15, which shows lower χ^2(avg) values overall in the best-fit region where R_g(avg) is between 37 and 41 Å. While this model shows only one of many possible configurations for the TM domain of OmpA, it does provide an example of an alternate tethering strategy, and thus a different region of conformation space, for the PP domain of OmpA than is represented by S3e24PP4 or S3e24PP5.

Plots of the data at each contrast compared to the calculated S3e24PP5 SANS curves for the best- and worst-fit structures, as well as the average SANS curve for the entire ensemble, are shown in Fig. 16. As explained earlier, the discrepancy between the calculated SANS curves and the SANS data in 98% D_2O is due to the use of the S3 model for Skp for both Skp–OmpW and Skp–OmpA. The portion of the Skp–OmpA data near

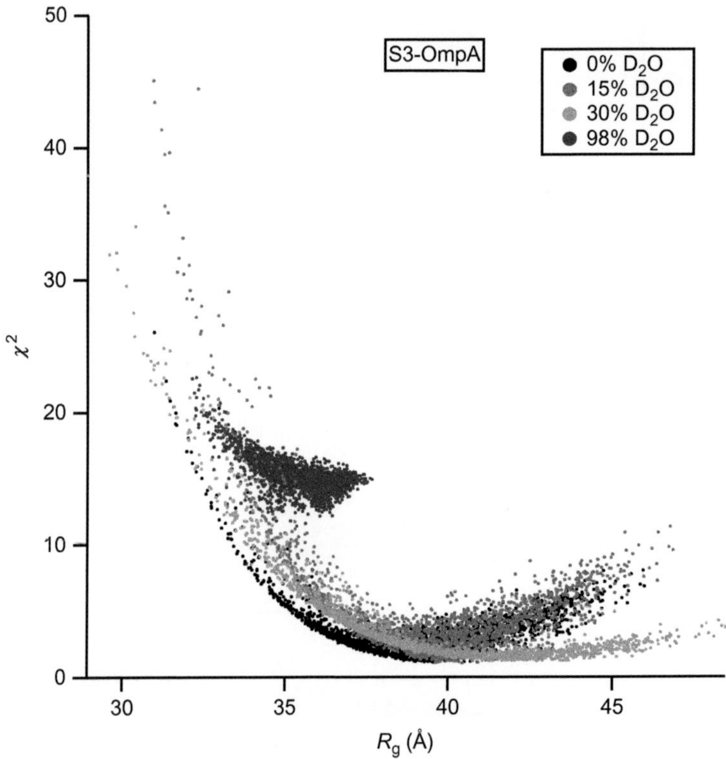

Figure 14 χ^2 versus R_g plots as a function of contrast from SASSIE runs exploring the conformation space of the OmpA PP domain for model S3-OmpA.

$q \approx 0.18$ Å^{-1} is not well matched since the peak observed in this region does not exist in the Skp–OmpW data. Additional MD simulations were not performed on Skp to try to find Skp–OmpA models that better fit the 98% D_2O data, since this would have only resulted in a shift in χ^2(avg) by a constant value in Figs. 13 and 15.

Figure 16 shows that the average SANS curve for the entire ensemble of S3e24PP5 structures fits the data almost as well as that for the best-fit single structure in each case, indicating that the entire generated ensemble of structures is also a reasonable fit to the SANS data. Similar plots are shown for the calculated S3-OmpA SANS curves compared to the data at each contrast in Fig. 17. Again, the average curve for the entire ensemble fits the data almost as well as that for the best-fit single structure in each case.

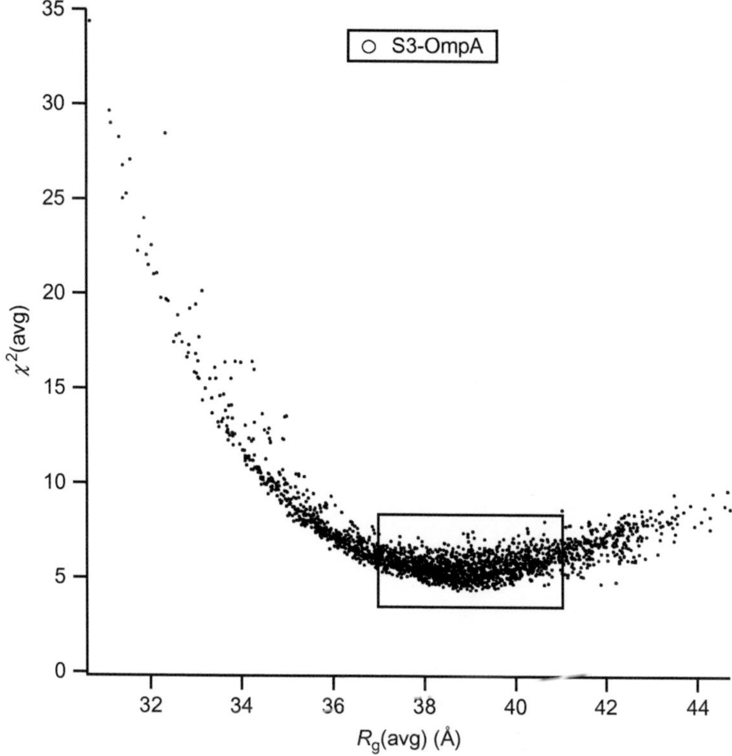

Figure 15 χ^2(avg) versus R_g(avg) plots from Fig. 14 for S3-OmpA.

The calculated SANS curves from the global best-fit individual structure are shown in Fig. 18 at each contrast for S3e24PP5 and in Fig. 19 for S3-OmpA, along with the corresponding global best- and worst-fit structures for comparison. The resultant R_g values as a function of contrast for the global best-fit individual structures are listed in Table 8 and the R_g values for the Skp and OmpA components, as well as the CM distance, D_{CM}, for the global best-fit individual structures are listed in Table 9. In agreement with the χ^2(avg) versus R_g(avg) curves, the parameters obtained from S3e24PP5 are in better agreement with the data than those obtained from S3e24PP4, although the D_{CM} value is still a little large compared to those obtained from the data. The parameters obtained from S3-OmpA are in very good agreement with those obtained from the data.

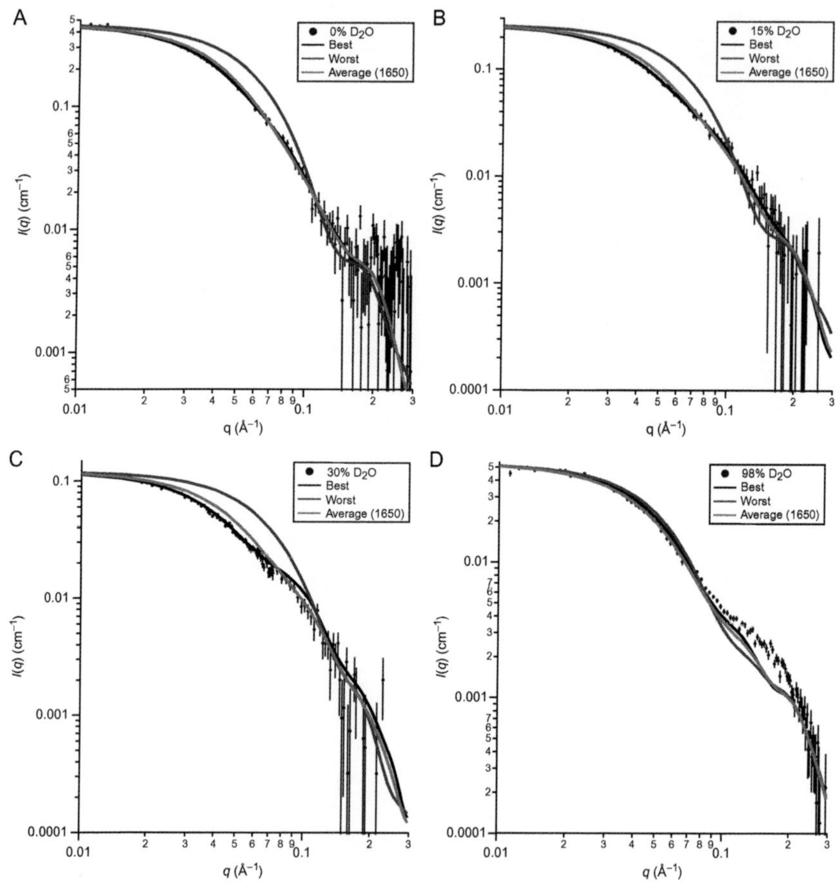

Figure 16 Model SANS data calculated from the best- and worst-fit single S3e24PP5 structures, as well as the average SANS curve calculated from the entire S3e24PP5 ensemble, along with the SANS data on an absolute scale for Skp–OmpA in (A) 0% D_2O, (B) 15% D_2O, (C) 30% D_2O, and (D) 98% D_2O. Error bars represent the standard error of the mean with respect to the number of pixels used in the data averaging.

Surface density plots for S3e24PP5 and S3-OmpA showing the configuration space covered by the PP domain for the ensemble (gray) and the best-fit ensemble (green) are presented in Fig. 20. Skp is shown in black and the TM domain of OmpA is in red. These plots show that the PP domain can assume multiple positions and still produce structures that fit the data well. The PP domain in S3e24PP5 was tethered such that it could

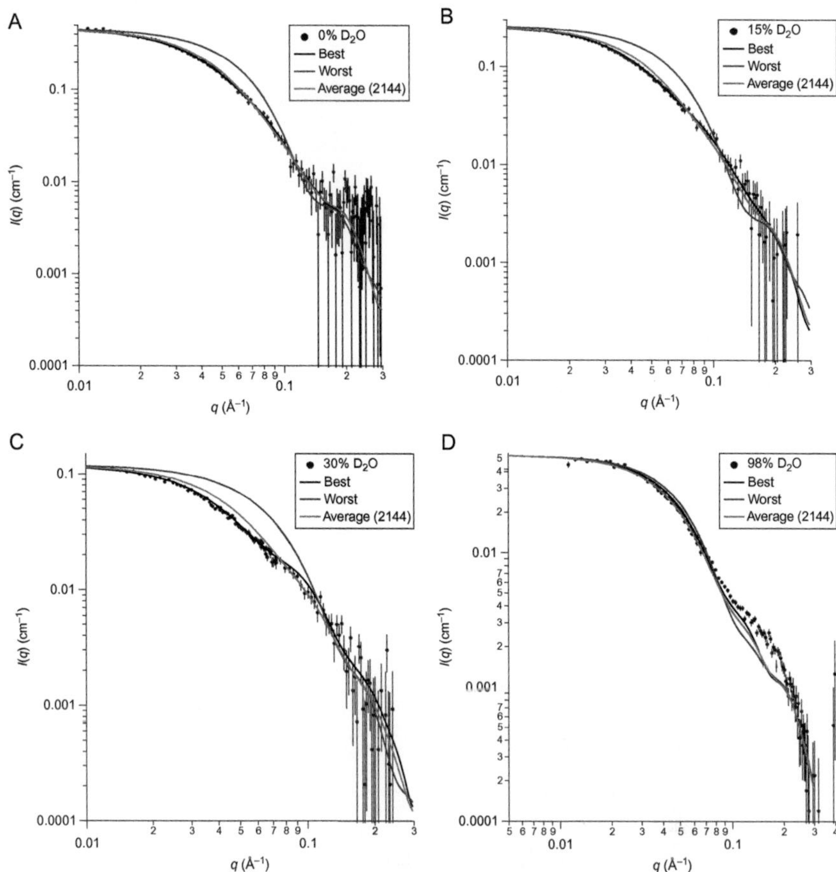

Figure 17 Model SANS data calculated from the best- and worst-fit single S3-OmpA structures, as well as the average SANS curve calculated from the entire S3-OmpA ensemble, along with the SANS data on an absolute scale for Skp–OmpA in (A) 0% D_2O, (B) 15% D_2O, (C) 30% D_2O, and (D) 98% D_2O. Error bars represent the standard error of the mean with respect to the number of pixels used in the data averaging.

occupy all positions shown in gray and those structures that fall in the rectangle in the χ^2(avg) versus R_g(avg) plot (Fig. 13) can take the positions shown in green. The gray areas that do not overlap with the green areas represent the conformations that do not fall in the rectangle. The PP domain was tethered higher within the Skp cavity in S3-OmpA, so a different part of conformation space was explored using SASSIE. But, the result is similar

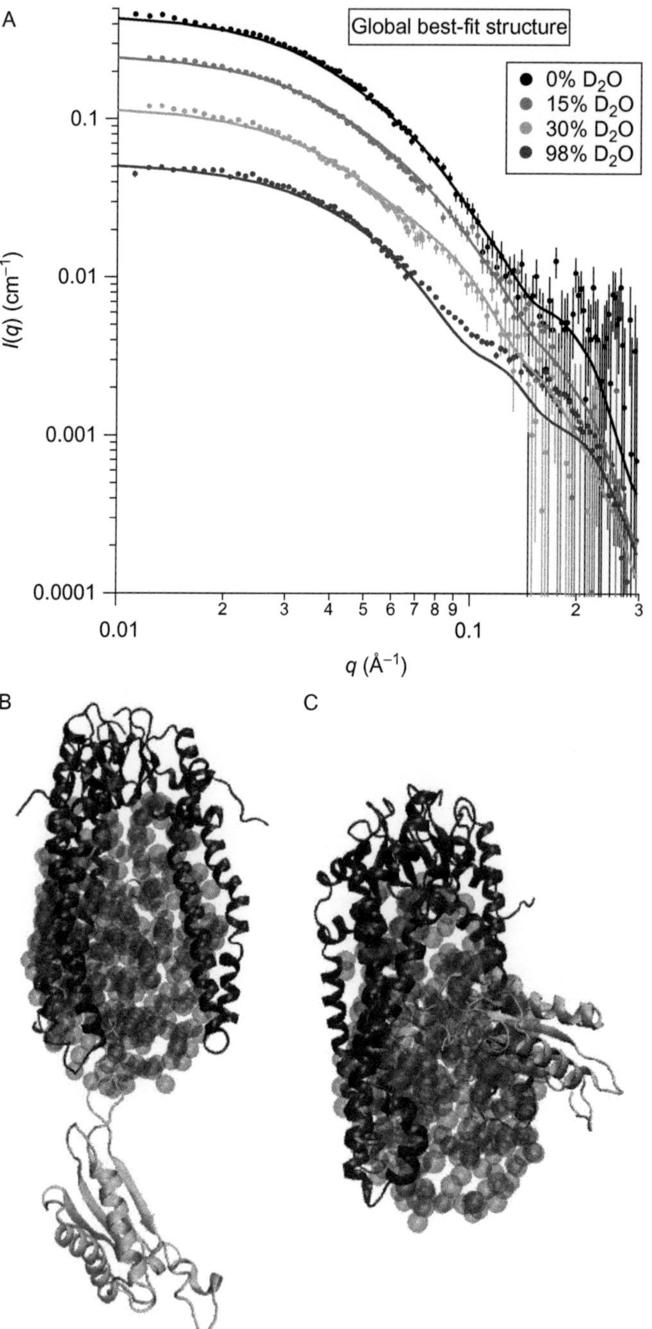

Figure 18 (A) Skp–OmpA SANS data on an absolute scale along with model SANS curves from the global single best-fit S3e24PP5 structure. Error bars represent the standard error of the mean with respect to the number of pixels used in the data averaging. The (B) best-fit and (C) worst-fit model structures are also shown for comparison. The ball representation of the ellipsoids is for clarity. The ellipsoids were represented by nonoverlapping points for SANS curve calculations, as described in Section 3. (See the color plate.)

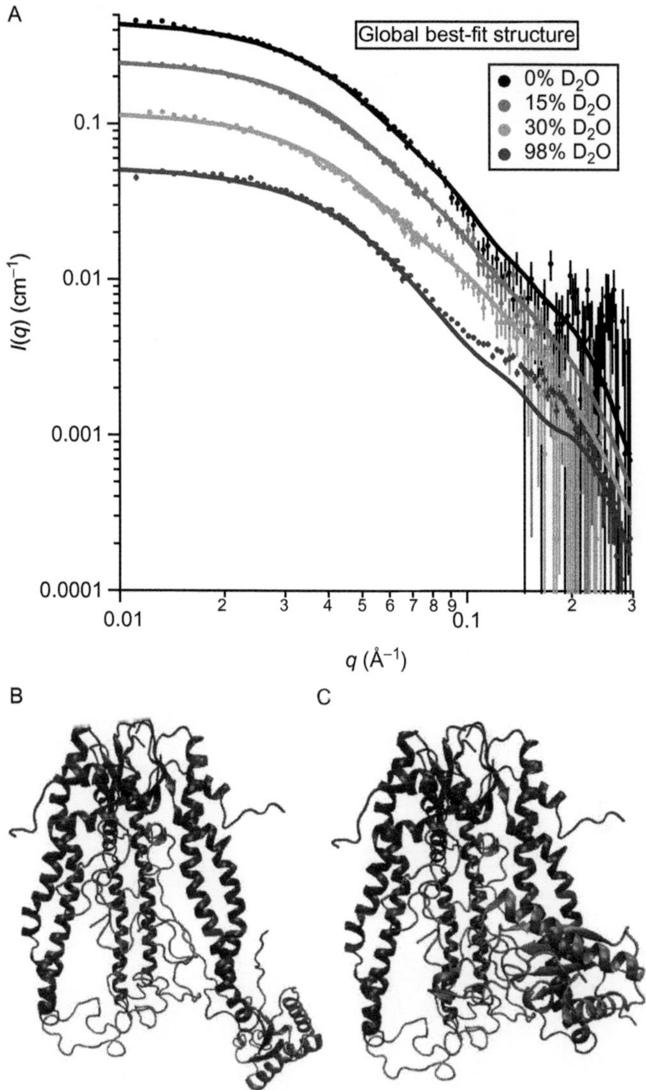

Figure 19 (A) Skp–OmpA SANS data on an absolute scale, along with model SANS curves from the global single best-fit S3-OmpA structure. Error bars represent the standard error of the mean with respect to the number of pixels used in the data averaging. The (B) best-fit and (C) worst-fit model structures are also shown for comparison. (See the color plate.)

to that of S3e24PP5 in that there is only a small region of conformation space (the gray that is not overlapping with the green) that represents the conformations that do not fall in the rectangle in the χ^2(avg) versus R_g(avg) plot (Fig. 15). Both surface density plots reveal that there is a portion

Table 8 Skp–OmpA Global Best-Fit Structure R_g as a Function of Contrast

SASSIE Run[a]	Structure Number[b]	0% D_2O R_g (Å)	15% D_2O R_g (Å)	30% D_2O R_g (Å)	98% D_2O R_g (Å)
S3e24PP4 (39.9)	442 (1650)	41.4	40.8	39.2	38.3
S3e24PP5 (40.8)	740 (1650)	42.2	41.8	41.1	38.2
S3-OmpA (38.7)	1188 (2144)	39.9	39.7	38.9	36.4
Guinier		39.7 ± 0.1	39.8 ± 0.3	39.1 ± 0.3	36.3 ± 0.7
P(*r*)		41.1 ± 0.4	42.0 ± 0.8	41.8 ± 0.3	38.1 ± 0.1

[a]Average R_g (Å) for the global best-fit structure is in parenthesis.
[b]Total number of structures is in parenthesis.

Table 9 Component R_g values for Skp–OmpA Global Best-Fit Structure

SASSIE Run	Structure Number	R_g (Skp) (Å)	R_g (OmpA) (Å)	D_{CM} (Å)
S3e24PP4	442	32.0	33.8	51.3
S3e24PP5	740	32.0	37.8	46.6
S3-OmpA	1188	31.7	36.6	40.5
Stuhrmann		33.5 ± 0.8	36.1 ± 0.8	38 ± 3
Parallel axis		32.7 ± 0.9	35.7 ± 0.8	40 ± 3

Values are from the coordinates only. Contrast was not taken into account.

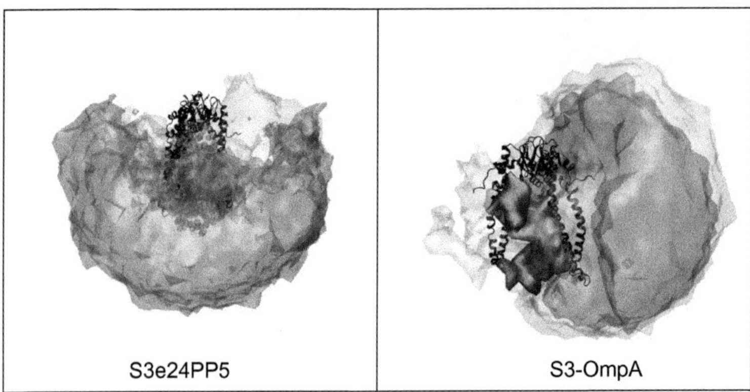

Figure 20 Structure density plots representing all of conformation space (gray) and the best-fit conformation space within the rectangles in Figs. 13 and 15 (green) explored by the OmpA PP domain using the S3e24PP5 and S3-OmpA models. The OmpA TM domain is shown in red and Skp is shown in black. (See the color plate.)

of conformation space around Skp that is not sampled by the PP domain of OmpA.

5. DISCUSSION

5.1 Skp Alone in Solution

The SANS data for Skp alone in solution are consistent with a structure similar to that of the Skp X-ray crystal structure (Walton et al., 2009), but with the helical tentacles more open resulting in a slightly larger R_g value than that for the crystal structure (Fig. 6). This splaying of Skp is not unexpected because the tips of the helical tentacles are positively charged and flexible, based on the fact that two of them are missing in the X-ray crystal structure (PDB ID: 1U2M). Also, recent NMR data from Skp–uOMP complexes demonstrated that residues 89–93, higher up the Skp tentacles closer to the body, display increased helicity on substrate binding and could therefore be at the pivot position for the splaying of the tentacles (Burmann et al., 2013). A crystal structure of the Tim 9–10 complex (PDB ID: 2BSK) also lacks density for the tentacles, possibly due to inherent flexibility (Beverly, Sawaya, Schmid, & Koehler, 2008; Webb, Gorman, Lazarou, Ryan, & Gulbis, 2006).

5.2 Skp–uOMP Structure and Binding Mechanism

Comparison of the SANS data for Skp alone in 98% D_2O and for the Skp–OmpA complex in 98% D_2O confirm that the conformation of Skp when bound to OmpA is similar to that when it is alone in solution, although it has a smaller R_g when bound to OmpA. When bound to OmpW, Skp adopts a conformation with a less symmetric configuration of the three tentacles and with an even smaller R_g. However, the R_g value for Skp in both complexes remains larger than that observed for the crystal structure. These data suggest that, in the absence of substrate, the Skp tentacles are in an open conformation (Fig. 6B); on binding an uOMP, Skp collapses in a clamp-like mechanism towards the more closed conformation represented by the crystal structure (Fig. 6A). This flexibility of the Skp tentacles may allow for different conformations of the Skp α-helices to accommodate different uOMPs with TM domains larger or smaller than those of OmpW and OmpA.

The SANS data revealed that the OmpA and OmpW TM domains are unfolded, but constrained in size, when bound to Skp. The R_g value of the TM region ($R_g \approx 24$ Å) is significantly larger in comparison to that calculated from the crystal structure of the folded TM domain ($R_g \approx 14$ Å) (Hong et al., 2006; Pautsch & Schulz, 1998). A similar radius (21 Å) was measured by NMR for paramagnetic spin-labeled OmpX that was bound to Skp (Burmann et al., 2013), in which case the volume occupied by OmpX was found to contain ≈50% water. Like OmpW, OmpX (16.5 kDa) folds into an eight-stranded beta-barrel integral membrane protein (Vogt & Schulz, 1999). Significantly, if these TM domains adopted a random coil state, as when fully chemically denatured, the theoretical R_g would be much larger (Fleming & Rose, 2005). For example, the radius of OmpX in 8 *M* urea was determined by NMR to be 45 Å (Burmann & Hiller, 2012). Additionally, an R_g of the unfolded TM domain of OmpA in solution under folding conditions can be estimated from the sedimentation coefficient to be 32 Å (Danoff & Fleming, 2011). This value is still considerably larger than the value determined by SANS, which means that Skp sequesters the unfolded TM domains of the uOMPs in a conformation that is smaller than they would naturally adopt in solution.

The SANS analysis of Skp–OmpA and Skp–OmpW complexes suggests a model for the binding of the uOMP TM domains near the tips of the Skp tentacles (S3e24 in Figs. 7 and 9; S3e24PP5 and S3-OmpA in Fig. 11). This region of Skp, which is also involved in the initial capture of OmpA (Lyu, Shao, Gao, & Zhao, 2012), is highly positively charged due to a large number of arginine residues. Conversely, the uOMP TM domains are typified by their overall negative charge with a p*I* ranging between 4 and 7 (Jarchow et al., 2008). The Skp–uOMP complex may consequently be stabilized by continued electrostatic interactions between the tentacle tips and the uOMP. Such a notion is supported by fluorescence studies that show weakening of observed binding in the presence of high salt (Qu et al., 2007).

A similar arrangement of the TM domain of OmpX was found from TROSY NMR data collected from a SurA–OmpX complex. The results support the notion that the OmpX TM domain has the same preferred conformational ensemble either in complex with Skp or with SurA (Burmann & Hiller, 2012). The similarity between uOMP spectra may be explained by the small surface area by which both these chaperones interact with uOMPs. In particular, SurA can bind short peptides (7–14 residues long) with a specific tripeptide motif (Bitto & McKay, 2003). Additionally, amino acid deletion experiments on the small Tim complexes (Vergnolle et al., 2005) and prefoldin (Siegert, Leroux, Scheufler, Hartl, & Moarefi, 2000) show that the tips of their tentacles are required for substrate binding. The SANS-derived structures show the OmpW and OmpA TM domains are bound to Skp in a position analogous to that of actin bound to prefoldin (Lundin et al., 2004). However, only the tips of the prefoldin tentacles interact with the unfolded actin and no electron density can be observed in the prefoldin cavity.

The model structures for Skp–OmpW and Skp–OmpA fit the SANS data best when the TM domain of OmpA is centered farther from the tips of the Skp tentacles, and thus closer to the body of Skp, than that of OmpW. In both cases, the closed conformation of Skp and the size and relative positions of the TM domains imply that a portion of the uOMP substrate is protruding outside the Skp cavity. This observation is consistent with interaction data from NMR analysis of Skp–uOMP complexes (Burmann et al., 2013; Callon et al., 2014; Walton et al., 2009), which demonstrated that binding is focused on the Skp tentacles, but there is no single preferred conformation of uOMP in this region of Skp. Because NMR only maps the presence of interactions, any substrate residues that extend away from the Skp cavity would be spectroscopically silent. Thus, the SANS-derived structure models implying that individual strands of uOMP migrate out of the Skp cavity are consistent with the NMR results. The combination of the inherent flexibility in the Skp tentacles and the lack of a specific structure for the bound uOMPs apparently insure that the uOMPs are sufficiently shielded from self-aggregation until the outer membrane is reached even if portions of the unfolded region protrude outside of the Skp cavity.

The best-fit structures to the Skp–OmpA SANS contrast variation data set using models with both ellipsoidal and all-atom representations of the TM region support a model for an elongated Skp–OmpA complex with a folded PP domain of OmpA protruding out near the bottom of the Skp tentacles. However, the ensembles of structures that represent the lowest $\approx$10% of χ^2(avg) values include those in which the PP domain of OmpA exits out to the side of Skp, and its location in any individual structure is not necessarily fixed. Furthermore, the calculated average SANS curve for the entire generated ensemble of structures fits the data almost as well as that from the best-fit structure. Thus, the solution can consist of a similar ensemble of structures in which the PP domain of OmpA assumes a number of different positions outside the Skp cavity and still be consistent with the SANS data.

5.3 Release Mechanism

The SANS data show that the Skp tentacles change conformation to accommodate different uOMP TM domains using a simple clamp-like mechanism used by jellyfish-like chaperones (Stirling et al., 2006). However, the fact that a portion of the uOMP polypeptide resides outside the Skp cavity implies that there is no requirement for a large-scale rearrangement of Skp for uOMP to be exchanged with an empty Skp trimer (Burmann et al., 2013) and other molecular chaperones (Schwalm et al., 2013; Sklar et al., 2007), or for uOMP to be presented to the Bam complex or to the bacterial outer membrane. While the PP domain of OmpA can exit from openings at either the base or the sides of the tentacles, there is an ample fraction of conformation space not occupied by the PP domain for any individual Skp–OmpA structure. Thus, the SANS data do not rule out the hypothesis that OmpA would exit the chaperone from its side. This is consistent with the capability of Skp to directly interact with membranes (De Cock et al., 1999), as well the idea that Skp may facilitate the folding of OmpA into negatively charged lipids or lipopolysaccharides (Bulieris et al., 2003; Patel, Behrens-Kneip, Holst, & Kleinschmidt, 2009). Even if the presence of the negatively charged uOMP counteracts the positive charges at the base of the Skp tentacles, a large section of the chaperone would still be available to bind lipids. In particular, in the SANS model of the Skp–OmpA complex, the three symmetrically related, highly conserved regions of Skp that are centered on residues E29, K77, and R88 (Burmann, Holdbrook, Callon, Bond, & Hiller, 2015) in the middle of each tentacle (Walton & Sousa, 2004) would remain accessible, and only one binding site would be lost during those times when the PP domain occupied that portion of conformation space.

6. CONCLUSIONS

Deuterium labeling of the uOMP component in Skp–OmpW and Skp–OmpA complexes coupled with SANS and contrast variation has enabled the measurement of individual R_g values of Skp and uOMP when in complex with each other. This experimental strategy also provided information on the distances between their centers of mass. These unique structural properties cannot be obtained using unlabeled complexes or other methodologies. Simultaneous analysis of the data allowed a determination of structure models consistent with the entire contrast variation data set. In conjunction with independent studies using other techniques, the SANS data were used to postulate how uOMPs are captured by Skp and subsequently released to the outer membrane.

Using the R_g information for Skp and the contrast variation data to constrain the model structures, it was determined that Skp can undergo conformational rearrangement to accommodate its client: Skp alone in solution is larger than Skp bound to OmpA, which is larger than Skp bound to OmpW, which is larger than Skp in protein crystals. Furthermore, the Skp tentacles are less ordered with respect to each other in the Skp–OmpW complex than in the Skp–OmpA complex. These findings would not have been revealed without deuterium labeling that allowed the separation of the Skp structure as it exists within the Skp–uOMP complexes. Skp dominated the scattering profile in 98% D_2O buffer, where differences were clearly observed between the Skp–OmpW data and the Skp–OmpA data. These differences were not readily observed in the 0% D_2O data, for instance, where Skp does not contribute as much to the scattering. This result suggests that the Skp tentacles have the ability to adjust to accommodate different uOMPs. Accordingly, because the uOMPs were labeled and distinguishable from the Skp component, it was possible to determine that the OmpW TM domain likely sits lower in the Skp cavity than the OmpA TM domain.

Finally, the resultant structure density plots mapping the conformation space of the OmpA PP domain revealed that it could exit from openings at either the base or the sides of the Skp tentacles. However, there is an ample fraction of conformation space not occupied by the PP domain for any individual Skp–OmpA structure. Given the flexibility of the Skp tentacles and the fact that a portion of the TM domain of uOMP resides outside the Skp cavity, it can be postulated that a low-energy path will always exist such that the TM domains of both OmpW and OmpA can be delivered to the outer membrane or other chaperones. Specifically, the SANS data do not rule out the premise that the TM domain can exit from the side of Skp, even when a PP domain is present as in the case of OmpA.

ACKNOWLEDGMENTS

This work utilized facilities supported in part by the National Science Foundation under Agreements DMR-0944772 and MCB-1412108. The National Institute of Standards and Technology, U.S. Department of Commerce provided the Biomolecular Labeling Laboratory facilities used in this work. This work was supported by grants from the National Institutes of Health: R01-079440 (KGF) and T32-GM008403. This work benefitted from CCP-SAS software developed through a joint EPSRC (EP/K039121/1) and NSF (CHE-1265821) grant.

REFERENCES

Berrow, N. S., Alderton, D., Sainsbury, S., Nettleship, J., Assenberg, R., Rahman, N., et al. (2007). A versatile ligation-independent cloning method suitable for high-throughput expression screening applications. *Nucleic Acids Research*, *35*, e45.

Beverly, K. N., Sawaya, M. R., Schmid, E., & Koehler, C. M. (2008). The Tim8–Tim13 complex has multiple substrate binding sites and binds cooperatively to Tim23. *Journal of Molecular Biology*, *382*, 1144–1156.

Bitto, E., & McKay, D. B. (2003). The periplasmic molecular chaperone protein SurA binds a peptide motif that is characteristic of integral outer membrane proteins. *The Journal of Biological Chemistry*, *278*, 49316–49322.

Bulieris, P. V., Behrens, S., Holst, O., & Kleinschmidt, J. H. (2003). Folding and insertion of the outer membrane protein OmpA is assisted by the chaperone Skp and by lipopolysaccharide. *The Journal of Biological Chemistry*, *278*, 9092–9099.

Burgess, N. K., Dao, T. P., Stanley, A. M., & Fleming, K. G. (2008). Beta-barrel proteins that reside in the *Escherichia coli* outer membrane in vivo demonstrate varied folding behavior in vitro. *The Journal of Biological Chemistry*, *283*, 26748–26758.

Burmann, B. M., & Hiller, S. (2012). Solution NMR studies of membrane-protein-chaperone complexes. *Chimia*, *66*, 759–763.

Burmann, B. M., Holdbrook, D. A., Callon, M., Bond, P. J., & Hiller, S. (2015). Revisiting the interaction between the chaperone Skp and lipopolysaccharide. *Biophysical Journal*, *108*, 1516–1526.

Burmann, B. M., Wang, C., & Hiller, S. (2013). Conformation and dynamics of the periplasmic membrane-protein-chaperone complexes OmpX-Skp and tOmpA-Skp. *Nature Structural & Molecular Biology*, *20*, 1265–1272.

Callon, M., Burmann, B. M., & Hiller, S. (2014). Structural mapping of a chaperone-substrate interaction surface. *Angewandte Chemie (International Ed in English)*, *53*, 5069–5072.

Chen, R., & Henning, U. (1996). A periplasmic protein (Skp) of *Escherichia coli* selectively binds a class of outer membrane proteins. *Molecular Microbiology*, *19*, 1287–1294.

Chen, J., Im, W., & Brooks, C. L. (2005). Application of torsion angle molecular dynamics for efficient sampling of protein conformations. *Journal of Computational Chemistry*, *26*, 1565–1578.

Clark, N. J., Zhang, H., Krueger, S., Lee, H. J., Ketchem, R. R., Kerwin, B., et al. (2013). Small-angle neutron scattering study of a monoclonal antibody using free-energy constraints. *The Journal of Physical Chemistry B*, *117*, 14029–14038.

Curtis, J. E., Raghunandan, S., Nanda, H., & Krueger, S. (2012). SASSIE: A program to study intrinsically disordered biological molecules and macromolecular ensembles using experimental scattering restraints. *Computer Physics Communications*, *183*, 382–389.

Danoff, E. J., & Fleming, K. G. (2011). The soluble, periplasmic domain of OmpA folds as an independent unit and displays chaperone activity by reducing the self-association propensity of the unfolded OmpA transmembrane β-barrel. *Biophysical Chemistry*, *159*, 194–204.

De Cock, H., Schäfer, U., Potgeter, M., Demel, R., Müller, M., & Tommassen, J. (1999). Affinity of the periplasmic chaperone Skp of *Escherichia coli* for phospholipids, lipopolysaccharides and non-native outer membrane proteins. Role of Skp in the biogenesis of outer membrane protein. *European Journal of Biochemistry/FEBS*, *259*, 96–103.

Denoncin, K., Schwalm, J., Vertommen, D., Silhavy, T. J., & Collet, J.-F. (2012). Dissecting the *Escherichia coli* periplasmic chaperone network using differential proteomics. *Proteomics*, *12*, 1391–1401.

Engelman, D., & Moore, P. B. (1975). Determination of quaternary structure by small-angle neutron scattering. *Quarterly Reviews of Biophysics*, *4*, 219–241.

Entzminger, K. C., Chang, C., Myhre, R. O., McCallum, K. C., & Maynard, J. A. (2012). The Skp chaperone helps fold soluble proteins in vitro by inhibiting aggregation. *Biochemistry*, *51*, 4822–4834.

Fleming, P. J., & Rose, G. D. (2005). Conformational properties of unfolded proteins. In J. Buckner & T. Kiefhaber (Eds.), *Protein folding handbook: vol. 2* (pp. 710–736). Weinheim, Germany: Wiley-VCH.

Glatter, O. (1977). A new method for the evaluation of small-angle scattering data. *Journal of Applied Crystallography*, *10*, 415–421.

Glatter, O., & Kratky, O. (1982). *Small-angle X-ray scattering*. New York: Academic Press.

Glinka, C. J., Barker, J. G., Hammouda, B., Krueger, S., Moyer, J. J., & Orts, W. J. (1998). The 30 m small-angle neutron scattering instruments at the national institute of standards and technology. *Journal of Applied Crystallography*, *31*, 430–445.

Grizot, S., & Buchanan, S. K. (2004). Structure of the OmpA-like domain of RmpM from *Neisseria meningitidis*. *Molecular Microbiology*, *51*, 1027–1037.

Guinier, A., & Fournet, G. (1955). *Small-angle scattering of X-rays*. New York: John Wiley & Sons, Inc.

Hansen, S. (1990). Calculation of small-angle scattering profiles using Monte Carlo simulation. *Journal of Applied Crystallography*, *23*, 344–346.

Harms, N., Koningstein, G., Dontje, W., Muller, M., Oudega, B., Luirink, J., et al. (2001). The early interaction of the outer membrane protein PhoE with the periplasmic chaperone Skp occurs at the cytoplasmic membrane. *The Journal of Biological Chemistry*, *276*, 18804–18811.

Heidorn, D. B., & Trewhella, J. (1988). Comparison of the crystal and solution structures of calmodulin and troponin C. *Biochemistry*, *27*, 909–915.

Heller, W. T. (2010). Small-angle neutron scattering and contrast variation: A powerful combination for studying biological structures. *Acta Crystallographica Section D: Biological Crystallography*, *66*, 1213–1217.

Hong, H., Patel, D. R., Tamm, L. K., & van den Berg, B. (2006). The outer membrane protein OmpW forms an eight-stranded beta-barrel with a hydrophobic channel. *The Journal of Biological Chemistry*, *281*, 7568–7577.

Ibel, K., & Stuhrmann, H. B. (1975). Comparison of neutron and X-ray scattering of dilute myoglobin solutions. *Journal of Molecular Biology*, *93*, 255–265.

Ieva, R., Tian, P., Peterson, J. H., & Bernstein, H. D. (2011). Sequential and spatially restricted interactions of assembly factors with an autotransporter beta domain. *Proceedings of the National Academy of Sciences of the United States of America*, *108*, E383–E391.

Jacques, D. A., & Trewhella, J. (2010). Small-angle scattering for structural biology—Expanding the frontier while avoiding the pitfalls. *Protein Science*, *19*, 642–657.

Jacrot, B. (1976). The study of biological structures by neutron scattering from solution. *Reports on Progress in Physics*, *39*, 911–953.

Jarchow, S., Lück, C., Görg, A., & Skerra, A. (2008). Identification of potential substrate proteins for the periplasmic *Escherichia coli* chaperone Skp. *Proteomics*, *8*, 4987–4994.

Kline, S. R. (2006). Reduction and analysis of SANS and USANS data using IGOR Pro. *Journal of Applied Crystallography*, *39*, 895–900.

Korndörfer, I. P., Dommel, M. K., & Skerra, A. (2004). Structure of the periplasmic chaperone Skp suggests functional similarity with cytosolic chaperones despite differing architecture. *Nature Structural & Molecular Biology*, *11*, 1015–1020.

Krueger, S., Gorshkova, I., Brown, J., Hoskins, J., McKenney, K. H., & Schwarz, F. P. (1998). Determination of the conformations of cAMP receptor protein and its T127L, S128A mutant with and without cAMP from small angle neutron scattering measurements. *Journal of Biological Chemistry*, *273*, 20001–20006.

Krueger, S., Shin, J.-H., Curtis, J. E., Rubinson, K. A., & Kelman, Z. (2014). The solution structure of full-length dodecameric MCM by SANS and molecular modeling: Structure of dodecameric MCM helicase. *Proteins: Structure, Function, and Bioinformatics*, *82*, 2364–2374.

Krueger, S., Shin, J.-H., Raghunandan, S., Curtis, J. E., & Kelman, Z. (2011). Atomistic ensemble modeling and small-angle neutron scattering of intrinsically disordered protein complexes: Applied to minichromosome maintenance protein. *Biophysical Journal, 101*, 2999–3007.

Li, J., Shi, C., Gao, Y., Wu, K., Shi, P., Lai, C., et al. (2012). Structural studies of *Mycobacterium tuberculosis* Rv0899 reveal a monomeric membrane-anchoring protein with two separate domains. *Journal of Molecular Biology, 415*, 382–392.

Lundin, V. F., Stirling, P. C., Gomez-Reino, J., Mwenifumbo, J. C., Obst, J. M., Valpuesta, J. M., et al. (2004). Molecular clamp mechanism of substrate binding by hydrophobic coiled-coil residues of the archaeal chaperone prefoldin. *Proceedings of the National Academy of Sciences of the United States of America, 101*, 4367–4372.

Lyu, Z.-X., Shao, Q., Gao, Y. Q., & Zhao, X. S. (2012). Direct observation of the uptake of outer membrane proteins by the periplasmic chaperone Skp. *PLoS One*, 7. e46068.

MacKerell, A. D., Bashford, D., Bellott, Dunbrack, Evanseck, R. L., Field, J. D., et al. (1998). All-atom empirical potential for molecular modeling and dynamics studies of proteins. *The Journal of Physical Chemistry B, 102*, 3586–3616.

Martín-Benito, J., Boskovic, J., Gómez-Puertas, P., Carrascosa, J. L., Simons, C. T., Lewis, S. A., et al. (2002). Structure of eukaryotic prefoldin and of its complexes with unfolded actin and the cytosolic chaperonin CCT. *The EMBO Journal, 21*, 6377–6386.

Martín-Benito, J., Gómez-Reino, J., Stirling, P. C., Lundin, V. F., Gómez-Puertas, P., Boskovic, J., et al. (2007). Divergent substrate-binding mechanisms reveal an evolutionary specialization of eukaryotic prefoldin compared to its archaeal counterpart. *Structure, 15*, 101–110.

McMorran, L. M., Bartlett, A. I., Huysmans, G. H. M., Radford, S. E., & Brockwell, D. J. (2013). Dissecting the effects of periplasmic chaperones on the in vitro folding of the outer membrane protein PagP. *Journal of Molecular Biology, 425*, 3178–3191.

Moon, C. P., Zaccai, N. R., Fleming, P. J., Gessmann, D., & Fleming, K. G. (2013). Membrane protein thermodynamic stability may serve as the energy sink for sorting in the periplasm. *Proceedings of the National Academy of Sciences of the United States of America, 110*, 4285–4290.

Moore, P. B. (1982). Small-angle scattering techniques for the study of biological macromolecules and macromolecular aggregates. In G. Ehrenstein & H. Lecar (Eds.), *Methods of experimental physics: vol. 20* (pp. 337–390). New York: Academic Press.

Patel, G. J., Behrens-Kneip, S., Holst, O., & Kleinschmidt, J. H. (2009). The periplasmic chaperone Skp facilitates targeting, insertion, and folding of OmpA into lipid membranes with a negative membrane surface potential. *Biochemistry, 48*, 10235–10245.

Pautsch, A., & Schulz, G. E. (1998). Structure of the outer membrane protein A transmembrane domain. *Nature Structural Biology, 5*, 1013–1017.

Peng, Y., Curtis, J. E., Fang, X., & Woodson, S. A. (2014). Structural model of an mRNA in complex with the bacterial chaperone Hfq. *Proceedings of the National Academy of Sciences of the United States of America, 111*, 17134–17139.

Phillips, J. C., Braun, R., Wang, W., Gumbart, J., Tajkhorshid, E., Villa, E., et al. (2005). Scalable molecular dynamics with NAMD. *Journal of Computational Chemistry, 26*, 1781–1802.

Putnam, C. D., Hammel, M., Hura, G. L., & Tainer, J. A. (2007). X-ray solution scattering (SAXS) combined with crystallography and computation: Defining accurate macromolecular structures, conformations and assemblies in solution. *Quarterly Reviews of Biophysics, 40*, 191–285.

Qu, J., Mayer, C., Behrens, S., Holst, O., & Kleinschmidt, J. H. (2007). The trimeric periplasmic chaperone Skp of *Escherichia coli* forms 1:1 complexes with outer membrane proteins via hydrophobic and electrostatic interactions. *Journal of Molecular Biology, 374*, 91–105.

Rambo, R. P., & Tainer, J. A. (2010). Bridging the solution divide: Comprehensive structural analyses of dynamic RNA, DNA, and protein assemblies by small-angle X-ray scattering. *Current Opinion in Structural Biology, 20*, 128–137.

Ramelot, T. A., Zhao, L., Hamilton, K., Maglaqui, M., Xiao, R., Liu, J., et al. (unpublished). Solution NMR structure of the folded C-terminal fragment of YiaD from *Escherichia coli*. Northeast structural genomics consortium target ER553.

REDUX. (n.d.). http://pages.jh.edu/pfleming/sw/redux/; retrieved 20 April 2015.

Rhodius, V. A., Suh, W. C., Nonaka, G., West, J., & Gross, C. A. (2006). Conserved and variable functions of the sigmaE stress response in related genomes. *PLoS Biology*, *4*, e2.

Sarachan, K. L., Curtis, J. E., & Krueger, S. (2013). Small-angle scattering contrast calculator for protein and nucleic acid complexes in solution. *Journal of Applied Crystallography*, *46*, 1889–1893.

SASSIE-web:Beta. (n.d.). https://sassie-web.chem.utk.edu/sassie2/; retrieved 19 June 2015.

Schäfer, U., Beck, K., & Müller, M. (1999). Skp, a molecular chaperone of gram-negative bacteria, is required for the formation of soluble periplasmic intermediates of outer membrane proteins. *The Journal of Biological Chemistry*, *274*, 24567–24574.

Schlapschy, M., Dommel, M. K., Hadian, K., Fogarasi, M., Korndörfer, I. P., & Skerra, A. (2004). The periplasmic *E. coli* chaperone Skp is a trimer in solution: Biophysical and preliminary crystallographic characterization. *Biological Chemistry*, *385*, 137–143.

Schwalm, J., Mahoney, T. F., Soltes, G. R., & Silhavy, T. J. (2013). Role for Skp in LptD assembly in *Escherichia coli*. *Journal of Bacteriology*, *195*, 3734–3742.

Schwede, T., Kopp, J., Guex, N., & Peitsch, M. C. (2003). SWISS-MODEL: An automated protein homology-modeling server. *Nucleic Acids Research*, *31*, 3381–3385.

Semenyuk, A. V., & Svergun, D. I. (1991). GNOM—A program package for small-angle scattering data processing. *Journal of Applied Crystallography*, *24*, 537–540.

Siegert, R., Leroux, M. R., Scheufler, C., Hartl, F. U., & Moarefi, I. (2000). Structure of the molecular chaperone prefoldin: Unique interaction of multiple coiled coil tentacles with unfolded proteins. *Cell*, *103*, 621–632.

Sklar, J. G., Wu, T., Kahne, D., & Silhavy, T. J. (2007). Defining the roles of the periplasmic chaperones SurA, Skp, and DegP in *Escherichia coli*. *Genes & Development*, *21*, 2473–2484.

Stirling, P. C., Bakhoum, S. F., Feigl, A. B., & Leroux, M. R. (2006). Convergent evolution of clamp-like binding sites in diverse chaperones. *Nature Structural & Molecular Biology*, *13*, 865–870.

The PyMOL Molecular Graphics System (version 1.7.4). (n.d.). Schrödinger, LLC.

Vergnolle, M. A. S., Baud, C., Golovanov, A. P., Alcock, F., Luciano, P., Lian, L.-Y., et al. (2005). Distinct domains of small Tims involved in subunit interaction and substrate recognition. *Journal of Molecular Biology*, *351*, 839–849.

Vogt, J., & Schulz, G. E. (1999). The structure of the outer membrane protein OmpX from *Escherichia coli* reveals possible mechanisms of virulence. *Structure (London, England: 1993)*, *7*, 1301–1309.

Walton, T. A., Sandoval, C. M., Fowler, C. A., Pardi, A., & Sousa, M. C. (2009). The cavity-chaperone Skp protects its substrate from aggregation but allows independent folding of substrate domains. *Proceedings of the National Academy of Sciences of the United States of America*, *106*, 1772–1777.

Walton, T. A., & Sousa, M. C. (2004). Crystal structure of Skp, a prefoldin-like chaperone that protects soluble and membrane proteins from aggregation. *Molecular Cell*, *15*, 367–374.

Webb, C. T., Gorman, M. A., Lazarou, M., Ryan, M. T., & Gulbis, J. M. (2006). Crystal structure of the mitochondrial chaperone TIM9.10 reveals a six-bladed alpha-propeller. *Molecular Cell*, *21*, 123–133.

Webb, C. T., Heinz, E., & Lithgow, T. (2012). Evolution of the β-barrel assembly machinery. *Trends in Microbiology*, *20*, 612–620.

Whitten, A. E., Cai, S., & Trewhella, J. (2008). MULCh: Modules for the analysis of small-angle neutron contrast variation data from biomolecular assemblies. *Journal of Applied Crystallography*, *41*, 222–226.

Whitten, A. E., Jacques, D. A., Hammouda, B., Hanley, T., King, G. F., Guss, J. M., et al. (2007). The structure of the KinA-Sda complex suggests an allosteric mechanism of histidine kinase inhibition. *Journal of Molecular Biology*, *368*, 407–420.

Whitten, A. E., & Trewhella, J. (2009). Small-angle scattering and neutron contrast variation for studying bio-molecular complexes. In R. S. Foote & J. W. Lee (Eds.), *Micro and nano technologies in bioanalysis: vol. 544* (pp. 307–323). Totowa, NJ: Humana Press.

Wu, S., Ge, X., Lv, Z., Zhi, Z., Chang, Z., & Zhao, X. S. (2011). Interaction between bacterial outer membrane proteins and periplasmic quality control factors: A kinetic partitioning mechanism. *Biochemical Journal*, *438*, 505–511.

Wu, X.-B., Tian, L.-H., Zou, H.-J., Wang, C.-Y., Yu, Z.-Q., Tang, C.-H., et al. (2013). Outer membrane protein OmpW of *Escherichia coli* is required for resistance to phagocytosis. *Research in Microbiology*, *164*, 848–855.

Zaccai, G. (2012). Straight lines of neutron scattering in biology: A review of basic controls in SANS and EINS. *European Biophysics Journal*, *41*, 781–787.

CHAPTER SEVEN

Deuteration in Biological Neutron Reflectometry

Frank Heinrich[*,†,1]
*Department of Physics, Carnegie Mellon University, Pittsburgh, Pennsylvania, USA
†Center for Neutron Research, National Institute of Standards and Technology, Gaithersburg, Maryland, USA
[1]Corresponding author: e-mail address: fheinrich@cmu.edu

Contents

Abstract

Neutron reflectometry (NR) is uniquely positioned in structural biology, because of its ability to characterize biomolecular interfacial architectures like lipid membranes and membrane-associated proteins nondestructively and in their native environment. Mimicking biological processes, samples can be manipulated and their structural response can be measured. Specific deuteration is an integral part of biological NR as it is essential for resolving, for example, individual components of membrane-bound protein–protein complexes. Data analysis techniques have been developed in the past decade that extract the maximum structural detail from reflectivity data obtained from samples with complex deuteration schemes while avoiding overinterpretation. This is achieved by employing robust methods for the determination of modeling uncertainties. Integrative modeling approaches for NR are emerging as an essential part of the technique.

Methods in Enzymology, Volume 566
ISSN 0076-6879
http://dx.doi.org/10.1016/bs.mie.2015.05.019

1. INTRODUCTION

1.1 Biological Neutron Reflectometry

Specular neutron reflectometry (NR) is a structural technique that is uniquely able to characterize biological interfaces under fully hydrated biomimetic conditions (Fragneto, 2009; Harroun, Kučerka, Nieh, & Katsaras, 2009; Junghans et al., 2015; Lakey, 2009; Majkrzak, Berk, Krueger, & Perez-Salas, 2006; Wacklin, 2010). One of the most prominent applications for biological NR is the determination of the structure of lipid bilayer membranes and associated proteins (see Fig. 1; Heinrich & Lösche, 2014; Kent et al., 2010; Le Brun, Holt, Shah, Majkrzak, & Lakey, 2011). This chapter will exclusively utilize this particular example when discussing methodical aspects, but all conclusions can be generally applied. Other applications of biological NR include, for example, the association of small molecules with lipid membranes (Benedetto et al., 2014), protein fouling at artificial interfaces (Al-Jawad, Fragneto, Liu, Chang, & Clarkson, 2009; Hollmann, Steitz, & Czeslik, 2008), protein denaturation at solid interfaces relevant for the storage of pharmaceuticals (Brouette et al., 2013), design of biosensors (Le Brun et al., 2011), or the characterization of whole cell adhesion to interfaces (Smith et al., 2010).

Using NR, biological systems can be studied destruction free and under conditions close to their native environment. For lipid bilayer membranes, this means that experiments can be performed on in-plane disordered, fluid-immersed membranes (Shenoy et al., 2010) with experimental control over environmental variables like chemical composition, temperature, and pH value of the buffer solution (Singh et al., 2013), or the electrical field strength across the lipid membrane (Gupta, Dura, Freites, Tobias, & Blasie, 2012). This is in stark contrast to the currently dominant structural techniques, such as protein crystallography, cryoelectron microscopy, and nuclear magnetic resonance. During a typical NR experiment, the biological system can be manipulated and its structural response to different stimuli can be recorded. This makes NR a unique tool to study the conformation and conformational changes of proteins at interfaces (Datta et al., 2011).

NR is a low-resolution technique with a smallest resolvable feature size of about 1–2 nm, and it provides structural information only along one direction, normal to the interface, while averaging in-plane parallel to the interface. In order to gain access to in-plane structural information, NR data have to be combined with external information, for example, from

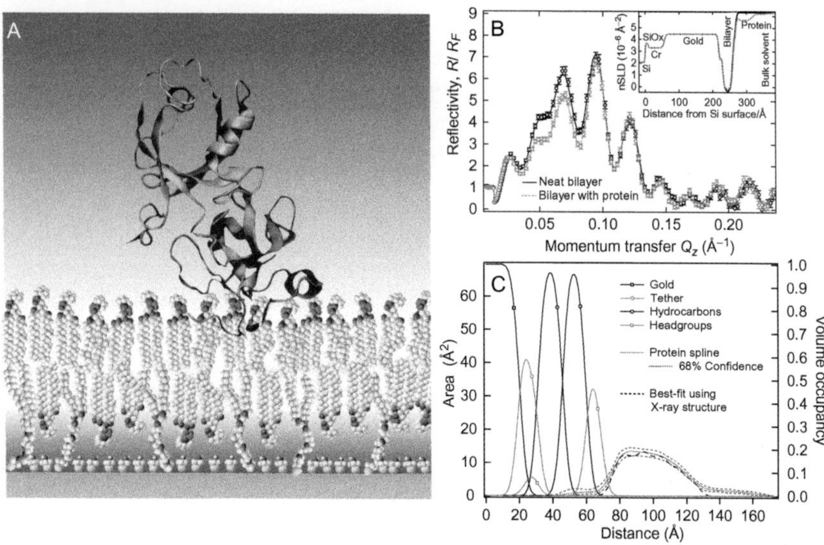

Figure 1 (A) Standard application for biological neutron reflectometry showing a sparsely tethered lipid bilayer membrane (stBLM) on a gold support and a membrane-associated protein (GRASP55). Neutron reflectometry was used to determine the bilayer structure, and the location and orientation of the protein using an integrative modeling approach (Heinrich et al., 2014) utilizing an existing high-resolution structure of the protein (Truschel et al., 2011). (B) Typical neutron reflectometry curves for a stBLM (9:1 DOPC:DOGS-NTA) before and after protein (GRASP55) association, measured using D_2O-based bulk solvent. Error bars represent 68% confidence limits. The momentum transfer Q_z increases with the incident angle of the neutron beam. The reflectivity (i.e., the flux ratio of reflected and incident neutrons) is normalized to the Fresnel reflectivity RF (i.e., the ideal reflectivity between the two bulk media, silicon- and D_2O-based bulk solvent). Inset: Modeled nSLD profiles along the membrane normal that describes the experimental data. Changes in the nSLD profile due to the associated protein are readily visible. (C) Composition space model of the GRASP55/ membrane complex. The lipid bilayer is parameterized using the continuous distribution model (Shekhar, Nanda, Lösche, & Heinrich, 2011). For simplicity, certain submolecular components that were modeled separately have been combined in this representation. The six distinct distributions shown represent the outer 20 Å of the gold substrate layer, the hydrophilic region of the tether molecules and βME that form the hydrated submembrane space, the substrate–proximal lipid headgroups and (lipid and tether) hydrocarbon chains, the substrate–distal lipid hydrocarbon chains and headgroups. The protein's cross section (component volume occupancy profile, CVO(z)) has been determined using a monotonic Hermite spline. Uncertainties were quantified by a Monte Carlo Markov Chain method and are displayed for the protein envelope as 68% confidence bands (Heinrich et al., 2009; Kirby et al., 2012). Uncertainty intervals for the molecular distributions of the tethered bilayer and the substrate are omitted for clarity, and only best-fit distributions are displayed. *This research was originally published in the Journal of Biological Chemistry (Heinrich et al., 2014), © The American Society for Biochemistry and Molecular Biology.*

computational or auxiliary experimental sources. Therefore, integrative modeling approaches play an important role during data analysis. It is already standard to use existing (partial) high-resolution structures of proteins or peptides obtained by X-ray crystallography or nuclear magnetic resonance during NR data analysis. Under the assumption that the high-resolution structure is invariant, the position and orientation of the molecule within the biomolecular architecture can be determined (rigid body modeling, see Fig. 2; Heinrich et al., 2014; McGillivray et al., 2009; Nanda et al., 2010; Yap et al., 2015). A large number of examples can be found in which molecular dynamics (MD) simulations complement NR (Pfefferkorn et al., 2012; Shenoy, Nanda, & Lösche, 2012). Mostly, those examples do not truly qualify as integrative modeling, because the MD simulations are not biased by the existing NR data during runtime, but rather time-averaged MD structures are compared to the experimental data after the simulation has finished. Consequently, integrative NR-steered MD simulations are currently under development.

NR structurally averages not only in plane but also in time. This has the consequence that structural information, for example, about a membrane-bound protein will always be obtained as an ensemble average. While this feature might prove disadvantageous for some applications, it makes NR an excellent tool to study disordered or conformationally flexible molecules at interfaces (Pfefferkorn et al., 2012; Shenoy, Shekhar, et al., 2012).

Finally and to the core of this chapter, the use of deuteration allows NR to distinguish molecular species that would otherwise be indistinguishable based on their scattering properties. Most notably, specific deuteration of components of a protein complex allows resolving their individual contributions to the structure (Clifton, Neylon, & Lakey, 2013; Yap et al., 2015). Therefore, NR is uniquely positioned to structurally characterize protein–protein complexes at interfaces and their structural response to external stimuli.

1.2 Deuteration in Biological NR

Selective deuteration, or the replacement of protium with deuterium, is an effective tool in neutron scattering to flexibly change the scattering properties of parts of an interfacial architecture with the aim of increasing or decreasing the sensitivity to selected molecules. In detail, deuteration changes the coherent neutron scattering length density (nSLD), ρ, of those molecules. To achieve highest sensitivity in resolving selected molecules

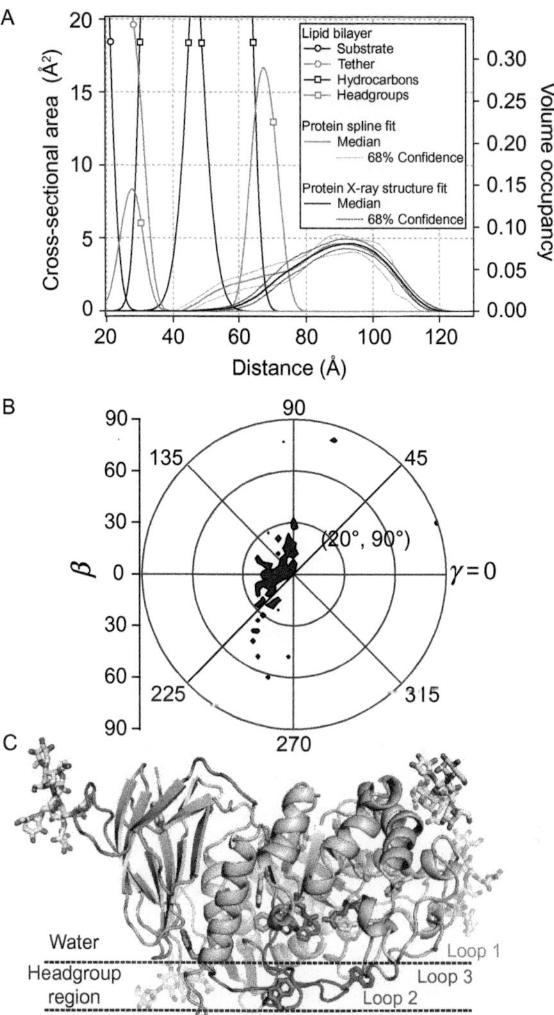

Figure 2 Example of a rigid body modeling of a protein (GCase) associated with a tethered lipid membrane composed of 50:50 molar fraction POPC:POPS (Yap et al., 2015). (A) Composition space modeling result. A difference between results obtained from a free-form modeling approach using a Hermite spline and an integrative modeling approach using an available X-ray crystal structure is apparent at the substrate–distal headgroup–hydrocarbon interface. It is likely that the protein adopts a different configuration when membrane-bound compared to its crystalline form with a deeper penetration of flexible loops into the hydrocarbon region of the substrate–distal lipid leaflet. (B) Probability plot of two Euler angles (β and γ) defining the protein orientation at the interface. Robust methods to determine modeling uncertainties are essential for complex modeling approaches. Here, a Monte Carlo Markov Chain algorithm was employed (Kirby et al., 2012). (C) The orientation of the protein at the membrane in agreement with the NR data and further supported by additional fluorescence data (Yap et al., 2015). *This research was originally published in the Journal of Biological Chemistry (Yap et al., 2015), © The American Society for Biochemistry and Molecular Biology.* (See the color plate.)

within the architecture, generally, the largest difference (contrast) in nSLD to their neighborhood is desirable. In the other extreme, to completely blindfold the experiment to contributions from a certain molecule, its nSLD has to be equal to that of its surrounding (contrast matching).

In the case of an interfacially stabilized lipid bilayer membrane with membrane-penetrating protein (see Fig. 3), the first possible target for deuteration is the lipid bilayer, and in particular, the lipid hydrocarbon chains. The nSLD of protiated lipid hydrocarbon chains is roughly $\rho \approx -0.5 \times 10^{-6}$ Å^{-2}, whereas, depending on the lipid type, perdeuterated hydrocarbon chains have an nSLD of $\rho \approx 7 \times 10^{-6}$ Å^{-2}. Protiated protein has an average nSLD of $\rho \approx 2 \times 10^{-6}$ Å^{-2}. Therefore, to structurally resolve protiated transmembrane proteins, a lipid bilayer with perdeuterated hydrocarbon chains provides a better contrast than their protiated equivalent. The second obvious deuteration target, the protein, has an nSLD of $\rho \approx 7 \times 10^{-6}$ Å^{-2} in its perdeuterated form. Therefore, in combination with a protiated lipid bilayer, it exceeds the previous contrast and yields an even greater sensitivity. In practice, however, factors like the availability of deuterated material, the molar fraction of the molecule of interest in the architecture, and the precise location contribute significantly to the choice of the deuteration scheme.

One of the currently most exciting applications of deuteration in NR is the structural characterization of membrane-associated protein complexes. Because there is little or no scattering contrast between individual protiated components of a protein complex, two parts of the complex have to be labeled differently. Preparing one part in its protiated and the other in its deuterated form allows them to be structurally separated during data analysis. It will be discussed in detail how this is technically achieved.

There is a third application of deuteration, which has received little attention so far. In cases where a molecule of interest is present only at low molar or volume fraction, a combined analysis of reflectivity from a protiated and a deuterated sample can significantly improve the sensitivity. Hereby, either the scattering length density of the molecule of interest or its surrounding, or both, can be varied.

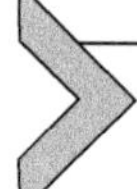

2. EXPERIMENTAL METHODS

2.1 Instrumentation

Experimental infrastructure for biological NR is in place at all major neutron scattering facilities world-wide (Dura et al., 2006; Teixeira et al., 2008). In

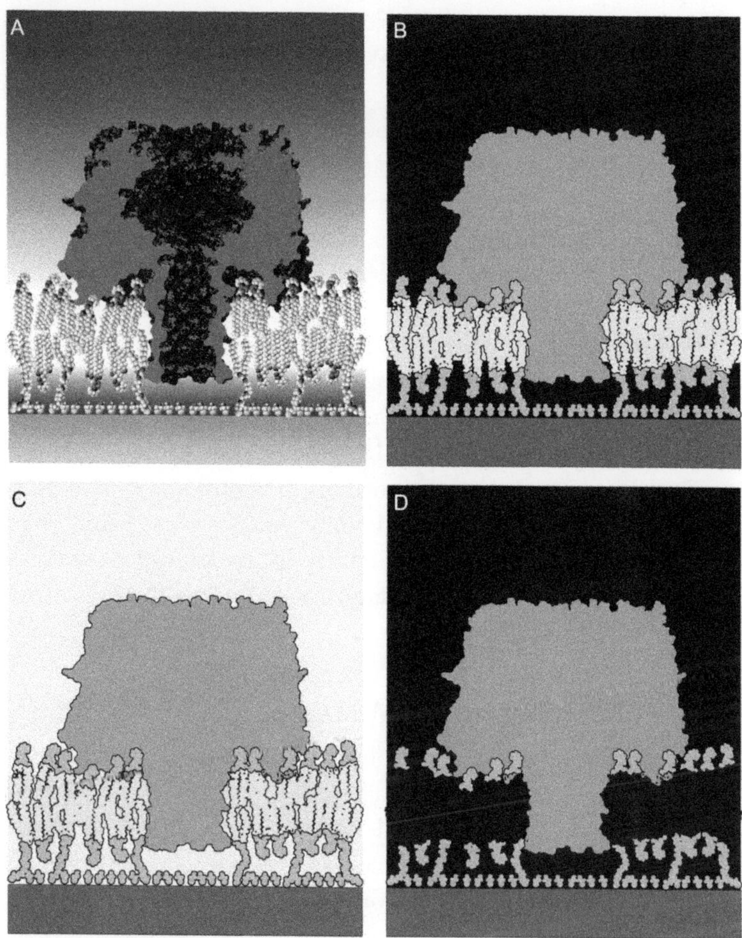

Figure 3 Deuteration schemes in biological NR. (A) Membrane structure and penetration depth of the toxin α-hemolysin as determined by NR (McGillivray et al., 2009) using rigid body modeling based on the X-ray structure of the protein (Song et al., 1996). (B–D) Illustration of scattering contrasts obtained by employing various deuteration schemes. The nSLD is color-coded ranging from light gray ($\approx -0.5 \times 10^{-6}$ Å^{-2}; H_2O; protiated hydrocarbons) to black ($\approx 6.5 \times 10^{-6}$ Å^{-2}; D_2O; deuterated hydrocarbons). The choice of deuteration scheme allows for increasing the sensitivity to certain molecular groups by maximizing their nSLD contrast with the surrounding. (B) Standard solvent contrast using D_2O-based buffer, a protiated lipid bilayer, and protiated protein. This contrast shows high sensitivity to the lipid hydrocarbon chains and the protein profile in the bulk solvent region adjacent to the lipid bilayer. (C): Standard complimentary contrast by exchanging the buffer for an H_2O-based buffer. The sensitivity to all solvent-containing regions is greatly improved when (C) and (B) are analyzed simultaneously. (C) An exemplary deuteration scheme using deuterated lipid hydrocarbon chains. This scheme is particularly useful to increase sensitivity to transmembrane protein and to protiated lipid headgroups.

most cases, biological NR does not require special instrumentation that is not available at scattering facilities, and it can be performed on a typical horizontal or vertical cold neutron reflectometer in standard configuration. Biological sample environment tends to be small in comparison to that of other research fields and fits most sample stations without substantial adjustment. In the case of membrane-associated proteins, the main challenge in terms of instrumentation lies in the accuracy of the measured reflectivity and the available neutron flux. Very often, the amount of protein binding to a membrane is low, contributing very little to the measured neutron reflectivity, which is dominated by the solid substrate and the lipid bilayer. It is therefore essential that systematic errors originating, for example, from the beam optics, the initial characterization of the incident neutron beam, or the neutron detection are minimized, while the neutron flux on the sample is maximized. These requirements are even more severe when one attempts to structurally resolve individual components of protein complexes using specific deuteration. In many cases, these types of problems expose the effective resolution limits of current instrumentation and motivate further instrumental development (Majkrzak, Carpenter, Heinrich, & Berk, 2011).

2.2 Deuteration of Biological Molecules

Obtaining deuterated materials for biological NR typically involves the synthesis of small molecules or harvesting small molecules and proteins from cells cultured in deuterated media (de Ghellinck et al., 2014; Pfefferkorn et al., 2012). Since this poses a significant entry barrier for many researchers in neutron scattering, all major neutron scattering centers provide access to biomolecular labeling facilities that compliment each other in that they specialize in different types of molecules (Teixeira et al., 2008). Typically, access is granted using a peer-review user program. Depending on the facility, the required deuterated molecules will then be produced with a varying degree of participation by the researcher.

2.3 Sample Requirements for Biological NR

Independent of the use of specific deuteration, samples for current instrumentation in NR are ideally prepared on a relatively large (several cm^2), flat, and smooth surfaces that are as transparent as possible for neutrons. This requirement limits the choice of the substrate to polished, single-crystalline silicon, sapphire, or quartz. Fortunately, substrates of this type are commercially available at high quality.

In many cases, the crystalline surface is further modified by adding thin films of various materials using thin film deposition methods or by chemical functionalization. The deposition of a buried magnetic thin film, for example, enables a magnetic reference layer technique that can be used to increase the effective resolution of the neutron experiment, albeit currently at the cost of measurement time (Holt et al., 2009; Le Brun et al., 2011). Alternative to solid–liquid interfaces, liquid–air interfaces using a Langmuir trough play an important role (Kent et al., 2010) but require a horizontal reflectometer. Solid–air interfaces can be studied under tight control of the H_2O and D_2O humidity at many facilities using specialized sample chambers.

For our chosen example of the characterization of lipid membrane-associated proteins, a versatile membrane model that fulfills all requirements for NR is the sparsely tethered bilayer lipid membrane (stBLM, see Fig. 1; Budvytyte et al., 2013; Heinrich et al., 2009; McGillivray et al., 2007). As a first preparation step, a self-assembled monolayer (SAM) of tether molecules is grafted onto a silicon wafer that has been terminated with a 10–20-nm thick gold film. Thereafter, a single lipid bilayer membrane is formed, typically via vesicle fusion to the tether layer. stBLMs are excellent mimics of natural membranes with respect to lipid fluidity and structure (Shenoy et al., 2010). They can be prepared using a large variety of lipids, including zwitterionic (phosphatidylcholine, phosphatidylethanolamine) or anionic lipids (phosphatidylserine, phosphatidylglycerol, phosphatidylphosphate) with saturated or unsaturated chains, sterols, and phosphatidylinositol or phosphatidylinositol phosphate lipids. Spacing of the synthetic tether lipids, grafted to the terminal gold film of the substrate through thiol chemistry and (typically) an oligo(ethylene oxide) linker, is achieved by coadsorption with β-mercaptoethanol (βME). stBLMs can be prepared virtually defect free, which prohibits unspecific protein adsorption to exposed substrate that would interfere with structural characterization of the membrane-associated protein (Valincius, Meškauskas, & Ivanauskas, 2012). Of similar importance, stBLMs are stable for the time scale of NR experiments (Valincius et al., 2008). stBLMs exhibit a low interfacial roughness, because of their proximity to the substrate ($\approx$20 Å). From a scattering point of view, this is important for achieving a high resolution (Majkrzak et al., 2011). On the other hand, the proximity of the substrate to the membrane and interference with the tethering chemistry—typically 50 mol% of the lipids located in the inner lipid leaflet are tether lipids—may limit the reconstitution of membrane proteins with large extramembrane domains. Alternative lipid membrane model systems in use are solid-supported (Benedetto et al., 2014), floating

(Fragneto et al., 2001), and polymer-cushioned lipid bilayer membranes (Burgess et al., 2004; Majewski et al., 1998; Smith et al., 2009).

A typical NR experiment with the aim to characterize a membrane-associated protein, and many other biological NR experiments, use a flow-through cell (Majkrzak et al., 2000) that enables buffer exchange *in situ*. This allows one to study a sample under different isotopic buffer compositions or to determine the response of the sample to external manipulations. Every condition is typically measured at least twice and in isotopically distinct bulk solvents (contrasts), for example, in H_2O- and D_2O-based buffer solutions. This greatly improves the ability to separate scattering contributions from the solvent and from interfacial components. For membrane-associated proteins this means that, typically, the as-prepared lipid bilayer is characterized first with at least two contrasts. Thereafter, the protein is either incubated using at least two contrasts, or it is incubated in H_2O-based buffer, and two different contrasts are measured after rinsing off the protein. The same principle of measuring several contrasts is applied to all following sample manipulations. Co-refinement of these different sets of reflectivity data from one unique sample measured under a variety of conditions then allows determining structural profiles with high confidence (Kirby et al., 2012). It also permits the precise quantification of changes in the interfacial structure after manipulation. The exact experimental procedure depends on a number of factors such as binding affinities, stability of the sample in solution, stability in D_2O, and sample availability. Typical flow cell volumes are about 1 mL, which makes NR competitive in terms of the required amount of sample compared to other structural methods. Sufficient statistics for a single contrast are typically collected within a couple to a few hours, depending on the neutron source and instrument, which makes NR a rather low-throughput technique at the moment. Upcoming next-generation NR instrumentation is poised to improve counting times by at least one order of magnitude and make high-throughput studies possible.

3. DATA ANALYSIS

3.1 General Data Analysis

Neutron reflectivity, like most scattering techniques, suffers from the phase problem, which is the loss of the phase information of the scattered neutron (wave) during detection. The loss of the phase information in most cases prohibits direct inversion of the collected neutron reflectivity data to obtain the one-dimensional structural (scattering length density) profile. Apart from a few special cases for which direct inversion strategies have been developed (Majkrzak & Berk, 1995, 2003), NR typically deals with this situation by using a modeling approach. Since the inverse problem of calculating the neutron reflectivity from a given nSLD profile has a unique solution, a parameterized trial profile is created and the calculated reflectivity is compared to the experimental data. Using an iterative process, all model parameters are optimized to obtain the best fit to the reflectivity. Whether it is possible to obtain a good fit to the measured reflectivity, therefore, critically depends on the choice of the model and the performance of the used optimization algorithm.

In the past, and frustratingly until today, neutron reflectivity data analysis has been dominated by the slab model (Ankner & Majkrzak, 1992), which describes the interfacial structure in terms of stratified slabs of constant nSLD. The slab model was greatly successful because of its close relationship with the optical matrix method that allows for an efficient calculation of the parent reflectivity spectrum. However, the slab model is impractical when applied to complex molecular architectures that consist of spatially intermixing submolecular groups. A transition to structure-based composition space modeling (Schalke, Krüger, Weygand, & Lösche, 2000; Shekhar et al., 2011; Wiener & White, 1992), which is standard in other structural techniques such as X-ray diffraction of membranes (Kučerka et al., 2008; Nagle & Tristram-Nagle, 2000), allows one to (a) build detailed molecular models, (b) directly parameterize relevant structural features of the biomolecular film, and (c) include external information about chemical and physical properties of the molecular constituents in the model.

We implemented a composition space approach that uses error functions for the modeling of component volume occupancy profiles, CVO(z), along the surface normal, z (continuous distribution model; Heinrich & Lösche, 2014; Shekhar et al., 2011). Its implementation for lipid bilayer membranes was validated against distributions of submolecular components obtained from MD simulations. This modeling strategy provides high flexibility in arranging molecular groups in space while intrinsically maintaining physically valid boundary conditions. For example, the continuous distribution model achieves volume filling of complex molecular architectures without over or under-filling. This feature makes the new parameterization particularly suited for complex molecular architectures such as stBLM with associated proteins, and it can also be readily applied to lipid bilayers physisorbed to solid supports, SAMs, or Langmuir monolayers at the air–water interface.

The composition space approach as implemented in the continuous distribution model is readily applicable to interfaces such as surface supported lipid bilayer membranes, whose architecture is in principal known and a model refinement only determines specific values for parameters such as bilayer thickness, lipid composition, and surface coverage. However, if parts of the molecular architecture are of unknown structure, the model needs to be extended by components that are more flexible and can assume different functional forms. Those strategies typically involve quasi free-form modeling approaches (Berk & Majkrzak, 1995; Kunz, Reiter, Götzelmann, & Stamm, 1993; Laub & Kuhl, 2006; van der Lee, Salah, & Harzallah, 2007). In combination with the continuous distribution model, we implemented a Hermite spline to describe unknown structural elements, in particular, membrane-associated proteins (Shenoy, Shekhar, et al., 2012). The Hermite spline has the advantage that it is a locally parameterized real-space object, in contrast to free-form implementations using truncated expansions of basis functions such as Chebychev polynomials (Laub & Kuhl, 2006). Hermite splines have the advantage over other established spline functions like parametric B-splines (Berk & Majkrzak, 1995) in that they require fewer parameters, that they are less prone to yield physically implausible profiles, and that the spline function passes through the defining control points, while still maintaining enough flexibility to describe realistic protein profiles. It is not straightforward to combine continuous distribution model components and Hermite splines in a way that satisfies all physical constraints. With regard to membrane-associated proteins, we achieved a satisfying technical solution that has been discussed in a comprehensive review (Heinrich & Lösche, 2014). This review also details the use of high-resolution atomic structures of known components of the interfacial architecture within a rigid body analysis.

It is of importance to keep in mind that all modeling strategies discussed so far take place in one dimension. While this approach appears to be sufficient for the current statistical quality of reflectivity data, fundamentally, it is clear that projections of molecular distributions onto one dimension cannot be expected to be generally well described by relatively simple functional forms, such as pairs of error functions used here. One particular example for this is the distortion of a local bilayer structure by a protein closely interacting with surrounding lipid molecules. There are currently no modeling strategies available to account for this apart from free-form methods. If the need arises triggered by next-generation instrumentation, existing pioneering work can serve a starting point for further model development (Politsch, 2001, 2003; Politsch & Cevc, 2000, 2002).

3.2 Deuteration-Specific Data Analysis

Specific deuteration requires a modeling strategy that can flexibly parameterize specific parts of the biomolecular interfacial architecture. While conventional slab models are able to describe lipid bilayers with deuterated hydrocarbon chains sufficiently well (Wacklin, 2010, 2011), they will intrinsically fail to resolve, for example, the deuterated part of a protein complex that might spatially overlap with the protiated part, or other components of the interface. Fortunately, it is straightforward to apply the composition space modeling approach discussed in section 3.1 to interfacial architectures with specific deuteration. The scattering length density of submolecular groups is readily accessible in the continuous distribution model as a fixed or fittable quantity. Therefore, biomolecular architectures can easily be constructed and parameterized including specifically deuterated components.

Some modification, however, is needed to extend the discussed free-form spline modeling approach to include the characterization of parts of the biomolecular architecture that are of unknown structure and that contain specific deuteration. One example for this is the characterization of a membrane-bound protein complex in which one component has been perdeuterated (see Fig. 4). To separate the structure of both components, we added an nSLD component to the free-form spline modeling approach (Yap et al., 2015). The traditional protein spline describes only a component volume occupancy profile, CVO(z), as a function of position, z, along the surface normal, and the nSLD of the material represented by the spline is fixed. The extended spline adds a second variable component, nSLD(z), which in itself is a spline function. It shares the position of the control points of the CVO spline to reduce the number of fitting parameters. The scattering length density at each control point is allowed to vary between that of the protiated component and that of the deuterated component, taking into account proton exchange of the protein with the buffer solution in which the biomolecular architecture is immersed. Individual profiles for the deuterated and the protiated part of the complex are obtained during postprocessing of the fit using a linear decomposition of the CVO(z) and nSLD(z) splines, based on the known scattering length densities of the two parts.

3.3 Uncertainty Analysis

In particular, free-form modeling of unknown structures requires a relatively large number of parameters when compared to slab models or to continuous

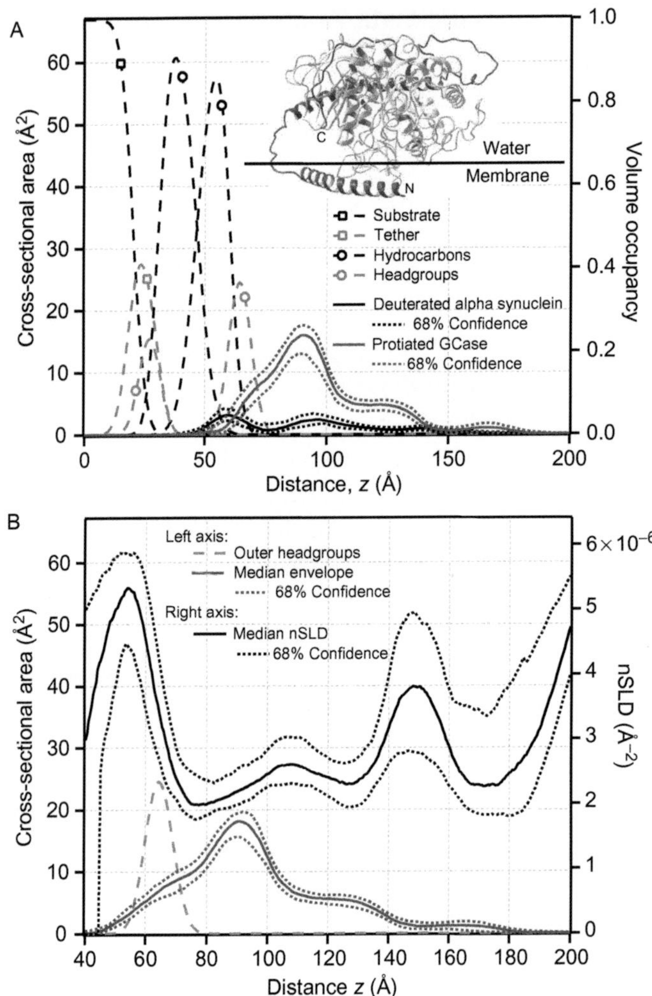

Figure 4 Characterization of a membrane-bound protein–protein complex of α-synuclein and GCase (see Fig. 2) at the lipid membrane using NR and specific deuteration (Yap et al., 2015). (A) Modeling result for the lipid bilayer using a continuous distribution model and for the volume occupancy profile of the protein complex using Hermite splines. Within 68% confidence limits, the volume occupancy profile of deuterated α-synuclein in the complex shows a bimodal envelope, which is in contrast to membrane-bound-free α-synuclein that exhibits only one maximum at the headgroup/hydrocarbon interface. Inset: A molecular model of the complex that satisfies the constraints set by NR and auxiliary data from fluorescence spectroscopy. In order to fit the NR profile, the second α-helix has to lift off the bilayer and is most likely closely associated with the GCase protein. (B) Implementation of the free-form Hermite spline modeling for protein–protein complexes with specific deuteration. Two Hermite splines model the volume occupancy profile and the nSLD profile independently,

distribution models. It is therefore absolutely necessary to employ a robust method to calculate uncertainties of and correlations between model fit parameters. We prefer using either a Monte Carlo simulation technique (Heinrich et al., 2009) or a Monte Carlo Markov Chain-based algorithm (Kirby et al., 2012) to determine those quantities. Besides being able to report accurate confidence intervals for fit parameters, volume occupancy profiles, or nSLD profiles, accurate confidence intervals play an important role in model optimization. While it is not possible to determine the validity of a particular model with those methods, it is very well possible, using an iterative process, to determine the optimal number of parameters for a given model that are supported by the data. This way, overparameterization as well as under-parameterization can be avoided. While it is well accepted in the community that overparameterization should be refrained from, it is generally not recognized that under-parameterization is equally, if not more, problematic. Having a robust uncertainty analysis in place, over-parameterization can be easily recognized by inflated error bars. Under-parameterization, however, tends to give tight error bars on an average quantity that easily could be separated into two parameters. Experience shows that under-parameterization affects fit parameters that are structurally not related to the feature that is under-parameterized. This is due to the nature of reflectivity modeling taking place in real-space while the comparison between model reflectivity and data takes place inverse space. Therefore, it is essential during modeling to find the optimal number of fit parameters using an iterative process of expanding and contracting the parameter space, thus, identifying a good balance between model detail and parameter uncertainties. The goodness of fit, or the χ^2 value, is a valuable tool to determine under and overparameterization but takes second place to a detailed analysis of parameter uncertainties and correlations. While it is technically possible to automate the model optimization process using methods from information theory, there is no functioning implementation of this, yet.

---

sharing the position of control points that are in this example on average 20 Å apart. The volume occupancy profile can vary between 0% and 100% volume filling, the nSLD profile can vary between values for protiated GCase (1.67×10^{-6} Å^{-2}) and deuterated α-synuclein (6.12×10^{-6} Å^{-2}). Individual volume occupancy profiles for the two components as shown in (A) are computed from the two Hermite splines and confidence limits are propagated. *This research was originally published in the Journal of Biological Chemistry (Yap et al., 2015), © The American Society for Biochemistry and Molecular Biology.* (See the color plate.)

4. CONCLUSION

Specific deuteration is an integral part of biological NR. Apart from optimizing scattering contrasts to obtain the highest sensitivity to the structural detail of interest within biomolecular architecture, specific deuteration is essential for resolving individual components of, for example, protein–protein complexes at lipid membranes. NR is uniquely positioned to address scientific questions in this exciting area of research, because of its ability to structurally characterize those systems nondestructively in their native fluid-like, disordered state. Measurements can be carried out in a biomimetic setting under full control of all environmental variables and while manipulating the sample to mimic biological processes. Biological NR experiments do not pose any particular technical challenge from a scattering perspective. The largest technical obstacle is the availability and production of deuterated material, which is addressed by all major neutron scattering facilities with the establishment of deuteration facilities with user programs. Data analysis techniques have been developed in the past decade to extract the maximum structural detail from the reflectivity while avoiding over-interpretation of the data by employing robust methods to determine modeling uncertainties. Integrative modeling approaches, which combine neutron reflectivity data with experimental data from other techniques or simulation, are emerging.

5. ACKNOWLEDGMENTS

I thankfully acknowledge Rana Ashkar and Christopher Bertrand for helpful comments. This research was supported by the Department of Commerce (MSE 70NANB11H8139 and 70NANB13H009) and by the NIST IMS program "Precision Measurements for Integral Membrane Proteins." Research was performed in part at the National Institute of Standards and Technology (NIST) Center for Nanoscale Science and Technology. Certain commercial materials, equipment, and instruments are identified in this work to describe the experimental procedure as completely as possible. In no case does such an identification imply a recommendation or endorsement by NIST, nor does it imply that the materials, equipment, or instrument identified are necessarily the best available for the purpose.

REFERENCES

Al-Jawad, M., Fragneto, G., Liu, J., Chang, S. R., & Clarkson, B. (2009). Fibronectin adsorption studied using neutron reflectometry and complementary techniques. *The European Physical Journal. E, Soft Matter*, *30*(2), 175–179.

Ankner, J. F., & Majkrzak, C. F. (1992). Subsurface profile refinement for neutron specular reflectivity. In *SPIE proceedings—International society for optics and photonics: Vol. 1738* (pp. 260–269).

Benedetto, A., Heinrich, F., Gonzalez, M. A., Fragneto, G., Watkins, E., & Ballone, P. (2014). Structure and stability of phospholipid bilayers hydrated by a room-temperature ionic liquid/water solution: A neutron reflectometry study. *The Journal of Physical Chemistry B*, *118*(42), 12192–12206.

Berk, N., & Majkrzak, C. F. (1995). Using parametric B splines to fit specular reflectivities. *Physical Review B: Condensed Matter*, *51*(17), 11296–11309.

Brouette, N., Fragneto, G., Cousin, F., Moulin, M., Haertlein, M., & Sferrazza, M. (2013). A neutron reflection study of adsorbed deuterated myoglobin layers on hydrophobic surfaces. *Journal of Colloid and Interface Science*, *390*(1), 114–120.

Budvytyte, R., Valincius, G., Niaura, G., Voiciuk, V., Mickevicius, M., Chapman, H., et al. (2013). Structure and properties of tethered bilayer lipid membranes with unsaturated anchor molecules. *Langmuir: The ACS Journal of Surfaces and Colloids*, *29*(27), 8645–8656.

Burgess, I., Li, M., Horswell, S. L., Szymanski, G., Lipkowski, J., Majewski, J., et al. (2004). Electric field-driven transformations of a supported model biological membrane—An electrochemical and neutron reflectivity study. *Biophysical Journal*, *86*(3), 1763–1776.

Clifton, L. A., Neylon, C., & Lakey, J. H. (2013). Examining protein-lipid complexes using neutron scattering. In N. J. Clifton (Ed.), *Methods in molecular biology: Vol. 974*. (pp. 119–150). Totowa, NJ: Humana Press, chapter 7.

Datta, S. A. K., Heinrich, F., Raghunandan, S., Krueger, S., Curtis, J. E., Rein, A., et al. (2011). HIV-1 Gag extension: Conformational changes require simultaneous interaction with membrane and nucleic acid. *Journal of Molecular Biology*, *406*(2), 205–214.

de Ghellinck, A., Schaller, H., Laux, V., Haertlein, M., Sferrazza, M., Maréchal, E., et al. (2014). Production and analysis of perdeuterated lipids from Pichia pastoris cells. *PLoS One*, *9*(4), e92999.

Dura, J. A., Pierce, D. J., Majkrzak, C. F., Maliszewskyj, N. C., McGillivray, D. J., Lösche, M., et al. (2006). AND/R: Advanced neutron diffractometer/reflectometer for investigation of thin films and multilayers for the life sciences. *Review of Scientific Instruments*, 77(7), 74301–7430111.

Fragneto, G. (2009). Biological physics near surfaces/interfaces: A perspective. *The European Physical Journal E, Soft Matter*, *30*(2), 239–243.

Fragneto, G., Charitat, T., Graner, F., Mecke, K., Perino-Gallice, L., & Bellet-Amalric, E. (2001). A fluid floating bilayer. *Europhysics Letters*, *53*, 100–106.

Gupta, S., Dura, J. A., Freites, J. A., Tobias, D. J., & Blasie, J. K. (2012). Structural characterization of the voltage-sensor domain and voltage-gated K+-channel proteins vectorially oriented within a single bilayer membrane at the solid/vapor and solid/liquid interfaces via neutron interferometry. *Langmuir: The ACS Journal of Surfaces and Colloids*, *28*(28), 10504–10520.

Harroun, T. A., Kučerka, N., Nieh, M.-P., & Katsaras, J. (2009). Neutron and X-ray scattering for biophysics and biotechnology: Examples of self-assembled lipid systems. *Soft Matter*, *5*(14), 2694–2703.

Heinrich, F., & Lösche, M. (2014). Zooming in on disordered systems: Neutron reflection studies of proteins associated with fluid membranes. *Biochimica et Biophysica Acta*, *1838*(9), 2341–2349.

Heinrich, F., Nanda, H., Goh, H. Z., Bachert, C., Lösche, M., & Linstedt, A. D. (2014). Myristoylation restricts orientation of the GRASP domain on membranes and promotes membrane tethering. *The Journal of Biological Chemistry*, *289*(14), 9683–9691.

Heinrich, F., Ng, T., Vanderah, D. J., Shekhar, P., Mihailescu, M., Nanda, H., et al. (2009). A new lipid anchor for sparsely tethered bilayer lipid membranes. *Langmuir: The ACS Journal of Surfaces and Colloids*, *25*(7), 4219–4229.

Hollmann, O., Steitz, R., & Czeslik, C. (2008). Structure and dynamics of alpha-lactalbumin adsorbed at a charged brush interface. *Physical Chemistry Chemical Physics (PCCP)*, *10*(10), 1448–1456.

Holt, S. A., Le Brun, A. P., Majkrzak, C. F., McGillivray, D. J., Heinrich, F., Lösche, M., et al. (2009). An ion-channel-containing model membrane: Structural determination by magnetic contrast neutron reflectometry. *Soft Matter*, *5*(13), 2576–2586.

Junghans, A., Watkins, E. B., Barker, R. D., Singh, S., Waltman, M. J., Smith, H. L., et al. (2015). Analysis of biosurfaces by neutron reflectometry: From simple to complex interfaces. *Biointerphases*, *10*(1), 019014.

Kent, M. S., Murton, J. K., Sasaki, D. Y., Satija, S., Akgun, B., Nanda, H., et al. (2010). Neutron reflectometry study of the conformation of HIV Nef bound to lipid membranes. *Biophysical Journal*, *99*(6), 1940–1948.

Kirby, B. J., Kienzle, P. A., Maranville, B. B., Berk, N. F., Krycka, J., Heinrich, F., et al. (2012). Phase-sensitive specular neutron reflectometry for imaging the nanometer scale composition depth profile of thin-film materials. *Current Opinion in Colloid & Interface Science*, *17*(1), 44–53.

Kučerka, N., Nagle, J. F., Sachs, J. N., Feller, S. E., Pencer, J., Jackson, A., et al. (2008). Lipid bilayer structure determined by the simultaneous analysis of neutron and X-ray scattering data. *Biophysical Journal*, *95*(5), 2356–2367.

Kunz, K., Reiter, J., Götzelmann, A., & Stamm, M. (1993). Model-free analysis of neutron reflectivity data from polymer thin-films with the simulated annealing technique. *Macromolecules*, *26*(16), 4316–4323.

Lakey, J. H. (2009). Neutrons for biologists: A beginner's guide, or why you should consider using neutrons. *Journal of the Royal Society Interface*, *6*, S567–S573.

Laub, C. F., & Kuhl, T. L. (2006). Fitting a free-form scattering length density profile to reflectivity data using temperature-proportional quenching. *Journal of Chemical Physics*, *125*(24), 244702.

Le Brun, A. P., Holt, S. A., Shah, D. S. H., Majkrzak, C. F., & Lakey, J. H. (2011). The structural orientation of antibody layers bound to engineered biosensor surfaces. *Biomaterials*, *32*(12), 3303–3311.

Majewski, J., Wong, J. Y., Park, C. K., Seitz, M., Israelachvili, J. N., & Smith, G. S. (1998). Structural studies of polymer-cushioned lipid bilayers. *Biophysical Journal*, *75*(5), 2363–2367.

Majkrzak, C. F., & Berk, N. (1995). Exact determination of the phase in neutron reflectometry. *Physical Review B: Condensed Matter*, *52*(15), 10827–10830.

Majkrzak, C. F., & Berk, N. F. (2003). Phase sensitive reflectometry and the unambiguous determination of scattering length density profiles. *Physica B: Condensed Matter*, *336*, 27–38.

Majkrzak, C. F., Berk, N. F., Krueger, S., Dura, J. A., Tarek, M., Tobias, D. J., et al. (2000). First-principles determination of hybrid bilayer membrane structure by phase-sensitive neutron reflectometry. *Biophysical Journal*, *79*, 3330–3340.

Majkrzak, C. F., Berk, N. F., Krueger, S., & Perez-Salas, U. (2006). Structural investigation of membranes in biology by neutron reflectometry. In J. Fitter, T. Gutberlet, & J. Katsaras (Eds.), *Neutron scattering in biology* (pp. 225–263). Berlin, Heidelberg, New York: Springer.

Majkrzak, C. F., Carpenter, E., Heinrich, F., & Berk, N. F. (2011). When beauty is only skin deep; Optimizing the sensitivity of specular neutron reflectivity for probing structure beneath the surface of thin films. *Journal of Applied Physics*, *110*(10), 102212.

McGillivray, D. J., Valincius, G., Heinrich, F., Robertson, J. W. F., Vanderah, D. J., Febo-Ayala, W., et al. (2009). Structure of functional *Staphylococcus aureus* alpha-hemolysin channels in tethered bilayer lipid membranes. *Biophysical Journal*, *96*(4), 1547–1553.

McGillivray, D. J., Valincius, G., Vanderah, D. J., Febo-Ayala, W., Woodward, J. T., Heinrich, F., et al. (2007). Molecular-scale structural and functional characterization of sparsely tethered bilayer lipid membranes. *Biointerphases*, *2*(1), 21–33.

Nagle, J. F., & Tristram-Nagle, S. A. (2000). Structure of lipid bilayers. *Biochimica et Biophysica Acta*, *1469*(3), 159–195.

Nanda, H., Datta, S. A. K., Heinrich, F., Lösche, M., Rein, A., Krueger, S., et al. (2010). Electrostatic interactions and binding orientation of HIV-1 matrix studied by neutron reflectivity. *Biophysical Journal*, *99*(8), 2516–2524.

Pfefferkorn, C. M., Heinrich, F., Sodt, A. J., Maltsev, A. S., Pastor, R. W., & Lee, J. C. (2012). Depth of α-synuclein in a bilayer determined by fluorescence, neutron reflectometry, and computation. *Biophysical Journal*, *102*(3), 613–621.

Politsch, E. (2001). String fit: A new structurally oriented X-ray and neutron reflectivity evaluation technique. *Journal of Applied Crystallography*, *34*, 239–251.

Politsch, E. (2003). Structure-related simultaneous X-ray and neutron reflectivity analysis method for Langmuir films. *Langmuir: The ACS Journal of Surfaces and Colloids*, *19*(4), 1221–1226.

Politsch, E., & Cevc, G. (2000). Determination of the molecular arrangement from neutron or X-ray reflectivity. *Physica B: Condensed Matter*, *276*, 390–391.

Politsch, E., & Cevc, G. (2002). Unbiased analysis of neutron and X-ray reflectivity data by an evolution strategy. *Journal of Applied Crystallography*, *35*, 347–355.

Schalke, M., Krüger, P., Weygand, M., & Lösche, M. (2000). Submolecular organization of DMPA in surface monolayers: Beyond the two-layer model. *Biochimica et Biophysica Acta*, *1464*(1), 113–126.

Shekhar, P., Nanda, H., Lösche, M., & Heinrich, F. (2011). Continuous distribution model for the investigation of complex molecular architectures near interfaces with scattering techniques. *Journal of Applied Physics*, *110*(10), 102216–10221612.

Shenoy, S., Moldovan, R., Fitzpatrick, J., Vanderah, D. J., Deserno, M., & Lösche, M. (2010). In-plane homogeneity and lipid dynamics in tethered bilayer lipid membranes (tBLMs). *Soft Matter*, *2010*(6), 1263–1274.

Shenoy, S. S., Nanda, H., & Lösche, M. (2012). Membrane association of the PTEN tumor suppressor: Electrostatic interaction with phosphatidylserine-containing bilayers and regulatory role of the C-terminal tail. *Journal of Structural Biology*, *180*(3), 394–408.

Shenoy, S., Shekhar, P., Heinrich, F., Daou, M.-C., Gericke, A., Ross, A. H., et al. (2012). Membrane association of the PTEN tumor suppressor: Molecular details of the protein-membrane complex from SPR binding studies and neutron reflection. *PLoS One*, 7(4), e32591.

Singh, S., Junghans, A., Tian, J., Dubey, M., Gnanakaran, S., Chlistunoff, J., et al. (2013). Polyelectrolyte multilayers as a platform for pH-responsive lipid bilayers. *Soft Matter*, *9*(37), 8938–8948.

Smith, H. L., Hickey, J., Jablin, M. S., Trujillo, A., Freyer, J. P., & Majewski, J. (2010). Mouse fibroblast cell adhesion studied by neutron reflectometry. *Biophysical Journal*, *98*(5), 793–799.

Smith, H. L., Jablin, M. S., Vidyasagar, A., Saiz, J., Watkins, E., Toomey, R., et al. (2009). Model lipid membranes on a tunable polymer cushion. *Physical Review Letters*, *102*(22), 228102.

Song, L., Hobaugh, M., Shustak, C., Cheley, S., Bayley, H., & Gouaux, J. (1996). Structure of staphylococcal alpha-hemolysin, a heptameric transmembrane pore. *Science, 274*(5294), 1859–1866.

Teixeira, S. C. M., Ankner, J., Bellissent-Funel, M. C., Bewley, R., Blakeley, M. P., Coates, L., et al. (2008). New sources and instrumentation for neutrons in biology. *Chemical Physics, 345*(2–3), 133–151.

Truschel, S. T., Sengupta, D., Foote, A., Heroux, A., Macbeth, M. R., & Linstedt, A. D. (2011). Structure of the membrane-tethering GRASP domain reveals a unique PDZ ligand interaction that mediates Golgi biogenesis. *The Journal of Biological Chemistry, 286*(23), 20125–20129.

Valincius, G., Heinrich, F., Budvytyte, R., Vanderah, D. J., McGillivray, D. J., Sokolov, Y., et al. (2008). Soluble amyloid beta-oligomers affect dielectric membrane properties by bilayer insertion and domain formation: Implications for cell toxicity. *Biophysical Journal, 95*(10), 4845–4861.

Valincius, G., Meškauskas, T., & Ivanauskas, F. (2012). Electrochemical impedance spectroscopy of tethered bilayer membranes. *Langmuir: The ACS Journal of Surfaces and Colloids, 28*(1), 977–990.

van der Lee, A., Salah, F., & Harzallah, B. (2007). A comparison of modern data analysis methods for X-ray and neutron specular reflectivity data. *Journal of Applied Crystallography, 40*, 820–833.

Wacklin, H. P. (2010). Neutron reflection from supported lipid membranes. *Current Opinion in Colloid & Interface Science, 15*(6), 445–455.

Wacklin, H. P. (2011). Composition and asymmetry in supported membranes formed by vesicle fusion. *Langmuir: The ACS Journal of Surfaces and Colloids, 27*(12), 7698–7707.

Wiener, M. C., & White, S. H. (1992). Structure of a fluid dioleoylphosphatidylcholine bilayer determined by joint refinement of x-ray and neutron diffraction data. III. Complete structure. *Biophysical Journal, 61*(2), 434–447.

Yap, T. L., Jiang, Z., Heinrich, F., Gruschus, J. M., Pfefferkorn, C. M., Barros, M., et al. (2015). Structural features of membrane-bound glucocerebrosidase and α-synuclein probed by neutron reflectometry and fluorescence spectroscopy. *The Journal of Biological Chemistry, 290*(2), 744–754.

CHAPTER EIGHT

Deuterium Labeling Strategies for Creating Contrast in Structure–Function Studies of Model Bacterial Outer Membranes Using Neutron Reflectometry

Anton P. Le Brun*[,1], Luke A. Clifton[†], Stephen A. Holt[‡], Peter J. Holden*, Jeremy H. Lakey[§]

*National Deuteration Facility, Bragg Institute, Australian Nuclear Science and Technology Organization, Lucas Heights, New South Wales, Australia
[†]ISIS Neutron Facility, STFC Rutherford Appleton Laboratory, Didcot, Oxfordshire, United Kingdom
[‡]Bragg Institute, Australian Nuclear Science and Technology Organisation, Lucas Heights, New South Wales Australia
[§]Institute for Cell and Molecular Biosciences, Newcastle University, Framlington Place, Newcastle upon Tyne, United Kingdom
[1]Corresponding author: e-mail address: abn@ansto.gov.au

Contents

Abstract

Studying the outer membrane of Gram-negative bacteria is challenging due to the complex nature of its structure. Therefore, simplified models are required to undertake structure–function studies of processes that occur at the outer membrane/fluid interface. Model membranes can be created by immobilizing bilayers to solid supports such as gold or silicon surfaces, or as monolayers on a liquid support where the surface pressure and fluidity of the lipids can be controlled. Both model systems are amenable to having their structure probed by neutron reflectometry, a technique that provides a one-dimensional depth profile through a membrane detailing its thickness and composition. One of the strengths of neutron scattering is the ability to use contrast

Methods in Enzymology, Volume 566
ISSN 0076-6879
http://dx.doi.org/10.1016/bs.mie.2015.05.020

matching, allowing molecules containing hydrogen and those enriched with deuterium to be highlighted or matched out against the bulk isotopic composition of the solvent. Lipopolysaccharides, a major component of the outer membrane, can be isolated for incorporation into model membranes. Here, we describe the deuteration of lipopolysaccharides from rough strains of *Escherichia coli* for incorporation into model outer membranes, and how the use of deuterated materials enhances structural analysis of model membranes by neutron reflectometry.

1. INTRODUCTION

The cell envelope of Gram-negative bacteria is a complex multicomponent structure which can be generally divided into three different sections (Ruiz, Kahne, & Silhavy, 2006). The first is the inner membrane consisting of a phospholipid bilayer with membrane proteins. The proteins can be either integral membrane proteins containing α-helical transmembrane domains or lipoproteins anchored to a leaflet of the membrane through fatty acid modification. The next layer is the periplasm which contains cross-linked chains of peptidoglycan that maintain cell shape and turgidity. The periplasm is a hydrophilic environment that contains enzymes for disulphide bond formation, protein folding and trafficking, and small molecule transport. The final section is the outer membrane (OM). The OM is highly asymmetric with the inner leaflet composed of phospholipids and the outer leaflet composed of lipopolysaccharides (LPS) (Nikaido, 2003). There are two types of protein found in the OM—integral outer membrane proteins and lipoproteins. The former generally have a β-barrel structure and mainly serve as channels for the uptake of nutrients and excretion of waste molecules but some also act as enzymes and adhesion molecules (Bishop, 2008; Delcour, 2009; Krishnan & Prasadarao, 2012). The OM lipoproteins are exemplified by Braun's lipoprotein which anchors the OM to the peptidoglycan in the periplasm (Dramsi, Magnet, Davison, & Arthur, 2008). LPS, the main component of the outer leaflet of the OM, is also a highly complex molecule consisting of lipid A, a core polysaccharide and the O-antigen (Fig. 1; Wilkinson, 1996). Lipid A is a phosphorylated glucosamine disaccharide acylated with four to seven fatty acids and anchors LPS into the OM (Raetz, Reynolds, Trent, & Bishop, 2007). Covalently linked to lipid A is the core polysaccharide, which is conserved across bacterial species and can generally be divided into an inner core consisting of 2-keto-3-deoxyoctonoic acid (Kdo) and heptoses and an outer core

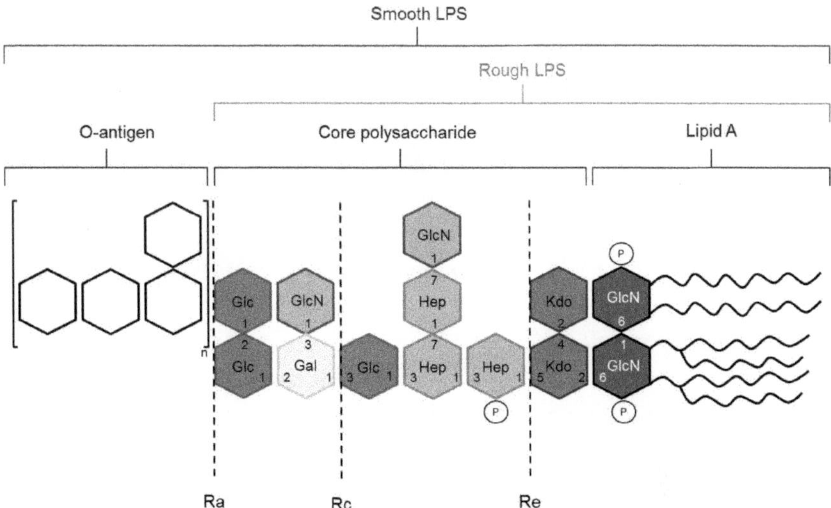

Figure 1 The general structure of LPS from B strains of *E. coli* showing the R3 core. The numbers in the colored hexagons indicate the carbons through which the glycosidic bonds are formed between each sugar moiety and the abbreviations are Kdo, 2-keto-3-deoxyoctonic acid; Hep, L-glycero-D-manno heptose; Glc, glucose; Gal, galactose; GlcN, 2-amino-2-deoxyglucose; P (in circle), phosphate. The lipid A tails are represented by the wavy lines and consist of four (R)-3-hydroxy-mystic acids, one myristic acid, and one lauric acid. (See the color plate.)

consisting of hexoses and hexosamines (Raetz & Whitfield, 2002). The final part of LPS is the O-antigen which is a long chain of polysaccharide that extends out into the environment. The O-antigen is highly variable and differs between different bacterial strains (Liu et al., 2014; Stenutz, Weintraub, & Widmalm, 2006). Bacterial strains which express LPS containing the O-antigen are termed smooth strains, whereas strains with LPS consisting of just lipid A and the core polysaccharide (or a portion of the core) are termed rough strains (Fig. 1).

Structure–function studies of the OM of Gram-negative bacteria are of interest due to the OM's important role in adhesion, infection, antibiotic uptake or resistance, and adjustment to environmental stresses (Nikaido, 2003). However due to the small size, complex structure and dynamic nature of the OM, a reductionist approach is needed to embark upon such structure–function studies and simplified models of the OM need to be created. There are a number of different methods for creating model biological membranes that mimic a particular aspect or property of a native membrane, with many of the models confining the membrane to a solid or liquid support (Le Brun, Darwish & James, 2013). Membranes in the form of bilayers on silicon, gold, quartz, or other surfaces or a lipid monolayer on a liquid surface are amenable to neutron reflectometry, a technique which probes the structure of surfaces, thin films, and buried interfaces at nanoscale dimensions by providing a one-dimensional depth profile through the sample. Each section is described as a different layer defined by its thickness and composition (Wacklin, 2010). Neutrons are suited for studying biological systems as they are (i) highly penetrating, enabling model membranes to be studied under different sample environments and under physiological conditions; (ii) nondestructive; and (iii) have the ability to discriminate between hydrogen and its isotope deuterium facilitating identification of the different components in complex systems (Clifton, Neylon & Lakey, 2013; Clifton et al., 2012; Wacklin, 2010). Enriching molecules with deuterium can be done biologically by growing microorganisms in deuterium oxide (commonly known as heavy water or D_2O) and then purifying the molecule of interest. This has been achieved for proteins (Blakeley et al., 2008; Chen et al., 2012; Clifton et al., 2012) and nucleic acids (Shotton et al., 1997). An alternative method is to label small precursor molecules with deuterium using hydrothermal exchange under pressure in the presence of a catalyst, followed by organic synthesis to produce the desired molecule—as has been described for phospholipids (Darwish et al., 2013; Yepuri et al., 2014). LPS is a major component of the OM covering approximately 75% of the bacterial surface (Nikaido & Vaara, 1987), making it an essential component in any model of the OM. In this chapter, the basic principles of neutron reflectometry and the studies of monolayers and bilayers containing LPS are discussed and methods for producing deuterated rough LPS for such studies are described.

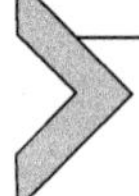

2. BRIEF INTRODUCTION TO NEUTRON REFLECTOMETRY AND DATA ANALYSIS

Specular neutron reflectometry measures the reflection of a beam of neutrons from an interface. A collimated beam of neutrons from either a reactor source or a spallation source is directed at a surface at glancing angles (typically less than 5°). Above the critical angle (below which total external reflection occurs), some of the neutrons will be transmitted through the sample, some will be absorbed by the sample and some will be reflected by the sample. The reflectivity (simply the number of reflected neutrons divided by the number of neutrons incident on the sample) is measured as a function of momentum transfer, Q, defined as:

$$Q = \frac{4\pi \sin \theta}{\lambda}$$

where θ is the angle of incidence and λ is the neutron wavelength. Thus, there are two ways to measure reflectivity: either by using a monochromatic neutron wavelength and varying the angle of incidence or by using a white beam of neutrons and measuring at a few of angles of incidence to cover the desired Q range. Neutron reflectometry can be conducted at a number of different instruments at various neutron facilities around the world, and a list of instruments can be found at: http://en.wikipedia.org/wiki/Neutron_reflectometry.

Reflectivity decreases with increasing momentum transfer. For example, a clean liquid surface with no surface film results in a featureless reflectivity profile which decreases by Q^4 (Fig. 2). However, if a film is deposited onto the liquid surface, there will be constructive and destructive interference between the reflected beams from the liquid/film interface and the film/air interface. This results in fringes known as Kiessig fringes (Fig. 2). The width of the fringes, ΔQ_f, is related to the thickness of the film, d, by

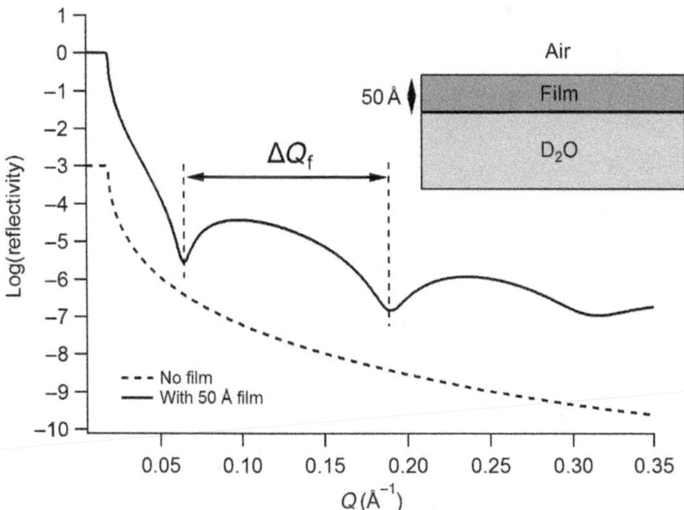

Figure 2 A simulation comparing the reflectivity data of a D_2O surface at the air–liquid interface with no film (black trace that is offset for clarity) with that of a 50-Å single layer film deposited (red trace (black in the print version)) showing Kiessig fringes appearing at 0.06–0.19 and 0.19–0.31 Å^{-1}.

$$\Delta Q_f = \frac{2\pi}{d}$$

and the intensity of the fringe is determined by the difference between the neutron scattering power of the film compared with the substrate.

The structural information provided by neutron reflectometry is in the form of a one-dimensional scattering length density profile through the sample, in essence providing a depth profile. A neutron scattering length density (nSLD) can be considered analogous to a neutron's refractive index and can be determined from the physical properties of a material using the following equation:

$$\text{nSLD} = N_A \sum_i \frac{p_i}{A_i} b_i$$

where N_A is Avogadro's number, p_i is the physical density, A_i is the atomic mass, and b_i is the nuclear scattering length of component i. Nuclear scattering lengths vary from element to element and from isotope to isotope. The difference in nuclear scattering length between hydrogen, ^1H ($b_H = -3.74 \times 10^{-5}$ Å), and its stable isotope deuterium, ^2H ($b_D = +6.64 \times 10^{-5}$ Å) is particularly useful in biological systems. Lipids, which mainly contain hydrogen and carbon ($b_C = +6.65 \times 10^{-5}$ Å), will have a different nSLD to proteins which also contain oxygen ($b_O = +5.81 \times 10^{-5}$ Å) and nitrogen ($b_N = +9.36 \times 10^{-5}$ Å) in large proportion. This means that solvent contrast matching can be utilized. D_2O has a higher nSLD (6.35×10^{-6} Å^{-2}) than H_2O (-0.56×10^{-6} Å^{-2}) and by varying the ratio of D_2O to H_2O in an aqueous system the nSLD of the bulk solution can be adjusted to be the same as a particular component rendering that component invisible to neutrons against the bulk solution. For example, the hydrocarbon tails of the lipid dipalmitoylphosphatidylcholine (DPPC) ($\times 2$ $CH_3(CH_2)_{14}-$) have an nSLD of -0.39×10^{-6} Å^{-2} when in the fluid phase and will be closely matched to H_2O, whereas hydrogenous proteins typically have an nSLD of $\sim 2.2 \times 10^{-6}$ Å^{-2} and will match out in an isotopic contrast consisting of ~40% D_2O/60% H_2O. One useful contrast when studying monolayers at the air–liquid interface is using a mixture of 8.2% D_2O to 91.8% H_2O by volume resulting in an nSLD of zero which is termed null reflecting water (NRW) or air contrast matched water. This means that the measured reflectometry signal will be purely from the material at the air/water interface. Disentangling the details of complex biological systems requires not only having contrast against the bulk surrounding but also between the components that make up the complex. In this case, contrast can be achieved through enriching one of the components with deuterium, and thus increasing its nSLD against the other molecule (Fig. 3).

For modeling neutron reflectivity data, the system studied can be described using a slab model which entails dividing the interface into a series of layers each of which is defined by three parameters: thickness, nSLD, and roughness. From the initial model, the theoretical reflectivity is calculated and compared to the experimental reflectivity. The parameters for each layer are then varied until the theoretical reflectivity matches the experimental data. This can be done using a range of specialized software packages including MOTOFIT (Nelson, 2006), RasCAL (Hughes, 2010), and Reflfit (Kienzle, O'Donovan, Ankner, Berk, & Majkrzak, 2000), all of which

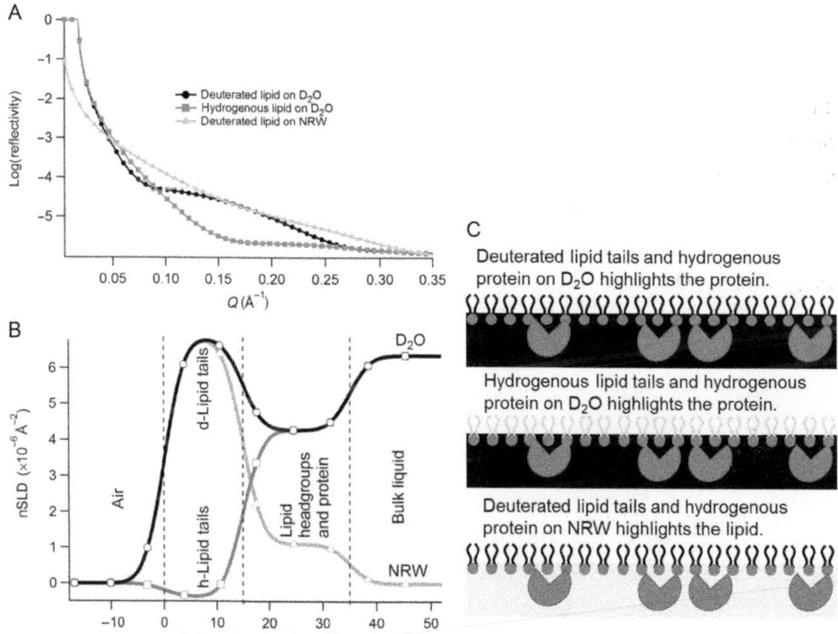

Figure 3 Contrast matching in neutron reflectometry. (A) Simulated reflectometry data of three different contrasts: tail deuterated lipid with hydrogenous protein on D_2O, hydrogenous lipid with hydrogenous protein on D_2O, and tail deuterated lipid with hydrogenous protein on NRW. (B) The corresponding nSLD profile with the vertical dashed lines delineating the different layers. (C) A cartoon describing which parts of the membrane system are highlighted in the different contrasts used.

utilize least-square fitting routines to minimize χ^2 (Chi-squared) values to achieve the best fit by using Levenburg–Marquardt and/or genetic algorithm procedures. In the vast majority of cases, a single neutron reflectivity profile cannot provide a unique solution to the final structural model. This is because the phase information is lost in the reflectometry data. A way to alleviate this problem is to simultaneously fit multiple data sets from the same system but using different isotopic contrasts, since an incorrect structural model is unlikely to suitably fit all contrasts. When simultaneously fitting multiple contrast data, it is assumed that the thickness and roughness for each layer in the model is the same for each contrast and that only the nSLD varies with changing isotopic contrast. The final resulting real-space nSLD profile can then be interpreted in terms of molecular structure with the layers of the nSLD profile each describing a different part of the system (Fig. 3). For instance, the acyl chains in phospholipids have a distinct nSLD from the polar headgroups, and thus can be modeled as separate layers. In Fig. 3B, the peak of nSLD at distances of 0–15 Å in the NRW contrast can be interpreted as originating from the deuterated lipid tails as the nSLD of this layer can only correspond to the nSLD of deuterated acyl chains. Additional evidence for this layer consisting of lipid tails is that lipid tails have no hydrogens that are exchangeable with the surrounding liquid, so the nSLD of the lipid tails should remain the same regardless of the nSLD of the liquid. If protein has not penetrated into the lipid tails region and there is complete coverage of lipid, the acyl chain layer nSLD will remain unchanged. The next layer observed is consistent with a protein layer bound to the lipid headgroups. Proteins contain hydroxyl, amine, and thiol groups that all have hydrogens that are exchangeable with solution hydrogens and deuteriums. The level of deuteration of these exchangeable protons is proportional to the amount of D_2O in the solvent. Consequently, a protein's nSLD will vary with the nSLD of the solution. In Fig. 3B, it can be seen that the nSLD of the protein layer from 15 to 25 Å changes when the solution is varied from D_2O to NRW, revealing this layer to be the protein containing layer. If the structure and dimensions of the protein are known, then the orientation of the bound protein can possibly be inferred from the thickness of this layer and this has been done using computationally advanced methods utilizing known X-ray crystallography or NMR structures, and molecular dynamics simulations for determining accurate orientation and penetration of proteins binding to tethered lipid bilayers (Heinrich & Lösche, 2014; Nanda, Heinrich, & Lösche, 2015).

Some form of error analysis should be completed on the data when the final real-space nSLD profile has been determined. This can be done using statistical techniques such as Monte Carlo resampling (Heinrich et al., 2009; Holt et al., 2009), where approximately 1000 synthetic data sets are obtained. The parameter ranges obtained from the best fit are used in the resampling procedure, where 1000 separate datasets are "synthesized" from the original dataset by adding Gaussian noise, weighted by the counting statistics of the real data, to each data point. These synthesized datasets are fitted individually using a computing cluster. The fits to the synthetic data are analyzed by binning the fitted values for each parameter into intervals and plotting a histogram of the frequencies of each bin. The distribution of each parameter can then be statistically analyzed with the parameter value taken as the midpoint of the 95% confidence interval and its uncertainty from the standard deviation of the distribution.

3. ISOLATION OF DEUTERATED LPS FROM ROUGH STRAINS OF *ESCHERICHIA COLI*

LPS is a major component of the OM and there are many proteins and other ligands that associate with the OM through LPS. Therefore, in some neutron reflectometry studies, it can be beneficial to use a deuterated version of LPS to provide contrast between such protein and ligand molecules in neutron reflectometry experiments. In most cases, perdeuteration, that is where 100% of the nonexchangeable hydrogens are replaced with deuterium, is not necessary as perdeuterated molecules can have a higher nSLD than D_2O, as is the case with perdeuterated proteins and lipids. This results in the molecule being no longer matched out in the D_2O contrast. In most neutron reflectometry experiments, having an nSLD close to D_2O is not always necessary as deuteration to the point where the nSLD of the deuterated molecule is significantly higher than that of the hydrogenous components in the system is sufficient. Therefore, partial deuteration is sufficient where only a certain percentage of the nonexchangeable hydrogens are exchanged for deuterium. This has the advantage of achieving the necessary contrast for the neutron reflectometry experiment while reducing the amount of costly deuterated starting materials needed. This section describes the procedure to obtain a partially deuterated rough LPS from *E. coli*. The strain used is *E. coli* J5 (ATCC # 43745), which is a rough mutant of *E. coli* O111:B4 (Elbein & Heath, 1965). The original rough mutants were observed in *Salmonella minnesota* and we use the equivalent chemotypes to denote *E. coli* rough mutants (Raetz, 1990), and hence we use the term Rc-LPS to describe the LPS obtained from *E. coli* J5. Figure 1 shows the generalized configuration of the core polysaccharide from *E. coli* J5, which was originally determined by NMR analysis (Müller-Loennies, Holst, & Brade, 1994; Müller-Loennies, Holst, Lindner, & Brade, 1999).

After streaking the bacteria onto LB agar plates, the colonies can be picked off to inoculate a small scale culture (10 mL) in LB medium (Bertani, 1951) incubated at 37 °C with shaking overnight. The next stage is to adapt the *E. coli* J5 to ModC1 medium (Chen et al., 2012; Middelberg, O'Neill, Bogle, & Snoswell, 1991). ModC1 can be prepared by dissolving the following to final concentrations in either H_2O or D_2O: 40 g L^{-1} glycerol, 4.16 g L^{-1} Na_2HPO_4, 2.58 g L^{-1} NH_4Cl, 2.54 g L^{-1} KH_2PO_4, and 1.94 g L^{-1} K_2SO_4. The solution can then be autoclaved if the bulk salts are made in pure H_2O or if any amount of D_2O is used, the solution should be filter-sterilized (0.22 μm) as the vapor from autoclaving potentially lowers the D_2O content. Four trace element stock solutions need to be made at 1000× concentrations as follows:

Solution A: 8.8 g per 100 mL sodium citrate and 0.912 g per 100 mL $FeCl_2$.

Solution B: 0.437 g per 100 mL $MnCl_2$, 0.408 g per 100 mL $ZnCl_2$, and 0.049 g per 100 mL $CuSO_4$.

Solution C: 4.8 g per 100 mL Thiamine–HCl.

Solution D: 32.7 g per 100 mL $MgSO_4$ (anhydrous).

The trace element solutions need to be 0.22 μm filtered regardless of whether they contain D_2O or H_2O and solution C should be stored in the dark. When making up the final medium, the four trace element solutions should be added to a final concentration of 1× immediately before use. To adapt the cells grown in the overnight LB culture, 10 mL of ModC1 made with H_2O is inoculated with 250 μL of the LB culture and grown to an $OD_{600} > 1.0$ at 37 °C with shaking at 220 rpm. This culture is then used to inoculate 100 mL of ModC1 made with 100% D_2O. This is then grown to $OD_{600} > 1.0$ in a sealed conical flask at 37 °C with shaking. A conical flask with large volume (e.g., 2 L) should be used at this point to increase the surface area of the culture for maximum aeration. The D_2O culture is then used to inoculate a bioreactor containing 1 L of ModC1 made with 100% D_2O to a starting OD_{600} of at least 0.1. The culture should then be grown at 37 °C until the dissolved oxygen content reaches 30% at which point the temperature should be reduced to 20 °C and the culture grown further until the carbon source is exhausted. This is indicated by an increase in the pH and usually occurs between OD_{600} of 20 and 30.

The cells are harvested by centrifugation and then washed by resuspending in H_2O and recentrifuging—this wash step is repeated twice. The cells are then flash frozen in liquid nitrogen and freeze dried to remove all moisture from the cells. At this point, the dried cells can be stored at −80 °C.

The method used to purify the Rc-LPS is based on one originally described by Galanos, Lüderitz, and Westphal (1969). To extract the Rc-LPS, the thawed cells need to be homogenized in phenol/chloroform/ petroleum spirit (2:5:8 by volume) using 10 mL g^{-1} of dried cells. This needs to be done gently so that the cells are not ruptured. This can be done with either a homogenizer (Potter-Elvehjem or motorized) or by placing the cells and organic solvent in a stoppered round bottomed flask with a magnetic stirrer bar set to stir gently for 3 h. After homogenization, the cell mass is removed from the solvent by a low speed centrifugation at 4000 × *g* for 15 min in glass centrifuge tubes. The cell mass can be rehomogenized twice more to achieve maximum yield. The Rc-LPS-containing solvent should be a golden brown color and large particulates can be removed by vacuum filtration using course filter paper. The chloroform and petroleum spirit should be removed by rotary evaporation. The LPS can then be precipitated by the drop wise addition of acetone/diethyl ether (1:1) until a white precipitate forms. Leave the solution to fully precipitate overnight at room temperature in stoppered glass tubes. Centrifuge at 4000 × *g* for 30 min and carefully remove and discard supernatant. Wash the pellet in 80% phenol to remove any protein contamination and then in diethyl ether to remove any residual phenol with a centrifugation step of 4000 × *g* for 30 min between each wash. The pellet can then be resuspended in ultra-pure water with heating to 45 °C and sonication. The solution should then be centrifuged at 100,000 × *g* for 4 h and the resulting pellet should contain pure Rc-LPS.

The Rc-LPS can be resuspended in pure water or buffer for analysis. Protein contamination can be monitored by running a sample on SDS–PAGE with staining in Coomassie Blue. Alternatively, measuring the absorption at 280 nm can estimate protein content, while nucleic acid contamination can be checked by measuring the absorption of samples at 260 nm. The purified Rc-LPS can also be further verified through assaying for the presence of the Kdo saccharide. This is done by using an assay first described by Karkhanis, Zeltner, Jackson, and Carlo (1978) and later modified by Proskurnin et al. (2010) (by replacing toxic arsenic sulfate with much safer sodium sulfate). The assay oxidizes Kdo released from LPS by acid hydrolysis to formylpyruvic acid, which is then converted to a chromophore with absorption at 548 nm. Determination of deuteration can be done

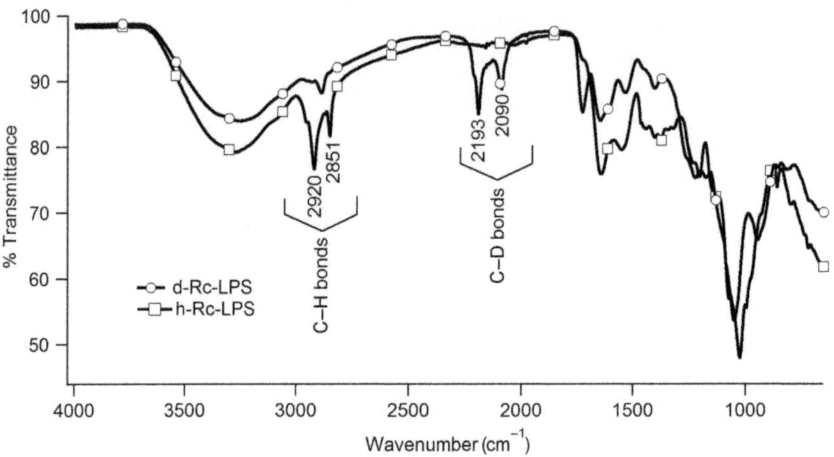

Figure 4 FTIR spectra of hydrogenous Rc-LPS (black) and deuterated Rc-LPS (red (black in the print version)) from *E. coli* J5. The peaks at 2193 and 2090 correspond to C–D bonds and the peaks at 2920 and 2851 correspond to C–H bonds. *Reproduced from Le Brun, Clifton, et al. (2013) with permission, copyright © 2013 American Chemical Society.*

qualitatively by conducting FTIR measurements on a few milligrams of solid material. A double peak corresponding to carbon–deuterium bonds will appear at wavenumbers ~2193 and ~2089 cm^{-1} as opposed to carbon–hydrogen which will appear at ~2919 and ~2851 cm^{-1} (Fig. 4).

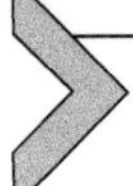

4. FORMATION OF ROUGH LPS MONOLAYERS AT THE AIR–LIQUID INTERFACE

Rough LPS from *E. coli* can be deposited at the air–liquid interface (Clifton et al., 2013; Le Brun et al., 2013). The LPS has to be dissolved in chloroform/methanol/water at 1 mg mL^{-1} to create a cloudy suspension (the solution may need sonication to fully dissolve). It is best to form the monolayers on a liquid surface contained within a Langmuir trough so that surface pressure can be controlled. The trough and the barriers should be thoroughly cleaned with a suitable organic solvent such as chloroform or propan-2-ol before filling with the liquid substrate. The rough LPS can be deposited onto liquid surfaces containing a variety of buffers and salts as well as EDTA. Using a Hamilton syringe, which has been extensively cleaned with the same chloroform/methanol/water mixture the LPS is dissolved in, the LPS solution is added to the liquid surface drop wise, making sure that the height the drop has to fall is less than 1 cm. If depositing onto a liquid surface that is contained within a Langmuir trough, the mobile barrier should be set to approximately half of the trough area. Once a surface pressure of ~5 mN m^{-1} has been reached the barriers should be opened until the surface pressure has reduced to 0 mN m^{-1}. If at this point, the barriers are not fully open then more LPS can be deposited onto the surface. If this is the case, LPS should be added drop wise again until ~5 mN m^{-1} is reached and then the barriers reopened. The deposition can continue until the barriers are fully open and should ideally have a surface pressure of 0 mN m^{-1}. The LPS monolayer film can be compressed at slow speeds (<10 mN min^{-1}) to a collapse pressure of 30–50 mN m^{-1} depending on the buffer used in the liquid (Fig. 5A). The compression and relaxation of the film can be repeated a number of times; however, a small amount of the LPS material is lost to the bulk solution on each compression (Fig. 5B).

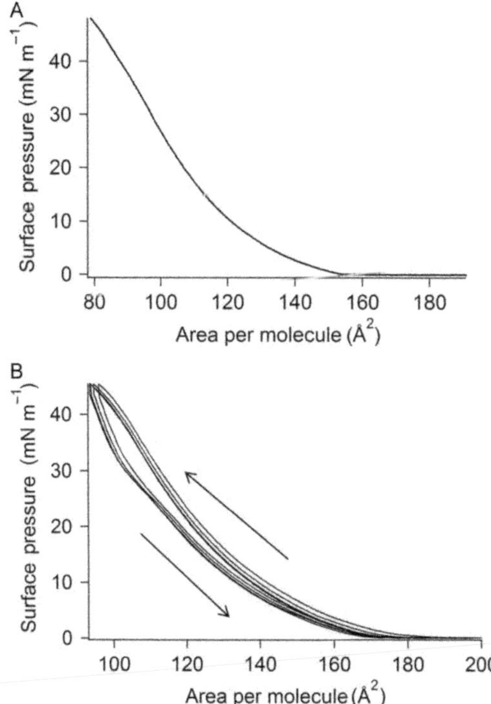

Figure 5 Pressure–area isotherms on monolayers of Rc-LPS on a liquid surface of 20 m*M* sodium phosphate, pH 7.0 in a Langmuir trough. (A) A single compression of a monolayer of Rc-LPS. (B) Multiple compressions and expansions of the Langmuir trough barriers of a monolayer of Rc-LPS. The arrows show the direction of the surface pressure upon compression and expansion. *Reproduced from Le Brun, Clifton, et al. (2013) with permission, copyright © 2013 American Chemical Society.*

5. MODELING MONOLAYERS OF Rc-LPS FROM NEUTRON REFLECTOMETRY DATA

For Rc-LPS and other lipid monolayer systems, three different contrasts can be used. Deuterated Rc-LPS can be deposited onto a subphase of NRW, which means that the reflectometry signal will be purely from the LPS monolayer and the maximum signal possible is achieved by depositing deuterated LPS in this case. The reverse contrast of hydrogenous Rc-LPS deposited onto a D_2O substrate should be done and a third additional contrast of deuterated Rc-LPS on D_2O can also be used to provide a unique solution to the structure. The three contrasts can be simultaneously fitted together (Le Brun, Clifton, et al., 2013; Fig. 6). Although the three monolayers are separate samples the thickness and roughness between each sample should be the same and only the scattering length density different. The Rc-LPS monolayer can be divided into three different layers (Fig. 6B), the first layer being the hydrocarbon tails which will have a low nSLD in the hydrogenous monolayer and a high nSLD in the deuterated monolayer. The second layer is composed of the sugar moieties of the Rc-LPS. As this layer contains solvent, the nSLD of the deuterated monolayer on either NRW or D_2O will vary between contrasts. From this difference, the amount of solvent in this layer can be calculated. The third and final layer is an outer layer of sugar headgroups with a higher solvent content then the inner sugar layer. The volume fraction of different components in Rc-LPS monolayers can be determined from neutron reflectometry using data from the different contrasts. If we take the case of a pure Rc-LPS monolayer, then the overall nSLD of the layer will consist of contributions from the LPS itself and any solvent that maybe be present in the layer:

$$\rho_{\text{layer}} = (\rho_{\text{LPS}}\varphi_{\text{LPS}}) + (\rho_{\text{water}}\varphi_{\text{water}}),$$

where ρ is nSLD of each component and φ is the volume fraction of each component. If using NRW then the water term becomes zero and it is straightforward to calculate the volume fraction of LPS. For the D_2O contrasts, the volume fraction can be calculated:

$$\varphi_{\text{LPS}} = \frac{\rho_{\text{layer}} - \rho_{\text{water}}}{\rho_{\text{LPS}} - \rho_{\text{water}}}.$$

From the volume fraction, the area per molecule (A) can be calculated as:

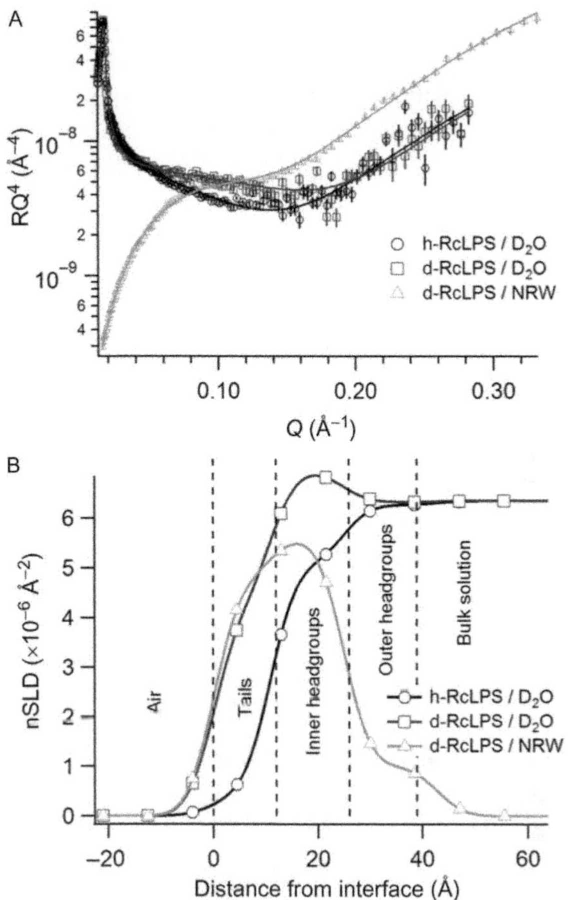

Figure 6 (A) Neutron reflectivity data (symbols with error bars) and fits (solid lines) of Rc-LPS at the air–liquid interface under pressure control at 20 mN m^{-1}. Three contrasts were fitted simultaneously with h-Rc-LPS on D_2O (black), d-Rc-LPS on D_2O (red (dark gray in the print version)), and d-Rc-LPS on NRW (green (light gray in the print version)). (B) The corresponding real-space nSLD profile with colors corresponding to those of the reflectivity profile and the vertical dashed lines showing the different layers. *Adapted from Le Brun, Clifton, et al. (2013) with permission, copyright © 2013 American Chemical Society.*

$$A = \frac{V_{\mathrm{m}}}{\varphi_{\mathrm{LPS}}\tau},$$

where τ is the thickness and V_{m} is the molecular volume. The molecular volume can be obtained from drawing out the molecule on web tools such as Molinspiration (http://www.molinspiration.com/cgi-bin/properties), from X-ray diffraction data for LPS, phospholipids, and sugars (Ferrier, 1960; Kastowsky, Gutberlet, & Bradaczek, 1993; Nagle & Tristram-Nagle, 2000; Snyder, Kim, & McIntosh, 1999), molecular dynamics simulations (Wu et al., 2013), or densitometry measurements (Sasaki & White, 2008). This results in a volume of 4255 Å^3 for Rc-LPS. The surface excess (amount per unit area), Γ, of the LPS on the liquid surface can then be calculated from:

$$\Gamma = \frac{1}{N_A A}.$$

From simultaneously fitting the multiple contrasts and calculating other values a detailed picture of the Rc-LPS monolayer structure can be developed, which is a starting point for investigating any changes in Rc-LPS structure upon binding other proteins, peptides, or ligands. These materials can be bound to the monolayer by injecting the material into the liquid subphase. Care must be taken while doing this not to disturb the monolayer so it is best to inject on the side of the barrier without the monolayer. Binding of analytes can be measured on the Langmuir trough by measuring the change in surface pressure with time as the analytes bind to the monolayer.

6. FORMING ASYMMETRIC BILAYERS USING PHOSPHOLIPIDS AND ROUGH LPS

The bacterial outer membrane is an asymmetric bilayer composed mainly of phospholipids in the inner leaflet and LPS in the outer leaflet. Langmuir–Blodgett and Langmuir–Schaefer techniques can be used to create such an asymmetric bilayer on silicon wafers (Fig. 7; Clifton, Skoda,

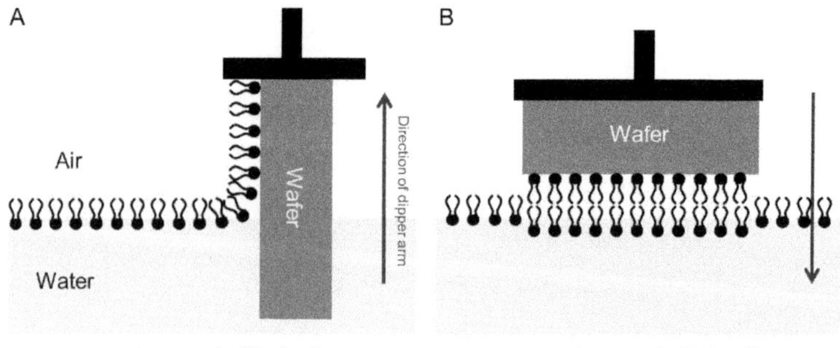

Figure 7 A generalized diagram of a monolayer deposition onto a solid support by (A) Langmuir–Blodgett dipping and (B) Langmuir–Schaefer dipping.

et al., 2013; Clifton et al., 2015). A Langmuir–Blodgett dip is where a solid wafer will be raised or lowered vertically through a monolayer on a liquid surface depositing the monolayer on the surfaces of the solid support as it passes through the monolayer (Fig. 7A). In Langmuir–Schaefer dipping the solid wafer is pushed horizontally through the monolayer to transfer the monolayer to the solid surface (Fig. 7B). This section will describe how to use dipping techniques and rough LPS monolayers to create asymmetric bilayers.

The first step is to prepare a silicon wafer with a polished surface which needs to be extensively cleaned. This can be done with extensive washes of ultra-pure water followed by UV-ozone treatment. Strongly acidic washes such as acid piranha solution (70% H_2SO_4 and 30% H_2O_2 by volume) which forms highly oxidizing peroxymonosulfuric acid (H_2SO_5, Caro's acid) can be used for particularly unclean wafers, but care needs to be taken as the procedure is dangerous and the acid solution needs to be neutralized before disposal. Due to the safety concerns of using acid piranha usually a dilute acid piranha solution (50% H_2O, 40% H_2SO_4, and 10% H_2O_2 by volume) is sufficient to clean wafers. A specialized Langmuir trough which has a well that is wide enough and deep enough to fit the silicon wafer when submerged has to be used. Typically, the silicon wafers will have the dimensions of $40 \times 80 \times 15\ mm^3$ so the troughs tend to be custom made to accommodate them. The trough barriers, pressure sensor, and dipping arm require calibration according to the manufacturer's instructions before use. The customized troughs may also have a laser scanner that can auto-adjust the level of the wafer with respect to the level of the water during the Schaefer dip and this should be calibrated also. The silicon block then needs to be fixed to the dipper arm and submerged below the liquid surface. The backing plate that is to be used to hold the wafer during neutron reflectometry experiments should be extensively cleaned in ethanol and be placed at the bottom of the well as the wafer will be driven into the backing plate once the Langmuir–Schaefer dip is complete.

The buffer used for the liquid surface should contain 5 m*M* $CaCl_2$ and be cooled to 10 °C to make sure that the best possible transfers are achieved. The phospholipid should then be deposited from a chloroform solution drop wise and then compressed slowly to 35 mN m^{-1}. When the target pressure of 35 mN m^{-1} is reached the Langmuir–Blodgett layer, deposition can commence and the wafer can be raised slowly (2 mm min^{-1}) up through the monolayer while maintaining a constant surface pressure of 35 mN m^{-1}. The polar headgroups of the phospholipids will transfer onto the hydrophilic surface of the silicon wafers leaving a monolayer of phospholipid with the hydrophobic acyl tails facing the air interface. Software that controls the dipping procedure can usually calculate a transfer ratio to determine how much lipid has been deposited onto the silicon wafer. However, this software usually assumes that a thin silicon wafer is being used in the dip, and therefore there are only two faces being covered with lipid. In this case, large wafers are being used resulting in four sides of the wafer being lifted through the lipid monolayer. Thus, the transfer ratio cannot be relied upon for absolute values; however, the slope of the pressure versus dipper position gives a good indication of monolayer transfer.

Once the Langmuir–Blodgett dip is complete, the phospholipid monolayer should be removed and a monolayer of rough LPS deposited on the same liquid surface and compressed to 35 mN m^{-1}. The laser scan should then be used to scan the water surface and the flat surface of the wafer so that level between the water and wafer can be auto-adjusted. The wafer with the deposited phospholipid should then be pushed down horizontally through the rough LPS monolayer. Once the dip has been completed, the wafer should be moved down into the backing plate (therefore, it is important that the wafer is also aligned with the position of the backing plate). It is important at this point that the formed bilayer is kept under liquid and not exposed to air otherwise the bilayer will be damaged. As the silicon wafer moves through the monolayer during the Schaefer dip, a jump in surface pressure will occur with the magnitude of the jump depending on composition and surface pressure of the monolayer.

When measuring the formed bilayer using neutron reflectometry, three solvent isotopic contrasts should be used: H_2O, D_2O, and 38% D_2O/62% H_2O (silicon matched water, this means that the reflectometry signal is purely from the deposited bilayer). When modeling the data, the bilayer can be divided into inner headgroups, inner tails, outer tails, and outer headgroups (Fig. 8). Lipid A, Rc-LPS, and Ra-LPS can be used to create stable asymmetric bilayers allowing for the core polysaccharide structure of the LPS to be modified to the needs of the experimental system to be studied (Clifton, Skoda, et al., 2013). Figure 8 shows an asymmetric model outer membrane created using Ra-LPS, which contains lipid A plus the complete core polysaccharide generating a realistic Gram-negative bacterial surface. Using hydrogenous phospholipid and hydrogenous rough, LPS enables determination of the structure of the total bilayer. However, an internal contrast can be used using a deuterated phospholipid. This will show the asymmetry in the bilayer as the lower phospholipid leaflet should have a high nSLD and the upper leaflet of the rough LPS a low nSLD. If there is any loss

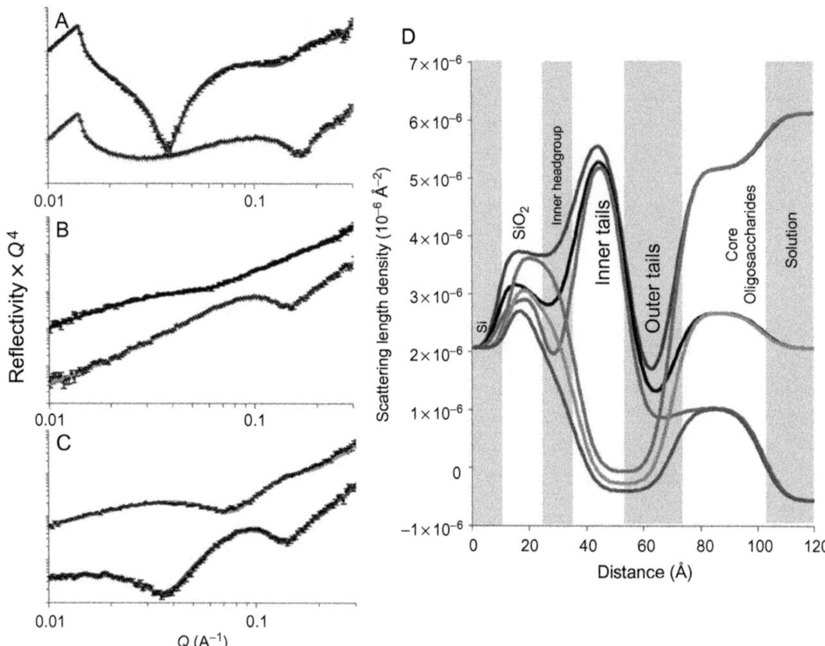

Figure 8 Neutron reflectometry profile and model data fits (A–C) and the scattering length density profiles these fits describe (D) for asymmetrically deposited DPPC (inner leaflet): Ra-LPS (outer leaflet) bilayer in the presence of 20 m*M* HEPES, pH 7.2 buffer with 5 m*M* $CaCl_2$. The six simultaneously fitted isotopic contrasts shown are (A) d-DPPC/Ra-LPS in D_2O (red line), h-DPPC/Ra-LPS in D_2O (blue line); (B) d-DPPC/Ra-LPS in 38% D_2O (black line), h-DPPC/Ra-LPS in 38% D_2O (gray line); and (C) d-DPPC/Ra-LPS in H_2O (green line), h-DPPC/Ra-LPS in H_2O (purple line). *Reproduced from Clifton et al. (2015) with permission, copyright © 2014 American Chemical Society.* (See the color plate.)

of asymmetry during the deposition process, then the nSLD of the tails will average out (Clifton, Skoda, et al., 2013; Clifton et al., 2015).

7. FINAL REMARKS

Neutron reflectometry is a powerful tool with which to study the structure of model biological membranes and is complementary to many other structural and biophysical techniques. The usefulness of the technique is a function of being able to discriminate between components with different nSLD, particularly if one has been enriched with deuterium, using contrast matching. As the structures to be studied using neutron reflectometry move from "simple" phospholipid bilayers and monolayers to structures that are more complex, the molecules that require deuteration become more complex. Purified LPS can be used to mimic the surface and asymmetry of model Gram-negative bacterial outer membranes and deuteration of rough LPS from *E. coli* allows for structure–function studies of such complex model membranes to occur.

ACKNOWLEDGMENTS

The National Deuteration Facility is partly funded by the National Collaborative Research Infrastructure Strategy of the Australian Government. The authors thank ANSTO's Bragg Institute, ISIS Pulsed Neutron and Muon Source, and Oak Ridge National Laboratory Spallation Neutron Source for the award of beam time. A.P.L.B. thanks the Australian Research Council for a Discovery Early Career Researcher Award (DE140101788). A.P.L.B. and S.A.H. thank the Access to Major Research Facilities Programme and the Australian Institute for Nuclear Science and Engineering (AINSE Ltd.) for travel support.

REFERENCES

Bertani, G. (1951). Studies on lysogenesis. I.: The mode of phage liberation by lysogenic *Escherichia coli*. *Journal of Bacteriology*, *62*, 293–300.

Bishop, R. E. (2008). Structural biology of membrane-intrinsic β-barrel enzymes: Sentinels of the bacterial outer membrane. *Biochimica et Biophysica Acta - Biomembranes*, *1778*, 1881–1896.

Blakeley, M. P., Ruiz, F., Cachau, R., Hazemann, I., Meilleur, F., Mitschler, A., et al. (2008). Quantum model of catalysis based on a mobile proton revealed by subatomic X-ray and neutron diffraction studies of h-aldose reductase. *Proceedings of the National Academy of Sciences*, *105*, 1844–1848.

Chen, X., Wilde, K. L., Wang, H., Lake, V., Holden, P. J., Middelberg, A. P. J., et al. (2012). High yield expression and efficient purification of deuterated human protein galectin-2. *Food Bioproduction and Processing*, *90*, 563–572.

Clifton, L. A., Johnson, C. L., Solovyova, A. S., Callow, P., Weiss, K. L., Ridley, H., et al. (2012). Low resolution structure and dynamics of a colicin-receptor complex determined by neutron scattering. *Journal of Biological Chemistry*, *287*, 337–346.

Clifton, L. A., Neylon, C., & Lakey, J. H. (2013). Examining protein–lipid complexes using neutron scattering. In J. H. Kleinschmidt (Ed.), *Lipid-Protein Interactions: Vol. 974.* (pp. 119–150). New York: Humana Press.

Clifton, L. A., Skoda, M. W. A., Daulton, E. L., Hughes, A. V., Le Brun, A. P., Lakey, J. H., et al. (2013). Asymmetric phospholipid: Lipopolysaccharide bilayers; a gram-negative bacterial outer membrane mimic. *Journal of the Royal Society, Interface*, *10*, 20130810.

Clifton, L. A., Skoda, M. W. A., Le Brun, A. P., Ciesielski, F., Kuzmenko, I., Holt, S. A., et al. (2015). Effect of divalent cation removal on the structure of gram-negative bacterial outer membrane models. *Langmuir*, *31*, 404–412.

Darwish, T. A., Luks, E., Moraes, G., Yepuri, N. R., Holden, P. J., & James, M. (2013). Synthesis of deuterated [d32]oleic acid and its phospholipid derivative [d64]dioleoyl-sn-glycero-3-phosphocholine. *Journal of Labelled Compounds and Radiopharmaceuticals*, *56*, 520–529.

Delcour, A. H. (2009). Outer membrane permeability and antibiotic resistance. *Biochimica et Biophysica Acta - Proteins and Proteomics*, *1794*, 808–816.

Dramsi, S., Magnet, S., Davison, S., & Arthur, M. (2008). Covalent attachment of proteins to peptidoglycan. *FEMS Microbiology Reviews*, *32*, 307–320.

Elbein, A. D., & Heath, E. C. (1965). The biosynthesis of cell wall lipopolysaccharide in *Escherichia coli*: I. The biochemical properties of a uridane diphosphate galactose 4-epimeraseless mutant. *Journal of Biological Chemistry*, *240*, 1919–1925.

Ferrier, W. G. (1960). Crystal structure of (beta)-d-glucose. *Acta Crystallographica*, *13*, 678–679.

Galanos, C., Lüderitz, O., & Westphal, O. (1969). A new method for the extraction of R lipopolysaccharides. *European Journal of Biochemistry*, *9*, 245–249.

Heinrich, F., & Lösche, M. (2014). Zooming in on disordered systems: Neutron reflection studies of proteins associated with fluid membranes. *Biochimica et Biophysica Acta - Biomembranes*, *1838*, 2341–2349.

Heinrich, F., Ng, T., Vanderah, D. J., Shekhar, P., Mihailescu, M., Nanda, H., et al. (2009). A new lipid anchor for sparsely tethered bilayer lipid membranes. *Langmuir*, *25*, 4219–4229.

Holt, S. A., Le Brun, A. P., Majkrzak, C. F., McGillivray, D. J., Heinrich, F., Loesche, M., et al. (2009). An ion-channel-containing model membrane: Structural determination by magnetic contrast neutron reflectometry. *Soft Matter*, *5*, 2576–2586.

Hughes, A. V. (2010, 2013). RasCAL. 2015, from http://sourceforge.net/projects/rscl/.

Karkhanis, Y. D., Zeltner, J. Y., Jackson, J. J., & Carlo, D. J. (1978). A new and improved microassay to determine 2-keto-3-deoxyoctonate in lipopolysaccharide of gram-negative bacteria. *Analytical Biochemistry*, *85*, 595–601.

Kastowsky, M., Gutberlet, T., & Bradaczek, H. (1993). Comparison of X-ray powder-diffraction data of various bacterial lipopolysaccharide structures with theoretical model conformations. *European Journal of Biochemistry*, *217*, 771–779.

Kienzle, P. A., O'Donovan, K. V., Ankner, J. F., Berk, N. F., & Majkrzak, C. F. (2000). Ncnr reflectometry software. 2015, from http://www.ncnr.nist.gov/reflpak.

Krishnan, S., & Prasadarao, N. V. (2012). Outer membrane protein A and OPRF: Versatile roles in gram-negative bacterial infections. *FEBS Journal*, *279*, 919–931.

Le Brun, A. P., Clifton, L. A., Halbert, C. E., Lin, B., Meron, M., Holden, P. J., et al. (2013). Structural characterization of a model gram-negative bacterial surface using lipopolysaccharides from rough strains of *Escherichia coli*. *Biomacromolecules*, *14*, 2014–2022.

Le Brun, A. P., Darwish, T. A., & James, M. (2013). Studies of biomimetic cellular membranes using neutron reflection. *Journal of Chemical and Biological Interfaces*, *1*, 3–24.

Liu, B., Knirel, Y. A., Feng, L., Perepelov, A. V., Senchenkova, S. N., Reeves, P. R., et al. (2014). Structural diversity in salmonella o antigens and its genetic basis. *FEMS Microbiology Reviews*, *38*, 56–89.

Middelberg, A. P. J., O'Neill, B. K., Bogle, I. D. L., & Snoswell, M. A. (1991). A novel technique for the measurement of disruption in high-pressure homogenization: Studies on E. coli containing recombinant inclusion bodies. *Biotechnology and Bioengineering*, *38*, 363–370.

Müller-Loennies, S., Holst, O., & Brade, H. (1994). Chemical structure of the core region of *Escherichia coli* j-5 lipopolysaccharide. *European Journal of Biochemistry*, *224*, 751–760.

Müller-Loennies, S., Holst, O., Lindner, B., & Brade, H. (1999). Isolation and structural analysis of phosphorylated oligosaccharides obtained from *Escherichia coli* j-5 lipopolysaccharide. *European Journal of Biochemistry*, *260*, 235–249.

Nagle, J. F., & Tristram-Nagle, S. (2000). Structure of lipid bilayers. *Biochimica et Biophysics Acta - Reviews in Biomembranes*, *1469*, 159–195.

Nanda, H., Heinrich, F., & Lösche, M. (2015). Membrane association of the PTEN tumor suppressor: Neutron scattering and MD simulations reveal the structure of protein–membrane complexes. *Methods*, *77–78*, 136–146.

Nelson, A. (2006). Co-refinement of multiple-contrast neutron/X-ray reflectivity data using motofit. *Journal of Applied Crystallography*, *39*, 273–276.

Nikaido, H. (2003). Molecular basis of bacterial outer membrane permeability revisited. *Microbiology and Molecular Biology Reviews*, *67*, 593–656.

Nikaido, H., & Vaara, M. (1987). Outer membrane. In C. Neidhardt, J. L. Ingraham, K. Brooks Low, B. Magasanik, M. Schaechter, & H. E. Umbarger (Eds.), *Vol. 1. Escherichia coli and Salmonella typhimurium. Cellular and molecular biology* (pp. 7–22). Washington DC: American Society for Microbiology.

Proskurnin, M. A., Ageeva, E. V., Senyuta, V. V., Orlova, N. V., Fokin, A. V., Ovchinnikov, O. B., et al. (2010). Comparison of optoacoustic and photothermal-lens determination of lipopolysaccharides. *Talanta*, *81*, 377–384.

Raetz, C. R. H. (1990). Biochemistry of endotoxins. *Annual Review of Biochemistry*, *59*, 129–170.

Raetz, C. R. H., Reynolds, C. M., Trent, M. S., & Bishop, R. E. (2007). Lipid a modification systems in gram-negative bacteria. *Annual Review of Biochemistry*, *76*, 295–329.

Raetz, C. R. H., & Whitfield, C. (2002). Lipopolysaccharide endotoxins. *Annual Review of Biochemistry*, *71*, 635–700.

Ruiz, N., Kahne, D., & Silhavy, T. J. (2006). Advances in understanding bacterial outer-membrane biogenesis. *Nature Reviews. Microbiology*, *4*, 57–66.

Sasaki, H., & White, S. H. (2008). Aggregation behavior of an ultra-pure lipopolysaccharide that stimulates TLR-4 receptors. *Biophysical Journal*, *95*, 986–993.

Shotton, M. W., Pope, L. H., Forsyth, T., Langan, P., Denny, R. C., Giesen, U., et al. (1997). A high-angle neutron fibre diffraction study of the hydration of deuterated a-DNA. *Biophysical Chemistry*, *69*, 85–96.

Snyder, S., Kim, D., & McIntosh, T. J. (1999). Lipopolysaccharide bilayer structure: Effect of chemotype, core mutations, divalent cations, and temperature. *Biochemistry*, *38*, 10758–10767.

Stenutz, R., Weintraub, A., & Widmalm, G. (2006). The structures of *Escherichia coli* o-polysaccharide antigens. *FEMS Microbiology Reviews*, *30*, 382–403.

Wacklin, H. P. (2010). Neutron reflection from supported lipid membranes. *Current Opinion in Colloid and Interface Science*, *15*, 445–454.

Wilkinson, S. G. (1996). Bacterial lipopolysaccharides—Themes and variations. *Progress in Lipid Research*, *35*, 283–343.

Wu, E. L., Engström, O., Jo, S., Stuhlsatz, D., Yeom, M. S., Klauda, J. B., et al. (2013). Molecular dynamics and nmr spectroscopy studies of E. coli lipopolysaccharide structure and dynamics. *Biophysical Journal*, *105*, 1444–1455.

Yepuri, N. R., Holt, S. A., Moraes, G., Holden, P. J., Hossain, K. R., Valenzuela, S. M., et al. (2014). Stereoselective synthesis of perdeuterated phytanic acid, its phospholipid derivatives and their formation into lipid model membranes for neutron reflectivity studies. *Chemistry and Physics of Lipids*, *183*, 22–33.

CHAPTER NINE

Essential Strategies for Revealing Nanoscale Protein Dynamics by Neutron Spin Echo Spectroscopy

David J.E. Callaway[1], Zimei Bu[1]
Department of Chemistry and Biochemistry, City College of New York, City University of New York, New York, USA
[1]Corresponding authors: e-mail address: dcallaway@ccny.cuny.edu; zbu@ccny.cuny.edu

Contents

Abstract

Determining the internal motions of a protein on nanosecond-to-microsecond timescales and on nanometer length scales is challenging by experimental biophysical techniques. Neutron spin echo spectroscopy (NSE) offers a unique opportunity to determine such nanoscale protein domain motions. However, the major hurdle in applying NSE to determine nanoscale protein motion is that the time and length scales of internal protein motions tend to be comparable to that of the global motions of a protein. The signals detected by NSE tend to be dominated by rigid-body translational and rotational diffusion. Using theoretical analyses, our laboratory showed that selective deuteration of a protein domain or a subunit can enhance the capability of NSE to reveal the internal motions in a protein complex. Here, we discuss the essential theoretical analysis and experimental methodology in detail. Protein nanomachines are far more complex than

Methods in Enzymology, Volume 566
ISSN 0076-6879
http://dx.doi.org/10.1016/bs.mie.2015.05.011

any molecular motors that have been artificially constructed, and their skillful utilization likely represents the future of medicine. With selective deuteration, NSE will allow us to see these nanomachines in motion.

1. INTRODUCTION

Proteins MOVE! Protein dynamics is essential for protein function (Frauenfelder, Sligar, & Wolynes, 1991; Miyashita, Onuchic, & Wolynes, 2003; Shoemaker, Portman, & Wolynes, 2000). Proteins adapt their conformations through protein motions in order to bind a variety of ligands or to form signaling complexes (Boehr, Nussinov, & Wright, 2009; Daniel, Dunn, Finney, & Smith, 2003; English et al., 2006; Hammes-Schiffer & Benkovic, 2006; Mobley & Dill, 2009). An abundance of experimental and theoretical studies have shown that functionally important protein motions are hierarchical, occurring on timescales ranging from femtoseconds to longer than seconds, and on length scales from angstroms to micrometers (Daniel et al., 2003; Frauenfelder et al., 1991; Ha et al., 1999; Mukhopadhyay, Krishnan, Lemke, Lindquist, & Deniz, 2007; Palmer, 2004; Zaccai, 2000). Protein motions on picosecond to nanosecond timescales, and from 50 microsecond to millisecond timescales can typically be characterized by NMR at atomic resolution (Mittermaier & Kay, 2006). Single-molecule biophysics has allowed the dynamics of biological macromolecules to be observed on timescales from milliseconds to seconds (Deniz, Mukhopadhyay, & Lemke, 2008; English et al., 2006; Greenleaf, Woodside, & Block, 2007). However, protein motions on nanosecond-to-microsecond timescales and on nanometer length scales are difficult to access by existing experimental biophysical techniques, including NMR and single-molecule detection. Currently, there is a spatial–temporal dynamic gap, on nanosecond-to-microsecond timescales and on nanometer length scales, where we cannot effectively determine the dynamics of proteins and large protein complexes. Here, we refer to protein motions on nanosecond-to-microsecond timescales and on nanometer length scales as *nanoscale protein motions*.

Neutron spin echo spectroscopy (NSE) is unique in its capacity to determine nanoscale motions. NSE is one of the quasielastic neutron scattering techniques that measures the difference in velocities between the incident and the scattering neutrons, in order to determine the energy exchange between neutrons and the molecules with motions, thus obtaining the dynamic information about the molecules (Bee, 1988; Higgins & Benoit, 1994). NSE employs the Larmor precession of neutron spins in a magnetic guide field as a clock to measure extremely small changes in velocities of scattering neutrons (Mezei, 1980; Mezei, Pappas, & Gutberlet, 2003), and thus enables the detection of very small energy changes in the scattering neutrons of $\delta E \sim 10^3$–10^{-2} μeV corresponding to nanosecond-to-microsecond dynamics. NSE thus fills an important spatial–temporal dynamic gap to determine nanoscale protein motions.

NSE thus has the potential to reveal protein domain motions because of the length scales and timescales probed by NSE. However, the major hurdle in applying NSE to determine nanoscale protein motion is that the time and length scales of internal protein motions are comparable to that of the global motions of a protein. The NSE measured correlation functions from a fully hydrogenated protein tend to be dominated by global motions of a protein (Hong et al., 2014; Farago, Li, Cornilescu, Callaway, & Bu, 2010). We have found that selective deuteration of a subunit can aid NSE to reveal effectively the internal domain motions in a reconstituted protein complex. Here, we first present a theoretical analysis, using nonequilibrium statistical mechanics, in order to reveal the advantage of selective deuteration. We also present the experimental scheme of selective deuteration, complex reconstitution, and sample preparation for NSE experiments. These theoretical, analytical and experimental schemes can be applied to determine nanoscale protein domain motions in a single protein, with selective deuteration of a protein domain or specific residues.

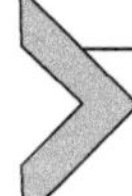

2. ESSENTIALS FOR DETERMINING NANOSCALE PROTEIN INTERNAL MOTION

2.1 Theory

2.1.1 Nonequilibrium Statistical Mechanics and the Mobility Tensor

NSE measures the intermediate scattering function $I(Q,t)$, which is the spatial Fourier transformation of the space–time van Hove correlation function $G(r,t)$ (Mezei, 1980),

$$I(Q,t) = \int_V G(r,t)\exp(-iQ\cdot r)\,dr$$

with Q the magnitude of the scattering vector, t the time, and r the position of a scattering center. The designation "intermediate" arises precisely because only one of the variables of $G(r,t)$ is Fourier transformed. Because of the short wavelength of neutrons, NSE measures $I(Q,t)$ on nanometer to submicron length scales and can reveal the nanoscale fluctuations in a protein. NSE thus measures motions on nanometer to micron length scales, and on nanosecond to microsecond timescales (Ewen & Richter, 1997; Farago et al., 2010; Mezei, 1980).

For a protein in solution, $I(Q,t)$ can typically be fit to a single exponential in time (and is difficult to fit to more exponentials) at a given Q. A natural way to interpret the NSE data is to examine the effective diffusion constant $D_{eff}(Q)$ as a function of Q, which is determined by the normalized intermediate scattering function $I(Q,t)/I(Q,0)$:

$$\begin{aligned} \Gamma(Q) &= -\lim_{t\to 0}\frac{\partial}{\partial t}\ln\left[I(Q,t)/I(Q,0)\right] \\ D_{eff}(Q) &= \frac{\Gamma(Q)}{Q^2} \end{aligned} \tag{1}$$

where $I(Q,0)$ is the static form factor.

In order to describe the dynamics of a protein in solution, we utilize the Akcasu–Gurol approach originally developed to describe the dynamics of random coil polymers (Akcasu & Gurol, 1976), generalized to include rotational motion (Bu, Biehl, Monkenbusch, Richter, & Callaway, 2005; Callaway & Bu, 2015):

$$D_{eff}(Q) = \frac{k_B T}{Q^2}\frac{\sum_{jl}\left\langle b_j b_l\left(Q\cdot H_{jl}^T\cdot Q + L_j\cdot H_{jl}^R\cdot L_l\right)e^{iQ\cdot(r_j-r_l)}\right\rangle}{\sum_{jl}\left\langle b_j b_l e^{iQ\cdot(r_j-r_l)}\right\rangle} \tag{2}$$

where b_j is the coherent scattering length of a subunit j, H^T is the translational mobility tensor, H^R is the rotational mobility tensor, and $k_B T$ is the usual temperature factor. The structural coordinates of the macromolecule, taken relative to the center of friction of the protein are given by r_j (note that $\Sigma r_j = 0$). In practice, the structural coordinates can be atoms, protein domains in a multidomain protein, or subunits in a multimeric protein complex, and may be obtained from high-resolution crystallography or NMR, or from low-resolution electron microscopy and small-angle X-ray and neutron scattering (SAXS and SANS). In Eq. (2), $L_j = r_j \times Q$ is the torque vector for each coordinate. The brackets $\langle\rangle$ denote an orientational average over the vector Q. The translational mobility tensor H^T in Eq. (2) is defined by the velocity response $v = H^T F$ to an applied force F. The rotational mobility tensor H^R is defined by the angular velocity response $\omega = H^R \tau$ to an applied torque τ.

It is necessary to include the contributions from rotational diffusion when the timescale of the measurement is short enough. The relevant timescale is the inverse of the rotational diffusion constant. This constant can be estimated from the fact that globular proteins tumble in solution at frequencies close to that for rigid spheres. Such frequencies are usually determined via the *rotational correlation time* $\varphi = V\eta/k_B T$, where V is the molecular volume, η is the viscosity of the medium, k_B is the Boltzmann's constant, and T is the temperature. Assuming the Stokes–Einstein relation $D_R = k_B T/(8\pi\eta R^3)$ for the rotational diffusion constant of a sphere with radius R, and using a spherical molecular volume $V = (4/3)\pi R^3$, one finds a relation between the rotational diffusion time $(D_R)^{-1} = 6\varphi$ and the molecular weight M_w of the protein at ambient temperatures:

$$(D_R)^{-1}(\text{nanoseconds}) \approx 3M_w(\text{kiloDaltons})$$

Most multidomain proteins of interest are of order 50 kDa or greater, while the NSE Fourier time is at most only a few hundred nanoseconds. Thus, rotational diffusion must typically be considered in performing NSE experiments.

Equation (2) is valid for either rigid bodies or rigid-body subunits connected by soft spring linkers (Bu et al., 2005; Farago et al., 2010; Ho et al., 2004). For a completely flexible body, the rotational diffusion term (involving H^R) is absent. For a rigid body composed of N identical beads, the translational mobility tensor H^T is a matrix with N^2 identical 3×3 elements since H^T yields the same velocity response of, e.g., subunits B and C to a force applied to subunit A.

For an object with internal flexibility, the elements of the mobility tensor will not be equal, so forces applied to a given bead would result in different velocities for other beads and the body would not remain rigid. This reflects itself in an effective diffusion constant with a different Q dependence for rigid bodies or bodies with internal motion. Comparing models of the mobility tensor from Eq. (2) to experimental $D_{eff}(Q)$ from NSE experiments allows one to extract the internal dynamics of a protein or protein complex (Bu et al., 2005; Bu & Callaway, 2011; Farago et al., 2010). The mobility tensor provides a direct indication of the existence of internal degrees of freedom (Bu et al., 2005; Bu & Callaway, 2011; Farago et al., 2010). The relationship between force, velocity, and the translational mobility tensor is illustrated in Fig. 1. These considerations allow a wide variety of results to be easily derived, such as the fact that $D_{eff}(Q)$ at infinite Q is twice its value

Mobility tensor $\vec{v} = \overleftrightarrow{H}\,\vec{F}$

Figure 1 The relationship between force and the mobility tensor. The translational mobility tensor gives the velocity response (speed and direction) of a given protein domain to a force applied to itself or to another domain.

at zero Q for a uniform rigid body (Callaway, Farago, & Bu, 2013; Farago et al., 2010).

The above simple theoretical approach does not require complicated molecular dynamics simulations, elastic network models, fits to rotational expansions in spherical harmonics, or Navier–Stokes hydrodynamics. We do not have to fit NSE data, but can directly predict the outcome of an NSE experiment from the structural coordinates in order to test models of the mobility tensor.

2.2 Partial Deuteration

Why does partial deuteration of a protein complex enhance the ability of NSE to reveal internal motions? Initially, this seems counterintuitive, as one is using contrast variation to eliminate the scattering signal from some of the sample, and one expects naively that a smaller sample should evince a smaller signal. But one must remember that one is observing the effective diffusion constant, as per Eq. (2), and not merely the total scattering cross-section.

The partial deuteration affects $D_{\mathrm{eff}}(Q)$ in two ways. To see this, consider the following *gedanken* experiment: Take a uniform disk and fix its center. At zero Q the effective diffusion constant $D_{\mathrm{eff}}(Q=0)=0$, since rotational diffusion is absent because $\Sigma\mathbf{r}=0$, and translational diffusion is absent because the center of the disk is fixed. Now deuterate half the disk and employ contrast matching to render this half invisible. Both translational and rotational diffusion (as seen by NSE) now appear at $Q=0$ because the coherent scattering length is no longer uniform and both therefore contribute to D_{eff}. Thus, D_{eff} is actually *larger* in the partially deuterated contrast matched case, since $\Sigma(b\mathbf{r})$ is nonzero and depends on time, even though $\Sigma\mathbf{r}=0$ and is independent of time because the center of the disk is fixed. Thus, even for a rigid body, partial deuteration increases the value of the NSE measured diffusion constant $D_{eff}(Q)$. This effect is even more dramatic if there is internal motion, as we will see below.

Second, selective deuteration of a member of a protein complex or a protein domain can also highlight the internal modes involving hydrogenated parts of the system. From the following example, we show that partial deuteration increases the sensitivity of NSE to reveal protein internal motion.

2.3 Activation of Nanoscale Internal Motion in NHERF1: A Specific Example

NHERF1 plays essential roles in modulating the intracellular trafficking and assembly of a number of receptors and ion transport proteins. NHERF1 is a multidomain protein that has two modular domains, PDZ1, PDZ2, and a disordered but compact C-terminal domain, with three domains connected by unstructured linkers (Bhattacharya et al., 2010; Li, Dai, Jana, Callaway, & Bu, 2005; Li et al., 2007). The C-terminal domain binds to the FERM domain of Ezrin with high affinity, $K_d = 19$ nM (Reczek, Berryman, & Bretscher, 1997). We have shown that binding to FERM to the C-terminal domain of NHERF1 allosterically increases the binding affinity of both PDZ1 and PDZ2 domains of NHERF1 for the cytoplasmic tail of CFTR (Li, Callaway, & Bu, 2009; Li et al., 2005). The PDZ1 and PDZ2 domains are 110 and 80 Å away, respectively, from the FERM binding site in the CT domain. The NHERF1·FERM complex thus suggests long-range allosteric transmission of binding signals on nanometer length scales (Li et al., 2009). Our NSE experiments revealed the activation of interdomain motions of the PDZ domains in NHERF1 on submicrosecond timescales upon binding to FERM. A dynamic protein can recognize more binding partner proteins and bind to one partner more tightly than a rigid homolog (Bhattacharya et al., 2013). We thus correlate the activated domain motions with the increased binding capabilities of the PDZ domains in the complex, and thus the propagation of allosteric signals from the Ezrin-binding site to the remote PDZ domains that are located as far as 110 Å away.

For NHERF1 alone in solution, the calculated rigid body $D_{eff}(Q)$ agrees with the NSE experimental data quite well, see Fig. 2. The rigid-body calculation uses as input only the translational diffusion coefficient D_0 of NHERF1 obtained from pulsed-field gradient NMR and the "dummy atom" structural coordinates (Svergun, 1999) reconstructed from solution SAXS and SANS (Li et al., 2007, 2009).

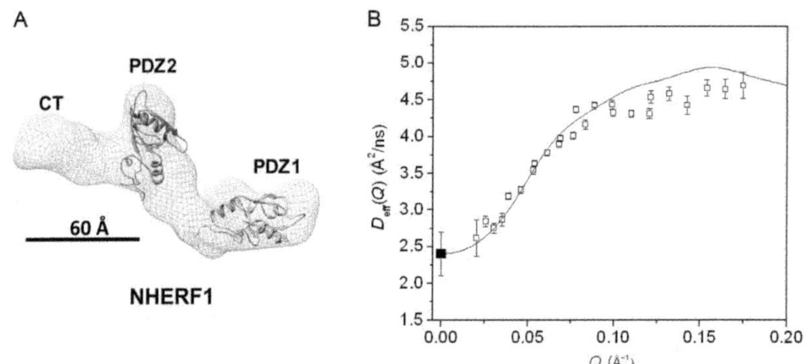

Figure 2 NHERF1 alone behaves as a rigid-body in solution as shown from NSE experiments. (A) The 3-D shape of NHERF1 reconstructed from SAXS (Li et al., 2009) using the *ab initio* program DAMMIN (Svergun, 1999). The known high-resolution structures of the PDZ1 (PDB code: 1I92) and PDZ2 (PDB code: 2KJD) domains are docked into the 3-D shape, using UCSF chimera (Pettersen et al., 2004). EBD, which overlaps with the last 13 amino acid residues that interact with PDZ2 is not marked in the graph. (B) Comparing the experimental D_{eff} (Q) of NHERF1 (black open square) with the rigid-body calculation (black solid line). The overall translational diffusion constant D_0 (filled black square) at $Q=0$ Å^{-1} is $D_0=2.4$ Å^2/ns from pulsed-PFG NMR measurements.

We have compared our calculations with the NSE experimental results on two types of complexes of NHERF1 bound to FERM, see Fig. 3. One complex is the hydrogenated NHERF1 in complex with the hydrogenated FERM (NHERF1·^{h}FERM), and the other complex is hydrogenated NHERF1 bound to deuterium labeled FERM (NHERF1·^{d}FERM). As we have pointed out that, at low Reynolds number, the dynamics of a protein as seen by NSE should not depend upon its mass, but rather upon its size. In our calculations, we thus always impose the constraint that the dynamics and therefore the mobility tensors of the hydrogenated and deuterated components are the same. When calculating $D_{eff}(Q)$ for the NHERF1·^{d}FERM complex, the scattering from the deuterated component is treated as "invisible" in Eq. (2) because of contrast matching, i.e., the neutron scattering length density of the deuterated component contrast matches that of the D_2O buffer background. We used D_0 of the deuterated complex or the hydrogenated complex obtained from PFG NMR or from dynamic light scattering and the structural coordinates obtained from SANS.

As shown in Fig. 3A, the agreement between the experimental NSE data and rigid-body calculations is poor for both the NHERF1·^{d}FERM and the

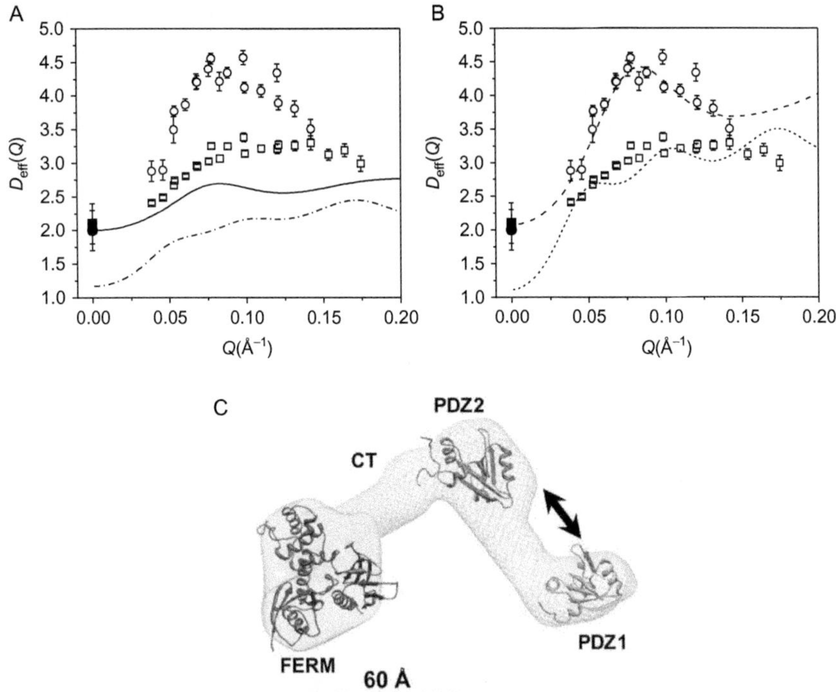

Figure 3 Activation of interdomain motion in NHERF1 upon binding to the FERM domain of Ezrin. (A) Comparing experimental $D_{eff}(Q)$ of NHERF1·dFERM and NHERF1·hFERM with rigid-body calculations. Open circle are the NSE data from NHERF1·dFERM. Open squares are the NSE data from NHERF1·hFERM. Solid circle and square are the self-diffusion constants D_0 of NHERF1·dFERM and NHERF1·hFERM obtained from PFG NMR, respectively. The solid line is from rigid-body model calculations of the NHERF1·dFERM complex. The dot dash line is from rigid-body model calculations of the NHERF1·hFERM complex. (B) Comparing experimental $D_{eff}(Q)$ of deuterated complex NHERF1·dFERM and hydrogenated complex NHERF1·hFERM with calculations incorporating interdomain motion between PDZ1 and PDZ2. The symbols for the experimental data are the same as in (A). The dashed curve is calculated from model incorporating domain motion between PDZ1 and PDZ2 for the NHERF1·dFERM complex. The short dashed curve is calculated from model incorporating domain motion between PDZ1 and PDZ2 for the NHERF1·hFERM complex. The comparisons in (A) and (B) show that deuteration of the FERM domain amplifies the effects of protein internal motions detected by NSE. (C) A model representing domain motion between PDZ1 and PDZ2 in the complex. The 3-D shape of the complex is reconstructed from SANS (Li et al., 2009). The known high-resolution structure fragments of PDZ1, PDZ2, and the FERM domain (PDB code: 1NI2) are docked into the envelope using UCSF chimera (Pettersen et al., 2004). The arrows represent translational motion between PDZ1 and PDZ2. A length-scale bar of 60 Å is shown. (See the color plate.)

NHERF1·^{h}FERM complexes. We have then incorporated domain motions in our calculations, with the mobility tensor with an internal mode between the PDZ1 and PDZ2 domains (Fig. 3B). The calculated $D_{eff}(Q)$ with internal motion agrees quite well with the NSE results for the NHERF1·^{d}FERM complex. Nevertheless, for the NHERF1·^{h}FERM complex, the computed D_0 at $Q=0$ is not close to the experimental values from PFG NMR measurements. We attribute this discrepancy to large conformational variations in NHERF1 by the unfolding of the CT domain upon binding to FERM. Such complications are minimal in the NHERF1·^{d}FERM complex because the deuterated ^{d}FERM is "invisible" to neutrons.

2.4 A Simple Four-Point Model Shows How Selective Deuteration Can Enhance the Effects of Internal Motion

The simple calculations we presented above require only the structural coordinates and a single constraint, the diffusion constant at $Q=0$ Å^{-1} for the deuterated complex, which can be measured by PFG NMR, to generate the computed $D_{eff}(Q)$. We further introduce an even more simplified model that yields the same effect, and serves to explain the $D_{eff}(Q)$ observed by NSE experiments. The simplified model is taken by extracting four points that represent the coordinates of the center-of-friction of domains obtained from the SANS data of the NHERF1·FERM complex. These points form a triangle, as shown in Fig. 4A, with the distances FERM–PDZ2 = 80 Å, PDZ2–PDZ1 = 59 Å, and FERM–PDZ1 = 110 Å. The CT domain is taken as being halfway between the FERM and PDZ2 domains. We include the point representing the FERM domain with a weight factor of three to account for its larger size relative to the other domains. Because it is possible to obtain the center-of-friction distances between the domains with confidence even with low-resolution SAXS or SANS data, this model possesses fewer uncertainties than a model based upon the molecular shape. More details of the four-point calculations are described previously (Farago et al., 2010).

Figure 4B is the $D_{eff}(Q)$ of the four-point rigid-body model, without incorporating internal domain motion between PDZ1 and the rest of the complex. Figure 4C is the $D_{eff}(Q)$ of the four-point model incorporating internal domain motion between PDZ1 and the rest of the complex. After incorporating internal motion, the overall $D_{eff}(Q)$ from the four-point model agrees well with the experimental data for both the partially deuterated and the hydrogenated complexes. The comparison between calculation and experimental data improves after including the form factor of a 20 Å

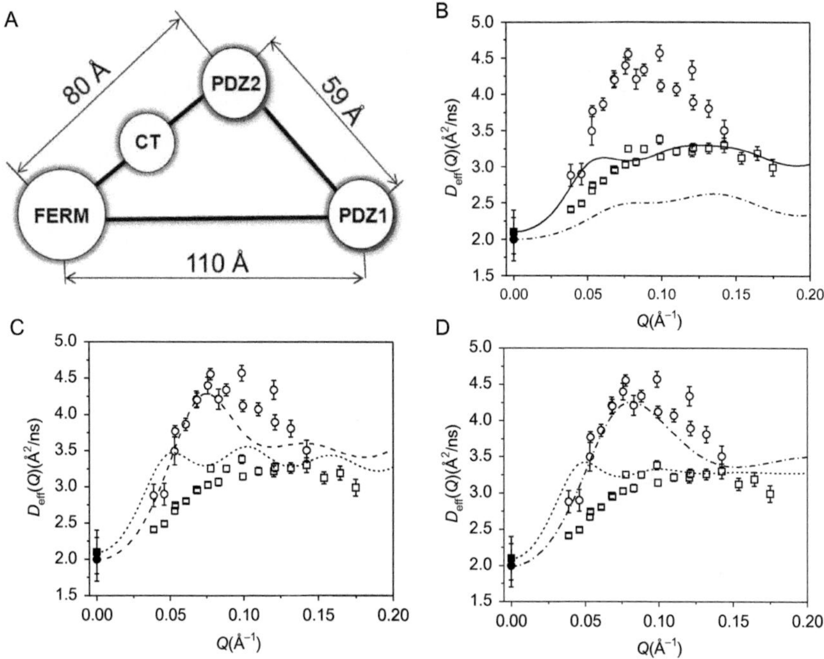

Figure 4 A simple four-point model can well describe domain motion in the complex. (A) The four-point model represents the NHERF1·FERM complex, with the centers of PDZ1, PDZ2, CT, and FERM domain taken from Fig. 3A. (B) Comparing the experimental NSE data with the four-point rigid-body calculations for NHERF1·hFERM (open squares are the experimental data and blue solid line is the calculated data) and for NHERF1·dFERM (open circles are experimental data and dot dash line is the calculated data). D_0 of NHERF1·dFERM (solid circle) squares) and NHERF1·dFERM (solid square) are from PFG NMR are shown. (C) Comparing the experimental data with calculations assuming interdomain motion between PDZ1 and PDZ2 in NHERF1·dFERM (dash line) and NHERF1·hFERM (blue (dark gray in the print version) short dash line). The experimental symbols are the same as in (B). (D) Comparing the experimental data with calculations incorporating interdomain motion between PDZ1 and PDZ2, as well as assuming finite size form factor of spheres of 20 Å radius for the FERM domain and for both PDZ domains in NHERF1·dFERM (dash dot line) and in NHERF1·hFERM (blue (dark gray in the print version) short dash line).

radius sphere for the FERM domain and both PDZ domains in the calculation (Fig. 4D). Thus, the NSE data are better represented by the four-point model that includes PDZ1–PDZ2 interdomain motion than by a model that assumes the complex is a rigid body. Further improvement likely requires the use of methods of evaluating the mobility tensors for proteins with high accuracy.

Moreover, from the four-point model calculations, we note that $D_{\mathrm{eff}}(Q)$ for the hydrogenated rigid complex and the hydrogenated complex with internal motion are nearly indistinguishable (Fig. 5A). For the deuterated complex, $D_{\mathrm{eff}}(Q)$ obtained from the interdomain motion model is significantly different from that of the rigid-body model (Fig. 5B). This can be explained as due to the relatively large contribution to Eq. (2) of the effects of rotational diffusion of the overall object, which dominates and obscures the effects of internal motion when no deuteration is performed. For the partially deuterated complex, both the docked domain calculation (Fig. 3B) and the four-point model (Fig. 5B) show that $D_{\mathrm{eff}}(Q)$ of the rigid-body complex is significantly different from that of the complex with internal domain motion. Thus, deuteration of a

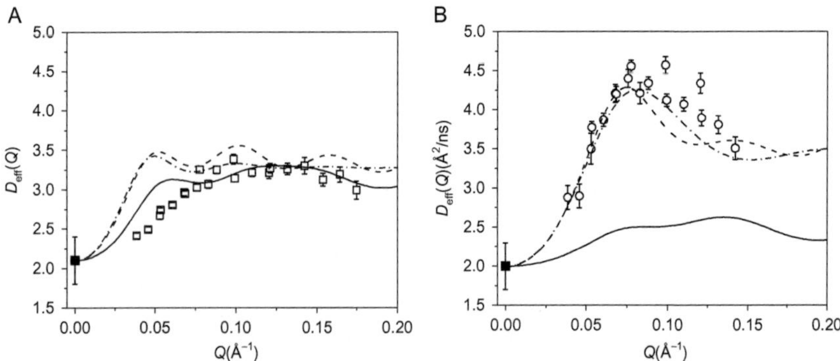

Figure 5 For the hydrogenated NHERF1·hFERM complex, the difference in $D_{\mathrm{eff}}(Q)$ between the rigid-body model and domain-motion models is very small, but is significantly increased in the deuterated complex. (A) Comparing the rigid-body calculation with the domain-motion calculation in the four-point model in the hydrogenated NHERF1·hFERM complex. NSE data from the NHERF1·hFERM (open squares), the four-point rigid-body model (solid line), four-point model incorporating domain motion between PDZ1 and PDZ2 (dash line), four-point model incorporating domain motion between PDZ1 and PDZ2 and finite size form factor of 20 Å radius for the FERM domain, PDZ1, and PDZ2 (dash dot line). D_0 at $Q=0$ Å^{-1} as measured from PFG NMR is shown in solid square. (B) Comparing the rigid-body calculation with the domain-motion calculation in the four-point model in the deuterated NHERF1·dFERM complex. NSE data from the NHERF1·dFERM (open circles), the four-point rigid-body model (solid line), four-point model incorporating domain motion between PDZ1 and PDZ2 (dash line), four-point model incorporating domain motion between PDZ1 and PDZ2 and finite size form factor of 20 Å radius for the FERM domain, PDZ1, and PDZ2 (dash dot line). D_0 at $Q=0$ Å^{-1} as measured from PFG NMR is shown in red (light gray in the print version) solid square.

domain or subunit in a protein complex can amplify the effects of internal protein dynamics as detected by NSE.

3. PREPARATION OF PARTIALLY DEUTERATED PROTEIN SAMPLES FOR NSE EXPERIMENTS

In the discussion above, we showed that deuteration of a part of a protein or protein complex can amplify the effects of internal motion observed by NSE. Essential to this analysis is the careful preparation of deuterated proteins. The readers can refer to a thoughtful paper by Meilleur et al., which gives a comprehensive description about the methodology of deuterium labeling of proteins for neutron scattering experiments (Meilleur, Weiss, & Myles, 2009). Here, we only describe the protocol that we used to produce deuterated proteins and to reconstitute protein complexes for NSE experiments.

3.1 Protein Expression and Purification

The proteins used for our NSE experiments were all expressed in bacteria. The pET151/D-TOPO vector (Invitrogen, Inc.) or pET32a (EMD Biosciences) was used to express the FERM domain of human Ezrin (FERM, amino acid residues 1–298), the full-length human NHERF1 (residues 11–358). The protein expressed by the pET151/D-TOPO vector contains an N-terminal V5 epitope plus a hexa-histidine fusion tag. All plasmids were subjected to DNA sequencing to verify the DNA sequence.

The protein expression plasmids were transformed into Rosetta 2(DE3) cells (Novagen). The cells were grown in Luria *broth* (LB) medium until the optical density at 600 nm reaches 0.8–0.9. The cells were induced with 0.5 m*M* Isopropyl β-D-1-thiogalactopyranoside (IPTG) for 2 h. The harvested cells were resuspended and lysed in buffer containing 20 m*M* sodium phosphate buffer, 150 m*M* NaCl, 0.1 m*M* phenylmethylsulfonyl fluoride, and 10 m*M* imidazole, pH 7.5. The protein extracts were first further purified by Ni^{2+} HiTrap chelating column (Amersham Biosciences). The proteins were then purified and analyzed by size-exclusion chromatography, using a Superdex 200 10/30 column (Amersham Biosciences). The N-terminal fusion tag was cleaved using Tobacco etch virus protease (Invitrogen) after purification.

3.2 Producing Deuterated Protein Subunit that is Contrast Matched in 100% D_2O Buffer Solution

As described in Meilleur et al. (2009), proteins with 65–70% deuteration can be contrast matched in 100% D_2O. A growth medium containing 85% D_2O can usually produce proteins of 65–70% deuteration content for SANS and NSE experiments.

3.2.1 Materials

M9 medium: 4 g glucose, 0.5 g NaCl, 1 g NH_4Cl, 3 g KH_2PO_4, 3.18 g Na_2HPO_4, 2 ml of 1 *M* $MgSO_4$, 0.1 ml $CaCl_2$. The salts and glucose are dissolved in 1 l of 85% D_2O (v/v of D_2O/H_2O).

3.2.2 Procedure

The following protocol is for growing 1.0 l of cell culture. The transformed Rosetta 2 (DE3) cells are grown at 37 °C in sterile LB medium. After overnight growth, 35 ml of the cell culture is centrifuged at 5000 rpm using a table-top centrifuge. The supernatant is discarded. The cell pellet is washed in 20 ml M9 medium containing 85% D_2O (Cambridge Isotope Laboratories). The cells are pelleted by centrifugation at 3000 rpm.

The cells are then resuspended in 150 ml M9 medium containing 85% D_2O and grown at 37 °C for overnight. The $O.D._{600nm}$ of this overnight culture is about 3.5–4. The overnight culture was added to 850 ml 85% D_2O M9 medium, and the cells grow at 37 °C until $O.D._{600nm}$ reaches 0.7–0.8. This process will take about 3–5 h. The cells were then grown until $O.D._{600nm}$ reaches 0.8–0.9. The cells were then induced with 0.25 m*M* IPTG at 37 °C overnight. Protein purification and fusion tag removal for the deuterated proteins were the same as described above. After protein purification, the deuterium content of the protein is measured by MALDI-TOF to determine the exact deuterium content.

Using the above protocol, our method does not need to adapt the cells on D_2O agarose plates. This can save considerable time. However, with this method, it is important to maintain $O.D._{600nm}$ at above 0.4 after the overnight culture is added to 850 ml of 85% D_2O M9 medium. We find that it is helpful to first add the 150 ml of 85% D_2O M9 overnight culture to 500–600 ml of 85% D_2O M9 and grow the cells until the $O.D._{600nm}$ reaches 0.7–0.8. The rest of the medium is then added, and the cells are grown until the $O.D._{600nm}$ reaches 0.7–0.8 before IPTG induction.

3.3 Reconstitution of the Partially Deuterated Protein Complex

The concentration of the purified protein is measured by ultraviolet light absorbance at 280 nm, using the theoretical extinction coefficient calculated from the amino acid sequence of the protein (Gasteiger et al., 2005). The deuterated protein is then mixed with hydrogenated protein at the stoichiometry ratio of their interaction. For example, for NHERF1 binding to FERM at 1:1 molar ratio, and for CFTRct binding to NHERF1 at a 2:1 molar ratio. The protein complex is purified by size-exclusion chromatography.

The above method of using size-exclusion chromatography to separate the complex is useful to purify a protein complex with high affinity, with dissociation constant K_d from nanomolar to submicromolar. For protein complexes with weak interactions, we recommend incubating the deuterated protein at a higher molar ratio to the hydrogenated protein than the stoichiometry of the protein complex and not to use size-exclusion to resolve the complex. This is because after exchanging the protein into buffer, the deuterated component is contrast matched. NSE will measure the dynamics of the hydrogenated component in complex to the deuterated component.

3.4 Exchanging the Proteins and Protein Complex into D_2O Buffer

The protein or protein complex needs to be completely exchanged into D_2O buffer before the NSE experiment. The D_2O used for making the buffer is 99.9%. The buffer components should also be deuterated. If phosphate buffer is used, D_2O can be added to the phosphate salts, and the mixture then lyophilized. This process will be repeated for three to four times to remove the H content in the salt. If Tris buffer is used, it is recommended to use deuterated Tris, and to use DCl or NaOD to adjust the pH.

The buffer exchange uses a protein concentrating device, such as VivaSpin (Vivaproducts, Inc., Littleton, MA) with a molecular mass cutoff below the protein or protein complex. The procedures are as follows:

1. Concentrating the protein to 1/10th of original volume. Dilute the protein 10 times with D_2O buffer. Repeat the process three times.
2. Dilute the protein 10 times with D_2O buffer. Leave the solution for 12–24 h. Concentrate the protein to 1/10th volume. Add D_2O buffer to adjust the protein concentration. For our experiments, we have obtained good results with a 15–20 mg/ml protein concentration.

4. SUMMARY

We have shown that it is now possible to observe directly internal motions in proteins on nanosecond timescales and nanometer length scales. These motions have been hypothesized to be the most important dynamic modes for protein function, in long-range allostery and enzymatic catalysis. Much work remains to be done, but we are confident that our work will eliminate several formidable roadblocks to further progress. Protein nanomachines are far more complex than any molecular motors that have been artificially constructed, and their skillful utilization likely represents the future of medicine. With appropriate partial deuteration, NSE will allow us to see these nanomachines in motion.

REFERENCES

Akcasu, Z., & Gurol, H. (1976). Quasi-elastic scattering by dilute polymer-solutions. *Journal of Polymer Science, Part B: Polymer Physics*, *14*, 1–10.

Bee, M. (1988). *Quasielastic neutron scattering: Principles and applications in solid state chemistry, biology and materials science*. Bristol and Philadelphia: Adam Hilger.

Bhattacharya, S., Dai, Z., Li, J., Baxter, S., Callaway, D. J. E., Cowburn, D., et al. (2010). A conformational switch in the scaffolding protein NHERF1 controls autoinhibition and complex formation. *Journal of Biological Chemistry*, *285*, 9981–9994.

Bhattacharya, S., Ju, J. H., Orlova, N., Khajeh, J. A., Cowburn, D., & Bu, Z. (2013). Ligand-induced dynamic changes in extended PDZ domains from NHERF1. *Journal of Molecular Biology*, *425*, 2509–2528.

Boehr, D. D., Nussinov, R., & Wright, P. E. (2009). The role of dynamic conformational ensembles in biomolecular recognition. *Nature Chemical Biology*, *5*, 789–796.

Bu, Z., Biehl, R., Monkenbusch, M., Richter, D., & Callaway, D. J. (2005). Coupled protein domain motion in Taq polymerase revealed by neutron spin-echo spectroscopy. *Proceedings of the National Academy of Sciences of the United States of America*, *102*, 17646–17651.

Bu, Z., & Callaway, D. J. (2011). Proteins MOVE! Protein dynamics and long-range allostery in cell signaling. *Advances in Protein Chemistry and Structural Biology*, *83*, 163–221.

Callaway, D. J., Farago, B., & Bu, Z. (2013). Nanoscale protein dynamics: A new frontier for neutron spin echo spectroscopy. *The European Physical Journal E, Soft Matter*, *36*, 76.

Callaway, D. J., & Bu, Z. (2015). Nanoscale protein domain motion and long-range allostery in signaling proteins- a view from neutron spin echo sprectroscopy. *Biophysical Reviews*, *7*, 165–174.

Daniel, R. M., Dunn, R. V., Finney, J. L., & Smith, J. C. (2003). The role of dynamics in enzyme activity. *Annual Review of Biophysics and Biomolecular Structure*, *32*, 69–92.

Deniz, A. A., Mukhopadhyay, S., & Lemke, E. A. (2008). Single-molecule biophysics: At the interface of biology, physics and chemistry. *Journal of the Royal Society Interface*, *5*, 15–45.

English, B. P., Min, W., van Oijen, A. M., Lee, K. T., Luo, G., Sun, H., et al. (2006). Ever-fluctuating single enzyme molecules: Michaelis-Menten equation revisited. *Nature Chemical Biology*, *2*, 87–94.

Ewen, B., & Richter, D. (1997). Neutron spin echo investigations on the segmental dynamics of polymers in melts, networks and solutions. In *Neutron spin echo spectroscopy viscoelasticity, rheology. Advances in polymer science: Vol. 134* (pp. 1–129). Berlin Heidelberg: Springer. ISBN: 978-3-540-62713-5. http://dx.doi.org/10.1007/3-540-68449-2_1.

Farago, B., Li, J., Cornilescu, G., Callaway, D. J., & Bu, Z. (2010). Activation of nanoscale allosteric protein domain motion revealed by neutron spin echo spectroscopy. *Biophysical Journal, 99*, 3473–3482.

Frauenfelder, H., Sligar, S., & Wolynes, P. (1991). The energy landscapes and motions of proteins. *Science, 254*, 1598–1603.

Gasteiger, E., Hoogland, C., Gattiker, A., Wilkins, M. R., Appel, R. D., & Bairoch, A. (2005). Protein identification and analysis tools on the ExPASy server. In *The proteomics protocols handbook* (pp. 571–607). Springer.

Greenleaf, W. J., Woodside, M. T., & Block, S. M. (2007). High-resolution, single-molecule measurements of biomolecular motion. *Annual Review of Biophysics and Biomolecular Structure, 36*, 171.

Ha, T., Ting, A. Y., Liang, J., Caldwell, W. B., Deniz, A. A., Chemla, D. S., et al. (1999). Single-molecule fluorescence spectroscopy of enzyme conformational dynamics and cleavage mechanism. *Proceedings of the National Academy of Sciences of the United States of America, 96*, 893–898.

Hammes-Schiffer, S., & Benkovic, S. J. (2006). Relating protein motion to catalysis. *Annual Review of Biochemistry, 75*, 519–541.

Higgins, J. S., & Benoit, H. C. (1994). *Polymers and neutron scattering*. Oxford: Clarendon Press.

Ho, D. L., Byrnes, W. M., Ma, W. P., Shi, Y., Callaway, D. J., & Bu, Z. (2004). Structure-specific DNA-induced conformational changes in Taq polymerase revealed by small angle neutron scattering. *Journal of Biological Chemistry, 279*, 39146–39154.

Hong, L., Sharp, M. A., Poblete, S., Biehl, R., Zamponi, M., Szekely, N., et al. (2014). Structure and dynamics of a compact state of a multidomain protein, the mercuric ion reductase. *Biophysical Journal, 107*, 393–400.

Li, J., Callaway, D. J., & Bu, Z. (2009). Ezrin induces long-range interdomain allostery in the scaffolding protein NHERF1. *Journal of Molecular Biology, 392*, 166–180.

Li, J., Dai, Z., Jana, D., Callaway, D. J., & Bu, Z. (2005). Ezrin controls the macromolecular complexes formed between an adapter protein Na+/H+ exchanger regulatory factor and the cystic fibrosis transmembrane conductance regulator. *Journal of Biological Chemistry, 280*, 37634–37643.

Li, J., Poulikakos, P. I., Dai, Z., Testa, J. R., Callaway, D. J., & Bu, Z. (2007). Protein kinase C phosphorylation disrupts Na+/H+ exchanger regulatory factor 1 autoinhibition and promotes cystic fibrosis transmembrane conductance regulator macromolecular assembly. *Journal of Biological Chemistry, 282*, 27086–27099.

Meilleur, F., Weiss, K. L., & Myles, D. A. (2009). Deuterium labeling for neutron structure-function-dynamics analysis. *Methods in Molecular Biology, 544*, 281–292.

Mezei, F. (1980). *The principles of neutron spin echo. Neutron spin echo: Proceedings of a Laue-Langevin Institut workshop*. Heidelberg: Springer.

Mezei, F., Pappas, C., & Gutberlet, T. (2003). *Neutron spin echo spectroscopy: Basics, trends and applications: Vol. 601*. Springer.

Mittermaier, A., & Kay, L. E. (2006). New tools provide new insights in NMR studies of protein dynamics. *Science, 312*, 224–228.

Miyashita, O., Onuchic, J. N., & Wolynes, P. G. (2003). Nonlinear elasticity, proteinquakes, and the energy landscapes of functional transitions in proteins. *Proceedings of the National Academy of Sciences of the United States of America, 100*, 12570–12575.

Mobley, D. L., & Dill, K. A. (2009). Binding of small-molecule ligands to proteins: "What you see" is not always "what you get". *Structure, 17*, 489–498.

Mukhopadhyay, S., Krishnan, R., Lemke, E. A., Lindquist, S., & Deniz, A. A. (2007). A natively unfolded yeast prion monomer adopts an ensemble of collapsed and rapidly fluctuating structures. *Proceedings of the National Academy of Sciences of the United States of America, 104*, 2649–2654.

Palmer, A. G., 3rd. (2004). NMR characterization of the dynamics of biomacromolecules. *Chemical Reviews, 104*, 3623–3640.

Pettersen, E. F., Goddard, T. D., Huang, C. C., Couch, G. S., Greenblatt, D. M., Meng, E. C., et al. (2004). UCSF chimera—A visualization system for exploratory research and analysis. *Journal of Computational Chemistry, 25*, 1605–1612.

Reczek, D., Berryman, M., & Bretscher, A. (1997). Identification of EBP50: A PDZ-containing phosphoprotein that associates with members of the ezrin-radixin-moesin family. *Journal of Cell Biology, 139*, 169–179.

Shoemaker, B. A., Portman, J. J., & Wolynes, P. G. (2000). Speeding molecular recognition by using the folding funnel: The fly-casting mechanism. *Proceedings of the National Academy of Sciences of the United States of America, 97*, 8868–8873.

Svergun, D. I. (1999). Restoring low resolution structure of biological macromolecules from solution scattering using simulated annealing. *Biophysical Journal, 76*, 2879–2886.

Zaccai, G. (2000). How soft is a protein? A protein dynamics force constant measured by neutron scattering. *Science, 288*, 1604–1607.

SECTION III

Mass Spectrometry

CHAPTER TEN

Method for the Determination of ^{15}N Incorporation Percentage in Labeled Peptides and Proteins

Eric L. Kilpatrick[1]
Bioanalytical Science Group, Biomolecular Measurement Division, Material Measurement Laboratory, National Institute of Standards and Technology, Gaithersburg, Maryland, USA
[1]Corresponding author: e-mail address: eric.kilpatrick@nist.gov

Contents

Abstract

Use of labeled ^{15}N proteins and peptides as internal standards in isotope-dilution mass spectrometry for the quantification of proteins has been increasing and is now accepted as a gold standard for this analysis. As a necessary reagent in this process, stable heavy isotope-labeled internal standards must be rigorously characterized in a number of ways including identity, concentration, purity, and structure. Additionally, the degree of the incorporation of the heavy isotope is a critical feature to consider. For proteins that are ^{15}N labeled, the percentage of incorporation is a valid measurement used to assess the fitness-to-purpose of the material. This measurement should be objective, repeatable, and based on empirical analysis. One means of assigning this value is to compare a mass spectrum of the isotopic profile of a peptide against a series of theoretical profiles containing different enrichment rates. This comparison can be made using the Pearson product-moment correlation coefficient (r) to find the best match between the empirical and theoretical profiles. Theoretical profiles can be generated using probability

Methods in Enzymology, Volume 566
ISSN 0076-6879
http://dx.doi.org/10.1016/bs.mie.2015.09.027

multinomial analysis but are computationally intensive and require the use of computers for practical use. The method described in this chapter describes the development and use of a computer program to calculate the percentage of ^{15}N enrichment of a labeled internal standard. Additionally, methods will be described for the empirical determination of an isotopic profile using a variety of mass spectrometry techniques.

1. INTRODUCTION

Isotopically labeled proteins and peptides have become indispensable to the analysis of proteins using modern techniques such as mass spectrometry (MS) and nuclear magnetic resonance. In quantitative analysis, the labeled form is often an internal standard (IS) in isotope dilution mass spectrometry (IDMS) (Pritchard et al., 2004; Singh et al., 2007). The advantage of the technique is to reduce variation in the measurement by normalizing and accounting for losses in the various stages of analysis from sample preparation, chromatography, ionization, and instrumental drift. IDMS uses linear regression of the ratios of both the response and mass of the sample and calibrants to the IS rather than the absolute values of the response and mass. This normalizing ability is particularly important to proteins which typically require enzymatic processing before MS analysis. Cogeneration of the sample, or calibrant, and IS products used in analysis provides the optimum technique to normalize this procedure.

In labeled proteins and peptides, the natural elemental abundance of an element is replaced with the enriched abundance of a stable heavy isotope of that element. In general, the stable isotopes commonly available are ^{13}C, ^{2}H, ^{15}N, ^{17}O, and ^{18}O; however, for proteins and peptides, the most commonly encountered forms are those incorporating ^{13}C and ^{15}N isotopes. Enriched protein forms with discrete labeled amino acids in the protein or peptide can be generated using techniques such as sequential residue synthesis, *in vitro* transcription and translation synthesis or heterologous expression. Proteins and peptides may also be generated by heterologous expression in cells using ^{15}N as the only nitrogen component. This technique, using *Escherichia coli*, is frequently utilized due to the widely available reagents, technology, and lower costs for this expression platform. For intact proteins, heavy isotope nitrogen (^{15}N) enrichment is the most common labeling form available. This type of labeling has also been utilized to generate quantitative concatamers (Qconcat), synthetic protein-like sequences of multiple tryptic peptides, which may also be recombinantly expressed (Beynon, Hoherty, Pratt, & Gaskell, 2005; Cheung, Anderson, Wang, & Turko, 2015; Scott,

Turko, & Phinney, 2015). The degree of incorporation depends on the purity of the added isotope and the residual level of the naturally more abundant isotope within the culture reagents. This chapter will focus on the method to estimate the ^{15}N incorporation in a labeled peptide and, subsequently, to the protein from which it was isolated.

There are several reasons to determine the level of label incorporation in MS. First, as an IS, the advantage of labeling is to provide a molecule of different mass with otherwise identical properties. However, the difference in mass must be of sufficient magnitude to prevent overlap of isotopic peak spectra between the target analyte and the IS. In cases of nearly complete incorporation (98–99%), provided the peptide is of sufficient length (5–6 residues), this separation is virtually assured but an objective quantification is still desired. However, it is often necessary to sacrifice incorporation in order to attain sufficient yield. In these cases, it is necessary to establish minimum levels of incorporation to remain sufficient for purpose. For labeled proteins used as reference materials, an objective and rigorous estimation method can enable comparison between lots, establish technical requirements, and contribute to uncertainty analysis.

2. MATERIALS

Sequencing grade modified trypsin was obtained from Promega (Madison, WI). MS grade acetonitrile and water were obtained from Honeywell/Burdick and Jackson. Other reagents were purchased from Sigma-Aldrich (St. Louis, MO). Certain commercial equipment or materials are identified in order to adequately describe the experimental procedure. Such identification does not imply recommendation or endorsement by the National Institute of Standards and Technology, nor does it imply that the materials or equipment identified are necessarily the best available for the purpose.

3. BACKGROUND

The isotopic distribution of a peptide can be used as a means for estimating labeled nitrogen content as this distribution is dependent upon the isotopic composition of the compound. Mass spectrometers are instruments suitable for obtaining isotopic profiles, provided that a scan of the spectrum can be produced over the appropriate range. A comparison of the experimentally obtained peaks with theoretically derived peaks is performed with the best match reported as the enrichment level. This concept is straightforward but is computationally challenging, requiring the use of software to accomplish the task.

The size of the protein or peptide to be analyzed by MS is limited by the mass-to-charge (*m/z*) range of the instrument which produces acceptable isotopic peak resolution from approximately *m/z* 600 to *m/z* 2500, typically. This limits analysis to peptides of 5–20 residues in length and precludes the direct analysis of proteins. However, proteins can be broken down into smaller parts which will be within the MS analytical range. Enzymes are useful for this because there are predictable fragmentation patterns. Proteolysis via trypsin is well suited for this purpose because of its specificity for cleaving after lysine or arginine residues. This allows the prediction of "tryptic peptides" of known sequence allowing for calculation of mass and isotopic distributions. Other endoproteases may be used as well including chymotrypsin, elastase, thermolysin, pepsin, or glutamyl endopeptidase. Each of these have known hydrolysis sites and will lead to generation of predicted peptides. Which enzyme may be of best use is dependent upon the specific protein in question. Ultimately, the choice is one that provides an appropriate peptide(s) for evaluation. As lysine and arginine residues are frequent in proteins, trypsin may offer advantages in producing a greater number of unique peptides from representative regions of the protein.

The method described in this chapter estimates the ^{15}N incorporation within peptides released by digestion of intact proteins. Multiple peptides from the protein can be evaluated providing information regarding the uniformity of the labeling throughout the protein. The released peptides are analyzed via MS to produce a spectrum containing the isotopic *m/z* peaks associated with the peptide. Experimentally derived spectra are matched against peaks which are theoretically derived with a given level of ^{15}N incorporation. The comparison is made using the Pearson product-moment correlation coefficient. The incorporation percentage is established by the comparison which produces the maximum coefficient. This technique is restricted to proteins and peptides which are ionized via proton addition and in positive mode. Neither negative mode or ionization via adduct ions can be used with the technique described in this chapter.

4. TRYPSIN DIGESTION

Proteolysis can be performed either in solution or from in-gel digestion of bands excised from SDS-PAGE analysis depending largely on the degree of purification of the protein in question. Mixtures of proteins are frequently separated by SDS-PAGE first to isolate the band containing the labeled protein in order to increase the quality of the MS spectra. Additional considerations are the denaturation of the protein, reduction of disulfide bonds, and irreversible alkylation of cysteine residues to prevent disulfide bond reformation. Proteins with few cysteines may not require reduction or alkylation to achieve sufficient denaturation as long as a sufficient quantity of useful peptides is liberated for qualitative analysis. Furthermore, alkylations, or any other *in vitro* modifications, are likely to have normal isotopic distributions rather than the enriched composition of the expressed protein leading to an underestimation of enrichment results. Therefore, it would be best to avoid peptides containing cysteine, or any other modification not resulting from the recombinant expression. The trypsin should be of at least sequencing grade, having been modified to reduce autolysis which can lead to reduced digestion efficiency as well as inappropriate cleavage at sites other than arginine and lysine.

4.1 In-Gel Digest

This procedure begins with the performance of sodium deoxycholate sulfate polyacrylamide gel electrophoresis (SDS-PAGE) on the protein mixture to separate the constituent proteins (Laemmli, 1970). Specific protocols for this technique are not included in this chapter. However, selection of factors such as gel density, voltage, run time, etc., should be made to optimize the separation of the appropriate protein from others within the lane. SDS-PAGE also serves the purpose of denaturing the protein as well as reducing disulfide bonds. Using gloves and a clean razor blade, remove the band from the gel and cut into cubes with faces having areas of approximately 2–4 mm^2. Place the gel pieces into a clean microcentrifuge tube. Cover the gel with 500 μL of 50 mmol/L ammonium bicarbonate (NH_4HCO_3) and incubate 20 min. Replace the buffer with 150 μL of 50 mmol/L NH_4HCO_3 containing 5 mmol/L dithiothreitol (DTT) to maintain the reduction of disulfide bonds. Incubate at 60 °C for 15 min. Cool to room temperature and add 20 μL of 100 mmol/L iodoacetamide and incubate for 15 min in dark at room temperature. Discard the solvent and wash the gel cubes in 500 μL of 500 mL/L acetonitrile (ACN) with 50 mmol/L NH_4HCO_3 with shaking for 20 min. Remove the wash solution before adding 50 μL of ACN and dry the gel cubes. Drying may be accomplished via vacuum centrifugation or with a nitrogen stream. The quantity of trypsin to be used in the digestion is equal to one-tenth of the quantity of sample protein contained in the excised band and is solubilized with 25 mmol/L NH_4CO_3 in such a concentration that 10 μL contains the appropriate amount of trypsin. Digestion buffer/trypsin (10 μL) is added to the dried gel cubes and incubated for 10 min to allow absorption of the solution. Add 20–50 μL of digestion buffer as needed to completely cover the gel pieces. Incubate the samples at 37 °C with shaking a minimum of 4 h although overnight digestion is typical. Remove and keep the supernatant in fresh tubes. Extract additional peptides from gel by adding 50 μL of 600 mL/L ACN containing 1 mL/L trifluoroacetic acid (TFA) and sonicating for 20 min. Add supernatant to the initial reserved material. Repeat the sonicating extraction and dry the supernatant samples. Reconstitute the peptides in buffer appropriate to the MS technique to be utilized. For MS utilizing electrospray (ESI), add 50 μL of 1 mL/L formic acid. For matrix-associated laser desorption ionization mass spectrometry (MALDI) analysis, reconstitution in 1 mL/L (TFA) is recommended.

4.2 In-Solution Digestion

Samples of protein 3–5 μg are dried down and resolubilized in 25 mmol/L NH_4HCO_3. The simplest approach is to add trypsin directly to the sample and digest overnight at 37 °C. The amount of trypsin is reduced from the amount used in the in-gel digest by a factor of five producing a 50-fold excess of protein over trypsin. In the case where that does not generate sufficient tryptic peptides, addition of denaturing, reduction, or alkylation steps may be required. Addition of denaturants such as urea, guanidine, or acid-labile surfactants has been successfully used to denature proteins for MS as long as certain precautions are observed. Samples to which urea (8 mol/L) is added should not be subjected to elevated temperatures (>60 °C) in order to minimize carbamylation of the side-chain amino groups of lysine and arginine (Kollipara & Zahedi, 2013). Additionally, urea must be diluted at least eightfold prior to trypsin addition to avoid inactivation. Acid-labile surfactants will interfere with MS ionization and must be cleaved at a low pH and removed via centrifugation or solid phase extraction. If needed, reduction with DTT and alkylation with iodoacetamide can be performed using similar concentrations and conditions as for in-gel digestion. Digestion conditions are overnight incubation at 37 °C with shaking. Samples bound for ESI-MS analysis should be dried down and reconstituted in 1 mL/L formic acid (aqueous) while MALDI-MS samples should be reconstituted in 1 mL/L TFA (aqueous).

5. MASS SPECTROMETRY

The tryptic peptides released by digestion are to be analyzed by MS. There are far too many instruments and modes available on the market to attempt to develop specific methods in this chapter. However, some general remarks can be made to address the two principle modes of ionization which have been shown to produce appropriate spectra. In each case, peptides are ionized by proton acceptance on basic residues. Also, each ionization type may be used to provide ions to instruments that are able to generate scans through *m*/*z* ranges, leading to peptide isotopic profile spectra. The principle difference lies in the mechanism of the proton generation, whether by laser excitation or electrical charge. Additional types of ionization such as electron impact or fast atom bombardment are not widely used for peptide MS analysis and are not discussed.

5.1 MALDI

This type of ionization is produced when a peptide-containing sample is mixed with a matrix which can serve as a source of ionizing protons and is excited by laser. A common matrix for peptide work is α–cyano-4-hydroxycinnamic acid but the performance of others may be evaluated to find the optimum for a given analysis. These alternate matrices for peptides include 3,5-dimethoxy-4-hydroxycinnamic acid (sinapic acid), 2,5-dihydroxy benzoic acid. All of these matrices are organic acids and therefore may serve as proton donors. The mixture is prepared in solution and is spotted onto the instrument sample plate and allowed to dry. This will produce crystals of the matrix containing the sample peptides. For MALDI-MS, a portion of the digested sample is mixed in equal parts with 10 mg/mL of the appropriate matrix (very close to a saturated solution) in 400 mL/L acetonitrile containing 1 mL/L TFA and spotted onto the instrument target plate. The organic content of the matrix buffer serves to solubilize hydrophobic peptides and facilitate drying and subsequent crystal formation.

During the instrument operation, energy from a laser is directed onto the crystals to liberate the ionizing protons. The wavelength of the laser is established by the instrument maker but is chosen to maximize energy absorption by the matrix. Ultraviolet lasers are typically used in desorption and common wavelengths are 337 nm (nitrogen laser), 355 nm, or 266 nm (generated by harmonic frequencies of Nd:YAG lasers). In positive mode, peptides are ionized upon proton capture and are accelerated into the drift tube by a large positive electrical potential. The time until the ion strikes the detector is used to assign the *m/z* value to the ion and is described as a time-of-flight mass spectrometer (TOF-MS). Two other operating modes are available, linear or reflector, which differ in the length of the ion path. Reflector mode, with a longer flight path, generally provides the highest resolution at the expense of sensitivity and is usually required in order to generate separated isotopic peaks of a peptide. The isotopic profile of the peptide is described by the peak *m/z* and area (or intensity). MALDI-MS predominantly generates *m/z* peaks which are singly charged with occasional production of doubly charged peaks. The information regarding the singly charged profiles is used as input into the program described later in this chapter to determine ^{15}N isotopic enrichment.

5.2 ESI

This type of ionization is produced by spraying the solution of sample through a charged electrode following chromatographic separation via high-performance liquid chromatography. The solution contains formic acid and as it is charged will lead to proton transfer to the peptide. The charged peptide is separated from the accompanying solution by evaporative mechanisms and is subsequently ejected into the gas phase prior to entry into the MS. ESI predominantly produces multicharged ions with singly charged peaks being less abundant. Peptides with two or three charges are very frequently observed to have the greatest intensity. The type of mass spectrometer to be used with ESI could be either a TOF-MS or an ion trap MS. Ion trap MS can be used in a stand-alone arrangement or as a precursor to high mass accuracy instruments such as Fourier transform ion cyclotron resonance MS, Orbitrap or TOF instruments. Whichever instrument is used, it is important that the isotopic peaks are sufficiently resolved. This is particularly important for higher charge states where peaks may be separated by less than unit *m/z*.

The resulting spectra are analyzed and the isotopic peak profiles are found. The singly charged profile may be observed in some cases and may be used directly in the program. However, the most intense peaks are likely to be either doubly- or triply charged. In this case, the peak *m/z* must be converted to the equivalent singly charged value as described later in the chapter.

6. PROGRAM DESCRIPTION

The program reads the input peptide sequence and accumulates the number of constituent elements (C, H, N, O, S) in each residue to obtain a typical molecular formula in standard form, $C_a H_b N_c O_d S_e$; where a, b, c, d, and e represent the number of atoms of each element present in the peptide. The molecular formula is expanded to include the isotopic possibilities in the following manner: $^{12}C_j\ ^{13}C_k\ ^{1}H_l\ ^{2}H_m\ ^{14}N_n\ ^{15}N_o\ ^{16}O_p\ ^{17}O_q\ ^{18}O_r\ ^{32}S_s\ ^{33}S_t\ ^{34}S_u\ ^{36}S_v$. The subscripts for each elemental isotope will sum to be equal to the total number of atoms for each element as in the standard form. Specifically, the relationship between the subscripts is: $j+k=a$; $l+m=b$; $n+o=c$; $p+q+r=d$; and $s+t+u+v=e$.

A set of nested loops increments the number of isotopes for each element to create a unique formula for each combination of isotopic arrangements. The number of possible combinations is very large and is computationally intensive. Therefore, the program limits the number of each specific heavy isotope for ^{13}C, ^{2}H, ^{17}O, ^{18}O, ^{33}S, ^{34}S, and ^{36}S to manageable levels without user input. No such limits are placed on the total number of ^{15}N isotopes atoms included in the calculation. These limits are sufficient to capture the individual probabilities contributing to more than 99% of the total calculated probability. Use of larger values does not increase the calculated probability but does significantly increase the computation time. A point is quickly reached that will yield a probability smaller than the computational limit of many computer systems and the use of limits avoids these underflow situations. For each combination of isotopes generated by the program, a probability is calculated using the multinomial function defined as $M[f,g]=\frac{(f+g)!}{(f!\times g!)}$.

The probability of an elemental isotopic composition is the product of the multinomial function and all the isotopic abundances raised to the power of the corresponding isotope count (Demirev & Fenselau, 2002; Hsu, 1984; Margrave & Polansky, 1962; Wang et al., 2007). While a Monte Carlo method could be used to generate these distributions (Choudhary et al., 2006), the multinomial approach was chosen because it will find a unique solution for a given input. The expression of the full probability for any given isotopic combination is

$$\begin{aligned}\text{Probability} = & \left\{M[j,k] \times \left({}^{12}\mathrm{C}_{bio}\right)^{j} \times \left({}^{13}\mathrm{C}_{bio}\right)^{k}\right\} \\ & \times \left\{M[l,m] \times \left({}^{1}\mathrm{H}_{ab}\right)^{l} \times \left({}^{2}\mathrm{H}_{ab}\right)^{m}\right\} \\ & \times \left\{M[n,o] \times \left({}^{14}\mathrm{N}_{ab}\right)^{n} \times \left({}^{15}\mathrm{N}_{ab}\right)^{o}\right\} \\ & \times \left\{M[p,q,r] \times \left({}^{16}\mathrm{O}_{ab}\right)^{p} \times \left({}^{17}\mathrm{O}_{ab}\right)^{q} \times \left({}^{18}\mathrm{O}_{ab}\right)^{r}\right\} \\ & \times \left\{M[s,t,u,v] \times \left({}^{32}\mathrm{S}_{ab}\right)^{s} \times \left({}^{33}\mathrm{S}_{ab}\right)^{t} \times \left({}^{34}S_{ab}\right)^{u} \times \left({}^{36}S_{ab}\right)^{v}\right\}.\end{aligned}$$

The subscript *ab* represents the natural abundance of that associated elemental isotope with the exception of carbon. The abundance values (*bio*) of carbon from biological sources are used rather than the abundance of mineral carbon (Beavis, 1993).

Total probability is the sum all of the full probabilities for all the constituent elements in the peptide. However, to develop an isotopic distribution, the probabilities are grouped together according to heavy isotopic content, e.g., 0, 1, 2, etc. For each isotopic composition the corresponding mass is calculated and the probabilities are summed for each nominal isotopic interval. The nominal mass of each peak is based on the weighted average of the isotopic masses against the corresponding probability. In this stepwise manner, a theoretical probability distribution can be generated providing values for both mass and peak height. The program will generate a series of these distributions incrementing the labeled nitrogen incorporation. At each point, the theoretical distribution is compared to an empirical distribution and a goodness-of-fit analysis is performed. Several methods have been described for this comparison including graphical (Demirev & Fenselau, 2002), least squares (Snijders, de Koning, & Wright, 2005), and Pearson's *r* (MacCoss, Wu, Matthews, & Yates, 2005). The Pearson product-moment correlation coefficient, *r*, is used to determine the best fit between the two distributions where

$$r = \frac{\sum X_i Y_i - \frac{\sum X_i \sum Y_i}{N}}{\sqrt{\left(\sum X_i^2 - \frac{(\sum X_i)^2}{N}\right)\left(\sum Y_i^2 - \frac{(\sum Y_i)^2}{N}\right)}}.$$

The level of incorporation is assigned as the percentage of ^{15}N isotope content generating the maximum coefficient value.

6.1 Program Usage

The program is a Windows form application created using Visual Basic (Visual Studio 2010). A screen shot of the program is shown in Fig. 1. The program requires only the input of the known sequence of the peptide and the experimentally determined isotopic profile *m/z* peaks and areas/intensities. The sequence to be analyzed is placed in the peptide sequence input box located at the top of the center-top area of the program window (See Fig. 1).

Standard single letter residue codes for the 20 amino acid residues may be input along with additional codes for modified residues. These additional codes are shown by clicking the "Residue Codes" tab on the top menu bar of the program. Available modifications are alkylated cysteine, phosphorylated serine/threonine, pyroglutamate, and pyrrolysine. Program details and instructions can be viewed by clicking the "About" tab in the menu bar.

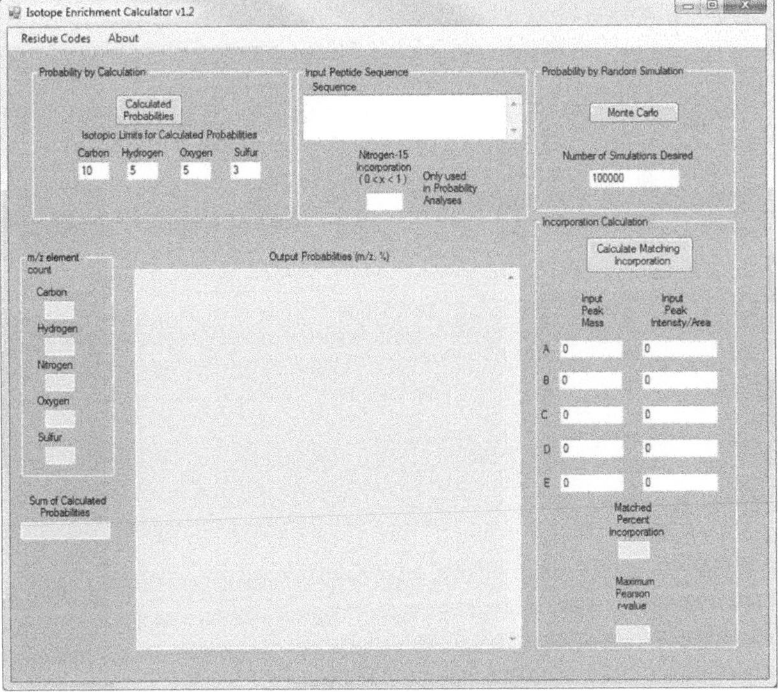

Figure 1 Isotope enrichment calculator. Screen shot of the program interface showing the input and output areas and the associated function buttons.

The isotopic profile data input area is located on the right side of the program area (see Fig. 1). The m/z and area/intensity of each of the peaks in the isotopic profile of the peptide are typed into the available boxes. Use of the areas of the peaks will generally provide more consistent results. However, the program will also function using the peak intensity. Up to five m/z—area pairs may be included in the calculation but a minimum of three are required. It is recommended to use as many peaks as possible. Use caution when identifying appropriate peaks as the m/z difference of any input peak with its immediately adjacent peaks must be less than m/z 1.1. Upon clicking of the "Calculate Matching Incorporation" button, the percent incorporation value and associated Pearson product-moment correlation coefficient (r) are output on the screen below the input fields.

The program is designed to use singly charged m/z but other charge states can be used if first converted to the equivalent singly charged state. The simplified equation defining m/z is given as

$$\frac{\text{mass}}{\text{charge}} = \frac{[M + z]}{z}.$$

In this equation, z represents the charge state (1, 2, 3, etc.) and M refers to the neutral mass of the peptide. This equation may be rearranged to solve for the neutral mass (M) of the higher charge state m/z which can then be used to calculate the equivalent singly charged state m/z. In application, the experimental m/z for a given higher charge state would be multiplied by that charge state and the mass would be reduced to an amount equivalent to the charge state. Conversion to singly charged m/z is accomplished by adding 1 to the neutral mass of the peptide. The combined equation for this process, where n is equal to the charge state (2, 3, etc.), is

$$\left(\frac{\text{mass}}{\text{charge}}\right)_1 = \left(\frac{\text{mass}}{\text{charge}}\right)_n \times n + (1 - n).$$

Once the m/z values have been converted, they may be used directly in the program with the corresponding peak area to find the incorporation value. It must be noted that this calculation produces a nominal value for m/z as the mass of the proton used to obtain the charge is not exactly one. The nominal value is of sufficient accuracy to use for the peak matching in the program. It may be of benefit to compare the incorporation values obtained from a singly charged isotopic profile with that of higher charge state as a back check of the analysis if the spectra are available.

The program has additional features that may be useful in some cases. Given a peptide sequence and desired ^{15}N abundance, the probability distribution of that sequence may be generated in either of two ways. The first is by determining the exact distribution using the multinomial analysis described above. For this function, the limits of the count of labeled isotopes for each element may be changed. The second provides an estimate generated by Monte Carlo analysis in which the number of simulations to be performed is specified by the user. The upper limit on simulation number is dependent on the memory resources of the user's computer. The Monte Carlo value, as a simulation, produces different probability distributions with each run of the program. To use these functions add both the sequence information and incorporation to the appropriate input boxes. Incorporation level is ratio of ^{15}N-to-^{14}N rather than the corresponding percentage and must be less than 1.

Initiating these functions will result in the probability distribution, in text form, in the output box. Additionally, the number of constituent atoms for each element of the charged peptide will be displayed. The sum of the calculated probabilities will be displayed only for the multinomial method and allows the effect of increasing isotopic limits on the quality of the output to be assessed.

7. RESULTS

Validation of the program can be accomplished by analyzing the wild-type distribution of a peptide from a protein having normal isotopic abundances with the program. The program was first tested with the peptide sequence from C-reactive protein (CRP), GYSIFSYATK as previously reported (Kilpatrick, Liao, Camara, Turko, & Bunk, 2012). A sample of natural abundance CRP was digested and analyzed by MALDI-MS. The resulting isotopic profile was input into the program and the abundance of ^{15}N reported matched the natural abundance with a Pearson product-moment correlation coefficient, $r=9991$.

Figure 2A shows the MALDI spectra from the analysis superimposed on the bar graph of the natural isotopic distribution generated by the program. Good matching is observed for both peak m/z and intensity demonstrating the ability of the program to reproduce a known enrichment. ^{15}N-labeled CRP was produced in *Pichia pastoris* and also digested and analyzed via MALDI-MS in the same study. The percentage of ^{15}N incorporation was calculated to be 98.5% with $r=0.9985$. As shown in Fig. 2B, the theoretical

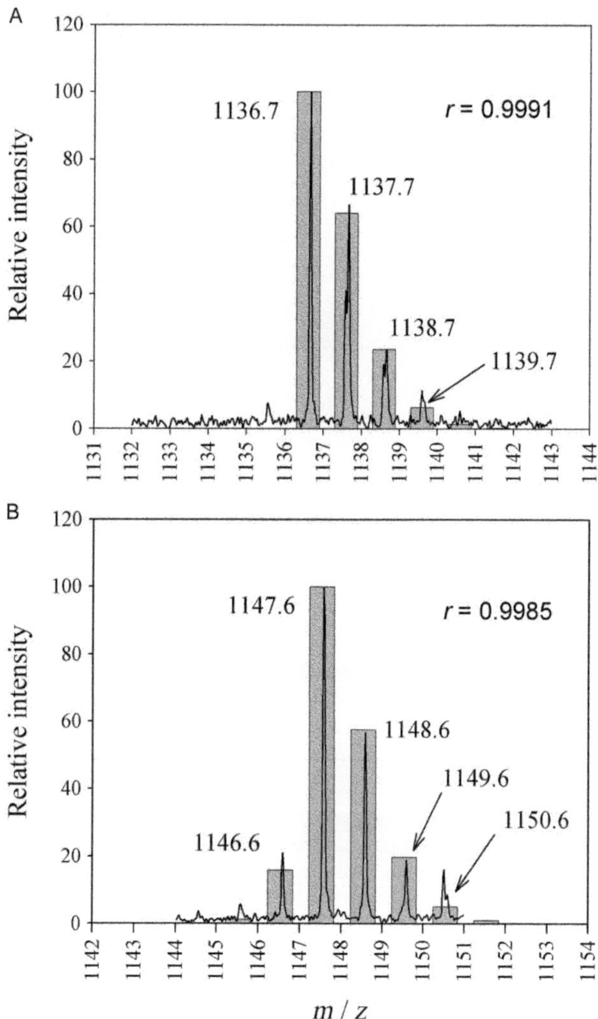

Figure 2 Comparison of theoretical isotopic probability distribution versus peak areas from MALDI-MS spectra for the peptide GYSIFSYATK. (A) MALDI-MS of unlabeled peptide is shown (solid line) and overlaid with the theoretical distribution generated (gray column) for naturally occurring ^{15}N incorporation. (B) MALDI-MS of ^{15}N-labeled peptide is shown (solid line) and overlaid with the theoretical distribution generated (gray column) for the ^{15}N-labeled peptide based on 98.5% incorporation rate. The corresponding Pearson product-moment correlation coefficient is shown in each panel. *Used with permission from Elsevier (Kilpatrick et al., 2012).*

Table 1 Incorporation Analysis for ^{15}N-Enriched CRP

Sequence	*m/z*	Incorporation Percentage	Pearson Correlation Coefficient	AA Positions
EAEAQTDMSR	1151.4	98.3	0.9999	−4 to 6
GYSIFSYATK	1147.5	98.5	0.9985	48 to 57
QDNEILIFWSK	1407.7	98.1	0.9971	59 to 69
YEVQGEVFTKPQLWP	1839.9	97.7	0.9977	192 to 206

Numbering is based on sequence of mature secreted native CRP.
Used with permission from Elsevier (Kilpatrick et al., 2012).

graph of 98.5% ^{15}N incorporation agrees very well with the MALDI spectra. It is beneficial to determine the incorporation with multiple peptides that would span the different parts of the protein. This increases the number of measurements that the average value is based upon leading to a more robust measurement. Also, as these same peptides may be used as the quantitative peptides in IDMS, knowing the incorporation differences of these peptides can be used to reduce potential measurement bias. Table 1 shows the incorporation analysis for four ^{15}N-labeled CRP peptides.

Each peptide shows a similar degree of incorporation and the Pearson product-moment correlation coefficient across the length of the protein.

8. CONCLUSIONS

The determination of the ^{15}N incorporation percentage of labeled proteins can be accomplished with a straightforward process beginning with protein digestion by appropriate enzymes followed by analysis with a variety of types of mass spectrometers. The resulting isotopic profiles of specific peptides can easily be analyzed using software developed for the purpose. This software has been demonstrated to faithfully reproduce a natural abundance incorporation and provide a consistent estimate across the length of a protein. The quantification of the degree of incorporation of ^{15}N-labeled proteins allows for the confident claim of fitness-for-purpose or batch comparison purposes of IS for IDMS for the quantification of protein. The program is available by request from the author or by download (http://www.nist.gov/mml/bmd/bioanalytical/isoenrichcalc.cfm).

REFERENCES

Beavis, R. C. (1993). Chemical mass of carbon in proteins. *Analytical Chemistry*, *65*, 496–497.

Beynon, R. J., Hoherty, M. K., Pratt, J. M., & Gaskell, S. J. (2005). Multiplexed absolute quantification in proteomics using artificial QCAT proteins of concatenated signature peptides. *Nature Methods*, *2*, 587–589.

Cheung, C. S., Anderson, K. W., Wang, M., & Turko, I. V. (2015). Natural flanking sequences for peptides included in a quantification concatamer internal standard. *Analytical Chemistry*, *87*, 1097–1102.

Choudhary, K., Spicer, V. L., Donald, L. J., Duckworth, H. W., Ens, W., Loewen, P. C., et al. (2006). Method for estimating the isotopic distributions of metabolically labeled proteins by MALDI-TOFMS: Application to NMR samples. *Analytical Chemistry*, *78*, 5419–5423.

Demirev, P. A., & Fenselau, C. F. (2002). Determination of isotope-enrichment ratios in proteins by high-resolution Fourier transform ion cyclotron resonance mass spectrometry. *European Journal of Mass Spectrometry*, *8*, 163–167.

Hsu, C. S. (1984). Diophantine approach to isotopic abundance calculations. *Analytical Chemistry*, *56*, 1356–1361.

Kilpatrick, E. L., Liao, W.-L., Camara, J. E., Turko, I. V., & Bunk, D. M. (2012). Expression and characterization of 15 N-labeled human C-reactive protein in *Escherichia coli* and *Pichia pastoris* for use in isotope-dilution mass spectrometry. *Protein Expression and Purification*, *85*, 94–99.

Kollipara, L., & Zahedi, R. P. (2013). Protein carbamylation: In vivo modification or in vitro artefact. *Proteomics*, *13*, 941–944.

Laemmli, U. K. (1970). Cleavage of structural proteins during the assembly of the head of bacteriophage T4. *Nature*, *227*, 680–685.

MacCoss, M. J., Wu, C. C., Matthews, D. E., & Yates, J. R., III. (2005). Measurement of the isotope enrichment of stable isotope-labeled proteins using high-resolution mass spectra of peptides. *Analytical Chemistry*, *77*, 7646–7653.

Margrave, J. L., & Polansky, R. B. (1962). Relative abundance calculations for isotopic molecular species. *Journal of Chemical Education*, *39*, 335–337.

Pritchard, C., Groves, K. J., Biesenbruch, S., O'Connor, G., Ashcroft, A. E., Aresene, C., et al. (2004). Quantification of human growth hormone in serum with a labeled protein as an internal standard: Essential considerations. *Analytical Chemistry*, *86*, 6525–6532.

Scott, K. B., Turko, I. V., & Phinney, K. W. (2015). Quantitative performance of internal standard platforms for absolute protein quantification using multiple reaction monitoring-mass spectrometry. *Analytical Chemistry*, *87*, 4429–4435.

Singh, R., Crow, F. W., Babic, N., Lutz, W. H., Lieske, J. C., Larson, T. S., et al. (2007). A liquid chromatography-mass spectrometry method for the quantification of urinary albumin using a novel 15 N-isotopically labeled albumin internal standard. *Clinical Chemistry*, *53*, 540–542.

Snijders, A. P. L., de Koning, B., & Wright, P. C. (2005). Pertubation and interpretation of nitrogen isotope distribution patterns in proteomics. *Journal of Proteome Research*, *4*, 2185–2191.

Wang, B., Sun, G., Anderson, D. R., Jia, M., Previs, S., & Anderson, V. E. (2007). Isotopologue distributions of peptide product ions by tandem mass spectrometry: Quantitiation of low levels of deuterium incorporation. *Analytical Biochemistry*, *367*, 40–48.

CHAPTER ELEVEN

QconCAT: Internal Standard for Protein Quantification

Kerry Bauer Scott*[,1], Illarion V. Turko*, Karen W. Phinney[†]

*Institute for Bioscience and Biotechnology Research, National Institute of Standards and Technology and the University of Maryland, Rockville, Maryland, USA

[†]Biomolecular Measurement Division, National Institute of Standards and Technology, Gaithersburg, Maryland, USA

[1]Corresponding author: e-mail address: kbauer@ibbr.umd.edu

Contents

Abstract

Protein quantification based on stable isotope labeling-mass spectrometry involves adding known quantities of stable isotope-labeled internal standards into biological samples. The internal standards are analogous to analyte molecules and quantification is achieved by comparing signals from isotope-labeled and analyte molecules. This methodology is broadly applicable to proteomics research, biomarker discovery and validation, and clinical studies, which require accurate and precise protein abundance measurements. One such internal standard platform for protein quantification is concatenated peptides (QconCAT). This chapter describes a protocol for the design, expression, characterization, and application of the QconCAT strategy for protein quantification.

Methods in Enzymology, Volume 566
ISSN 0076-6879
http://dx.doi.org/10.1016/bs.mie.2015.09.022

1. INTRODUCTION

QconCATs are artificial proteins comprised of concatenated peptides (Fig. 1). Usually, tryptic peptides are concatenated, but QconCATs can be used for the quantification of any peptide that is reproducibly produced by cleavage from any sample (Pratt et al., 2006). These peptides can originate from multiple different proteins allowing for the simultaneous quantification of many independent proteins. During expression of the QconCAT, a stable isotope (e.g., ^{13}C, ^{15}N) is incorporated into the protein product. The isotopically labeled QconCAT is added to the sample of interest prior to proteolysis to produce labeled and unlabeled peptides for detection by mass spectrometry. The mass spectrometric signal ratio of unlabeled to labeled peptide ions and the known concentration of the added QconCAT allows for quantification of the proteins in the sample.

The first QconCAT was used to quantify several proteins in chick skeletal muscle (Beynon, Doherty, Pratt, & Gaskell, 2005). QconCAT constructs have since been used to measure clinically relevant proteins in Alzheimer's disease (Chen, Wang, & Turko, 2012), study the stoichiometry of protein complexes (Al-Majdoub, Carroll, Gaskell, & Barber, 2014; Nanavati, Gucek, Milne, Subramaniam, & Markey, 2008), quantify

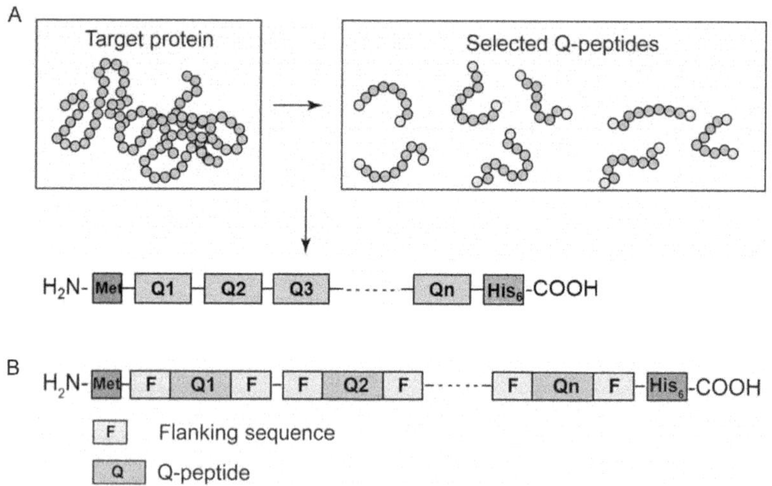

Figure 1 QconCAT schematic. (A) Selected peptides from a protein of interest concatenated into an artificial protein. (B) QconCAT with peptides carrying flanking sequences made up of their natural amino acid sequences on both sides of the peptides. *Adapted from Chen and Turko (2014). Reprinted with permission.* (See the color plate.)

posttranslational modifications (Carroll, Lanucara, & Eyers, 2011; Johnson, Eyers, Eyers, Beynon, & Gaskell, 2009), distinguish and quantify protein isoforms (Chen, Wang, & Turko, 2013), and interrogate the mammalian host response to pathogens (Bislev et al., 2012). These examples demonstrate the versatility of the QconCAT strategy.

Alternatives to the QconCAT internal standard are recombinant proteins and synthetic peptides. However, recombinant protein expression is often not feasible, especially for high-molecular weight proteins and membrane proteins. Whereas, synthetic peptides are broadly available, but their performance is affected by issues with resolubilization, degradation, peptide modification, and incomplete digestion of the analyte protein within the sample. Comparative studies between QconCATs and synthetic peptides or recombinant proteins have shown that a critical factor affecting equal release of the individual peptides comprising the QconCAT are proteolysis conditions (Mirzaei, McBee, Watts, & Aebersold, 2008). Moreover, the QconCAT peptides need to have similar proteolysis rates as their peptide counterparts from the native protein. Different digestion rates can result in overestimation or underestimation of the protein in the sample depending on whether the QconCAT is more or less susceptible to digestion (Brun et al., 2007; Scott, Turko, & Phinney, 2015).

This possible limitation of the QconCAT strategy has been shown to be combatted with the addition of natural flanking sequences into the QconCAT construct (Fig. 1). The direct concatenation of peptides into an artificial protein results in amino acid sequences surrounding the proteolytic cleavage sites that are not identical to the native proteins. During enzymatic digestion this can lead to different digestion rates as the efficiency of digestion can be influenced by amino acids surrounding the cleavage sites (flanking sequences). Natural flanking sequences include the amino acids (usually three to six) proximal to the cleavage sites of a peptide as they are present in the native protein of interest (Chen & Turko, 2014). QconCATs with natural flanking sequences have been shown to have improved accuracy over QconCAT constructs without natural flanking sequences (Cheung, Anderson, Wang, & Turko, 2015; Scott et al., 2015). These results, for QconCATs based on tryptic peptides, are due to the conserved primary sequence near the tryptic cleavage sites which reduce differences in digestion rates between the QconCAT and sample proteins.

The advantage of parallel quantification of multiple proteins makes the QconCAT an important and versatile tool for protein quantification. This tutorial provides practical information regarding the preparation and use of

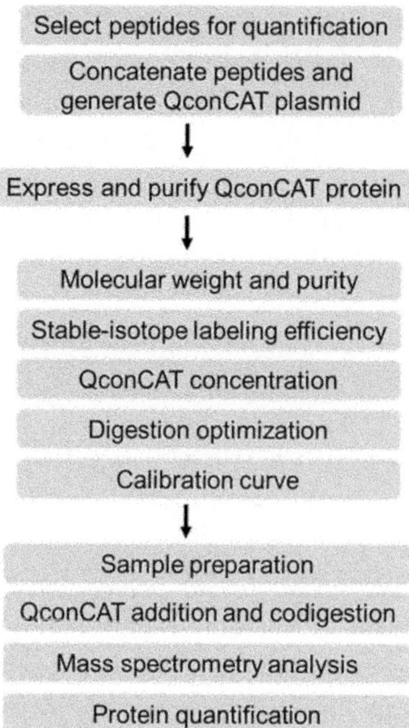

Figure 2 Flowchart for complete protocol including design, expression, characterization, and application of the QconCAT internal standard strategy.

QconCATs as a way of providing absolute or relative protein quantification. The QconCAT protocol presented (Fig. 2) here can be adapted for numerous applications and biological systems.

2. QconCAT DESIGN

QconCAT generation begins with the selection of a target list of proteins to measure. This is followed by the selection of representative peptides to incorporate into the QconCAT. Peptides for quantification should be 600–4000 Da in mass, unique to the protein of interest, and detectable by mass spectrometry. Standard peptides should lack residues that are prone to modification (e.g., methionine, cysteine, N-terminal glutamine) and lack sites with posttranslational modification in the native protein (e.g., phosphorylation, glycosylation). For tryptic peptides, dibasic residues and cleavage sites neighbored by acidic amino acids should also be avoided when possible as these sequence patterns promote miscleavages (Vandermarliere, Mueller, & Martens, 2013).

A peptide mapping experiment of the samples of interest using the cleavage strategy and mass spectrometer of choice for the study would ease peptide selection. This would enable comparisons of resultant peptide quality and quantity for selection of optimally performing peptides. Experimental analysis of the sample of interest would also provide insight into completeness of proteolysis, presence of posttranslational modifications, and incidence of isobaric and isomeric peptides derived from other proteins in the sample (Simpson & Beynon, 2012). In lieu of experimental data, there are many resources to aid in the selection of peptides including Blast protein search to determine peptide uniqueness and the National Institute of Standards and Technology (NIST) libraries of peptide tandem mass spectra and PeptideAtlas for mass spectrometry data from prior experiments (Deutsch, Lam, & Aebersold, 2008).

Despite thoughtful design of the QconCAT, some constructs can still result in difficult peptides that do not release well or peptides that do not show equimolarity upon digestion. To minimize these risks, several peptides for each protein of interest should be included in the QconCAT. This increases the robustness of the QconCAT. If one peptide proves unsuitable for proteolytic release or detection by mass spectrometry, another peptide can be used for quantification.

Depending on the number of proteins and corresponding peptides selected, multiple independent QconCAT proteins may be needed. Group peptides from proteins with similar abundances in the sample. Individual QconCAT peptides have a limited dynamic range (up to 2 orders of magnitude) where the mass spectrometric response is linear and protein quantification will be accurate. Incorporating peptides from proteins with similar dynamic ranges into the same QconCAT allows the linear response of these peptides to overlap the concentration range of the endogenous peptides in the sample. Otherwise, multiple samples have to be prepared with varying amounts of added QconCAT standard to accurately quantify all proteins of interest when the endogenous proteins have widely different abundances.

Selected peptides next need to be ordered in the QconCAT. For QconCATs without flanking sequences, peptides should be ordered to preserve the natural primary local sequence context when possible. For example, if a leucine immediately neighbors the cleavage site of a peptide, this peptide should be placed next to a peptide with an N-terminal leucine (Simpson & Beynon, 2012). Furthermore, as with trypsin, the three amino acids surrounding each site of a cleavage site can affect digestion efficiency (Vandermarliere et al., 2013). Preserving this much local primary sequence in the QconCAT may be impossible. Therefore, natural flanking sequences can be included to better mimic the local sequence of the native peptides and minimize the significance of peptide order.

In addition to the quantification peptides, other features should be included in the QconCAT construct. At the N-terminus, introduction of a short sequence (e.g., MGTK) provides an initiator methionine and a sacrificial peptide to protect the quantification peptides. The C-terminus should encode a hexahistidine purification tag (His-tag) for purification of the expressed QconCAT. The QconCAT gene can be optimized for codon usage for expression in *Escherichia coli* and mRNA secondary structure then cloned into a T7 expression vector such as pET-21a.

3. EXPRESSION AND PURIFICATION OF QconCAT

The following protocol is for 1 L of *E. coli* culture in minimal growth media with uniform ^{15}N isotopic labeling of the QconCAT. Necessary nitrogen essential for cell growth can be controlled by use of minimal growth media supplemented with isotopically labeled nutrients such as ^{15}N-ammonium chloride (^{15}N-NH_4Cl). Uniform labeling with ^{15}N-NH_4Cl is more cost effective compared to stable isotope labeling with specific amino acids. Uniform isotopic labeling also provides flexibility in cleavage strategy and transition selection. Incorporation of specific isotopic amino acids such as $^{13}C_6$-leucine would limit peptide selection to those containing a leucine residue. Even labeling with $^{13}C_6$-lysine and $^{13}C_6$-arginine limits cleavage to tryptic digestion; whereas, uniform isotopic labeling enables endopeptidases other than trypsin to be used. Uniform labeling also provides more freedom in transition selection for quantification by multiple reaction monitoring (MRM) assays. For instance, transitions would not be limited to y-ions due to a requirement of labeled $^{13}C_6$-lysine or $^{13}C_6$-arginine to be present in the fragment ion. If labeling with specific amino acids is desired for a consistent mass offset between every unlabeled and labeled peptide pair, cell-free expression systems offer a cost-effective alternative due to the small expression volumes (Chen, Wang and Turko, 2012, 2013). However, bacterial cell-free expression systems are not recommended for larger QconCAT proteins.

Transform the QconCAT plasmid into chemically competent BL21 (DE3) *E. coli* cells. Dissolve plasmid in TE buffer (10 mmol/L Tris–HCl, 1 mmol/L EDTA, pH 8.0) to a concentration of 10 ng/μL. Add 1 μL of resuspended plasmid to 25 μL of cells and incubate on ice for 10 min. Heat shock cells in a 42 °C water bath for 30 s and return to ice for 2 min. Add 250 μL SOC medium (2% tryptone, 0.5% yeast extract, 10 mmol/L NaCl, 2.5 mmol/L $MgCl_2$, 10 mmol/L $MgSO_4$, 20 mmol/L glucose) and incubate with shaking at 37 °C for 1 h. Plate 100 μL onto an LB agar plate with 100 μg/mL ampicillin and incubate inverted overnight at 37 °C.

Prepare 1 L of ^{15}N-enriched minimal growth media (M9 media; Table 1). Use a colony from the plate to inoculate a starter culture of 10 mL M9 media supplemented with 100 μg/mL ampicillin and grow in a shaker at 37 °C overnight. The following day inoculate main culture (remaining volume of M9 media supplemented with 100 μg/mL ampicillin) using overnight starter culture and grow until optical density at 600 nm reaches 0.6–0.8. Induce QconCAT expression with 400 μmol/L isopropyl β-D-1-thiogalactopyranoside and culture at 37 °C with shaking for 4 h before collecting the cells by centrifugation (4000 × *g*; 10 min).

Resuspend the bacterial cell pellet in lysis buffer (50 mmol/L Tris–HCl, 150 mmol/L sodium chloride, 1 mmol/L dithiothreitol, pH 7.4) and sonicate on ice. Centrifuge the solubilized cell pellet (15000 × *g*, 25 min, 4 °C) to recover the inclusion bodies. Solubilize the inclusion body pellet in equilibration buffer for the metal affinity chromatography (Table 2) and centrifuge (10000 × *g*, 10 min, 4 °C) to remove insoluble material. Although QconCATs resemble endogenous proteins in composition, unlike native analyte proteins, QconCATs do not have inherent higher order structure and usually aggregate in inclusion bodies. Verify QconCAT presence in

Table 1 ^{15}N-Enriched Minimal Media Composition

Component	Amount in 1 L
^{15}N-ammonium chloride ($^{15}NH_4Cl$)	1.0 g
Sodium phosphate (Na_2HPO_4)	6.8 g
Potassium phosphate (KH_2PO_4)	3.0 g
Sodium chloride (NaCl)	0.5 g
Glucose	4.0 g
Magnesium sulfate ($MgSO_4$)	240.0 mg
Calcium chloride ($CaCl_2$)	11.3 mg
Thiamine	1.0 mg
Iron chloride ($FeCl_3$)	3.0 mg

Weigh and dissolve components bring volume up to 1 L with water. Filter sterilize media before use.

Table 2 Metal Affinity Chromatography Equilibration, Wash, and Elution Buffer Compositions

Buffer	Component	Concentration (mol/L)
Equilibration	Sodium phosphate (Na_2HPO_4)	0.05
	Sodium chloride (NaCl)	0.30
	Guanidine hydrochloride	6.0
Wash	Sodium phosphate (Na_2HPO_4)	0.05
	Sodium chloride (NaCl)	0.30
	Guanidine hydrochloride	6.0
	Imidazole	0.01
Elution	Sodium phosphate (Na_2HPO_4)	0.05
	Sodium chloride (NaCl)	0.30
	Guanidine hydrochloride	6.0
	Imidazole	0.15

All buffers adjusted to a pH of 7.4. Guanidine hydrochloride can be substituted for urea (8 mol/L).

the inclusion body with sodium dodecyl sulfate–polyacrylamide gel electrophoresis (SDS-PAGE) and Coomassie Blue or silver staining of cell lysis supernatant and solubilized inclusion body fractions. Guanidine hydrochloride needs to be removed from inclusion body fractions with chloroform/methanol precipitation prior to SDS-PAGE analysis.

Prepare a conventional gravity flow chromatography column with cobalt or nickel resin and equilibrate with equilibration buffer (Table 2). Load the solubilized inclusion bodies by gravity flow onto the pre-equilibrated resin column. Wash the resin and elute His-tagged QconCAT protein with wash/elute buffer (Table 2). Eluent can be collected until the absorbance reading at 280 nm returns to baseline (A_{280} of elution buffer). The concentration of imidazole in the wash and elution buffers may need to be adjusted to decrease nonspecific binding and increase protein yield. Buffer exchange the purified QconCAT into 25 mmol/L ammonium bicarbonate (NH_4HCO_3) to remove contaminating reagents such as imidazole. The QconCAT protein may need to be solubilized in a buffer that includes a denaturing agent (e.g., urea or RapiGest surfactant) to keep the QconCAT in solution and prevent precipitation. This is especially important for larger QconCAT proteins which have a higher propensity for aggregation

(Brownridge et al., 2011). However, QconCATs have no requirement for functional folding allowing for flexibility in the final buffer composition.

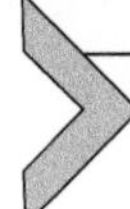

4. CHARACTERIZATION

4.1 Purity and Molecular Weight Evaluation

Protein purity and molecular weight can be assessed to evaluate the quality of the QconCAT. Evaluate purity of the QconCAT protein by SDS-PAGE and Coomassie Blue or silver staining. Perform intact protein analysis with matrix-assisted laser desorption ionization (MALDI) or electrospray ionization mass spectrometry to compare the calculated molecular weight to the theoretical molecular weight to ensure complete expression of the protein.

4.2 Determination of Stable Isotope Labeling Efficiency

Nearly, full stable isotope labeling is essential as the dynamic range of the experiment is largely ruled by the labeling efficiency. A QconCAT labeled at 90% would have a dynamic range of less than a 1 order of magnitude (Simpson & Beynon, 2012). Fortunately, high labeling efficiency (e.g., >98–99%) is usually obtainable for this uniform labeling approach when the isotopically labeled nutrient is greater than 98% pure. Determine ^{15}N incorporation (%) into the QconCAT at the peptide level. Digest the QconCAT and acquire precursor ion mass spectra through either MALDI or liquid chromatography–mass spectrometry (LC–MS) data collection. Calculate ^{15}N incorporation by the NIST Isotopic Enrichment Calculator which compares a series of theoretical isotopic distributions at incremental isotopic percentages for an input peptide sequence to the experimentally derived isotopic distribution profile (Kilpatrick, Liao, Camara, Turko, & Bunk, 2012). The software is freely available for download from: http://www.nist.gov/mml/bmd/bioanalytical/isoenrichcalc.cfm. For more information, see chapter "Method for the determination of ^{15}N incorporation percentage in labeled peptides and proteins" by Kilpatrick.

4.3 Concentration Determination

Accuracy in quantitative mass spectrometry requires accuracy in the concentration of the internal standard protein used. There are several approaches to quantify the purified QconCAT protein and each must be weighed in terms of accuracy and experimental needs. For experiments, where quantification will be relative or there is high biological variation, more imprecision in the accuracy of the QconCAT quantification can be tolerated. For absolute quantification experiments, a high degree of accuracy in the quantification of the QconCAT is required.

QconCAT quantification can be determined by a colorimetric protein assay such as the bicinchoninic acid assay. These assays offer low protein consumption and convenience, but variable accuracy depending on the amino acid composition of the QconCAT protein. The same pure protein should be used as a standard for the most accurate results. However, this is not usually possible and bovine serum albumin (BSA) is usually used as a standard. This may lead to over or underestimation of concentration depending on variation in the amount of types of amino acid residues between BSA and the QconCAT. Also, the presence of reducing and chelating agents can affect quantification accuracy and may need to be reduced by sample dilution or removed through dialysis or precipitation.

Spectrophotometric quantification from direct measurements of absorbance (280 nm) based on QconCAT amino acid composition provides convenience, low sample consumption, and high accuracy. The concentration of the purified QconCAT protein is determined based on the absorbance value at 280 nm and the protein-specific extinction coefficient in conjunction with Beer's law. The molecular weight and molar extinction coefficient of a novel QconCAT protein can be calculated based on the known amino acid composition using the ProtParam tool on the ExPASy server (http://web.expasy.org/protparam/; Wilkins et al., 1999). Absorptivity, and therefore accuracy, can be affected by buffer, ionic strength, and pH. In addition, non-protein components in the protein solution that absorb ultraviolet light (lipids, nucleic acids, and detergents) can interfere with the measurements and need to be accounted for or removed.

Amino acid analysis is the reference method to quantify proteins and can provide absolute quantification of pure proteins. Several factors can affect the accuracy of amino acid analysis measurements of concentration (Rutherfurd & Gilani, 2009). Incomplete hydrolysis of the protein into constituent amino acids or degradation of amino acids during hydrolysis will adversely affect the accuracy of quantification. The quality of the amino acid standards used for quantification will also determine the accuracy of the measurement.

4.4 Digestion Efficiency Evaluation

Accurate quantification using the QconCAT strategy relies on stoichiometric proteolysis of the QconCAT peptides and identical proteolytic release of the analyte and QconCAT peptides. A rigorous assessment of QconCAT digestion is required to determine digestion conditions that will diminish potential proteolysis challenges. Proteolytic rates and efficiency are highly impacted by protein higher order structure and primary sequence. QconCATs generally lack high order structure whereas native proteins can be tightly folded and resistant to digestion. Use of a denaturant (urea, guanidinium hydrochloride, RapiGest, or acetonitrile) and sufficient digestion time should diminish structural impediments to proteolysis, especially for the native proteins. Prior to proteolysis, the presence of these chaotropes would have to be reduced to prevent hindrance of enzyme activity.

Digestion conditions such as denaturant, enzyme concentration, and length of digestion incubation time greatly affect peptide release. These conditions can be tailored to best fit an individual QconCAT and analyte protein sample to allow for accurate quantification. Conduct a time course experiment to assess the rate and extent of peptide release by monitoring peptide signal over digestion time (Brownridge & Beynon, 2011). The optimal digestion protocol is indicated when relative signals between peptides from analyte protein and QconCAT reach a steady plateau. At earlier time points, the analyte and QconCAT peptides may have different rates of proteolysis. Longer digestion incubation times may be required until peptide digestion curves for both QconCAT and analyte plateau (Brownridge & Beynon, 2011). Quantification errors can occur if peptide release does not reach a plateau as complete digestion is not indicated, as is the case for slowly released peptides.

Each peptide in the QconCAT should be considered individually, since not every proteins and not all peptides within an individual protein behave the same under the same proteolytic conditions (Mirzaei et al., 2008; Scott et al., 2015). Individual peptides may require different digestion conditions for accurate quantification. In addition to proteolytic release, the digestion protocol may also require reduction and alkylation of cysteine residues if the analyte proteins contain disulfide bonds or the QconCAT contains cysteine residues.

4.5 Dynamic Range Determination

Quantification based on QconCATs requires the establishment of calibration curves prior to concentration determinations of any unknown sample. The calibration curve allows assessment of linearity, dynamic range, and limits of quantification. A standard curve is constructed relating peak ratio of labeled to nonlabeled peptide to amount of labeled QconCAT material. To reflect the intention of the QconCAT application, spike varying amounts of quantified QconCAT material into the sample, digest the sample, analyze by mass spectrometry, and construct a calibration curve for each QconCAT peptide. The spiked QconCAT concentration range needs to cover the concentration of analyte protein in the sample of interest and include at least five concentrations. In addition, spike the QconCAT material into a matrix similar in composition to the test samples that are going to be used experimentally to account for matrix effects.

Calibration curves for truly quantitative peptides will show low scatter and linearity over the entire concentration range. A comparison of the calibration curves for each peptide should show similar slope values and y-intercepts. Outlier slopes lower in value may indicate reduced digestion efficiency or peptide modification and variability in y-intercepts suggest unsuitability for use in quantification. The limit of quantification for each peptide can be determined by identifying the lowest calibration point at which the measured coefficient of variation is less than 20%.

5. APPLICATION OF QconCAT

QconCATs can be used to quantify proteins in any number of matrices from simple protein mixtures to complex tissue samples. The test samples may require preparation to homogenize tissue, lyse cells, and release protein content prior to QconCAT addition and subsequent co-proteolysis. One constraint of the QconCAT is that they do not have the same physicochemical properties as the native proteins they represent. QconCATs cannot be presumed to enrich, fractionate, chromatograph, or precipitate with the analyte proteins of interest. If samples of proteins are pre-treated in any of these ways, the QconCAT quantification may no longer reflect the absolute protein concentration in the original test sample. However, at the peptide level, the QconCAT peptides and analyte peptides do have the same physicochemical properties and will share identical chromatographic, ionization, and fragmentation behavior.

A known amount (in moles) of QconCAT protein is added to the test samples so that the analyte protein amount in the samples is within the dynamic linear range of the analogous QconCAT peptides. The mixed samples are digested according to the optimized digestion protocol. Following digestion, the samples are ready for analysis by mass spectrometry. Extracted ion chromatograms (EIC) from a high-resolution mass spectrometer can be extracted for a specific m/z (mass to charge) value with a defined mass tolerance window. Coeluting species may be able to be resolved as long as the species have independent isolation windows. Simultaneous tandem MS data acquisition can add sequence verification of the specific species. MRM assays on a triple quadrupole instrument can improve sensitivity and selectivity of a quantitative analysis. MRM assays can also be a solution to quantification of interfering isobaric and isomeric species may be able to be differentiated depending on different m/z fragmentation values.

To quantify the proteins, peak areas are determined from EICs or MRM assays for the QconCAT peptide ions and analyte peptide ions. The relative proportions of the labeled QconCAT and unlabeled sample peak areas are calculated. The experimentally determined peak area ratios are used to calculate the unknown amount of peptide and corresponding analyte protein in the sample:

$$\frac{\text{QconCAT peak area}}{\text{Sample peak area}} = \frac{\text{QconCAT amount}}{\text{Sample amonut}}$$

When the number of moles of QconCAT added to the sample is known, the amount of protein in the sample can be translated to the number of molecules per cell or mg of tissue or converted to a concentration, depending on the sample (Carroll et al., 2011).

Disclaimer. Certain commercial materials and equipment are identified in order to adequately specify the experimental procedure. Such identification neither implies recommendation or endorsement by the National Institute of Standards and Technology nor does it imply that the material or equipment identified is the best available for the purpose.

REFERENCES

Al-Majdoub, Z. M., Carroll, K. M., Gaskell, S. J., & Barber, J. (2014). Quantification of the proteins of the bacterial ribosome using QconCAT technology. *Journal of Proteome Research, 13*(3), 1211–1222. http://doi.org/10.1021/pr400667h.

Beynon, R. J., Doherty, M. K., Pratt, J. M., & Gaskell, S. J. (2005). Multiplexed absolute quantification in proteomics using artificial QCAT proteins of concatenated signature peptides. *Nature Methods*, *2*(8), 587–589. http://doi.org/10.1038/nmeth774.

Bislev, S. L., Kusebauch, U., Codrea, M. C., Beynon, R. J., Harman, V. M., Røntved, C. M., et al. (2012). Quantotypic properties of QconCAT peptides targeting bovine host response to *Streptococcus uberis*. *Journal of Proteome Research*, *11*(3), 1832–1843. http://doi.org/10.1021/pr201064g.

Brownridge, P., & Beynon, R. J. (2011). The importance of the digest: Proteolysis and absolute quantification in proteomics. *Methods (San Diego, CA, United States)*, *54*(4), 351–360. http://doi.org/10.1016/j.ymeth.2011.05.005.

Brownridge, P., Holman, S. W., Gaskell, S. J., Grant, C. M., Harman, V. M., Hubbard, S. J., et al. (2011). Global absolute quantification of a proteome: Challenges in the deployment of a QconCAT strategy. *Proteomics*, *11*(15), 2957–2970. http://doi.org/10.1002/pmic.201100039.

Brun, V., Dupuis, A., Adrait, A., Marcellin, M., Thomas, D., Court, M., et al. (2007). Isotope-labeled protein standards: Toward absolute quantitative proteomics. *Molecular & Cellular Proteomics*, *6*(12), 2139–2149. http://doi.org/10.1074/mcp.M700163-MCP200.

Carroll, K. M., Lanucara, F., & Eyers, C. E. (2011). Quantification of proteins and their modifications using QconCAT technology. *Methods in Enzymology*, *500*, 113–131. http://doi.org/10.1016/B978-0-12-385118-5.00007-4.

Chen, J., & Turko, I. V. (2014). Trends in QconCATs for targeted proteomics. *Trends in Analytical Chemistry*, *57*, 1–5. http://doi.org/10.1016/j.trac.2013.12.013.

Chen, J., Wang, M., & Turko, I. V. (2012). Mass spectrometry quantification of clusterin in the human brain. *Molecular Neurodegeneration*, 7, 41. http://doi.org/10.1186/1750-1326-7-41.

Chen, J., Wang, M., & Turko, I. V. (2013). Quantification of amyloid precursor protein isoforms using quantification concatamer internal standard. *Analytical Chemistry*, *85*(1), 303–307. http://doi.org/10.1021/ac3033239.

Cheung, C. S. F., Anderson, K. W., Wang, M., & Turko, I. V. (2015). Natural flanking sequences for peptides included in a quantification concatamer internal standard. *Analytical Chemistry*, *87*(2), 1097–1102. http://doi.org/10.1021/ac503697j.

Deutsch, E. W., Lam, H., & Aebersold, R. (2008). PeptideAtlas: A resource for target selection for emerging targeted proteomics workflows. *EMBO Reports*, *9*(5), 429–434. http://doi.org/10.1038/embor.2008.56.

Johnson, H., Eyers, C. E., Eyers, P. A., Beynon, R. J., & Gaskell, S. J. (2009). Rigorous determination of the stoichiometry of protein phosphorylation using mass spectrometry. *Journal of the American Society for Mass Spectrometry*, *20*(12), 2211–2220. http://doi.org/10.1016/j.jasms.2009.08.009.

Kilpatrick, E. L., Liao, W.-L., Camara, J. E., Turko, I. V., & Bunk, D. M. (2012). Expression and characterization of ^{15}N-labeled human C-reactive protein in *Escherichia coli* and *Pichia pastoris* for use in isotope-dilution mass spectrometry. *Protein Expression and Purification*, *85*(1), 94–99. http://doi.org/10.1016/j.pep.2012.06.019.

Mirzaei, H., McBee, J. K., Watts, J., & Aebersold, R. (2008). Comparative evaluation of current peptide production platforms used in absolute quantification in proteomics. *Molecular & Cellular Proteomics*, 7(4), 813–823. http://doi.org/10.1074/mcp.M700495-MCP200.

Nanavati, D., Gucek, M., Milne, J. L. S., Subramaniam, S., & Markey, S. P. (2008). Stoichiometry and absolute quantification of proteins with mass spectrometry using fluorescent and isotope-labeled concatenated peptide standards. *Molecular & Cellular Proteomics*, 7(2), 442–447. http://doi.org/10.1074/mcp.M700345-MCP200.

Pratt, J. M., Simpson, D. M., Doherty, M. K., Rivers, J., Gaskell, S. J., & Beynon, R. J. (2006). Multiplexed absolute quantification for proteomics using concatenated signature peptides encoded by QconCAT genes. *Nature Protocols*, *1*(2), 1029–1043. http://doi.org/10.1038/nprot.2006.129.

Rutherfurd, S. M., & Gilani, G. S. (2009). Amino acid analysis. In J. E. Coligan, et al. (Eds.), *Current protocols in protein science*. Chapter 11, Unit 11.9, http://doi.org/10.1002/0471140864.ps1109s58.

Scott, K. B., Turko, I. V., & Phinney, K. W. (2015). Quantitative performance of internal standard platforms for absolute protein quantification using multiple reaction monitoring-mass spectrometry. *Analytical Chemistry*, *87*(8), 4429–4435. http://doi.org/10.1021/acs.analchem.5b00331.

Simpson, D. M., & Beynon, R. J. (2012). QconCATs: Design and expression of concatenated protein standards for multiplexed protein quantification. *Analytical and Bioanalytical Chemistry*, *404*(4), 977–989. http://doi.org/10.1007/s00216-012-6230-1.

Vandermarliere, E., Mueller, M., & Martens, L. (2013). Getting intimate with trypsin, the leading protease in proteomics. *Mass Spectrometry Reviews*, *32*(6), 453–465. http://doi.org/10.1002/mas.21376.

Wilkins, M. R., Gasteiger, E., Bairoch, A., Sanchez, J. C., Williams, K. L., Appel, R. D., et al. (1999). Protein identification and analysis tools in the ExPASy server. *Methods in Molecular Biology (Clifton, N.J.)*, *112*, 531–552.

CHAPTER TWELVE

Production, Purification, and Characterization of ^{15}N-Labeled DNA Repair Proteins as Internal Standards for Mass Spectrometric Measurements

Prasad T. Reddy*[,1], Pawel Jaruga[†], Bryant C. Nelson[‡], Mark S. Lowenthal[†], Ann-Sofie Jemth[§], Olga Loseva[§], Erdem Coskun[†], Thomas Helleday[§], Miral Dizdaroglu[†,1]

*Institute for Bioscience and Biotechnology Research, National Institute of Standards and Technology and the University of Maryland, Rockville, Maryland, USA

[†]Biochemical Measurement Division, National Institute of Standards and Technology, Gaithersburg, Maryland, USA

[‡]Biosystems and Biomaterials Division, National Institute of Standards and Technology, Gaithersburg, Maryland, USA

[§]Science for Life Laboratory, Division of Translational Medicine and Chemical Biology, Department of Medical Biochemistry and Biophysics, Karolinska Institutet, Stockholm, Sweden

[1]Corresponding authors. e-mail address: prasad.reddy@nist.gov; miral@nist.gov

Contents

Methods in Enzymology, Volume 566
ISSN 0076-6879
http://dx.doi.org/10.1016/bs.mie.2015.06.044

Abstract

Oxidatively induced DNA damage is caused in living organisms by a variety of damaging agents, resulting in the formation of a multiplicity of lesions, which are mutagenic and cytotoxic. Unless repaired by DNA repair mechanisms before DNA replication, DNA lesions can lead to genomic instability, which is one of the hallmarks of cancer. Oxidatively induced DNA damage is mainly repaired by base excision repair pathway with the involvement of a plethora of proteins. Cancer tissues develop greater DNA repair capacity than normal tissues by overexpressing DNA repair proteins. Increased DNA repair in tumors that removes DNA lesions generated by therapeutic agents before they became toxic is a major mechanism in the development of therapy resistance. Evidence suggests that DNA repair capacity may be a predictive biomarker of patient response. Thus, knowledge of DNA–protein expressions in disease-free and cancerous tissues may help predict and guide development of treatments and yield the best therapeutic response. Our laboratory has developed methodologies that use mass spectrometry with isotope dilution for the measurement of expression of DNA repair proteins in human tissues and cultured cells. For this purpose, full-length ^{15}N-labeled analogs of a number of human DNA repair proteins have been produced and purified to be used as internal standards for positive identification and accurate quantification. This chapter describes in detail the protocols of this work. The use of ^{15}N-labeled proteins as internal standards for the measurement of several DNA repair proteins *in vivo* is also presented.

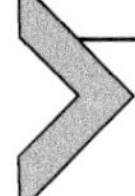

1. INTRODUCTION

Exogenous and endogenous sources such as free radicals and ionizing radiation generate oxidatively induced DNA damage by a variety of mechanisms, resulting in the formation of modified bases and sugars, DNA–protein cross-links, strand breaks, base-free sites, and tandem lesions such as 8,5′-cyclopurine-2′-deoxyribonucleosides and clustered damaged sites (reviewed in Dizdaroglu & Jaruga, 2012; Evans, Dizdaroglu, & Cooke, 2004). This type of DNA damage is thought to play an important role in disease processes such as carcinogenesis and aging (reviewed in Dizdaroglu, 2015; Friedberg et al., 2006). Elaborate DNA repair pathways exist in mammalian cells with approximately 150 different proteins involved (reviewed in Friedberg et al., 2006; Wood, Mitchell, & Lindahl, 2005). DNA repair deficiencies cause accumulation of DNA damage and mutations, leading to genomic instability, which is a major factor in carcinogenesis (Beckman & Loeb, 2005; Helleday, Petermann, Lundin, Hodgson, & Sharma, 2008; Hoeijmakers, 2001; Kelley, 2012; Liu, Yin, & Pu, 2007; Loeb, 2011; Madhusudan & Middleton, 2005). Oxidatively induced DNA damage is mainly repaired by base excision repair (BER) and also by nucleotide excision repair (NER), albeit to a lesser extent (Friedberg et al., 2006). In the first step of BER, DNA glycosylases hydrolyze the *N*-glycosidic bond releasing the damaged base and generating an abasic site, followed by the action of a series of other BER enzymes (Gros, Saparbaev, & Laval, 2002; Hegde, Hazra, & Mitra, 2008). Formamidopyrimidine DNA glycosylase (Fpg, also called MutM) is one of the main DNA glycosylases in *Escherichia coli*, which specifically excises 4,6-diamino-5-formamidopyrimidine (FapyAde), 2,6-diamino-4-hydroxy-5-formamidopyrimidine (FapyGua), and 8-hydroxyguanine (8-OH-Gua) from DNA containing multiple lesions (Boiteux, Gajewski, Laval, & Dizdaroglu, 1992; Dizdaroglu, 2005). In eukaryotes, 8-oxoguanine-DNA glycosylase (OGG1) including human OGG1 (hOGG1), which is a functional homolog of Fpg, exhibits a strong specificity for FapyGua and 8-OH-Gua, but does not act on FapyAde (Audebert, Radicella, & Dizdaroglu, 2000; Dherin, Radicella, Dizdaroglu, & Boiteux, 1999; Hu et al., 2005; Sidorenko, Grollman, Jaruga, Dizdaroglu, & Zharkov, 2009).

Mammalian NEIL1 protein, which is a *E. coli* Nei-like DNA glycosylase with an additional β,δ-elimination activity, has been discovered in eukaryotes (Bandaru, Sunkara, Wallace, & Bond, 2002; Hazra et al., 2002). This enzyme is unique in that it specifically removes FapyAde and FapyGua from DNA with multiple lesions, without exhibiting any significant activity for 8-OH-Gua (Bandaru et al., 2002; Chan et al., 2009; Doublie, Bandaru, Bond, & Wallace, 2004; Hazra et al., 2002; Hu et al., 2005; Jaruga, Birincioglu, Rosenquist, & Dizdaroglu, 2004; Liu et al., 2010; Muftuoglu et al., 2009; Rosenquist et al., 2003; Roy et al., 2007). The involvement of NEIL1 in NER has also been suggested on the basis of accumulation of 8,5′-cyclopurine-2′-deoxyribonucleosides in *neil1*$^{-/-}$ mice (Jaruga, Xiao, Vartanian, Lloyd, & Dizdaroglu, 2010). Homologs of *E. coli* Nth have also been found in yeast and mammals (Aspinwall et al., 1997; Roldán-Arjona, Anselmino, & Lindahl, 1996). Mammalian NTH1 mainly acted on pyrimidine-derived lesions, exhibiting a narrower substrate specificity than *E. coli* Nth (Karahalil, Roldan-Arjona, & Dizdaroglu, 1998; Roldán-Arjona et al., 1996); however, evidence has also been provided for purine-derived FapyAde to be the physiological substrate of NTH1 (Chan et al., 2009; Hu et al., 2005).

Other enzymes such as apurinic/apyrimidinic endonuclease 1 (APE1) and DNA polymerase β (Pol β) also play vital roles in BER. Abasic sites, which are left behind following removal of modified DNA bases by DNA glycosylases, are processed by APE1 (Demple, Herman, & Chen, 1991). APE1 provides over 95% of the total endonuclease function with some additional critical functions (Abbotts & Madhusudan, 2010; Barnes et al., 2009; Demple & Harrison, 1994; Demple et al., 1991; Fishel, Vascotto, & Kelley, 2013; Gros, Ishchenko, Ide, Elder, & Saparbaev, 2004; Tell, Quadrifoglio, Tiribelli, & Kelley, 2009; Wilson & Barsky, 2001; Xanthoudakis & Curran, 1992). The action of APE1 generates a one-nucleotide gap with a 3′-OH and a 5′-terminal 2′-deoxyribose phosphate (dRP) moiety. Pol β, which is found in all vertebrate species (Beard & Wilson, 2006, 2014; Braithwaite & Ito, 1993), binds to this gap and performs DNA synthesis with its DNA polymerase activity, filling in the gap. It then removes the blocking dRP moiety, paving the way for other enzymes such as DNA ligases to close the resulting gap to complete the repair of DNA (Beard & Wilson, 2000, 2006; Matsumoto & Kim, 1995; Srivastava et al., 1998).

Defects in BER enzymes such as those discussed above are associated with neurological disorders and cancer (Sampath, McCullough, & Lloyd, 2012; Wallace, Murphy, & Sweasy, 2012; Wilson, Kim, Berquist, & Sigurdson, 2011). Polymorphic variants of these enzymes have been found in human populations in connection to various cancer incidences (reviewed in Dizdaroglu, 2015). For example, *neil1*$^{-/-}$ mice developed obesity and symptoms of metabolic syndrome (Vartanian et al., 2006). Male and female *neil1*$^{-/-}$ and *nth1*$^{-/-}$ mice developed pulmonary and hepatocellular tumors to an extent up to ≈15%. A dramatic increase up to 75% in cancer incidence has been observed in *neil1*$^{-/-}$/*nth1*$^{-/-}$ male and female mice (Chan et al., 2009). In mice with both deleted alleles of *ape1*, early embryonic lethality has been observed; similarly, APE1 heterozygous mice exhibited increased oxidative stress, spontaneous mutagenesis and cancer incidences, and reduced survival of pups and embryos (Cabelof, 2012; Hakem, 2008; Huamani et al., 2004; Meira et al., 2001; Xanthoudakis, Smeyne, Wallace, & Curran, 1996). Thus, accumulated evidence unequivocally points to the crucial role of BER enzymes in the maintenance of genetic stability and disease prevention.

It is now well known that overexpression of DNA repair proteins that may increase the DNA repair capacity is common in cancer (Helleday et al., 2008; Kelley, 2012; Madhusudan & Middleton, 2005). Increased and effective DNA repair capacity in tumors that removes DNA lesions generated by ionizing radiation and/or chemotherapy before they became toxic is a major mechanism to develop therapy resistance. Evidence suggests that DNA repair capacity might be a predictive biomarker of patient response (Kelley, 2012; Perry, Sultana, & Madhusudan, 2012). In this respect, BER proteins have emerged as biomarkers for assessment of the effectiveness and outcome of therapy (Chan & Bristow, 2010; Curtin, 2013; Fishel et al., 2013; Helleday, 2008; Helleday et al., 2008; Illuzzi & Wilson, 2012; Kelley, 2012; Madhusudan & Middleton, 2005; Moeller, Arap, & Pasqualini, 2010; Perry et al., 2012). Targeting MTH1 protein, which is not a BER protein, has been also suggested to be beneficial in cancer therapy (Gad et al., 2014; Huber et al., 2014). This enzyme dephosphorylates modified 2′-deoxynucleoside triphosphates in the nucleotide pool so that they cannot be incorporated into DNA during DNA replication, preventing cell death and mutations (Maki & Sekiguchi, 1992; Mo, Maki, & Sekiguchi, 1992; Sakai et al., 2002; Sakumi et al., 1993; Sekiguchi & Tsuzuki, 2002; Tsuzuki et al., 2001). Its overexpression has been observed in many cancers (Gad et al., 2014; Huber et al., 2014; Kennedy, Cueto, Belinsky, Lechner, & Pryor, 1998; Okamoto et al., 1996; Speina et al., 2005).

The determination of the overexpression or underexpression of DNA repair proteins in normal and cancer tissues may help predict and guide development of treatments, potentially yielding the greatest therapeutic response (Kelley, 2012). In general, expression levels of DNA repair proteins have been estimated by Western blot methods. With these methods, proteins are separated by sodium dodecyl sulfate–polyacrylamide gel electrophoresis (SDS–PAGE) and identified by the use of an antibody specific to a target protein, meaning that they completely depend on a reliable and specific antibody resource. Antibodies may potentially exhibit some off-target binding, leading to false results. No positive identification and no absolute quantification are provided because of the lack of mass spectrometric evidence and suitable internal standards. Quantification is based on the estimation of stained area only. The molecular mass of a target protein is only estimated by the use of a molecular mass ladder of proteins. If DNA repair proteins are to be used as reliable biomarkers in cancer, their expression levels must be accurately measured in tissues, providing positive identification of a target protein and its absolute quantification. Mass spectrometry is the most suitable technique of choice for the measurement of proteins. The application of this technique would be essential for positive identification and accurate quantification of DNA repair proteins in human tissues using proper internal standards. Our laboratory has recently developed methodologies and stable isotope-labeled standards for the measurements of DNA repair proteins using liquid chromatography–tandem mass spectrometry (LC-MS/MS) with isotope dilution (Coskun et al., 2015; Dizdaroglu, Reddy, & Jaruga, 2011; Kirkali et al., 2013; Reddy, Jaruga, Nelson, Lowenthal, & Dizdaroglu, 2011; Reddy et al., 2013). Stable isotope-labeled analogs of DNA repair proteins, which are to be used as internal standards, were produced, purified, and characterized. This chapter describes the procedures used for this purpose and the application of the developed internal standards for the measurement of DNA repair proteins.

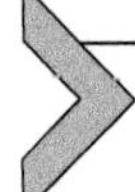

2. MATERIALS AND METHODS

2.1 Materials

Restriction endonucleases, DNA ligase, and calf intestinal alkaline phosphatase were obtained from New England Biolabs (MA). Isopropyl β-D-thiogalactoside (IPTG), acrylamide, bisacrylamide, and Tris(2-carboxyethyl)phosphine hydrochloride were purchased from Sigma Chemical Company (St. Louis, MO). Diethylaminoethyl (DEAE) cellulose (DE52) was from Whatman Laboratories. Shodex carboxymethyl cellulose HPLC preparative column (CM 2025) was from Phenomenex. YM10 membrane filters were from EMD Millipore Corporation (Bedford, MA). Ni-Sepharose column HisTrap HP and MonoQ anion-exchange column were obtained from GE Healthcare, Bio-Science (Uppsala, Sweden). BugBuster protein extraction reagent was from Millipore (Billerica, MA). Protease inhibitor cocktail was purchased from Roche (Indianapolis, IN). Benzonase nuclease was from Novagen (Denmark). $^{15}N\text{-}NH_4Cl$ was obtained from Sigma Chemical Company (St. Louis, MO).

2.2 *E. coli* Strains and Plasmids

The pET11a expression vector, for native protein production of *Ec*Fpg based on the IPTG induction, was obtained from Novagen, Madison, WI. The relevant strains for cloning and expression are *E. coli* Novablue (K12) and BL21(DE3), respectively. The human *ogg1* gene in the expression vector pET15b for His-tagged production of the protein was kindly provided by Dr. Dmitry Zharkov through Prof. Arthur Grollman's laboratory. The human *neil1* gene in the expression vector pET28 for His-tagged production of the protein was kindly provided by Dr. Stephen Lloyd (Oregon Health and Science University, Portland, Oregon). The human *ape1* gene was provided by Dr. David M. Wilson III of the National Institute on Aging, National Institutes of Health, Baltimore, Maryland, USA.

2.3 Preparation of Minimal Medium

The composition of the medium was: 6 g NaH_2PO_4, 3 g K_2HPO_4, 0.5 g NaCl, and 1 g ^{15}N-NH_4Cl, 5 g glucose, 246 mg $MgSO_4 \cdot 7H_2O$ per liter. Ampicillin or kanamycin was added to a final concentration of 50 μg/mL or 25 μg/mL, respectively.

2.4 DNA Procedures

E. coli Novablue (K12) harboring a recombinant plasmid was grown at 37 °C overnight in 10 mL Luria-Bertani medium (Miller, 1972), containing the appropriate antibiotic. Minipreparations of plasmid DNA were purified using a Qiagen Kit. Digestion of DNA with restriction enzymes was performed according to the manufacturer's recommendation. DNA fragments were separated by electrophoresis on SeaKem GTG agarose or NuSieve GTG agarose. DNA was purified using a Qiagen Kit. Ligation of DNA fragments was performed as described (Sambrook, Fritsch, & Maniatis, 1989). Competent cells of the *E. coli* strains used here were prepared by the Hanahan method (Hanahan, 1983; Sambrook et al., 1989).

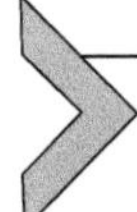

3. PRODUCTION AND PURIFICATION OF *E. COLI* ^{15}N-Fpg PROTEIN

3.1 Cloning of *fpg* Into pET11a Vector

The coding sequence for *E. coli fpg* gene from the plasmid pFpg was provided by Dr. Timothy R. O'Connor of the Beckman Research Institute of the City of Hope, Duarte, CA (O'Connor, Boiteux, & Laval, 1989). It was amplified by PCR with Pfu polymerase using a 5′-end primer (5′-GGAATTC CAT ATG CCT GAA TTA CCC GAA G-3′) and a 3′-end primer (5′-CCG CTC GAG TTA CTT CTG GCA CTG CCG AC-3′). The 5′-end primer sequence contained the *Nde*I restriction recognition sequence CATATG wherein ATG served as the initiation codon for protein expression. The 3′-end primer contained the TAA translation stop codon.

The amplified product was digested with *Nde*I to produce the gene with 5′-*Nde*I protruding end and 3′-blunt end. The DNA fragment was purified from a 1% agarose gel, and cloned into the *Nde*I and *Bam*HI (first digested with *Bam*HI, filled in with DNA polymerase to make this blunt end, then digested with *Nde*I) sites of the pET11a expression vector. A recombinant plasmid pET11a/Fpg was isolated from *E. coli* Nova Blue cells. *E. coli* BL21 (DE3) was transformed with the recombinant plasmid to induce protein expression with IPTG.

3.2 Production of ^{15}N-Fpg

E. coli BL21(DE3) harboring pET11a/Fpg recombinant plasmid was grown at 37 °C for 20 h on LB agar plate containing 100 μg/mL ampicillin. A colony was carefully (without touching into the LB medium) transferred to 10 mL minimal medium containing 1 mg ^{15}N-NH_4Cl/mL and 50 μg ampicillin/mL. Cells were grown for 1 h at 37 °C at 250 rpm in a 50 mL tube. This inoculum was transferred to 140 mL minimal medium, in a 500 mL baffled flask, containing ^{15}N-NH_4Cl and ampicillin as above. This culture was grown at 37 °C for 16 h. Next, 3×50 mL of this seed culture was transferred to 3 × 450 mL of minimal medium containing ^{15}N-NH_4Cl and ampicillin in 3 × 2 L baffled flasks. IPTG was added to a final concentration of 30 μmol/L to induce ^{15}N-Fpg production at 37 °C for 1 h. Next, the culture was shifted to 42 °C for 5 h to continue ^{15}N-Fpg production. Cells were harvested at 6000 × *g* for 20 min and washed with 25 mmol/L Tris–HCl buffer, pH 7.5. The wet weight of cells obtained in this procedure was ≈3 g.

3.3 Purification of ^{15}N-Fpg

Cell pellet was suspended in 40 ml of 50 mmol/L Tris–HCl, pH 8.0, 10 mmol/L β-mercaptoethanol, 1 mmol/L ethylenediaminetetraacetic acid (EDTA), and 50 mmol/L NaCl (lysis buffer 1), containing one tablet of protease inhibitors. Cell suspension was passed through a French press at 7.1×10^4 kPa twice. The cell-free extract was centrifuged at 10,000 × *g* for 20 min. The supernatant was mixed with 5 g of DE52 anion-exchange resin equilibrated with the lysis buffer 1 in a 500-mL bottle at 80 rpm for 1 h at 4 °C and then poured into a column. The flow through containing nearly all the ^{15}N-Fpg and fewer cellular proteins was collected. The resin was washed with 10 mL of the lysis buffer 1 and added to the flow through. The ^{15}N-Fpg enriched pool (50 mL) was dialyzed overnight against 1 L of 20 mmol/L potassium phosphate buffer, pH 7.4. The dialyzed pool was centrifuged at 100000 × *g* for 1 h to remove any particulate material.

The supernatant was chromatographed on a HPLC-Shodex carboxymethyl cellulose column (2 cm × 20 cm) equilibrated with 20 mmol/L potassium phosphate buffer, pH 7.4. The column was washed with 100 mL of 20 mmol/L potassium phosphate buffer, pH 7.4 until the A_{280} stabilized. Then ^{15}N-Fpg was eluted with a 0–0.5 mol/L potassium chloride gradient (250 mL each) in 20 mmol/L potassium phosphate buffer, pH 7.4. Pure ^{15}N-Fpg was eluted as a sharp peak at ≈0.3 mol/L KCl. Fractions containing ^{15}N-Fpg were pooled, concentrated on YM3 membrane filter, and dialyzed overnight against 500 mL of 50% glycerol, 20 mmol/L Tris–HCl, 1 mmol/L EDTA, and 10 mmol/L β-mercaptoethanol, pH 7.5. Protein concentration was determined by the Lowry method using bovine serum albumin (BSA) as standard (Lowry, Rosebrough, Farr, & Randall, 1951). Total yield of the protein was 1 mg. The SDS–PAGE and Coomassie staining of ^{15}N-Fpg (lane 5) and protein molecular mass standards (lane 6) are shown in Fig. 1, indicating high purity of the labeled protein (Reddy et al., 2011).

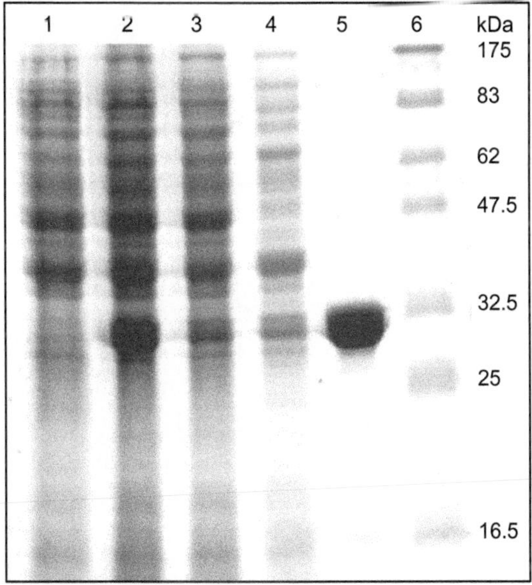

Figure 1 SDS (13%)–PAGE analysis of the purification steps of *E. coli* ^{15}N-Fpg: lane 1, 21 μg protein of the extract of BL21(DE3) cells harboring pET11a control plasmid; lane 2, 31 μg protein of the extract of BL21(DE3) cells harboring Fpg/pET11a plasmid induced with IPTG; lane 3, 26 μg of 10000 × *g* supernatant; lane 4, 8.6 μg flow through from DEAE cellulose column, 29 μg; lane 5, 8.5 μg ^{15}N-Fpg eluted from CM cellulose column, 26 μg; and lane 6, molecular mass markers. *Data from Reddy et al. (2011).*

4. PRODUCTION AND PURIFICATION OF ^{15}N-hOGG1

4.1 Production of ^{15}N-hOGG1

E. coli BL21(DE3) harboring pET15b/hOGG1 recombinant plasmid was grown at 37 °C for 20 h on LB agar plate containing 100 μg/mL ampicillin. A colony was carefully (without touching into the LB medium) transferred to 100 mL minimal medium containing 1 mg ^{15}N-NH_4Cl/mL and 50 μg ampicillin/mL. Cells were grown overnight (16 h) at 37 °C at 250 rpm in a 250-mL baffled flask. This inoculum was transferred to 1000 mL minimal medium, in a 2000-mL baffled flask, containing ^{15}N-NH_4Cl and ampicillin as above. This culture was grown at 37 °C for 1 h and briefly cooled in ice water to room temperature. Next, IPTG was added to a final concentration of 100 μmol/L to induce ^{15}N-hOGG1 production at 24 °C overnight. Cells were harvested at 6000 × *g* for 20 min and washed with 25 mmol/L Tris–HCl buffer, pH 7.5. The wet weight of cells obtained in this procedure was ≈2 g.

4.2 Purification of ^{15}N-hOGG1

The cell pellet was suspended in 20 mL of 50 mmol/L Tris–HCl, pH 8.0, 10 mmol/L β-mercaptoethanol, 300 mmol/L NaCl, and 10 mmol/L imidazole (lysis buffer 2) containing one tablet of protease inhibitors. Cell suspension was passed through a French press at 7.1 × 10^4 kPa. The cell-free extract was centrifuged at 40000 × *g* for 1 h. Meanwhile, 2 mL of nickel-agarose slurry (1 mL resin) was washed with the lysis buffer 2 in a 30 mL Bio-Rad polypropylene column. The supernatant was added to the resin and mixed on a rocker for 2 h at 4 °C. The flow through was collected, and the column was washed successively with three 10-mL aliquots of the lysis buffer 2. Next, the resin was washed three times with 3-mL aliquots of the lysis buffer 2 containing an additional 20 mmol/L imidazole (total 30 mmol/L). Next, the ^{15}N-hOGG1 was eluted with five 5-mL aliquots of the lysis buffer containing 100 mmol/L imidazole. The first 5 mL contained about 50% of ^{15}N-hOGG1 (≈95% pure) with minor contaminants as judged by SDS–PAGE. The subsequent four elutions contained the remaining ^{15}N-hOGG1 with higher purity (≈98%). All the five elutions containing ^{15}N-hOGG1 were pooled, concentrated on YM3 membrane filter, and dialyzed overnight against 500 mL of 50% glycerol, 20 mmol/L Tris–HCl, 1 mmol/L EDTA, and 10 mmol/L β-mercaptoethanol, pH 7.5. Mass measurements

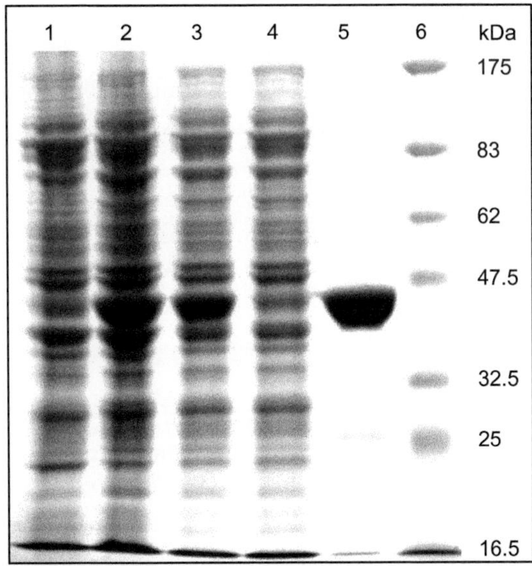

Figure 2 SDS (10%)–PAGE analysis of the purification steps of ^{15}N-hOGG1: lane 1, 20 μg protein of the extract of BL21(DE3) cells harboring pET11a control plasmid; lane 2, 21 μg protein of the extract of BL21(DE3) cells harboring hOGG1/pET15b plasmid induced with IPTG; lane 3, 17 μg of 48000 × *g* supernatant; lane 4, 15.5 μg from Ni-agarose flow through; lane 5, 8.3 μg ^{15}N-hOGG1 eluted from Ni-agarose column; and lane 6, molecular mass markers. *Data from Reddy et al. (2011).*

and activity analysis of ^{15}N-hOGG1 were carried out. Protein concentration was determined by the Lowry method using BSA as standard (Lowry et al., 1951). Total yield of the protein was 5.3 mg. Figure 2 shows the SDS–PAGE and Coomassie staining of ^{15}N-hOGG1 (lane 5) and protein molecular mass standards (lane 6), indicating high purity of the labeled protein (Reddy et al., 2011).

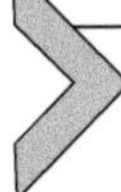

5. PRODUCTION AND PURIFICATION OF ^{15}N-hNEIL1

5.1 Production of ^{15}N-hNEIL1

E. coli BL21 (DE3) harboring pET28/human *neil1* recombinant plasmid was grown at 37 °C for 20 h on LB agar plate containing 100 μg ampicillin and 40 μg chloramphenicol/mL. A colony was carefully (without touching into the LB medium) transferred to 10 mL minimal medium containing 1 mg ^{15}N-NH_4Cl/mL, 50 μg ampicillin, and 20 μg chloramphenicol/mL. Cells were grown for 16 h at 37 °C at 250 rpm in a 50 mL tube. This inoculum was transferred to 90 mL minimal medium, in a 250 mL flask, containing

^{15}N-NH_4Cl, ampicillin, and chloramphenicol as above. This culture was grown at 37 °C for 9 h. Next, this seed culture was transferred to 900 mL of minimal medium containing ^{15}N-NH_4Cl, ampicillin, and chloramphenicol in a 2 L flask. This culture was grown at 37 °C for 4 h. Cells (A_{600} = 0.45) were briefly cooled in ice and ^{15}N-hNEIL1 production was induced with 100 μmol IPTG/mL at 25 °C for 16 h. Cells were harvested at 6000 × *g* for 20 min and washed with 25 mmol/L Tris–HCl buffer (pH 7.5). The wet weight of cells obtained in this procedure was 2.8 g/L culture.

5.2 Purification of ^{15}N-hNEIL1

Cell pellet was suspended at a rate of 1 g/10 mL of 50 mmol/L Na_2HPO_4–NaH_2PO_4 buffer (pH 8.0), 10 mmol/L β-mercaptoethanol, 10 mmol/L imidazole, and 300 mmol/L NaCl (lysis buffer). One tablet of protease inhibitors was added to the cell suspension. Cell suspension was passed through a French press at 7.1×10^4 kPa twice. The cell-free extract was centrifuged at 36000 × *g* for 30 min. Meanwhile, 1 mL of nickel-agarose slurry (0.5 mL resin) was washed with the lysis buffer in a 30-mL Bio-Rad polypropylene column. The supernatant was added to the resin and mixed on a rocker for 2 h at 4 °C. The flow through was collected, and the column was washed successively with three 10-mL aliquots of the lysis buffer. Next, the resin was washed successively two times with 10 mL aliquots of the lysis buffer containing an additional 30 mmol/L imidazole (total 40 mmol/L). Next, the resin was washed successively three times with 5-mL aliquots of the lysis buffer containing an additional 70 mmol/L imidazole (total 80 mmol/L). The first imidazole$_{80}$ wash did elute about 30% of ^{15}N-hNEIL1 with a number of contaminating proteins as judged by SDS–PAGE. The second and third imidazole$_{80}$ washes eluted about 30% of nearly homogeneous ^{15}N-hNEIL1. Next, ^{15}N-hNEIL1 was eluted successively with three 5 mL aliquots of the lysis buffer containing 150 mmol/L imidazole. The first 5 mL contained about 30% of ^{15}N-hNEIL1. The subsequent two elutions contained the remaining ^{15}N-hNEIL1. The second and third imidazole$_{80}$ elutions and all the imidazole$_{150}$ elutions were pooled and dialyzed overnight against 1 L of 50 mmol/L Na_2HPO_4–NaH_2PO_4 buffer (pH 8.0) and 300 mmol/L NaCl. Slight precipitate of ^{15}N-hNEIL1 appeared after dialysis even in the presence of 300 mmol/L NaCl. The precipitate was centrifuged off from the dialyzed pool, and the clear supernatant was concentrated on NMLW5 membrane in an Amicon cell. The above purification procedure developed for cells from 1 L culture was proportionately

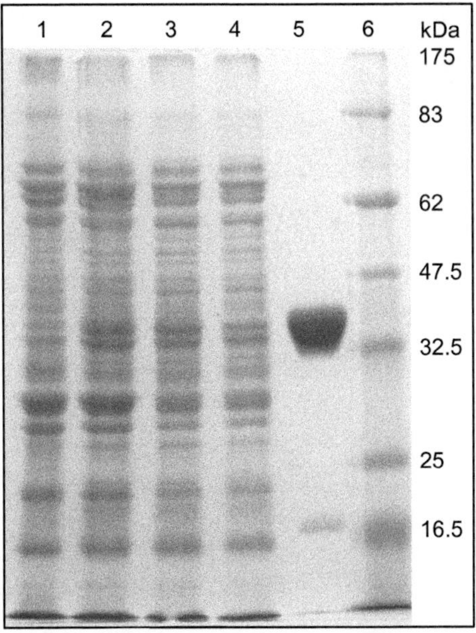

Figure 3 SDS (10%)–PAGE analysis of the purification steps of ^{15}N-hNEIL1: lane 1, uninduced cell extract, 24 μg; lane 2, induced cell extract, 28 μg; lane 3, 20000 × *g* supernatant, 22 μg; lane 4, flow through from nickel resin, 17 μg; lane 5, pure ^{15}N-NEIL1 eluted from nickel resin, 15 μg; and lane 6, molecular mass markers. *Data from Reddy et al. (2013).*

scaled up to 5 L as and when needed. Protein concentration was determined by the Lowry method using BSA as the standard (Lowry et al., 1951). The final yield of ^{15}N-hNEIL1 was ≈2 mg/L culture. The SDS–PAGE and Coomassie staining of ^{15}N-hNEIL1 (lane 5) and protein molecular mass standards (lane 6) are presented in Fig. 3. A high purity of the labeled protein can be seen with a minor contaminant at ≈17 kDa (Reddy et al., 2013).

6. PRODUCTION AND PURIFICATION OF ^{15}N-hAPE1

6.1 Production of ^{15}N-hAPE1

E. coli BL21 (DE3) harboring pET11/h*APE1* recombinant plasmid was grown at 37 °C for 20 h on LB agar plate containing 100 μg ampicillin/mL. A colony was carefully (without touching into the LB medium) transferred to 10 mL minimal medium containing 50 μg ampicillin/mL. Cells were grown for 3 h at 37 °C at 250 rpm in a 50 mL tube. This inoculum was transferred to 200 mL minimal medium with 50 μg ampicillin/mL in a 500 mL flask. This culture was grown at 37 °C for 14 h. Next, each

50 mL of this seed culture was transferred to 4 × 1000 mL of minimal medium containing $^{15}N\text{-}NH_4Cl$ and ampicillin in 4 × 2 L flasks. This culture was grown at 37 °C for 6 h and then at 20 °C for 90 min. Cell density at this stage was $A_{600} = 0.5$. The production of ^{15}N-hAPE1 was induced with 100 μmol IPTG/mL at 20 °C for 15 h. Cells were harvested at 6000 × *g* for 20 min and washed with 25 mmol/L Tris–HCl buffer (pH 7.5). The wet weight of cells obtained in this procedure was 2.25 g/L culture for a total of 9 g wet weight of cells.

6.2 Purification of ^{15}N-hAPE1

Nine grams of cells from 4-L culture were suspended in 90 mL of lysis buffer A: 50 mmol/L HEPES–KOH buffer, pH 7.5, containing 5% glycerol, 50 mmol/L KCl, and 1 mmol/L DTT. Cells were broken by passing through a French press at 7.1×10^4 kPa. The cell-free extract was centrifuged at 70000 × *g* for 1 h. The supernatant was mixed with 5 g of DE52 anion-exchange resin equilibrated with the buffer A. The supernatant/resin slurry in a 250-ml bottle was mixed at 178 rpm for 1 h and then poured into a column. The flow through containing nearly all the ^{15}N-hAPE1 and fewer cellular proteins was collected. The resin was washed with 10 mL of the buffer A and added to the flow through. The ^{15}N-hAPE1 enriched pool (100 mL) was chromatographed on a HPLC-Shodex carboxymethyl cellulose column (2.0 cm × 25 cm) equilibrated with buffer A. The column was washed with 100 mL of buffer A until the A_{280} stabilized. Then ^{15}N-hAPE1 was eluted with a potassium chloride gradient generated from buffer A and buffer A containing 0.65 mol/L KCl (250 mL each). Pure ^{15}N-hAPE1 was eluted as a sharp peak at ≈0.3 mol/L KCl. Fractions containing ^{15}N-hAPE1 were pooled, concentrated to ≈10 mg/mL on a YM10 membrane filter. The yield of ^{15}N-hAPE1 from 4 L culture of minimal medium was 60 mg. Aliquots of the protein was stored at −70 °C. Figure 4, lane 6, shows the SDS–PAGE analysis and Coomassie staining of ^{15}N-hAPE1 and protein molecular mass standards (lane 7), indicating high purity of the labeled protein (Kirkali et al., 2013).

7. PRODUCTION AND PURIFICATION OF ^{15}N-hMTH1

^{15}N-hMTH1 was produced according to the procedures used for the production of hMTH1 (Svensson et al., 2011). ^{15}N-hMTH1 was expressed from pET28a-hMTH1 in *E. coli* BL21(DE3) grown in a minimal medium (6 g Na_2HPO_4, 3 g KH_2PO_4, 0.5 g NaCl, 5 g glucose, 120.4 mg $MgSO_4$

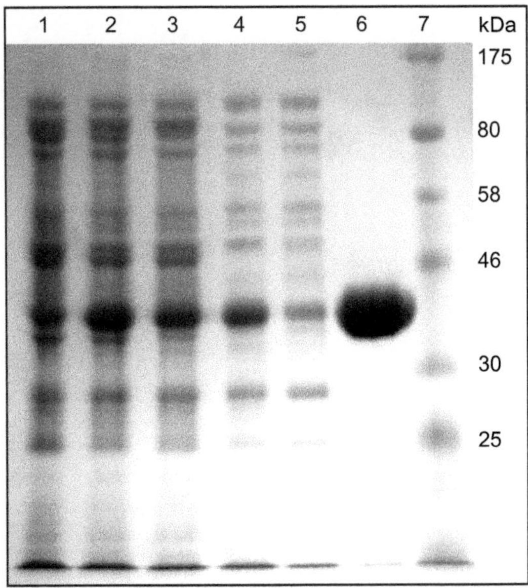

Figure 4 SDS (11%)–PAGE analysis of the purification steps of ^{15}N-hAPE1: lane 1, uninduced cell extract, 24 μg; lane 2, induced cell extract, 28 μg; lane 3, 70000 × *g* supernatant, 24 μg; lane 4, flow through from DEAE cellulose column, 17 μg; lane 5, flow through from CM cellulose column, 16 μg; lane 6, pure ^{15}N-APE1 eluted from CM cellulose column, 21 μg; and lane 7, molecular mass markers. *Data from Kirkali et al. (2013).*

per liter) containing 1 g ^{15}N-NH_4Cl. The production was induced by addition of 200 μmol/L isopropyl β-D-1-thiogalactopyranoside. Cells were harvested by centrifugation after 16 h at 20 °C, and the obtained pellet was dissolved in BugBuster protein extraction reagent supplemented with complete protease inhibitor cocktail and benzonase nuclease. The resulting suspension was centrifuged, and the cleared lysate was subjected to chromatography on Ni-Sepharose column HisTrap HP. Bound proteins were eluted using a linear gradient of imidazole (25–500 mmol/L) and analyzed by SDS–PAGE. Fractions containing ^{15}N-hMTH1 were after dialysis loaded onto a MonoQ anion-exchange column and eluted with a linear gradient of NaCl. Eluted fractions were analyzed using SDS–PAGE with Coomassie staining. Fractions containing ^{15}N-hMTH1 protein were pooled and stored in 20 mmol/L HEPES, 225 mmol/L NaCl, 10% glycerol, and 1 mmol/L Tris (2-carboxyethyl)-phosphine hydrochloride at −80 °C. The purity of the isolated ^{15}N-hMTH1 was checked using SDS–PAGE with Coomassie staining. Figure 5 illustrates the SDS–PAGE analysis of ^{15}N-hMTH1 along with hMTH1 and protein standards, indicating high purity of both proteins (Coskun et al., 2015).

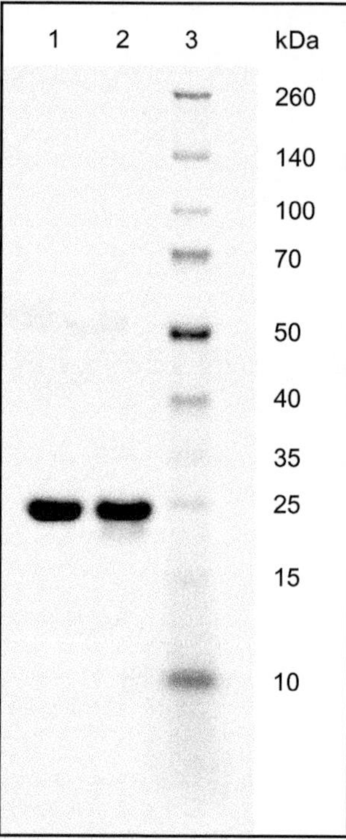

Figure 5 SDS–PAGE analysis of purified hMTH1 and ^{15}N-hMTH1. Lane 1, hMTH1; lane 2, ^{15}N-hMTH1; and lane 3, molecular mass markers. *Data from Coskun et al. (2015).*

8. MASS SPECTROMETRIC ANALYSIS OF ^{15}N-LABELED PROTEINS

Mass spectrometry was used to measure the molecular masses and to check the isotopic purities of ^{15}N-labeled DNA repair proteins that were produced as described above. MALDI-ToF mass spectrometry was applied to measure the molecular masses of *E. coli* ^{15}N-Fpg and ^{15}N-hOGG1 (Reddy et al., 2011). The protonated average molecular mass (MH$^+$) of *E. coli* ^{15}N-Fpg amounts to 30641 Da, as calculated using the *NIST Mass and Fragment Calculator.* The measurement yielded a value of 30442 Da. It is well known that the N-terminal Met is removed from proteins by peptide deformylase and methionineaminopeptidase as an essential process, when the penultimate residue is small and uncharged as Pro next to Met (Giglione, Boularot, & Meinnel, 2004). Without the N-terminal Met, the calculated value of the MH$^+$ of *E. coli* ^{15}N-Fpg amounted to

30509 Da. The measured value (30442 Da) is 99.8% of this calculated value. Using the same approach, the MH^+ of His-tagged ^{15}N-hOGG1 was found to be 41834 Da. The His-tag had a sequence of MGSSHHHHHHSSGLVPRGSHEL. The calculated MH^+ of this protein amounts to 41688 Da, which is 99.7% of the measured value. The measured MH^+ of His (KLAAALEHHHHHH)-tagged ^{15}N-hNEIL1 amounted to 45833 Da (unpublished results). The calculated MH^+ of this protein was 45785 Da, which is 99.9% of the measured value. Considering that the MALDI-ToF mass spectrometry provides a mass accuracy of 0.1%, these values represent more than 99% ^{15}N-labeling of *E. coli* Fpg, hOGG1, and hNEIL1 proteins.

An Orbitrap mass spectrometer coupled to an HPLC system, which has a mass accuracy of 0.001%, was used to measure the molecular mass of ^{15}N-hAPE1 (Kirkali et al., 2013). The calculated MH^+ of ^{15}N-hAPE1 with the loss of Met amounts to 35858 Da. The measured MH^+ was 35854 Da, which differs by 4 Da only from the calculated value and thus represents a 99.99% labeling. Within the mass accuracy of the Orbitrap mass spectrometer, this result showed that a complete ^{15}N-labeling of hAPE1 was achieved.

The molecular mass of ^{15}N-hMTH1 was measured using a high-resolution QToF mass spectrometer coupled to an HPLC system (Coskun et al., 2015). Without the N-terminal Met, the calculated MH^+ of ^{15}N-hMTH1 amounts to 20229.8 Da. The measured MH^+ of ^{15}N-hMTH1 was 20216.2 Da, which is 99.93% of the calculated value. Taken together, the measurement of the average molecular masses of the produced full-length ^{15}N-labeled proteins confirmed their identity and an almost complete ^{15}N-labeling, meaning that these labeled proteins can be used as excellent internal standards for the measurement of DNA repair proteins by LC-MS/MS or any other MS technique.

9. APPLICATIONS TO THE MEASUREMENT OF DNA REPAIR PROTEINS

9.1 Development of Methodologies

Following production of ^{15}N-labeled DNA repair proteins, we developed methodologies using these proteins as internal standards to positively identify and accurately quantify corresponding DNA repair proteins in human cultured cells and tissues with the use of LC-MS/MS. For this purpose, ^{15}N-labeled proteins and their unlabeled analogs were hydrolyzed with trypsin and then analyzed by LC-MS/MS. A number of tryptic peptides of each protein were identified on the basis of their full-scan mass spectra and product ion spectra. Identified tryptic peptides yielded protein scores of more than 100 with the use of the "Mascot" search engine (http://www.matrixscience.com), according to which protein scores greater than 70 are considered significant ($p < 0.05$) for positive identification. Moreover, using to the SwissProt database (http://prospector.ucsf.edu/prospector/cgi-bin/mssearch.cgi) and the taxonomy *Homo sapiens* (20,200 sequences), several combinations of only four out of many identified tryptic peptides provided a 100% match with studied proteins. This means that the simultaneous measurement of just four tryptic peptides would be sufficient to positively identify and quantify a target protein. As an example of full-scan mass spectra, Fig. 6 illustrates the mass spectrum of one of the tryptic peptides of hAPE1 and its ^{15}N-labeled analog (Kirkali et al., 2013). In general, mass spectra of such peptides contain an MH^+ ion of lower abundance (ion at m/z 1131 in Fig. 6) and a doubly protonated (charged) molecular ion [$(M+2H)^{2+}$] as the most abundant ion (m/z 566.1 in Fig. 6) (Kinter & Sherman, 2000). The mass spectrum of the ^{15}N-labeled peptide exhibited a shift of 14 Da in the mass of MH^+ of the unlabeled peptide, consistent with the fourteen ^{15}N atoms in the molecule. The shift in the mass of $(M+2H)^{2+}$ was also in agreement with the mass difference of 14 Da. To obtain product ion spectra, optimum collision energies using $(M+2H)^{2+}$ ions as the precursor ions were measured (Coskun et al., 2015; Reddy et al., 2013). These were in agreement with previously determined collision energies of peptides with similar masses (Kinter & Sherman, 2000). Product ion spectra of identified tryptic peptides were recorded using the determined collision energy for each peptide. As examples, product ion spectra of the peptides in Fig. 6 are illustrated in Fig. 7 (Kirkali et al., 2013). Such spectra are generally dominated by so-called y-series ions with b-series ions being of lesser intensity (Kinter & Sherman, 2000). In Fig. 7A, typical y-ions from the y_1-ion to the y_9-ion, with the most intense ion being the y_8-ion at m/z 904, are seen with only a few b-ions. The ^{15}N-labeled analog gave an essentially identical spectrum with mass shifts according to the ^{15}N-content of the fragments (Fig. 7B). The observed masses of all y-series ions and b-series ions in product ion spectra of analyzed tryptic peptides perfectly matched the corresponding theoretical masses calculated using *NIST Mass and Fragment Calculator* (www.nist.gov/mml/bmd/bioanalytical/massfragcalc.cfm).

Selected reaction monitoring (SRM) was used to analyze trypsin hydrolysates of proteins and their ^{15}N-labeled analogs to establish the sensitive and

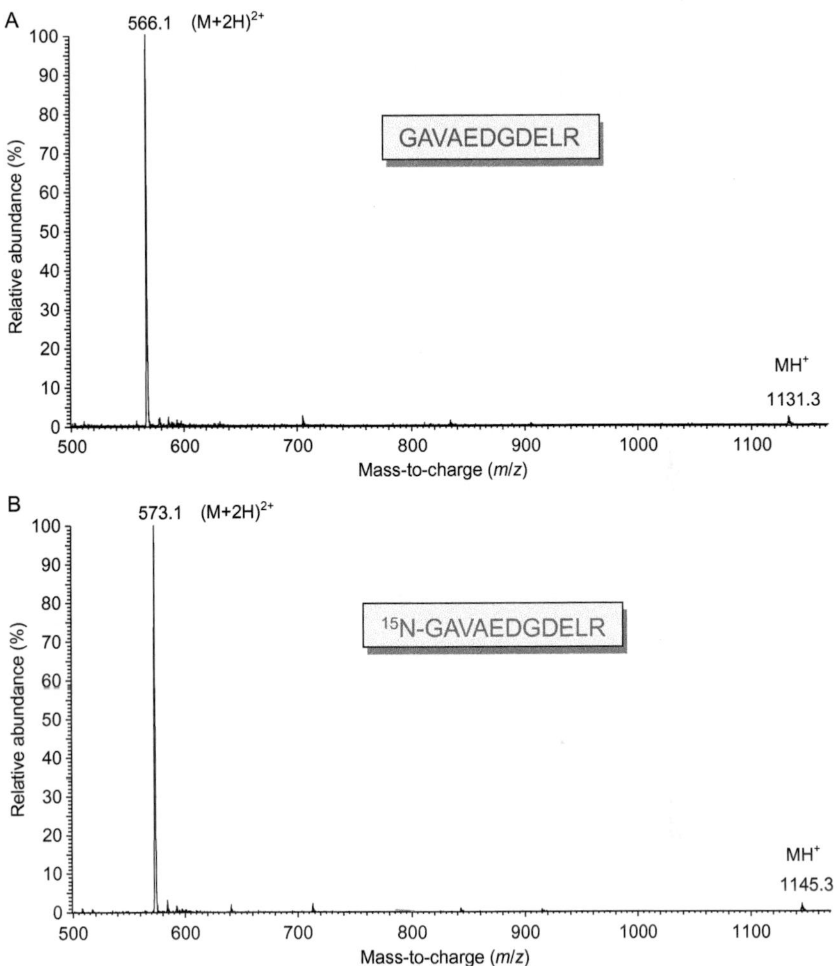

Figure 6 Full-scan mass spectra of the tryptic peptides GAVAEDGDELR (A) and ^{15}N-GAVAEDGDELR (B). *Data from Kirkali et al. (2013).*

accurate measurement of proteins *in vivo*. Mass transitions from $(M+2H)^{2+}$ to the most intense *y*-ion in product ion spectra were used. Ion-current profiles of the mass transitions of tryptic peptides exhibited excellent peak shapes and a baseline separation between peptides. As expected, each tryptic peptide and its respective ^{15}N-labeled analog coeluted at the same retention time (Coskun et al., 2015; Dizdaroglu et al., 2011; Kirkali et al., 2013; Reddy et al., 2013). To further ascertain the identity of a target peptide, multiple typical transitions were simultaneously monitored. All typical transitions of tryptic peptides lined up at their appropriate retention times,

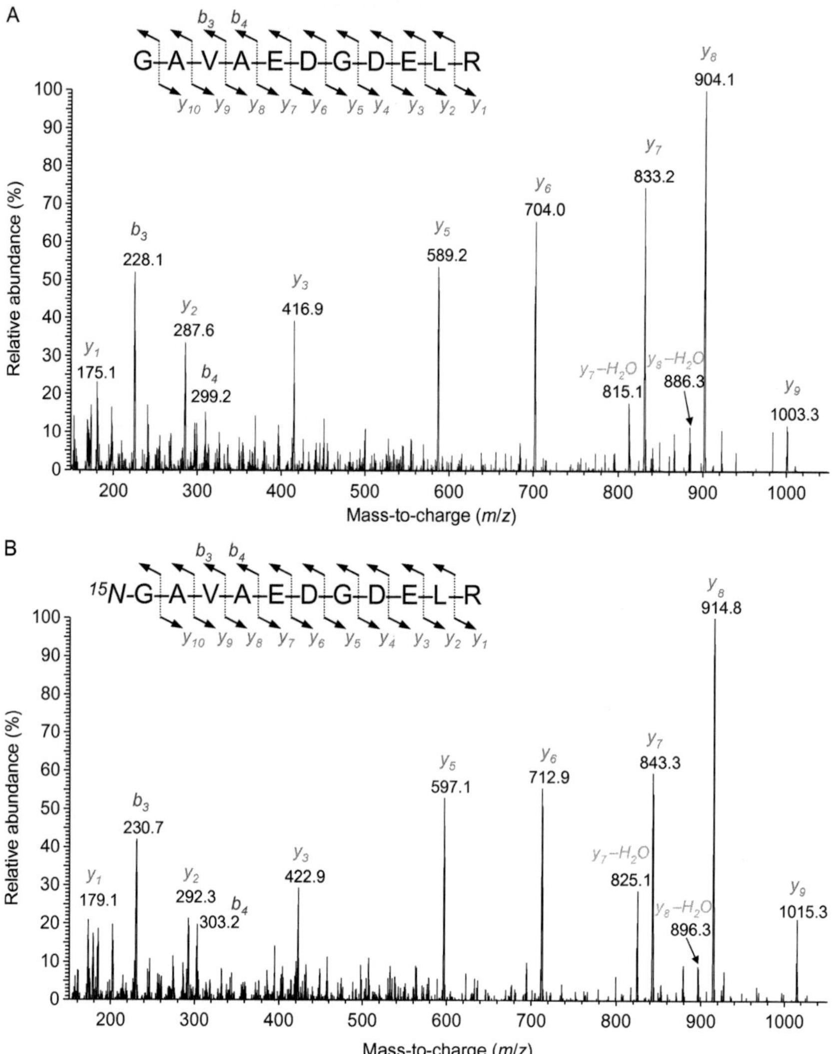

Figure 7 Product ion spectra of GAVAEDGDELR (A) and ^{15}N-GAVAEDGDELR (B). The $(M+2H)^{2+}$ ions *m/z* 566.3 (A) and *m/z* 573.3 (B) were used as the precursor ions. The insert shows the fragmentation pathways leading to the *b*- and *y*-ions. *Data from Kirkali et al. (2013).*

confirming their identity. The analytical sensitivity of the LC-MS/MS instrument was measured. Depending on the peptide, the on-column limit of detection varied from 10 to 20 fmol with a signal-to-noise ratio of at least 3, whereas the limit of quantification was approximately 50–100 fmol of target peptide with a signal-to-noise ratio of 10.

9.2 Measurement of DNA Repair Proteins *In Vivo*

hAPE1 has been positively identified and quantified in nuclear and cytoplasmic extracts of three human cell lines and in mouse liver (Kirkali et al., 2013), and in human cancerous and disease-free breast tissues (unpublished data). Different expression levels of hAPE1 were observed depending of the cell line and tissue types. As an example, Fig. 8 illustrates ion-current profiles of mass transitions of five tryptic peptides of hAPE1 and ^{15}N-hAPE1 obtained using the tryptic digest of a nuclear extract of human mammary gland epithelial adenocarcinoma (MCF-7) cells. The signals of five peptides

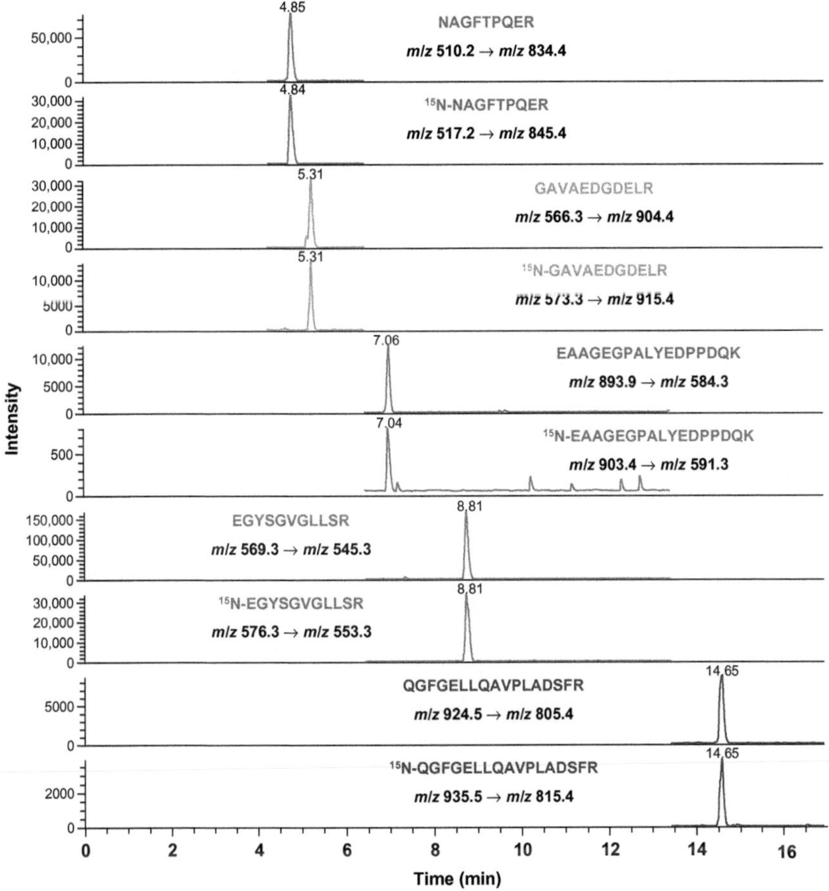

Figure 8 Ion-current profiles of mass transitions of five tryptic peptides of hAPE1 and ^{15}N-hAPE1 obtained using the tryptic digest of a nuclear extract of MCF-7 cells. The extract was spiked with an aliquot of ^{15}N-hAPE1 prior to separation by HPLC. *Data from Kirkali et al. (2013).*

of hAPE1 and ^{15}N-hAPE1, which was added to protein extracts as an internal standard, lined up at the corresponding retention times, unequivocally identifying hAPE1 in this human cell line. The expression levels of hAPE1 in cell lines and tissues were calculated by using the measured signals of mass transitions of each tryptic peptide and their ^{15}N-labeled analogs, the amount of the internal standard, and the protein amount in each protein extract. In addition, we demonstrated the detection of APE1 variants with single amino acid replacements, i.e., Gln51His and Gly241Arg, which occur in several cancers (Pieretti, Khattar, & Smith, 2001), and in the general population (Hadi, Coleman, Fidelis, Mohrenweiser, & Wilson, 2000; Mohrenweiser, Xi, Vazquez-Matias, & Jones, 2002).

hMTH1 has been unequivocally identified and quantified in human malignant and disease-free breast tissues and in four cultured human cell lines (Coskun et al., 2015). Extreme expression of MTH1 in malignant breast tumors was observed, suggesting that cancer cells are addicted to MTH1 for their survival. In agreement with this finding, cultured MCF-7 cells exhibited a much greater expression level of hMTH1 than their normal counterpart mammary gland epithelial (MCF-10A) cells.

10. CONCLUSIONS

In this chapter, we described the production and purification of full-length ^{15}N-labeled DNA proteins to be used as internal standards for the measurement of DNA repair proteins in human tissues *in vivo* by LC-MS/MS or by other mass spectrometric techniques. These internal standards are absolutely essential for positive identification and accurate quantification of target proteins. The use of these labeled proteins has been demonstrated for the measurement of several DNA repair proteins in human tissues and cultured cells. Such measurements will be of fundamental importance for the determination of DNA repair capacity, the use of DNA repair proteins as biomarkers, and the development of DNA repair inhibitors in cancer, and eventually in other diseases as well.

ACKNOWLEDGMENTS

We are grateful to Dr. Dimitry Zharkov (SB RAS Institute of Chemical Biology and Fundamental Medicine, Novosibirsk, Russia) for providing the *ogg1* gene in the expression vector of pET15b for the production of ^{15}N-hOGG1, and to Dr. Timothy R. O'Connor (Beckman Institute of the City of Hope, Duarte, CA) for the gift of the coding sequence for *E. coli fpg* gene from the plasmid pFpg for production of *E. coli* ^{15}N-Fpg. *E. coli* BL21 (λDE3) harboring pETApe recombinant of plasmid for the production of ^{15}N-hAPE1 was kindly provided by Dr. David M. Wilson III (National Institute on Aging, National Institutes of Health, Baltimore, Maryland, USA). The human *neil1* gene in the expression vector pET22b for the production of ^{15}N-hNEIL1 was a gift from Dr. Stephen Lloyd (Oregon Health and Science University, Portland, OR). This work on MTH1 protein was supported by the Knut and Alice Wallenberg Foundation (T.H.), Swedish Research Council (T.H.), Swedish Cancer Society (T.H.), the Swedish Pain Relief Foundation (T.H.), the Swedish Foundation for Strategic Research (T.H.), the Göran Gustafsson Foundation, and the Torsten and Ragnar Söderberg Foundation (T.H.).

Certain commercial equipment or materials are identified in this chapter in order to specify adequately the experimental procedure. Such identification does not imply recommendation or endorsement by the National Institute of Standards and Technology, nor does it imply that the materials or equipment identified are necessarily the best available for the purpose.

REFERENCES

Abbotts, R., & Madhusudan, S. (2010). Human AP endonuclease 1 (APE1): From mechanistic insights to druggable target in cancer. *Cancer Treatment Reviews*, *36*, 425–435.

Aspinwall, R., Rothwell, D. G., Roldan-Arjona, T., Anselmino, C., Ward, C. J., Cheadle, J. P., et al. (1997). Cloning and characterization of a functional human homolog of *Escherichia coli* endonuclease III. *Proceedings of the National Academy of Sciences of the United States of America*, *94*, 109–114.

Audebert, M., Radicella, J. P., & Dizdaroglu, M. (2000). Effect of single mutations in the OGG1 gene found in human tumors on the substrate specificity of the ogg1 protein. *Nucleic Acids Research*, *28*, 2672–2678.

Bandaru, V., Sunkara, S., Wallace, S. S., & Bond, J. P. (2002). A novel human DNA glycosylase that removes oxidative DNA damage and is homologous to Escherichia coli endonuclease VIII. *DNA Repair (Amsterdam)*, *1*, 517–529.

Barnes, T., Kim, W. C., Mantha, A. K., Kim, S. E., Izumi, T., Mitra, S., et al. (2009). Identification of Apurinic/apyrimidinic endonuclease 1 (APE1) as the endoribonuclease that cleaves c-myc mRNA. *Nucleic Acids Research*, *37*, 3946–3958.

Beard, W. A., & Wilson, S. H. (2000). Structural design of a eukaryotic DNA repair polymerase: DNA polymerase beta. *Mutation Research*, *460*, 231–244.

Beard, W. A., & Wilson, S. H. (2006). Structure and mechanism of DNA polymerase beta. *Chemical Reviews*, *106*, 361–382.

Beard, W. A., & Wilson, S. H. (2014). Structure and mechanism of DNA polymerase beta. *Biochemistry*, *53*, 2768–2780.

Beckman, R. A., & Loeb, L. A. (2005). Genetic instability in cancer: Theory and experiment. *Seminars in Cancer Biology*, *15*, 423–435.

Boiteux, S., Gajewski, E., Laval, J., & Dizdaroglu, M. (1992). Substrate specificity of the *Escherichia coli* Fpg protein (formamidopyrimidine-DNA glycosylase): Excision of purine lesions in DNA produced by ionizing radiation or photosensitization. *Biochemistry*, *31*, 106–110.

Braithwaite, D. K., & Ito, J. (1993). Compilation, alignment, and phylogenetic relationships of DNA polymerases. *Nucleic Acids Research*, *21*, 787–802.

Cabelof, D. C. (2012). Haploinsufficiency in mouse models of DNA repair deficiency: Modifiers of penetrance. *Cellular and Molecular Life Sciences*, *69*, 727–740.

Chan, N., & Bristow, R. G. (2010). "Contextual" synthetic lethality and/or loss of heterozygosity: Tumor hypoxia and modification of DNA repair. *Clinical Cancer Research*, *16*, 4553–4560.

Chan, M. K., Ocampo-Hafalla, M. T., Vartanian, V., Jaruga, P., Kirkali, G., Koenig, K. L., et al. (2009). Targeted deletion of the genes encoding NTH1 and NEIL1 DNA N-glycosylases reveals the existence of novel carcinogenic oxidative damage to DNA. *DNA Repair (Amsterdam)*, *8*, 786–794.

Coskun, E., Jaruga, P., Jemth, A. S., Loseva, O., Scanlan, L. D., Tona, A., et al. (2015). Addiction to MTH1 protein results in intense expression in human breast cancer tissue as measured by liquid chromatography-isotope-dilution tandem mass spectrometry. *DNA Repair (Amsterdam)*. in press.

Curtin, N. J. (2013). PARP and PARP inhibitor therapeutics. In S. Madhusudan & D. M. Wilson, III. (Eds.), *DNA repair and cancer from bench to clinic* (pp. 513–563). Boca Raton: CRC Press.

Demple, B., & Harrison, L. (1994). Repair of oxidative damage to DNA: Enzymology and biology. *Annual Review of Biochemistry*, *63*, 915–948.

Demple, B., Herman, T., & Chen, D. S. (1991). Cloning and expression of APE, the cDNA encoding the major human apurinic endonuclease: Definition of a family of DNA repair enzymes. *Proceedings of the National Academy of Sciences of the United States of America*, *88*, 11450–11454.

Dherin, C., Radicella, J. P., Dizdaroglu, M., & Boiteux, S. (1999). Excision of oxidatively damaged DNA bases by the human alpha-hOgg1 protein and the polymorphic alpha-hOgg1(Ser326Cys) protein which is frequently found in human populations. *Nucleic Acids Research*, *27*, 4001–4007.

Dizdaroglu, M. (2005). Base-excision repair of oxidative DNA damage by DNA glycosylases. *Mutation Research*, *591*, 45–59.

Dizdaroglu, M. (2015). Oxidatively induced DNA damage and its repair in cancer. *Mutation Research/Reviews in Mutation Research*, *763*, 212–245.

Dizdaroglu, M., & Jaruga, P. (2012). Mechanisms of free radical-induced damage to DNA. *Free Radical Research*, *46*, 382–419.

Dizdaroglu, M., Reddy, P. T., & Jaruga, P. (2011). Identification and quantification of DNA repair proteins by liquid chromatography/isotope-dilution tandem mass spectrometry using their fully 15N-labeled analogues as internal standards. *Journal of Proteome Research*, *10*, 3802–3813.

Doublie, S., Bandaru, V., Bond, J. P., & Wallace, S. S. (2004). The crystal structure of human endonuclease VIII-like 1 (NEIL1) reveals a zincless finger motif required for glycosylase activity. *Proceedings of the National Academy of Sciences of the United States of America*, *101*, 10284–10289.

Evans, M. D., Dizdaroglu, M., & Cooke, M. S. (2004). Oxidative DNA damage and disease: Induction, repair and significance. *Mutation Research*, *567*, 1–61.

Fishel, M. L., Vascotto, C., & Kelley, M. R. (2013). DNA base excision repair therapeutics: Summary of targets with a focus on APE1. In S. Madhusudan & D. M. Wilson, III. (Eds.), *DNA repair and cancer from bench to clinic* (pp. 233–287). Boca Raton: CRC Press.

Friedberg, E. C., Walker, G. C., Siede, W., Wood, R. D., Schultz, R. A., & Ellenberger, T. (2006). *DNA repair and mutagenesis* (2nd ed.). Washington, DC: ASM Press.

Gad, H., Koolmeister, T., Jemth, A. S., Eshtad, S., Jacques, S. A., Strom, C. E., et al. (2014). MTH1 inhibition eradicates cancer by preventing sanitation of the dNTP pool. *Nature*, *508*, 215–221.

Giglione, C., Boularot, A., & Meinnel, T. (2004). Protein N-terminal methionine excision. *Cellular and Molecular Life Sciences: CMLS*, *61*, 1455–1474.

Gros, L., Ishchenko, A. A., Ide, H., Elder, R. H., & Saparbaev, M. K. (2004). The major human AP endonuclease (Ape1) is involved in the nucleotide incision repair pathway. *Nucleic Acids Research*, *32*, 73–81.

Gros, L., Saparbaev, M. K., & Laval, J. (2002). Enzymology of the repair of free radicals-induced DNA damage. *Oncogene*, *21*, 8905–8925.

Hadi, M. Z., Coleman, M. A., Fidelis, K., Mohrenweiser, H. W., & Wilson, D. M., III. (2000). Functional characterization of Ape1 variants identified in the human population. *Nucleic Acids Research*, *28*, 3871–3879.

Hakem, R. (2008). DNA-damage repair; the good, the bad, and the ugly. *The EMBO Journal*, *27*, 589–605.

Hanahan, D. (1983). Studies on transformation of Escherichia coli with plasmids. *Journal of Molecular Biology*, *166*, 557–580.

Hazra, T. K., Izumi, T., Boldogh, I., Imhoff, B., Kow, Y. W., Jaruga, P., et al. (2002). Identification and characterization of a human DNA glycosylase for repair of modified bases in oxidatively damaged DNA. *Proceedings of the National Academy of Sciences of the United States of America*, *99*, 3523–3528.

Hegde, M. L., Hazra, T. K., & Mitra, S. (2008). Early steps in the DNA base excision/single-strand interruption repair pathway in mammalian cells. *Cell Research*, *18*, 27–47.

Helleday, T. (2008). Amplifying tumour-specific replication lesions by DNA repair inhibitors—A new era in targeted cancer therapy. *European Journal of Cancer*, *44*, 921–927.

Helleday, T., Petermann, E., Lundin, C., Hodgson, B., & Sharma, R. A. (2008). DNA repair pathways as targets for cancer therapy. *Nature Reviews. Cancer*, *8*, 193–204.

Hoeijmakers, J. H. (2001). Genome maintenance mechanisms for preventing cancer. *Nature*, *411*, 366–374.

Hu, J., de Souza-Pinto, N. C., Haraguchi, K., Hogue, B. A., Jaruga, P., Greenberg, M. M., et al. (2005). Repair of formamidopyrimidines in DNA involves different glycosylases: Role of the OGG1, NTH1, and NEIL1 enzymes. *Journal of Biological Chemistry*, *280*, 40544–40551.

Huamani, J., McMahan, C. A., Herbert, D. C., Reddick, R., McCarrey, J. R., MacInnes, M. I., et al. (2004). Spontaneous mutagenesis is enhanced in Apex heterozygous mice. *Molecular and Cellular Biology*, *24*, 8145–8153.

Huber, K. V., Salah, E., Radic, B., Gridling, M., Elkins, J. M., Stukalov, A., et al. (2014). Stereospecific targeting of MTH1 by (S)-crizotinib as an anticancer strategy. *Nature*, *508*, 222–227.

Illuzzi, J. L., & Wilson, D. M., III. (2012). Base excision repair: Contribution to tumorigenesis and target in anticancer treatment paradigms. *Current Medicinal Chemistry*, *19*, 3922–3936.

Jaruga, P., Birincioglu, M., Rosenquist, T. A., & Dizdaroglu, M. (2004). Mouse NEIL1 protein is specific for excision of 2,6-diamino-4-hydroxy-5-formamidopyrimidine and 4,6-diamino-5-formamidopyrimidine from oxidatively damaged DNA. *Biochemistry*, *43*, 15909–15914.

Jaruga, P., Xiao, Y., Vartanian, V., Lloyd, R. S., & Dizdaroglu, M. (2010). Evidence for the involvement of DNA repair enzyme NEIL1 in nucleotide excision repair of (5′*R*)- and (5′*S*)-8,5′-cyclo-2′-deoxyadenosines. *Biochemistry*, *49*, 1053–1055.

Karahalil, B., Roldan-Arjona, T., & Dizdaroglu, M. (1998). Substrate specificity of *Schizosaccharomyces pombe* Nth protein for products of oxidative DNA damage. *Biochemistry*, *37*, 590–595.

Kelley, M. R. (2012). Future directions with DNA repair inhibitors: A roadmap for disruptive approaches to cancer therapy. In M. R. Kelley (Ed.), *Molecular targets and clinical applications. DNA repair in cancer therapy* (pp. 301–310). Elsevier: Amsterdam.

Kennedy, C. H., Cueto, R., Belinsky, S. A., Lechner, J. F., & Pryor, W. A. (1998). Overexpression of hMTH1 mRNA: A molecular marker of oxidative stress in lung cancer cells. *FEBS Letters*, *429*, 17–20.

Kinter, M., & Sherman, N. E. (2000). *Protein sequencing and identification using tandem mass spectrometry*. New York: Wiley-Interscience.

Kirkali, G., Jaruga, P., Reddy, P. T., Tona, A., Nelson, B. C., Li, M., et al. (2013). Identification and quantification of DNA repair protein apurinic/apyrimidinic endonuclease 1 (APE1) in human cells by liquid chromatography/isotope-dilution tandem mass spectrometry. *PLoS One*, *8*, e69894.

Liu, M., Bandaru, V., Bond, J. P., Jaruga, P., Zhao, X., Christov, P. P., et al. (2010). The mouse ortholog of NEIL3 is a functional DNA glycosylase in vitro and in vivo. *Proceedings of the National Academy of Sciences of the United States of America*, *107*, 4925–4930.

Liu, R., Yin, L. H., & Pu, Y. P. (2007). Reduced expression of human DNA repair genes in esophageal squamous-cell carcinoma in china. *Journal of Toxicology and Environmental Health. Part A*, *70*, 956–963.

Loeb, L. A. (2011). Human cancers express mutator phenotypes: Origin, consequences and targeting. *Nature Reviews. Cancer*, *11*, 450–457.

Lowry, O. H., Rosebrough, N. J., Farr, A. L., & Randall, R. J. (1951). Protein measurement with the Folin phenol reagent. *Journal of Biological Chemistry*, *193*, 265–275.

Madhusudan, S., & Middleton, M. R. (2005). The emerging role of DNA repair proteins as predictive, prognostic and therapeutic targets in cancer. *Cancer Treatment Reviews*, *31*, 603–617.

Maki, H., & Sekiguchi, M. (1992). MutT protein specifically hydrolyses a potent mutagenic substrate for DNA synthesis. *Nature*, *355*, 273–275.

Matsumoto, Y., & Kim, K. (1995). Excision of deoxyribose phosphate residues by DNA polymerase β during DNA repair. *Science*, *269*, 699–702.

Meira, L. B., Devaraj, S., Kisby, G. E., Burns, D. K., Daniel, R. L., Hammer, R. E., et al. (2001). Heterozygosity for the mouse Apex gene results in phenotypes associated with oxidative stress. *Cancer Research*, *61*, 5552–5557.

Miller, J. H. (1972). *Experiments in molecular genetics*. Cold Spring Harbor, NY: Cold Spring Harbor Laboratory.

Mo, J. Y., Maki, H., & Sekiguchi, M. (1992). Hydrolytic elimination of a mutagenic nucleotide, 8-oxodGTP, by human 18-kilodalton protein: Sanitization of nucleotide pool. *Proceedings of the National Academy of Sciences of the United States of America*, *89*, 11021–11025.

Moeller, B. J., Arap, W., & Pasqualini, R. (2010). Targeting synthetic lethality in DNA damage repair pathways as an anti-cancer strategy. *Current Drug Targets*, *11*, 1336–1340.

Mohrenweiser, H. W., Xi, T., Vazquez-Matias, J., & Jones, I. M. (2002). Identification of 127 amino acid substitution variants in screening 37 DNA repair genes in humans. *Cancer Epidemiology, Biomarkers and Prevention*, *11*, 1054–1064.

Muftuoglu, M., de Souza-Pinto, N. C., Dogan, A., Aamann, M., Stevnsner, T., Rybanska, I., et al. (2009). Cockayne syndrome group B protein stimulates repair of formamidopyrimidines by NEIL1 DNA glycosylase. *Journal of Biological Chemistry*, *284*, 9270–9279.

O'Connor, T. R., Boiteux, S., & Laval, J. (1989). Repair of imidazole ring-opened purines in DNA: Overproduction of the formamidopyrimidine-DNA glycosylase of Escherichia coli using plasmids containing the fpg+gene. *Annali dell'Istituto Superiore di Sanità*, *25*, 27–31.

Okamoto, K., Toyokuni, S., Kim, W. J., Ogawa, O., Kakehi, Y., Arao, S., et al. (1996). Overexpression of human mutT homologue gene messenger RNA in renal-cell carcinoma: Evidence of persistent oxidative stress in cancer. *International Journal of Cancer*, *65*, 437–441.

Perry, C., Sultana, R., & Madhusudan, S. (2012). Personalized cancer medicine: DNA repair alterations are promising predictive biomarkers in cancer. In M. R. Kelley (Ed.), *Molecular targets and clinical applications. DNA repair in cancer therapy* (pp. 257–282). Elsevier: Amsterdam.

Pieretti, M., Khattar, N. H., & Smith, S. A. (2001). Common polymorphisms and somatic mutations in human base excision repair genes in ovarian and endometrial cancers. *Mutation Research*, *432*, 53–59.

Reddy, P. T., Jaruga, P., Kirkali, G., Tuna, G., Nelson, B. C., & Dizdaroglu, M. (2013). Identification and quantification of human DNA repair protein NEIL1 by liquid chromatography/isotope-dilution tandem mass spectrometry. *Journal of Proteome Research, 12*, 1049–1061.

Reddy, P. T., Jaruga, P., Nelson, B. C., Lowenthal, M., & Dizdaroglu, M. (2011). Stable isotope-labeling of DNA repair proteins, and their purification and characterization. *Protein Expression and Purification, 78*, 94–101.

Roldán-Arjona, T., Anselmino, C., & Lindahl, T. (1996). Molecular cloning and functional analysis of a *Schizosaccharomyces pombe* homologue of *Escherichia coli* endonuclease III. *Nucleic Acids Research, 24*, 3307–3312.

Rosenquist, T. A., Zaika, E., Fernandes, A. S., Zharkov, D. O., Miller, H., & Grollman, A. P. (2003). The novel DNA glycosylase, NEIL1, protects mammalian cells from radiation-mediated cell death. *DNA Repair (Amsterdam), 2*, 581–591.

Roy, L. M., Jaruga, P., Wood, T. G., McCullough, A. K., Dizdaroglu, M., & Lloyd, R. S. (2007). Human polymorphic variants of the NEIL1 DNA glycosylase. *Journal of Biological Chemistry, 282*, 15790–15798.

Sakai, Y., Furuichi, M., Takahashi, M., Mishima, M., Iwai, S., Shirakawa, M., et al. (2002). A molecular basis for the selective recognition of 2-hydroxy-dATP and 8-oxo-dGTP by human MTH1. *Journal of Biological Chemistry, 277*, 8579–8587.

Sakumi, K., Furuichi, M., Tsuzuki, T., Kakuma, T., Kawabata, S., Maki, H., et al. (1993). Cloning and expression of cDNA for a human enzyme that hydrolyzes 8-oxo-dGTP, a mutagenic substrate for DNA synthesis. *Journal of Biological Chemistry, 268*, 23524–23530.

Sambrook, J., Fritsch, E. F., & Maniatis, T. (1989). *Molecular cloning: A laboratory manual* (2nd ed.). Cold Spring Harbor, NY: Cold Spring Harbor Laboratory.

Sampath, H., McCullough, A. K., & Lloyd, R. S. (2012). Regulation of DNA glycosylases and their role in limiting disease. *Free Radical Research, 46*, 460–478.

Sekiguchi, M., & Tsuzuki, T. (2002). Oxidative nucleotide damage: Consequences and prevention. *Oncogene, 21*, 8895–8904.

Sidorenko, V. S., Grollman, A. P., Jaruga, P., Dizdaroglu, M., & Zharkov, D. O. (2009). Substrate specificity and excision kinetics of natural polymorphic variants and phosphomimetic mutants of human 8-oxoguanine-DNA glycosylase. *The FEBS Journal, 276*, 5149–5162.

Speina, E., Arczewska, K. D., Gackowski, D., Zielinska, M., Siomek, A., Kowalewski, J., et al. (2005). Contribution of hMTH1 to the maintenance of 8-oxoguanine levels in lung DNA of non-small-cell lung cancer patients. *Journal of the National Cancer Institute, 97*, 384–395.

Srivastava, D. K., Berg, B. J., Prasad, R., Molina, J. T., Beard, W. A., Tomkinson, A. E., et al. (1998). Mammalian abasic site base excision repair. Identification of the reaction sequence and rate-determining steps. *Journal of Biological Chemistry, 273*, 21203–21209.

Svensson, L. M., Jemth, A. S., Desroses, M., Loseva, O., Helleday, T., Hogbom, M., et al. (2011). Crystal structure of human MTH1 and the 8-oxo-dGMP product complex. *FEBS Letters, 585*, 2617–2621.

Tell, G., Quadrifoglio, F., Tiribelli, C., & Kelley, M. R. (2009). The many functions of APE1/Ref-1: Not only a DNA repair enzyme. *Antioxidants & Redox Signaling, 11*, 601–620.

Tsuzuki, T., Egashira, A., Igarashi, H., Iwakuma, T., Nakatsuru, Y., Tominaga, Y., et al. (2001). Spontaneous tumorigenesis in mice defective in the MTH1 gene encoding 8-oxo-dGTPase. *Proceedings of the National Academy of Sciences of the United States of America, 98*, 11456–11461.

Vartanian, V., Lowell, B., Minko, I. G., Wood, T. G., Ceci, J. D., George, S., et al. (2006). The metabolic syndrome resulting from a knockout of the NEIL1 DNA glycosylase.

Proceedings of the National Academy of Sciences of the United States of America, 103, 1864–1869.

Wallace, S. S., Murphy, D. L., & Sweasy, J. B. (2012). Base excision repair and cancer. *Cancer Letters, 327*, 73–89.

Wilson, D. M., III, & Barsky, D. (2001). The major human abasic endonuclease: Formation, consequences and repair of abasic lesions in DNA. *Mutation Research, 485*, 283–307.

Wilson, D. M., III, Kim, D., Berquist, B. R., & Sigurdson, A. J. (2011). Variation in base excision repair capacity. *Mutation Research, 711*, 100–112.

Wood, R. D., Mitchell, M., & Lindahl, T. (2005). Human DNA repair genes. *Mutation Research, 577*, 275–283.

Xanthoudakis, S., & Curran, T. (1992). Identification and characterization of Ref-1, a nuclear protein that facilitates AP-1 DNA-binding activity. *The EMBO Journal, 11*, 653–665.

Xanthoudakis, S., Smeyne, R. J., Wallace, J. D., & Curran, T. (1996). The redox/DNA repair protein, Ref-1, is essential for early embryonic development in mice. *Proceedings of the National Academy of Sciences of the United States of America, 93*, 8919–8923.

SECTION IV

Hydrogen-Deuterium Exchange Mass Spectrometry

CHAPTER THIRTEEN

Hydrogen Exchange Mass Spectrometry

Leland Mayne[1]
Johnson Research Foundation, Department of Biochemistry and Biophysics, Perelman School of Medicine, The University of Pennsylvania, Philadelphia, Pennsylvania, USA
[1]Corresponding author: e-mail address: mayne@mail.med.upenn.edu

Contents

Abstract

Hydrogen exchange (HX) methods can reveal much about the structure, energetics, and dynamics of proteins. The addition of mass spectrometry (MS) to an earlier fragmentation-separation HX analysis now extends HX studies to larger proteins at high structural resolution and can provide information not available before. This chapter discusses experimental aspects of HX labeling, especially with respect to the use of MS and the analysis of MS data.

Methods in Enzymology, Volume 566
ISSN 0076-6879
http://dx.doi.org/10.1016/bs.mie.2015.06.035

1. INTRODUCTION

Hydrogen–deuterium exchange mass spectrometry is increasingly being recognized as an important, information-rich, tool for protein studies (Englander, 2006; Kaltashov, Bobst, Nguyen, & Wang, 2013; Konermann, Tong, & Pan, 2008; Ling, Silva, Schriemer, & Schryvers, 2012; Pirrone, Iacob, & Engen, 2015; Zhang & Smith, 1993). The amide hydrogen (NH) of every residue (except proline) in every protein is available as a probe of its local environment. These probes can be used in many ways to inform us about the structural energetics and dynamics of proteins under native conditions, as well as how they respond to functional changes such as ligand binding and folding. Most hydrogen exchange experiments measure the exchange of H(^{1}H) and D(^{2}H) isotopes. This is reflected by the widespread use of the term hydrogen–deuterium exchange (HDX) rather than the more general term, hydrogen exchange (HX) that includes other isotopes. Mass spectrometry (MS) has taken an overwhelming lead since the late 1990s as the method of choice for HX measurements. While nuclear magnetic resonance (NMR) naturally achieves residue resolution, it requires $\sim 10^4$ times as much protein for each sample, is limited to fairly small proteins, $<\sim 200$ residues, that are soluble near m*M* concentrations, and often requires isotopic enrichment with ^{15}N.

This chapter will discuss HX-MS experiments in three sections: First HX labeling itself, beginning with exchange chemistry and the mechanisms of exchange in a protein. We then discuss the measurement of deuterated proteins by electrospray ionization MS. We will mostly deal with the fragment separation method (Englander & Englander, 1972; Englander & Kallenbach, 1983) in which labeled protein is enzymatically digested to yield peptide fragments that are in turn individually analyzed for D content based on the increase in mass (Δm) of deuterated peptides compared to unexchanged, all H, peptides. We finally discuss data processing, beginning with the identification of deuterated peptides and our method for moving toward residue resolution.

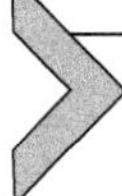

2. HX MECHANISM

2.1 Chemistry

The chemical exchange of a freely exposed amide NH is catalyzed by OH^- and H^+ and so is naturally a function of pH with a minimum rate at roughly pH 2.5. The rate is further influenced by inductive and steric effects from nearby amino acid side chains, temperature, and the isotopes involved. These effects are well calibrated and serve as reference rates when assessing structural protection (Bai, Milne, Mayne, & Englander, 1993; Connelly, Bai, Jeng, & Englander, 1993).

Careful control of pH is vital to successful HX measurements. Above pH 3 where proteins are usually studied, the rate increases by a factor of 10 with every pH unit. We take advantage of this pH dependence to control the rate of exchange for experimental convenience and most importantly to preserve the labeled state of the protein by quenching to the minimum rate. At pH 2.5 and 0 °C, the rate is near 0.01/min which allows time for the manipulations necessary for MS measurements.

2.2 HX Structural Physics

Stable hydrogen-bonded protein structure must open to expose the amide to solvent before it can exchange with its chemical rate as given by (Hvidt & Nielsen, 1966):

$$\text{closed (protected)} \underset{k_{cl}}{\overset{k_{op}}{\rightleftharpoons}} \text{open (exchangable)} \xrightarrow{k_{ch}} \text{exchanged}$$

This scheme leads to a measured exchange rate, k_{ex}, given by:

$$k_{ex} = \frac{k_{op}k_{ch}}{k_{cl} + k_{ch} + k_{op}} \tag{1}$$

here k_{ch} is the free peptide rate, a function of temperature, pH, and local sequence, and k_{op} and k_{cl} are the structural opening and closing rates that expose the amide.

For stable structure ($k_{op} \ll k_{cl}$), this general scheme reduces to two limiting cases depending on the relative rates of reclosing (k_{cl}) and the chemical exchange rate (k_{ch}).

When $k_{cl} \gg k_{ch}$, the protein opens and recloses many times before exchange occurs. The measured exchange rate then is k_{ch} during the fraction of time open. The equilibrium constant for opening, K_{op}, is given as k_{op}/k_{cl}, leading to a measured exchange rate as in Eq. (2).

$$k_{ex} = \frac{k_{ch}K_{op}}{K_{op} + 1} \cong k_{ch}K_{op} \tag{2}$$

In this case, referred to as EX2 or bimolecular exchange, exchange depends on k_{ch} and therefore on the pH. $1/K_{op}$ is often referred to as the protection factor. Protection is also often reported in free energy units as $\Delta G_{HX} = -RT\ln(K_{op})$.

When $k_{cl} \ll k_{ch}$, exchange is limited by the opening rate, k_{op}, and is independent of k_{ch}:

$$k_{ex} = k_{op} \tag{3}$$

This case, known as EX1, sometimes referred to as correlated exchange, is most likely at high pH, where k_{ch} is fast, or low protein stability where k_{cl} is slow.

When HX is measured by NMR, the exchange rate of each amide is measured individually and it is not easy to distinguish EX1 from EX2 in a single experiment. The most common test is to measure exchange at two or more pH values. If protein stability does not change with pH the exchange rate will change by a factor of 10 for each change of 1 pH unit under EX2 conditions and remain unaffected under EX1 conditions. If the exchange mechanism transitions between EX2 and EX1 as pH is raised, both the opening and closing rates can be measured by fitting the measured exchange rates as a function of pH to Eq. (1) allowing k_{op} and k_{cl} to float while using the known values of k_{ch} as a function of pH (Bedard, Mayne, Peterson, Wand, & Englander, 2008).

Measuring HX by MS can make the EX1, EX2 difference directly apparent and illustrates a significant advantage of using MS. MS measures exchange for groups of amino acids together whether peptides or the whole protein. If a segment of structure unfolds and stays unfolded for longer than the free peptide exchange rate before refolding, as in EX1 exchange, then all residues in that segment will fully exchange before it refolds. Under these conditions, at any point during the course of exchange the sample will contain peptides that are unexchanged (have not yet opened) and peptides that are fully exchanged (have opened at least once). As a result, it will display a bimodal isotopic profile. In EX2 exchange, individual amino acids are not correlated and so will always show a unimodal isotopic distribution that moves from starting to ending conditions (H to D or D to H) with longer exchange times.

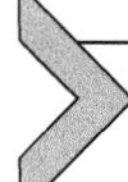

3. EXPERIMENTAL CONSIDERATIONS—EXCHANGE LABELING

Figure 1 outlines the two major classes of HX labeling experiments and the two major methods used for reading the state of labeling by MS.

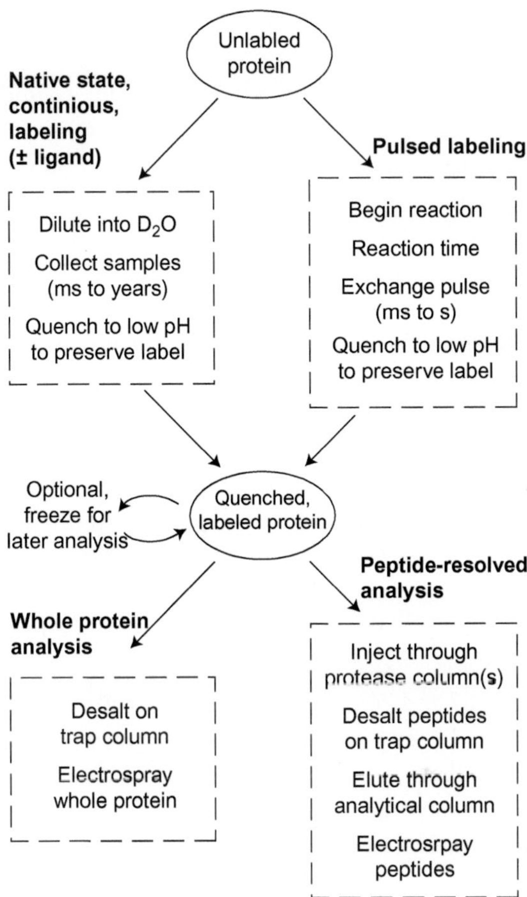

Figure 1 Classes of HX experiments.

Most experiments use continuous labeling followed by peptide level measurement. We will first discuss labeling protocols, then discuss MS methods for reading the label, and finally data analysis which includes choices depending on the nature of the labeling experiment.

3.1 Continuous Labeling (Native State)

Most HX experiments are some form of native state, continuous, labeling in which the exchange of the protein of interest is measured as a function of exchange time under conditions where the native state of the protein is stable. Experimental conditions are primarily determined by the protein being studied. The simplest way to begin exchange is to dilute a fully H protein sample into D_2O under experimental conditions. It is generally desirable to make a large dilution so as to increase the dynamic range of the measurement, though this may be limited by protein solubility or other considerations.

Two dilutions are required. First, protein is diluted into exchange conditions and then after some time, quench buffer is added to bring the solution to pH ~2.5 accompanied by rapid cooling or freezing for later MS analysis. Commonly, each exchange time point is mixed separately in individual microcentrifuge tubes. This can be done either by hand or robotically. With effort, labeling times as short as 10 s can be achieved by either method.

It is important to consider the range of time points chosen for exchange samples. We would like to measure the rates of each amino acid in our protein but individual amino acids within a native protein may have exchange rates that vary by factors of 10^8 or more, from free peptide rates of less than a second to residues that exchange only via transient unfolding of the entire protein, more than a year for a stable protein (~12 kcal/mol). In order to capture this dynamic range and extract as much information as is available from the experiment, it is clear that we cannot be limited to exchange measured by simple dilution experiments over a single day, a dynamic range of only ~1000. At pH 7 and 20 °C, many positions will be fully exchanged by the first time point, 10–30 s, while many amides may not have measurably exchanged in 24 h. The measurement range can be extended for many proteins by moving the intrinsic HX rate into a more convenient timescale by adjusting the temperature, pH, or both. If the protein under study does not change its stability, a change of 1 pH unit up or down moves the intrinsic exchange rate up or down by a factor of 10, increasing the measurable range of protection (Coales et al., 2010). The exchange temperature can be similarly adjusted; a change of 10 °C yields a change in rate of almost a factor of 3. It is important to verify that protein stability does not change in response to the change in conditions by overlapping the timescales so that some residues are measured under different exchange conditions and show the same protection factors.

Faster exchange rates can also be accessed by use of a flow system to perform the labeling. Two examples are illustrated in Fig. 2. Exchange times are determined by flow rates and the volumes between mixers. Times as short as 10 ms can be measured. With our BioLogic SFM400, we are, in addition, able to stop and restart the flow allowing for exchange times of minutes in the same apparatus. Wang, Abzalimov, Bobst, and Kaltashov (2013) have used a flow system to spray directly without proteolysis or HPLC.

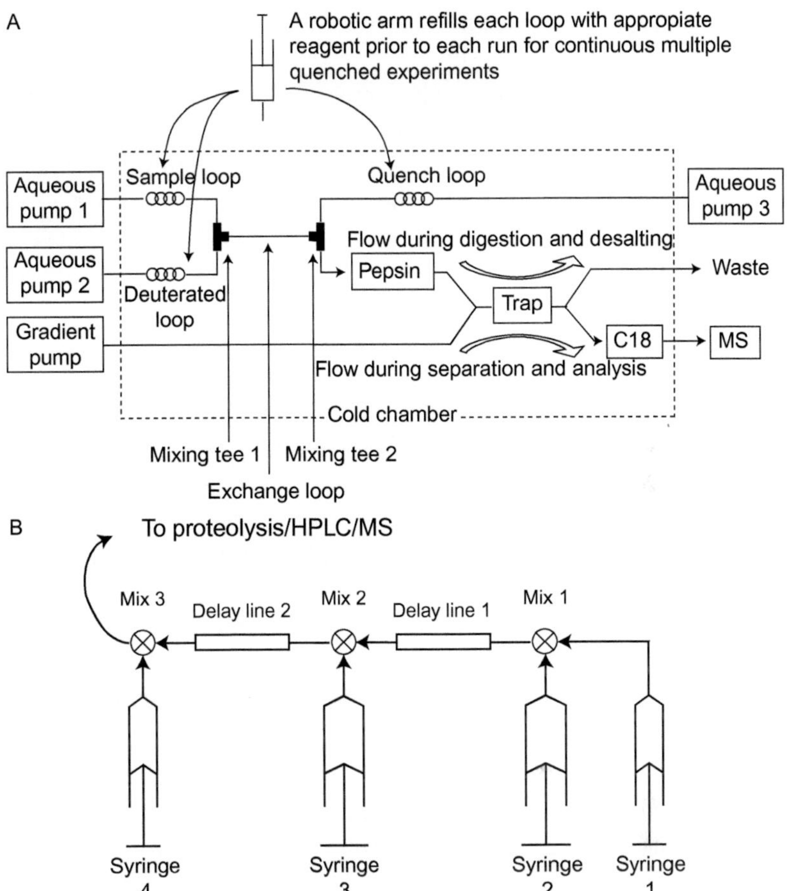

Figure 2 (A) Flow apparatus used for short labeling times as described by Coales et al. (2010). Exchange begins when the sample is mixed with D_2O buffer at mixing T1 pushed by pumps 1 and 2. After an exchange time, determined by the flow rate and the volume of the exchange loop, quench buffer is added. The sample flows through a pepsin column and onto a trap column. From the trap column, peptides are eluted to the MS by a gradient supplied by a separate pump. Not shown are valves that allow the reloading of the sample, D buffer, and quench loops, and that control flow through the trap column. (B) Our system using a BioLogic SFM400 stopped-flow mixer for labeling and a separate cooled chamber (Fig. 3) for proteolysis and chromatography. In the stopped-flow mixer, the speed of each syringe is separately controlled. For a simple exchange experiment, syringe 1 is not used. Unlabeled protein from syringe 2 is mixed with D_2O buffer from syringe 3. After an exchange time determined by the volume of delay line 2 and the speed of flow exchange is quenched by syringe 4. The quenched sample flows to the injection loop of our proteolysis/HPLC system. When used for refolding experiments, we start with unfolded protein in syringe 1. Folding is started with a dilution into refolding conditions at mixer 1. From here, the labeling pulse is as above.

Once a sample has exchanged, the label must be preserved for MS analysis. The sample is quenched by lowering the temperature and pH before making the measurement as quickly as possible. It is important to verify that the pH of the quenched sample after the mix is as expected. We verify the pH of the mixture for each new batch of buffers. The optimum pH seems to be 2.5 for the initial quench where the ionic strength is 10's or 100's of m*M* and 2.3 for the HPLC buffers where the ionic strength is <20 m*M* (Walters, Ricciuti, Mayne, & Englander, 2012). We use 0.1% formic acid and 0.05% trifluoroacetic acid for the aqueous HPLC buffers and 0.1% formic acid for the acetonitrile buffer.

The quenched sample can be flash frozen and stored at −80 °C for later MS analysis. It is important that frozen samples be thawed without excess heating. This is aided by the addition of antifreeze in the form of 10% glycerol or some other molecule. With the freezing point depressed, thawing can be done fairly quickly (1–2 min) in a 0 °C bath.

As discussed above, exchange of H-bonded NH's requires the transient opening of protecting structure. This can involve small-scale openings that only expose one amide NH at a time and larger scale openings that expose several amino acids together, all the way up to the global unfolding of the entire protein. Under EX2 conditions, the rate of a single residue will be determined by the fraction of time open, the sum of all opening reactions that expose that residue. Larger scale openings can be promoted by destabilizing conditions such as increased temperature or increasing concentrations of denaturant. This technique has been used to great effect by NMR for several proteins and has helped with the understanding of protein structural energetics and folding (Bai, Sosnick, Mayne, & Englander, 1995).

Some label will be unavoidably lost to back exchange during the workup for MS analysis. It is possible to measure the degree of back exchange with an "all D" control. In this experiment, a fully D protein sample subjected to the identical quench and analysis procedure used for exchange time points. The measured D occupancy for this sample can be used to calibrate the loss of label. The details of how back exchange is evaluated and corrected for depend on how the data are being analyzed and will be discussed in the analysis section.

Many people seem to have problems preparing all D samples. Given that some amides in stable proteins may be protected by 10^8 or more, simply waiting will not always work. We have had good success by heating a solution of protein in D_2O, at the same D/H ratio that will be present during the exchange experiment, to the beginning of the thermal melt transition, usually ~5 °C less than the midpoint of the melt. The protein is always cycling back and forth between folded and unfolded states but at room temperature the unfolded fraction may be only 10^{-8} or less. At a temperature near the beginning of the melt, this fraction will approach 10^{-2}, high enough to lower protection factors to only ~10^2–10^3 but not so high as to risk aggregation. In practice, we usually allow 10's of minutes to be sure that the whole sample has cycled through the unfolded state.

3.1.1 Binding Sites and Epitope Mapping

HDX is frequently used to localize the binding site of a drug or antibody (epitope mapping) by measuring exchange rates for free, unbound protein and for protein bound to the ligand of interest. Exchange rates are expected to slow for those residues that are affected by the binding. This can occur by more than one mechanism. HX slowing can occur due to direct hydrogen bonding between the ligand and exposed NH's on the protein. Binding interactions can stabilize a H-bond against exchange due to direct interaction of the ligand with either the H-bond donor or acceptor. Or interactions can stabilize H-bonded structure indirectly by stabilizing parts of the protein against larger scale openings (Mayne, Paterson, Cerasoli, & Englander, 1992; Paterson, Englander, & Roder, 1990; Wei et al., 2014).

It is of fundamental importance in binding studies to be sure that the protein binding site is saturated during the exchange time. The degree of saturation limits the extent to which HDX rate differences can be measured. If, for example, a ligand is bound only 90% of the time, then the maximal slowing will be a factor of 10 slower than for the unbound protein even in cases where exchange is completely stopped in the bound state. If this slowing affects only 1 or 2 residues in a peptide of 12 or more amino acids, the difference may be hard to distinguish.

Ligand bound experiments have to deal with the ligand in the MS analysis. The low pH quench will often unfold proteins and reduce binding. For small-molecule ligands, this alone may be enough if the ligand itself does not interfere with proteolysis or HPLC. In the case of protein–protein binding, e.g., antibody interactions, the ligand may be bigger than the protein being studied and is usually in molar excess. In this case, it is usually desirable to separate the protein from the antibody to avoid signal suppression and conflicts due to overlapping peptides. This is most easily accomplished by immobilizing the antibody on a column and binding the target protein to the antibody column. D_2O buffer is infused into the column already saturated with bound protein to start exchange. The protein is eluted and exchange quenched in a single step by infusing low pH quench buffer through the column directly into the analysis system (Paterson et al., 1990).

3.2 Pulsed Labeling

Protein conformational kinetics, e.g., protein folding, can be followed by HX pulsed labeling methods. In this experiment, rather than measuring exchange as a function of time under steady-state conditions, we start a reaction and after some time take a snapshot of the hydrogen-bonded structure with a short pulse of high pH where exchange is fast enough to fully exchange all freely exposed amides in a few ms. After a labeling time of usually 10's of ms, the pH is dropped to quench conditions and the sample analyzed.

While slower pulsed labeling experiments can be performed by manual mixing or by the same automated systems sometimes used for continuous labeling experiments, a flow system is needed for faster pulses. We use a Bio-Logic SFM400 stopped-flow mixer connected directly to the injection loop of our online proteolysis/HPLC system (Fig. 2). Three mixes are needed. The first, to start folding, will have to be large to dilute away denaturant. If exchange rates under folding conditions are slow relative to the folding time, then this dilution can also dilute the protein into D_2O, or H_2O if starting with unfolded D protein. The exchange pulse is started with an addition of high pH buffer. If the first dilution did not change the isotope (H to D or D to H), then this dilution will have to be large to move the protein to new isotopic solvent so that exchange can be measured. The pulse is quenched by the final addition of low pH buffer to bring the mix to pH 2.5.

There is an underappreciated feature of HX-MS in protein folding experiments. NMR measurements are done under native conditions. This is necessary in order to achieve the chemical shift dispersion necessary to resolve individual residues. In NMR-measured pulsed labeling experiments, the H/D pattern imposed by the pulse is "quenched" by native structural protection so only residues that are protected in the native protein will be measured, leaving open the possibility that nonnative structure could be present at the time of the pulse yet be unobserved if not protected by later native structure. MS-detected refolding experiments have so far not seen such nonnative protection, instead only stepwise accumulation of native structure has been observed (Hu et al., 2013; Walters, Mayne, Hinshaw, Sosnick, & Englander, 2013).

4. EXPERIMENTAL CONSIDERATIONS—MS MEASUREMENT

4.1 Whole Protein

Whole protein analysis is done without proteolysis and with little or no chromatography. Volatile buffers such as ammonium acetate and the addition of organic solvent allow quenched protein to be sprayed directly. Otherwise it is necessary to exchange the buffer to something more suitable for electrospray ionization. We do this by trapping the protein on a small C4 column, washing away buffer salts, and then eluting with a step to ~75% acetonitrile. The eluted protein is sprayed directly (Fig. 3).

4.2 Peptide Resolved

In order to better locate the sites of exchange, the quenched protein is usually digested enzymatically into peptides that are then washed of buffer salts,

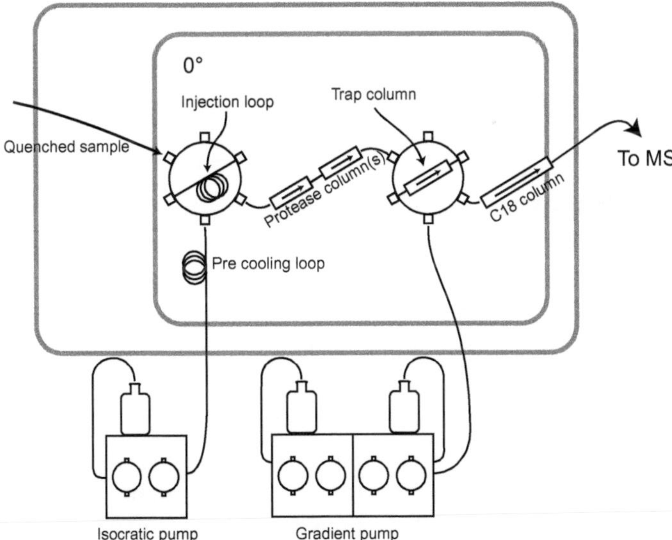

Figure 3 Our online digestion/HPLC system. The quenched sample is pushed through the protease column by a cold flow of pH 2.3 buffer (0.1% formic acid, 0.05% TFA). The produced peptides are bound by the small trap column and buffer salts are washed away. After switching the second valve peptides are eluted and separated by an H_2O/acetonitrile gradient through a C18 analytical column. The output of the C18 column goes directly to the electrospray. Whole protein can be desalted for ESI by omitting the protease and analytical columns.

and separated by rapid chromatography at 0 °C for electrospray ionization (Fig. 3) (Mayne et al., 2011). Quenched, labeled protein is loaded into the injection loop of a standard HPLC injection valve. A cold pH 2.3 aqueous buffer flow at 100–200 μl/min carries it through a protease column (sometimes more than one in tandem) and then via a second valve to a small reverse-phase "trap" column (typically C8 or C18, 1 mm × 10 mm, 5-μm beads). The peptides produced bind to the trap column while buffer salts are washed away. Switching the second valve directs the flow from a gradient pump through the trap column and on through an analytical column to the electrospray. Variations of this basic plan with such features as reversing the flow during the protease wash or swapping protease columns are possible.

4.2.1 Proteolysis

Ideally, we would like to produce many sequentially overlapping peptides that redundantly cover the entire protein. Most workers have adopted the use of immobilized protease columns in preference to solution digestion. The speed of digestion and ease of incorporation into a flow system greatly favor the column system. While commercial protease columns are available, we produce our own. We couple protease, either pepsin or Fungal Protease type XIII from Sigma, to POROS 20 AL media from Applied Biosystems. We couple according to the manufacturer's instructions at pH 4.4 with the addition of Na_2SO_4 (final concentration 750 m*M*) added over 2 h at room temperature. After several hours of additional reaction time, the beads are washed with pH 4.4 buffer for storage at 4 °C. A key step is to remove buffer salts from the commercial protease by gel filtration before beginning the reaction. Any amine containing buffer salts will compete with protein lysine side chains for binding sites. Once coupled, the beads are stable for months at 4 °C. We pack columns of either 1 or 2 mm diameter by 20 mm. We choose column sizes and flow rates according to the sizes and number of peptides we identify for each protein.

Digestion is often aided by the addition of denaturant to the quenched protein solution and is required for some proteins. We find our protease columns easily tolerate denaturant up to 2 *M* guanidinium chloride without damage. Addition of up to 200 m*M* TCEP (tris(2-carboxyethyl)phosphine) is included as necessary to reduce disulfides. Even with denaturant and TCEP, some proteins are resistant to proteolysis. There are some reports of successful electrochemical reduction of disulfide bonds (Mysling, Salbo, Ploug, & Jorgensen, 2014).

4.2.2 Chromatography

Peptides are eluted from the trap through an analytical HPLC column (C18, 0.3 mm × 50 mm, 3-μm beads) to the electrospray. Gradients are roughly 10–45% acetonitrile over 12–15 min at ~10 μl/min. We find that a shorter gradient is not particularly useful in reducing back exchange as it only speeds the measurement of late eluting peptides while reducing HPLC separation. While the gradient is running, we can wash the protease column with injections of guanidinium chloride to wash any protein that might carry over.

Peptides are identified using standard MS/MS methods (Mayne et al., 2011). We are sure to include in our sequence database all potential contaminants including potential impurities, the proteases used, and other proteins analyzed using the same system. As an extra check on correct peptide ID's, we confirm that sequentially overlapping peptides display consistent HD labeling.

5. MASS SPECTROMETER INSTRUMENT CONSIDERATIONS

HDX measurements can use many kinds of MS instruments. Everything we are discussing here assumes electrospray ionization. The combination of a large number of peptides, wide isotopic envelopes due to deuteration, and short HPLC gradients increases the probability of peptide overlap. High m/z resolution reduces the likelihood of individual isotopic peak overlap and high mass accuracy makes identification of peptides more certain. We use a Thermo Scientific LTQ Orbitrap XL usually operating at a resolution of 60,000 at m/z 400. Our data analysis program, ExMS, takes advantage of this higher resolution data to identify deuterated peptides.

Access to MS/MS capabilities is required for peptide identification but usually cannot be used to further localize D occupancy. Traditional collisional MS/MS fragmentation methods "scramble" amide D labels such that H's and D's equilibrate throughout the amides in a peptide before it fragments into daughter ions (Hamuro, Tomasso, & Coales, 2008). The increase in mass of daughter ions will reflect not the D occupancy levels of the daughter ion amino acids in the quenched protein but instead reflect the average fractional D occupancy of the peptide before fragmentation. Nonergodic fragmentation methods, electron-capture dissociation, and electron-transfer dissociation have been shown to cause minimal scrambling and provide the possibility of residue resolution of D occupancy. By analyzing the deuteration levels of a series of daughter ions differing by one amino acid at a time, the D occupancy of each new position can be determined by simple subtraction (Pan, Han, Borchers, & Konermann, 2009; Wang et al., 2013; Zehl, Rand, Jensen, & Jorgensen, 2008). In practice, enzymatic proteolysis may still be necessary for all but very short proteins to ensure coverage because not every possible site will fragment and the measurement of small mass differences in large fragments will limit accuracy. Care must still be taken to avoid scrambling in the spectrometer ion source. In some cases, parameters required to avoid ion scrambling severely reduce the signal.

6. DATA ANALYSIS

6.1 Whole Protein

Analysis of whole protein experiments is relatively straightforward. Continuous labeling EX2 experiments are characterized by the total D content as measured by the centroid shift. A back exchange correction can be applied by measuring the centroid of a fully D sample and correcting the measured Δm as discussed below.

Pulsed labeling experiments, EX1 exchange, or conformationally heterogeneous samples are expected to show multimodal isotopic profiles. If the m/z separation is sufficient, then the ratio of exchanged to unexchanged protein molecules can be readily measured by integrating the intensity of each envelope.

6.2 Peptide Analysis

Data analysis of HX experiments at the level of peptides can be an involved process depending on the nature of the exchange experiment and the peptide coverage. In all cases, it is important to remember that as the peptide bond is cut the N of the first amino acid of the resulting peptide becomes a free amine which exchanges very rapidly. In addition, the exchange rate of the second amino acid is greatly accelerated due to inductive effects on the chemical rate from the free amine (Bai et al., 1993). The result is that any D label present on the first amino acid is lost and, in general, label on the second residue is lost as well. In cases of very low back exchange if the first two side chains are isoleucine or valine, some label on the second residue might be retained.

6.2.1 Identification of Labeled Peptides

Identification of peptides with unknown levels of deuteration can be difficult. Several groups have developed software in an attempt to automate the identification and quantification of deuterated peptides, including (Pascal et al., 2012; Slysz et al., 2009) among others, as well as commercial software from Waters and Sierra Analytics. In general, one starts with a list of identified peptides and scans the MS chromatogram looking for each one while allowing for an unknown level of deuteration. The analysis usually starts with an all H sample where the *m/z* values are known in order to accurately measure the retention time window for each peptide. Once the retention time is accurately known, then deuterated samples are analyzed. A search over a narrow range of retention times simplifies the search over a larger range of *m/z* values.

We have developed the program ExMS (Kan, Mayne, Sevaugan Chetty, & Englander, 2011) which uses a list, with retention times, of peptides previously identified by MS/MS to identify deuterated peptides from MS data. In short, for each peptide on its list ExMS uses high-resolution accurate mass data to identify sets of isotopically resolved peaks that are consistent with deuterated samples of that peptide. A series of tests are applied to confirm the quality of the identification. The extracted data for each peptide are output as a table containing the retention time, the intensity of each isotopic peak, the overall intensity, and the centroid value for each peptide as well as information about each test.

At this point, various options are available depending on the nature of the experiment. One can simply plot average D occupancy, the increased centroid above the all H value, as a function of time as increased weight in Daltons or as percent exchanged. Percent D should be calculated as:

$$\frac{\Delta m}{((\mathrm{frac D}) \times (N - P - 2))}.$$

where Δm is the centroid mass above the centroid of the unlabeled peptide, frac D is the fraction D present during the exchange period, N is the number of amino acids in the peptide, and P is the number of proline residues in the peptide beyond the first two amino acids. As discussed above, no label is expected to be retained on the first two amino acids.

Whichever scale is used, it is helpful to indicate the size of the peptide involved so the reader can convert between the two scales. The use of a D scale rather than percent more readily relates to structural protection which operates on a residue level.

Protection provided by H-bonded structure directly influences the rate of exchange under EX2 conditions. Therefore, in order to measure the energetics of opening reactions that allow exchange, it is most useful to pay attention not to simple deuteration levels but instead to exchange rates (Fig. 4).

In pulsed labeling experiments or under EX1 conditions, analysis can involve a simple centroid shift measurement of average D occupancy. Alternatively, more information is available if multiple labeled populations are resolvable as in Fig. 5. In this case, the relative intensities of differently labeled populations yield information about the number of sites labeled (m/z axis) and the fraction of molecules in each resolved labeled population (relative intensities of each population).

6.2.2 Back Exchange Correction

The measured level of back exchange is an important measure of data quality. D recovery can be measured on a per-peptide basis by subjecting

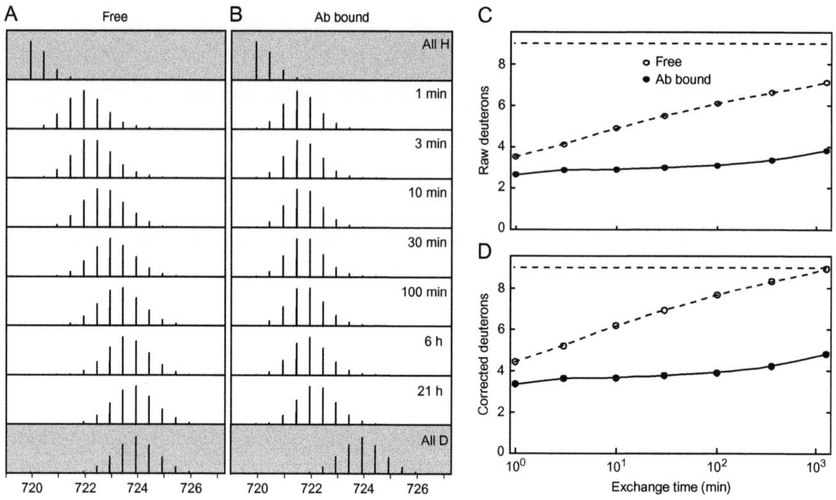

Figure 4 EX2 exchange. (A, B) Mass spectra of a peptide produced by pepsin digest (RVALTEDRLPRL) as a function of exchange time. The top spectra are for an unexchanged, all H, sample and the bottom are from a fully exchanged, all D, sample used to calibrate back exchange. Intermediate time points show the progress of exchange for (A) the protein free in solution and (B) the protein bound to an antibody. The increase in centroid above the all H sample uncorrected for back exchange is plotted in (C), corrected for back exchange in (D). In the absence of back exchange, we could expect nine deuterons for an all D sample. The measured centroid increase for the all D sample is 7 D's indicating 78% recovery. It can be seen that antibody binding apparently slows exchange for at least six positions by a factor of 1000 or more.

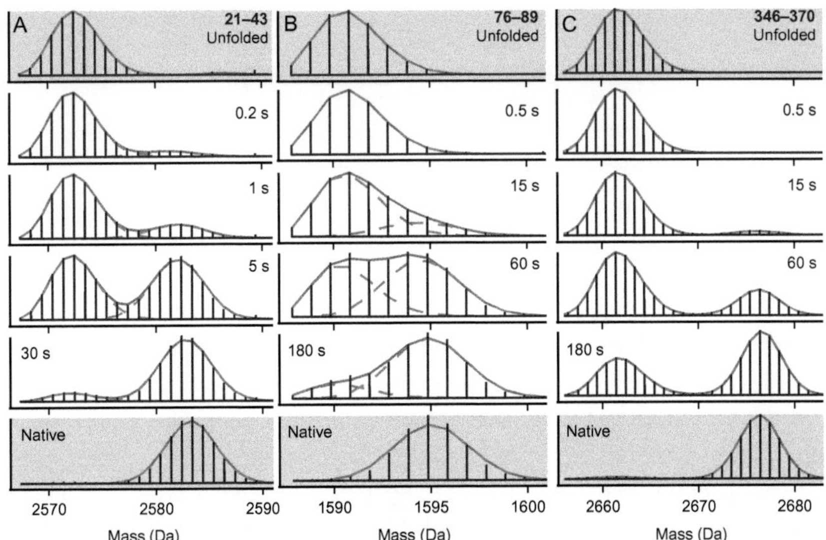

Figure 5 Example data from a pulsed labeling experiment following the folding of maltose binding protein (MBP) showing bimodal isotopic distributions. Three peptides are shown. Residues 21–43 (A), 76–89 (B), and 346–370 (C). Folding time before the labeling pulse is indicated. In this case, starting with all D unfolded protein, the heavy fraction reflects protein molecules that were protected from exchange during a 43-ms, pH 9 pulse. Different parts of the protein acquire protecting structure at different times after folding starts showing the progressive formation of the native structure. The data shown in gray are from unfolded and native control experiments and show that the protected and unprotected fractions behave like fully folded and unfolded protein. The fits are with binomials. A concerted EX1 exchange mechanism will yield similar data. *From Walters et al. (2013).*

an all D sample to the same proteolysis, HPLC, MS workflow used for experimental samples. The D recovery can then be calculated as:

$$\text{recovery} = \frac{\Delta m(\text{D, measured})}{\Delta m(\text{D, expected})}$$

where Δm(D, measured) is the measured shift for an "all D" sample. Δm(D, expected) is the expected shift for an all D sample in the absence of back exchange calculated as:

$$\Delta m(\text{D, expected}) = \text{frac}(\text{D})(N - P - 2)$$

here frac(D) is the fraction D present when the all D sample was prepared, N is the number of amino acids in the peptide, and P is the number of proline residues in the peptide beyond the first two amino acids.

A simple back exchange correction can be applied as:

$$\Delta m(\text{corrected}) = \frac{\Delta m(\text{measured})}{\text{recovery}}$$

It should be stressed that all back exchange corrections are at best approximate. The exchange correction suggested above implicitly assumes that labeled residues in the H/D experiment back exchange at the average exchange rate of the entire peptide as measured from the all D experiment. In reality, for a given peptide only some residues may have high D occupancy and may have exchange rates higher or lower than the average. A somewhat more sophisticated way of dealing with back exchange is discussed below in the context of moving toward residue-resolved analysis. In general, it is far better to reduce back exchange rather than try to correct for it. Even in experiments where back exchange seems like it can be ignored in analysis (ligand binding, EX1, pulsed labeling), the presence of back exchange still reduces the quality of the collected data, reducing the dynamic range available for resolving differences in exchange with and without ligand or for resolving folded and unfolded populations.

6.3 Toward Residue Resolution

For EX2 experiments, many groups have developed analysis methods that attempt to localize D occupancy beyond the peptide level. All such methods seek to use the information available from multiple sequentially overlapping peptides or fragments to localize D occupancy as much as possible given the cut sites. If nonergodic fragmentation methods are used or proteolysis provides sets of peptides with a common end differing by single residue steps at the other end, then simple subtraction can be used (Wang et al., 2013). We will not attempt to review the various published algorithms designed to deal with more general cases but will briefly describe the method we have developed, HDsite (Kan, Walters, Mayne, & Englander, 2013).

The HDsite algorithm takes advantage of not just the average, centroid, D occupancy for each peptide but also uses the information available in the shape of the isotopic distribution. Many different D occupancy patterns for a partially exchanged peptide can give rise to the same centroid value with different isotopic envelopes (Fig. 6). HDsite uses the set of D occupancies for each residue as adjustable parameters to simultaneously fit the measured isotopic envelopes for a set of overlapping peptides. While D sites cannot be localized beyond the information available from the cut sites in a set of

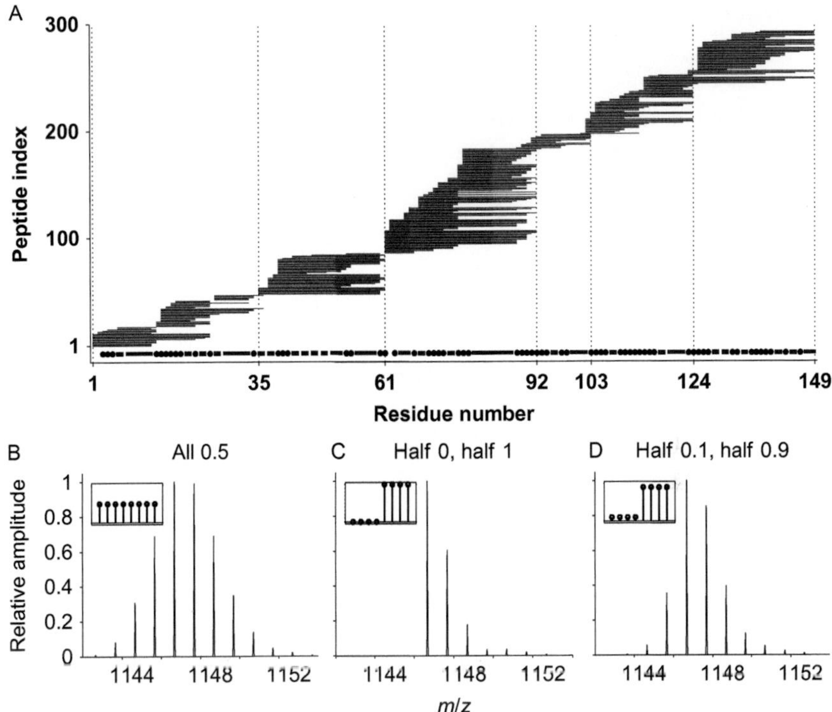

Figure 6 Toward residue resolution. (A) Bars represent the nearly 300 overlapping peptides obtained for staphylococcal nuclease. The staggered ends of overlapping peptides can in principle locate D occupancy to higher resolution than the peptide level. Given this set of peptides, the potential resolution is indicated by the dot and dashed line along the bottom. Small spots indicate residues that should be resolvable. Bars represent groups or two or more residues that always appear together and cannot be resolved. The HDsite algorithm can provide D occupancy information for each of these "switchable" residues but cannot assign D values to a particular residue in the group. (B–D) The basis for the HDsite algorithm. Simulated isotopic distributions for three different D distributions, all with the same centroid, yield different MS isotopic distributions. (B) All eight sites 50% D. (C) Four sites 100% D, four sites 0% D. (D) Four sites 10% D, four sites 90% D. *Adapted from Kan et al. (2013).*

overlapping peptides HDsite is able to resolve within "switchable" residues variable D occupancy levels without being able to assign which D level goes with which residue. The algorithm has been tested against NMR-measured rates for staphylococcal nuclease.

Back exchange can be dealt with as part of the fitting process. An effective back exchange time is measured for each peptide based on the measured centroid of an all D sample using the known free peptide rates under quench conditions for the residues of that peptide. For each iteration in the fitting procedure, the D occupancy levels are adjusted according to the effective back exchange time for that peptide. These adjusted D occupancies are used to calculate the envelope that is compared to experimental data.

We have since developed a version of the program that fits the entire exchange time course with a set of exchange rates, one for each residue rather than fitting D occupancy values for each time point. While computationally more intensive, this version improves the quality of the fit by reducing the number of adjustable parameters from one D occupancy per residue for each time point to one rate per residue for the whole time course.

6.4 Analysis of Pulsed Exchange Data

Data from pulsed labeling experiments, conformationally heterogeneous samples, and EX1 experiments are all expected to give rise to multimodal MS profiles. In these experiments, one is often more interested in the relative magnitude of the heavy and light populations than the degree of exchange (centroid) of each population. Analysis is greatly dependent on the quality of the MS data, particularly the separation of the labeled populations on the m/z axis. In favorable cases, the populations are separated well enough so that integration of the intensity of each is straightforward as is the calculation of the centroid of each. In other cases, separation is not sufficient to fully separate the isotopic envelopes of two populations. In this case, it is necessary to do some sort of curve fitting to extract the relative intensities and centroids of the labeled populations. This is not as simple a problem as it might seem because the shape of each isotopic distribution is not necessarily known. In practice, Gaussians or binomials can give good results except in cases where the overlap is particularly severe. The best quality data come from peptides with well-resolved protected and unprotected envelopes. Generally, this means peptides that are long enough so that ^{13}C envelopes are narrower than mass differences due to D but not so long that a given peptide spans more than one protein folding unit. We find that peptides of ~10–15 amino acids give good results. A high dynamic range of H/D during the pulse (ideally 0% D protein exchanging in 100% D or vice versa) and very low back exchange also improve separation of differentially labeled fractions.

7. CONCLUSION

HX methods continue to extend our understanding of protein structure, folding, and dynamics. HX methods are being applied to an increasing number of protein systems and sample handling techniques continue to advance. The use of MS has greatly extended the range of proteins amenable to HX analysis while vastly reducing the amount of protein required. As mass spectrometers with ever higher resolution and sensitivity become available combined with advances in data processing software, we can expect to extract more and more detailed information from HDX experiments.

REFERENCES

Bai, Y., Milne, J. S., Mayne, L., & Englander, S. W. (1993). Primary structure effects on peptide group hydrogen exchange. *Proteins: Structure, Function, and Genetics*, *17*, 75–86.

Bai, Y., Sosnick, T. R., Mayne, L., & Englander, S. W. (1995). Protein folding intermediates: Native-state hydrogen exchange. *Science*, *269*(5221), 192–197.

Bedard, S., Mayne, L. C., Peterson, R. W., Wand, A. J., & Englander, S. W. (2008). The foldon substructure of staphylococcal nuclease. *Journal of Molecular Biology*, *376*(4), 1142–1154. http://dx.doi.org/10.1016/j.jmb.2007.12.020.

Coales, S. J., Sook Yen, E., Lee, J. E., Ma, A., Morrow, J. A., & Hamuro, Y. (2010). Expansion of time window for mass spectrometric measurement of amide hydrogen/deuterium exchange reactions. *Rapid Communications in Mass Spectrometry*, *24*(24), 3585–3592. http://dx.doi.org/10.1002/rcm.4814.

Connelly, G. P., Bai, Y., Jeng, M. F., & Englander, S. W. (1993). Isotope effects in peptide group hydrogen exchange. *Proteins*, *17*(1), 87–92.

Englander, S. W. (2006). Hydrogen exchange and mass spectrometry: A historical perspective. *Journal of the American Society for Mass Spectrometry*, *17*(11), 1481–1489.

Englander, S. W., & Englander, J. J. (1972). Hydrogen tritium exchange methods. *Methods in Enzymology*, *26c*, 406–413.

Englander, S. W., & Kallenbach, N. R. (1983). Hydrogen exchange and structural dynamics of proteins and nucleic-acids. *Quarterly Reviews of Biophysics*, *16*, 521–655.

Hamuro, Y., Tomasso, J. C., & Coales, S. J. (2008). A simple test to detect hydrogen/deuterium scrambling during gas-phase peptide fragmentation. *Analytical Chemistry*, *80*(17), 6785–6790. http://dx.doi.org/10.1021/ac800645f.

Hu, W., Walters, B. T., Kan, Z. Y., Mayne, L., Rosen, L. E., Marqusee, S., et al. (2013). Stepwise protein folding at near amino acid resolution by hydrogen exchange and mass spectrometry. *Proceedings of the National Academy of Sciences of the United States of America*, *110*(19), 7684–7689. http://dx.doi.org/10.1073/pnas.1305887110.

Hvidt, A., & Nielsen, S. O. (1966). Hydrogen exchange in proteins. *Advances in Protein Chemistry*, *21*, 287–386.

Kaltashov, I. A., Bobst, C. E., Nguyen, S. N., & Wang, S. (2013). Emerging mass spectrometry-based approaches to probe protein-receptor interactions: Focus on overcoming physiological barriers. *Advanced Drug Delivery Reviews*, *65*(8), 1020–1030. http://dx.doi.org/10.1016/j.addr.2013.04.014.

Kan, Z. Y., Mayne, L., Sevaugan Chetty, P., & Englander, S. W. (2011). ExMS: Data analysis for HX-MS experiments. *Journal of the American Society for Mass Spectrometry*, *22*(11), 1906–1915. http://dx.doi.org/10.1007/s13361-011-0236-3.

Kan, Z. Y., Walters, B. T., Mayne, L., & Englander, S. W. (2013). Protein hydrogen exchange at residue resolution by proteolytic fragmentation mass spectrometry analysis. *Proceedings of the National Academy of Sciences of the United States of America*, *110*(41), 16438–16443. http://dx.doi.org/10.1073/pnas.1315532110.

Konermann, L., Tong, X., & Pan, Y. (2008). Protein structure and dynamics studied by mass spectrometry: H/D exchange, hydroxyl radical labeling, and related approaches. *Journal of Mass Spectrometry*, *43*(8), 1021–1036. http://dx.doi.org/10.1002/jms.1435.

Ling, J. M., Silva, L., Schriemer, D. C., & Schryvers, A. B. (2012). Hydrogen-deuterium exchange coupled to mass spectrometry to investigate ligand-receptor interactions. *Methods in Molecular Biology*, *799*, 237–252. http://dx.doi.org/10.1007/978-1-61779-346-2_15.

Mayne, L., Kan, Z. Y., Sevugan Chetty, P., Ricciuti, A., Walters, B. T., & Englander, S. W. (2011). Many overlapping peptides for protein hydrogen exchange experiments by the fragment separation-mass spectrometry method. *Journal of the American Society for Mass Spectrometry*, *22*(11), 1898–1905.

Mayne, L., Paterson, Y., Cerasoli, D., & Englander, S. W. (1992). Effect of antibody binding on protein motions studied by hydrogen-exchange labeling and two-dimensional NMR. *Biochemistry*, *31*(44), 10678–10685.

Mysling, S., Salbo, R., Ploug, M., & Jorgensen, T. J. (2014). Electrochemical reduction of disulfide-containing proteins for hydrogen/deuterium exchange monitored by mass spectrometry. *Analytical Chemistry*, *86*(1), 340–345. http://dx.doi.org/10.1021/ac403269a.

Pan, J., Han, J., Borchers, C. H., & Konermann, L. (2009). Hydrogen/deuterium exchange mass spectrometry with top-down electron capture dissociation for characterizing structural transitions of a 17 kDa protein. *Journal of the American Chemical Society*, *131*(35), 12801–12808. http://dx.doi.org/10.1021/ja904379w.

Pascal, B. D., Willis, S., Lauer, J. L., Landgraf, R. R., West, G. M., Marciano, D., et al. (2012). HDX workbench: Software for the analysis of H/D exchange MS data. *Journal of the American Society for Mass Spectrometry*, *23*(9), 1512–1521. http://dx.doi.org/10.1007/s13361-012-0419-6.

Paterson, Y., Englander, S. W., & Roder, H. (1990). An antibody binding site on cytochrome c defined by hydrogen exchange and two-dimensional NMR. *Science*, *249*(4970), 755–759.

Pirrone, G. F., Iacob, R. E., & Engen, J. R. (2015). Applications of hydrogen/deuterium exchange MS from 2012 to 2014. *Analytical Chemistry*, *87*(1), 99–118. http://dx.doi.org/10.1021/ac5040242.

Slysz, G. W., Baker, C. A., Bozsa, B. M., Dang, A., Percy, A. J., Bennett, M., et al. (2009). Hydra: Software for tailored processing of H/D exchange data from MS or tandem MS analyses. *BMC Bioinformatics*, *10*, 162. http://dx.doi.org/10.1186/1471-2105-10-162.

Walters, B. T., Mayne, L., Hinshaw, J. R., Sosnick, T. R., & Englander, S. W. (2013). Folding of a large protein at high structural resolution. *Proceedings of the National Academy of Sciences of the United States of America*, *110*(47), 18898–18903. http://dx.doi.org/10.1073/pnas.1319482110.

Walters, B. T., Ricciuti, A., Mayne, L., & Englander, S. W. (2012). Minimizing back exchange in the hydrogen exchange—Mass spectrometry experiment. *Journal of the American Society for Mass Spectrometry*, *23*, 2132–2139.

Wang, G., Abzalimov, R. R., Bobst, C. E., & Kaltashov, I. A. (2013). Conformer-specific characterization of nonnative protein states using hydrogen exchange and top-down mass spectrometry. *Proceedings of the National Academy of Sciences of the United States of America*, *110*(50), 20087–20092. http://dx.doi.org/10.1073/pnas.1315029110.

Wei, H., Mo, J., Tao, L., Russell, R. J., Tymiak, A. A., Chen, G., et al. (2014). Hydrogen/deuterium exchange mass spectrometry for probing higher order structure of protein therapeutics: Methodology and applications. *Drug Discovery Today*, *19*(1), 95–102. http://dx.doi.org/10.1016/j.drudis.2013.07.019.

Zehl, M., Rand, K. D., Jensen, O. N., & Jorgensen, T. J. D. (2008). Electron transfer dissociation facilitates the measurement of deuterium incorporation into selectively labeled peptides with single residue resolution. *Journal of the American Chemical Society*, *130*(51), 17453–17459.

Zhang, Z., & Smith, D. L. (1993). Determination of amide hydrogen exchange by mass spectrometry: A new tool for protein structure elucidation. *Protein Science*, *2*(4), 522–531.

CHAPTER FOURTEEN

Mapping Protein–Ligand Interactions with Proteolytic Fragmentation, Hydrogen/Deuterium Exchange-Mass Spectrometry

Elyssia S. Gallagher*[,1,2], Jeffrey W. Hudgens[†,‡,2]
*Institute for Bioscience and Biotechnology Research, National Institute of Standards and Technology and the University of Maryland, Rockville, Maryland, USA
†Institute for Bioscience and Biotechnology Research, Rockville, Maryland, USA
‡Biomolecular Measurement Division, National Institute of Standards and Technology, Rockville, Maryland, USA
[2]Corresponding authors: e-mail address: Elyssia_Gallagher@baylor.edu; jeffrey.hudgens@nist.gov

Contents

[1] Current address: Department of Chemistry and Biochemistry, Baylor University, Waco, Texas, USA.

Methods in Enzymology, Volume 566
ISSN 0076-6879
http://dx.doi.org/10.1016/bs.mie.2015.08.010

Abstract

Biological processes are the result of noncovalent, protein–ligand interactions, where the ligands range from small organic and inorganic molecules to lipids, nucleic acids, peptides, and proteins. Amide groups within proteins constantly exchange protons with water. When immersed in heavy water (D_2O), mass spectrometry (MS) can measure the change of mass associated with the hydrogen to deuterium exchange (HDX). Protein–ligand interactions modify the hydrogen exchange rates of amide protons, and the measurement of the amide exchange rates can provide rich information regarding the dynamical structure of the protein–ligand complex. This chapter describes a protocol for conducting bottom-up, continuous uptake, proteolytic fragmentation HDX-MS experiments that can help identify and map the interacting peptides of a protein–ligand interface. This tutorial outlines the fundamental theory governing hydrogen exchange; provides practical information regarding the preparation of protein samples and solutions; and describes the exchange reaction, reaction quenching, enzymatic digestion, chromatographic separation, and peptide analysis by MS. Tables list representative combinations of fluidic components used by HDX-MS researchers and summarize the available HDX-MS analysis software packages. Additionally, two HDX-MS case studies are used to illustrate protein–ligand interactions involving: (1) a continuous sequence of interacting residues and (2) a set of discontinuously numbered residues, residing spatially near each other.

1. INTRODUCTION

Biological processes are facilitated by noncovalent, protein–ligand interactions. Ligands range from small organic and inorganic molecules to lipids, nucleic acids, peptides, and proteins. As examples, in the respiration cycle reversible interactions of O_2 and carbamino compounds at distinct sites within hemoglobin enable transport of O_2 and CO_2 (Nelson & Cox, 2012). In blood plasma, progesterone, propranolol, and other small therapeutic molecules are carried by α1-acid glycoprotein (Fournier, Medjoubi-N, & Porquet, 2000; Huang & Hudgens, 2013). Additionally, a plethora of small molecule–protein (Kirschke, Goswami, Southworth, Griffin, & Agard,

2014), peptide–protein (Artimo et al., 2012; Das, Sharma, Kumar, Krishna, & Mathur, 2013), and protein–protein interactions mediate signal transduction across the plasma membrane and within living cells (Chen, Chattopadhyay, Bergen, Gadd, & Tannery, 2007).

Hydrogen/deuterium exchange-mass spectrometry (HDX-MS) is an important tool for characterizing protein–ligand interactions (Katta & Chait, 1991; Zhang & Smith, 1993). HDX-MS measures deuterium uptake of amide N–H groups along the protein backbone. This information is useful for identifying portions of the protein engaged in binding and the dynamics of the protein–ligand interface. While once quite challenging and labor intensive, advances in instrumentation and software have enabled research groups to rapidly complete insightful, conclusive HDX-MS studies. As a result, more laboratories are conducting analyses that utilize HDX-MS methods (Pirrone, Iacob, & Engen, 2015).

This chapter describes protocols for conducting studies with continuous uptake, proteolytic fragmentation HDX-MS, which has demonstrated utility for identifying the interacting peptides in protein–ligand interactions. The power of this HDX-MS protocol for mapping protein–protein interactions is demonstrated in a study by Malito et al. (2013), who used multiple techniques to map the epitope of complement factor H-binding protein (fHbp) against a monoclonal antibody (mAb 12C1). Malito et al. studied the interaction of fHbp with bactericidal mAb 12C1 with diverse mapping techniques, including peptide arrays, phage display, X-ray crystallography, and HDX-MS (Fig. 1A–D). The map obtained with each technique was compared to the epitope of fHbp (Fig. 1E), determined from modeling the fHbp:fH crystal structure (Schneider et al., 2009). Their HDX-MS study accounted for nearly the entire ≈ 1000 Å^2 contact interface between mAb 12C1 and fHbp. They concluded that "…hydrogen/deuterium exchange-mass spectrometry was the most effective method to rapidly supply near-complete information about epitope structure" (Malito et al., 2013).

2. H/D EXCHANGE THEORY

2.1 H/D Exchange

Within proteins, hydrogen atoms manifesting ionic character exchange with the labile protons of the surrounding solvent. When the solvent contains an isotope of hydrogen (deuterium or tritium), exchange can be used to

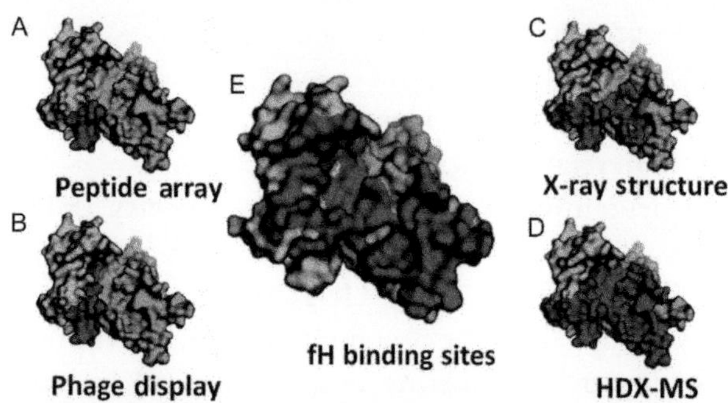

Figure 1 Binding maps of the fHbp antigenic epitope plotted by different techniques (Malito et al., 2013). The N- and C-terminal domains of fHbp are presented in green and cyan respectively. (A) Red denotes the fHbp epitope found by synthetic peptide arrays. (B) Red denotes the fHbp epitope found by phage display. (C) Red denotes the 23 residues of fHbp contacted by the Fab of mAb 12C1, as determined from the X-ray structure of the fHbp:Fab 12C1 complex. (D) Red denotes the peptide regions of fHbp contacted by the Fab of mAb 12C1 within the fHbp:12C1 complex, as determined by HDX-MS measurements. (E) Red denotes 52 fHbp surface residues bound with surface residues of domains 6 and 7 in human fH, as prepared by Malito et al. by modeling the crystal structure of fHbp:fH complex (Schneider et al., 2009). *Adapted images used with the permission of Malito et al. (2013).* (See the color plate.)

characterize protein structure and dynamics via an analytical method that detects a difference between the isotopes. Since hydrogen and deuterium have masses of 1.008 and 2.014 Da, respectively, MS can be utilized to monitor exchange by detecting the mass increase of the protein.

Each hydrogen bond type has a distinct range of H/D exchange rates (Fig. 2). For instance, covalently bound C–H groups do not undergo solvent-mediated exchange. Covalently bound hydrogen atoms in amine, indole, and hydroxyl functional groups of amino acid side chains and the N-terminal amide group of the protein backbone undergo rapid H/D exchange with half-lives on the order of $t_{1/2} \approx 0.01$–1 ms (Bai, Milne, Mayne, & Englander, 1993; Liepinsh & Otting, 1996). The other amides in the protein backbone exhibit exchange rates with half-lives ranging from $t_{1/2} = 5$ s to 60 days. These rates can be measured using proteolytic fragmentation HDX-MS. Since a backbone amide hydrogen is present at every amino acid (except proline) throughout the protein, monitoring the rates of exchange associated with these functional groups allows the entire protein structure to be sampled. In addition to the functional group, H/D exchange rates vary with solution pH, temperature, solvent accessibility, and protein

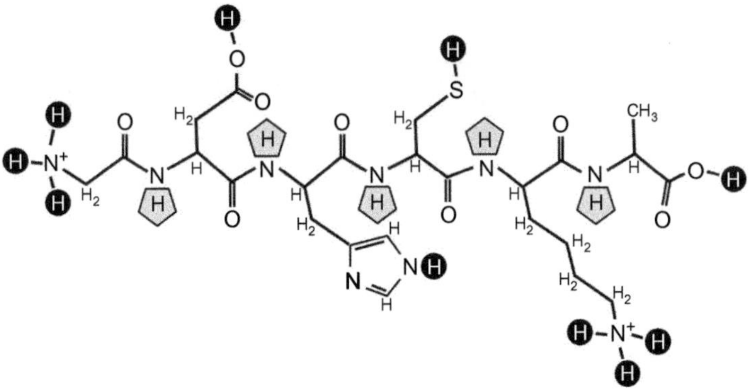

Gly – Asp – His – Cys – Lys – Ala

Figure 2 Hydrogens classified by exchangeability in an example peptide: glycine-aspartic acid-histidine-cysteine-lysine-alanine. HDX-MS measures the backbone amide hydrogens shown in green (light gray in the print version) pentagons. Hydrogens that are not monitored with HDX-MS due to their very rapid exchange rates are shown in red (dark gray in the print version) circles. The numerous hydrogens that do not exchange due to their covalent binding with carbon are designated by black letters.

structure (Balasubramaniam & Komives, 2013; Englander & Kallenbach, 1984; Mandell, Falick, & Komives, 1998).

2.2 Acid and Base Catalysis

A completely unfolded protein has essentially no protection against proton exchange between the solvent and each amide group of the protein backbone, that is, each amide is in the exchange-competent state, $N\text{–}H_{open}$. Proton exchange between an unprotected amide group and aqueous solvent is a chemical reaction catalyzed by acid, base, or water. Accordingly, the exchange rate (k_{ch}) is written (Bai et al., 1993; Englander, 2006; Englander & Kallenbach, 1984):

$$k_{ch} = k_{acid}^{int}[H_3O^+] + k_{base}^{int}[OH^-] + k_{water}^{int}[H_2O] \quad (1)$$

The k_{acid}^{int} and k_{base}^{int} are rate coefficients for the acid- and base-catalyzed exchange, and k_{water}^{int} is the intrinsic rate coefficient for the water-catalyzed exchange reaction (Englander, 2006). At physiological pH (7.0–8.0) the base-catalyzed mechanism is dominant. The acid-catalyzed mechanism becomes dominant as pH decreases below 3. Between pH 2.0 and 3.0, the exchange rate, k_{ch}, attains a minimum with the acid- and base-catalyzed mechanisms contributing equally to exchange, and the water-catalyzed mechanism being dominant. At the pH corresponding to minimum k_{ch}, the exchange rate for backbone amides is two orders of magnitude less than any other functional group (except for the guanidinium group of arginine). The small contribution associated with k_{water}^{int} is generally ignored, as it accounts for ≈38% of k_{ch} at the pH minimum and becomes negligible as the acid or the base rates increase (Bai et al., 1993; Gregory, Crabo, Percy, & Rosenberg, 1983). Of great consequence to the HDX-MS experiment, the exchange rate for an average backbone amide is reduced by 10,000-fold at pH 2.5 compared to pH 7.0 (Dempsey, 2001).

2.3 Temperature

The chemical exchange rate, k_{ch}, shows strong temperature dependence as described by the Arrhenius equation. A rate decrease of ≈10-fold is observed when the temperature is reduced from 20 to 0 °C (Englander, 2006). Decreasing the temperature lowers the kinetic energy of the molecules in solution, reducing the rate of diffusion and leading to fewer collisions and reactions between the protein, catalyst, and protium or deuterium donor. When the protein is under quench conditions (pH 2.5 and 0 °C), the average half-life for H/D exchange of a backbone amide hydrogen is between 30 and 120 min (Engen & Smith, 2000; Morgan & Engen, 2009; Thevenon-Emeric, Kozlowski, Zhang, & Smith, 1992), providing sufficient time for rapid digestion and separation of the resulting peptides prior to MS. To ensure highly reproducible analyses, it is necessary to precisely control solution pH and temperature at every step in the HDX-MS experiment, including protein equilibration, exchange, quench, and processing prior to MS.

2.4 Protein Structure

The exchange rate of a backbone amide N–H is dependent on the primary, secondary, tertiary, and quaternary protein structure. For an unfolded protein or peptide, the identities of the neighboring amino acids affect the magnitude of k_{ch}. Englander (Bai et al., 1993; Connelly, Bai, Jeng, & Englander, 1993; Englander, 2015) provides tables that can predict k_{ch} for unfolded peptides based on pH, temperature, and neighboring amino acids.

The Linderstrøm-Lang model of hydrogen exchange describes exchange by localized unfolding. Amides can reside in either exchange-incompetent ($N\text{-}H_{closed}$) or exchange-competent ($N\text{-}H_{open}$) states (Hvidt & Linderstrøm-Lang, 1954). Any hydrogen-bonded proton is exchange-incompetent because it is sterically inaccessible to exchange (Englander, Mayne, Bai, & Sosnick, 1997). Generally, secondary structural features (e.g., α-helices, β-sheets) in a folded protein are comprised of amides residing in exchange-incompetent states since they are stabilized through intramolecular hydrogen bonds. The Linderstrøm-Lang model posits that an amide N–H group becomes exchange-competent through unspecified transient events. The Linderstrøm-Lang model is defined kinetically by the structural opening and reclosing rates, k_{op} and k_{cl}, of an amide:

$$N-H_{closed} \underset{k_{cl}}{\overset{k_{op}}{\rightleftharpoons}} N-H_{open} \xrightarrow[D_2O]{k_{ch}} \text{exchanged} \tag{2}$$

The exchange process proceeds during transient, exchange-competent intervals when the hydrogen bond donor and acceptor are not bound ($N\text{-}H_{open}$) and are solvent accessible. The Linderstrøm-Lang model attributes the variation of k_{op} and k_{cl} to the presence and absence of secondary structure, and these dynamical variables connect protein structure to thermodynamic properties (Englander et al., 1997; Hvidt & Linderstrøm-Lang, 1954; Hvidt & Nielsen, 1966). Protection against amide N–H exchange, conferred by secondary structure through variation of k_{op} and k_{cl}, can reduce the H/D exchange rate at amide groups by as much as 10^8 (Fleming & Rose, 2005; Skinner, Lim, Bédard, Black, & Englander, 2012).

When a hydrogen bond is broken, a kinetic competition ensues between the rates of chemical exchange, k_{ch}, and structural reclosing, k_{cl} (Baldwin, 2011; Englander & Kallenbach, 1984; Hvidt & Nielsen, 1966). Under most physiological conditions, reclosing is faster than chemical exchange ($k_{cl} \gg k_{ch}$), requiring the same hydrogen bond to break multiple times before a successful exchange event. The hydrogen exchange rate then depends on the fraction of time the exchange-competent (open) form exists and therefore the structural opening equilibrium constant (K_{op}). The kinetic expression for this EX2 condition is written with a preequilibrium step:

$$k_{HDX} = \frac{k_{op}k_{ch}}{\left(k_{op} + k_{ch} + k_{cl}\right)} \approx K_{op}k_{ch} \tag{3}$$

where k_{HDX} is the observed exchange rate. The equilibrium term, K_{op}, represents the fraction of time that the amide is exchange-competent. The approximation shown in Eq. (3) holds for any significant level of structural protection ($k_{op}/k_{cl} \approx K_{op} < 1$). The Gibb's energy change, $\Delta_{op}G^{\circ}$, is the opening energy for exchange at amide N–H:

$$\Delta_{\text{op}} G^{\circ} = -RT \ln K_{\text{op}} = -RT \ln \left(\frac{k_{\text{op}}}{k_{\text{cl}}} \right) \quad (4)$$

As the chemical rate is made progressively faster, for example, by raising the pH, the amide exchange rate will increase linearly with the solvent catalyst concentration in the EX2 range. Ultimately, when $k_{\text{ch}} > k_{\text{cl}}$, the HX exchange rate will reach the asymptotic EX1 limit at which the deuteration rate coefficient, k_{HDX}, becomes

$$k_{\text{HDX}} = k_{\text{op}} \quad (5)$$

EX1 behavior can be documented by its insensitivity to pH or by observation of correlated D-uptake behavior, where multiple amides simultaneously exchange and exhibit the same EX1 exchange rate (Englander, 2006).

Many studies have been devoted to determining how protein structure governs amide hydrogen exchange patterns. These studies have explored solvent and catalyst penetration models. Analysis of model-based expectations for hydrogen exchange indicate that local unfolding pathways manifest activation energies of $E_a = 70$–145 kJ/mol and penetration mechanisms manifest $E_a > 165$ kJ/mol (Englander & Kallenbach, 1984). Studies of protein amide N–H exchange dynamics have successfully accounted for HDX-MS and HDX-NMR data with local unfolding and foldon models (Englander, Mayne, & Krishna, 2007; Hu et al., 2013; Krishna, Hoang, Lin, & Englander, 2004; Krishna, Maity, Rumbley, Lin, & Englander, 2006).

2.5 Effects of Ligand-Binding Interactions upon H/D Exchange Rates

Most commonly, protein–ligand-binding interactions reduce the observed deuterium uptake at the binding site; however, such binding is sometimes accompanied by allosteric effects. Sowole and Konermann offered scenarios where H/D exchange rates are reduced (type 1), unaffected (type 0), or enhanced (type 2) by ligand binding (Sowole & Konermann, 2014). Accounting for these scenarios requires no modification of the Linderstrøm-Lang model.

The complex NL formed by the binding of protein N and ligand L is characterized by its dissociation constant, K_d, defined as:

$$K_d = \frac{\{N\}\{L\}}{\{NL\}} \quad (6)$$

where curly braces denote the chemical potential of each species in solution, as referenced to a standard state (e.g., 1 mol/L) (Gilson, Given, Bush, & McCammon, 1997). (Formally, K_d is unitless; however, since activity coefficients ≈ 1 in dilute solution, the effects of reference states are nil, allowing direct use of species concentrations in Eq. (6). Hence, K_d values are customarily listed in concentration units.) Formation of complex NL is accompanied by its stabilization energy:

$$\Delta_d G^\circ = -RT \ln K_d \tag{7}$$

The magnitude of $\Delta_d G^\circ$ is governed by an interplay of entropic and enthalpic factors originating from the protein, ligand, and solvent (Bissantz, Kuhn, & Stahl, 2010; Tzeng & Kalodimos, 2009). Contributions to $\Delta_d G^\circ$ may originate from any structural element of the protein. Consequently, structural reorganization in the protein arising from complexation can affect N–H groups local to the binding interface, and it can allosterically affect remote N–H groups. With knowledge of the stabilization energy of NL, $\Delta_d G^\circ$, statistical mechanics can be used to reveal the spectrum of effects that complexation may have on amide hydrogen exchange rate coefficients.

In the statistical mechanics terminology of Sowole and Konermann a backbone amide N-H exchanging under EX2 conditions has the H/D exchange rate coefficient (Sowole & Konermann, 2014):

$$k_{\text{HDX}} = p_{\text{op}} k_{\text{ch}} \tag{8}$$

where p_{op} (as with K_{op}) is the fraction of time that the N–H bond spends in the exchange-competent (N–H_{open}) state (ref. Eq. 2) (Englander et al., 2007; Wooll, Wrabl, & Hilser, 2000). From Boltzmann statistics, the p_{op} coefficient is computed by

$$p_{\text{op}} = \frac{e^{-\Delta_{\text{op}} G^\circ / RT}}{Z} \tag{9}$$

where $\Delta_{\text{op}} G^\circ$ is the opening energy and Z is the partition function, defined as $Z = 1 + \exp(-\Delta_{\text{op}} G^\circ / RT)$, describing the manifold of all states (ground and excited) for the protein or protein complex. Stable proteins conform to the condition, $k_{\text{cl}} \gg k_{\text{op}}$, such that $\Delta_{\text{op}} G^\circ \gg 0$ and $Z \approx 1$. For convenience of illustration, Gibbs energy in the following text is expressed in RT units, $\Delta j = RT \Delta_{\text{op}} G^\circ$, so that

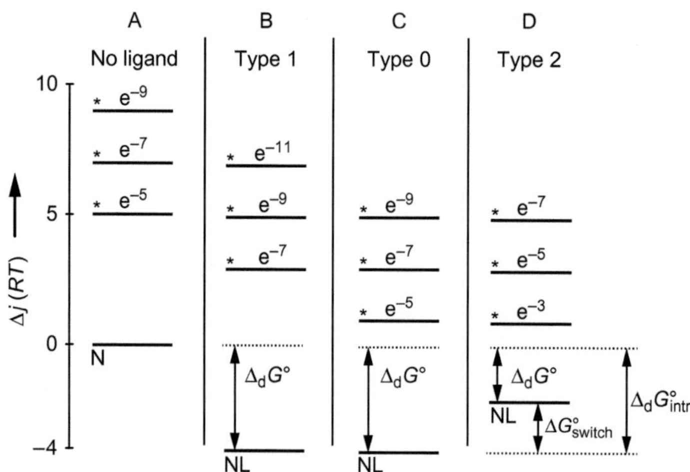

Figure 3 Free energy level diagram of protein N and protein–ligand complex NL, where energy is plotted in *RT* units. All amide N–H's of the lowest energy (ground) state are exchange-incompetent (i.e., $N–H_{closed}$). This simplified example depicts only three of a large ensemble of excited states. Higher energy states, denoted by asterisks, are partially unfolded states manifesting one or more exchange-competent amide N–H's (i.e., $N–H_{open}$). Adjacent to each asterisk, the Boltzmann occupancy is noted, and this occupancy determines k_{HDX} of N. (A) The uncomplexed protein molecule. Excited states are presumed to be $\Delta j = 5RT$, $7RT$, and $9RT$ above the ground state. (B) Type 1 scenario, where ligand binding lowers the free energy of the ground state by $\Delta_d G° (\equiv -4RT)$, and the excited-state energies are lowered (arbitrarily) by -2*RT*. The net stabilization increases the energy gap between the ground and excited states, which reduces populations of the excited states relative to those in (A); thus, k_{HDX} decreases. (C) Type 0 scenario, where the ground state and excited-state energies are lowered equally by $\Delta_d G° (\equiv -4RT)$. Formation of the protein–ligand complex NL does not alter the Boltzmann occupancies among the manifold of excited states; thus, k_{HDX} is unchanged. (D) Type 2 scenario, where the intrinsic protein–ligand binding affinity, $\Delta_d G^{\circ}_{intr}$, is countered by the expense of a conformational change in structure, $\Delta_d G^{\circ}_{switch}$. The energy gap between the ground and excited states is diminished, which increases the Boltzmann occupancies among the excited-states relative to those in (A); thus, k_{HDX} increases. The overall binding affinity, K_d, is determined by two competing contributions (Eqs. 7 and 11). *Adapted figure used by permission from Sowole and Konermann (2014).*

$$k_{HDX} = e^{-\Delta j} k_{ch} \tag{10}$$

Each complex NL has an energy state manifold, comprising an exchange-incompetent ground state (i.e., $N–H_{closed}$) and a manifold of sparsely populated, higher energy states (Fig. 3). The higher energy states represent the multitude of subglobal states experiencing partial or full unfolding through foldon fluctuations and local dynamics (Englander et al., 2007; Gu, Zitzewitz, & Matthews, 2007; Kaltashov, Bobst, & Abzalimov, 2013; Weinkam, Zimmermann, Romesberg, & Wolynes, 2010). Each higher energy state has one or more amide N–H's that is exchange-competent. In Fig. 3, each higher energy level is labeled with its Boltzmann occupancy, $e^{-\Delta j}$. The stable uncomplexed protein molecule N (Fig. 3A) has excited states, which are arbitrarily set to $\Delta j = 5RT$, $7RT$, and $9RT$ above the ground state for this discussion. The formation of the protein–ligand complex NL can result in three scenarios.

In the type 1 scenario (Fig. 3B), formation of the NL complex lowers the ground state energy by $\Delta_d G^\circ$. This lowering is expressed as $\Delta_d G^\circ$ (which is arbitrarily set to -4RT). Formation of the NL complex may also lower the energy of the excited-state manifold, which is shown (arbitrarily) by a change of -2RT. As a result, the net energy separation between the exchange-incompetent ground state and lowest exchange-competent higher state is greater for NL than for N (ref. Fig. 3A). Consequently, the thermal populations of the excited states in NL are diminished relative to those of the uncomplexed protein N; thus, the k_{HDX} observed in NL is also diminished. Predominately, the HDX-MS literature comprises observations of NL complexes exhibiting type 1 behavior.

The type 0 scenario (Fig. 3C) is characterized by energy lowering of the ground and excited states by the binding affinity, $\Delta_d G^\circ$. Since the relative energies among the ground and excited states are unchanged, excited-state fractional occupancies are also unchanged; thus, k_{HDX} remains unchanged. Such behavior is difficult to detect. Sowole and Konermann (2014) suggest that the type 0 scenario is implicated in the study of some mAb complexes (Pan, Salas-Solano, & Valliere-Douglass, 2014).

The noncanonical type 2 scenario results in enhanced k_{HDX}. Here, the protein possesses a binding site that has a large intrinsic affinity, $\Delta_d G^\circ_{intr}$ ($\equiv$-4RT arbitrarily). However, the ligand is accommodated in the NL ground state only after an unfavorable structural change has taken place. The energy penalty of this conformational switching event is termed ΔG°_{switch}. The overall binding affinity in this scenario is the sum of two opposing contributions

$$\Delta_d G^\circ = \left|\Delta_d G^\circ_{intr}\right| - \left|\Delta G^\circ_{switch}\right| \tag{11}$$

The NL ground state remains thermodynamically favored for $\Delta_d G^\circ > 0$. Ligand-induced distortion of the protein implies that NL conformers with exchange-competent, open N-H sites are more readily accessible than in the uncomplexed N. Thus, ligand binding lowers Δj values, resulting in increased HDX rates throughout the protein (Eq. 4, Fig. 3D). Sowole and Konermann have demonstrated that oxygenation of myoglobin results in a NL complex that exhibits type 2 behavior at residues displaced from the binding site (Sowole & Konermann, 2014). The observed behavior is consistent with cooperative T→R switching, where part of the intrinsic O_2 binding energy ($\Delta_d G^{\circ}_{intr}$) is reinvested for destabilization of the ground state ($\Delta G^{\circ}_{switch}$). This destabilization increases the Boltzmann occupancy of unfolded conformers, thereby enhancing HDX rates.

3. THE HDX-MS EXPERIMENT

3.1 Synopsis

As documented above, H/D exchange is dependent on any property affecting the molecular Gibb's energy, which most commonly spans pH, temperature, solvent accessibility, and structure. Thus, the pH and temperature of HDX experiments are optimized for measuring the forward exchange rates with minimal perturbing back exchange. Further, since these variables are controlled, the collected data can be used to infer the structural dynamics and solvent accessibility of each backbone amide within the protein.

The continuous-labeling, proteolytic fragmentation HDX-MS experiment, as depicted in Fig. 4, begins when a protein sample is diluted in D_2O, initiating exchange at exposed amides and side chains. After exposure in D_2O for selected time intervals (t_{HDX}), aliquots of deuterated protein are extracted and diluted into a quench buffer, which is maintained at pH 2.5 (H_2O) and 0 °C. These conditions minimize amide exchange rates, preventing loss of deuterium at labeled backbone amides. After the solution is quenched, the protein is digested into peptides by an acidic protease, and the resulting peptides are trapped on a C18 guard column where buffer exchange removes salt. Using reverse-phase (RP) chromatography, the peptides elute from an analytical column and are analyzed by electrospray ionization mass spectrometry (ESI-MS). The quench conditions (pH 2.5, ≈1 °C) are maintained throughout the digestion, buffer exchange, and chromatographic stages to minimize back exchange loss of deuterium labels, which proceeds in H_2O solvent via Eq. (2). From the MS data, the average mass change of each peptide is calculated and correlated to the presence of deuterium. Since side chains have more rapid exchange rates than backbone amides at pH 2.5 and 0 °C, the deuterium labels at these positions are

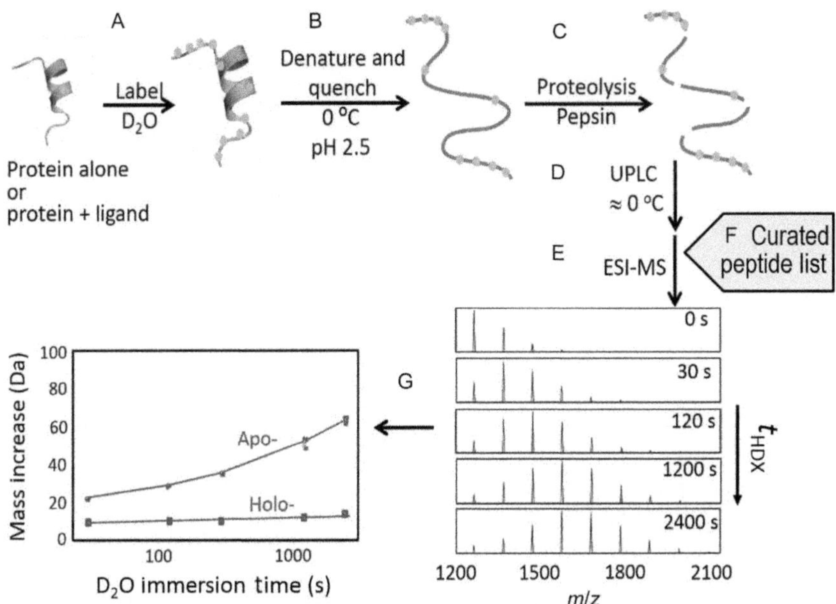

Figure 4 Schematic of the HDX-MS experiment showing the key steps: (A) deuterium labeling of the protein sample (apo) or protein + ligand samples (holo) by immersion in D_2O for selected periods (e.g., $t_{HDX} = 0$, 30, 120, 1200, 2400 s), (B) quenching and (optional) denaturing of the sample by immersion in H_2O (pH 2.5, ≈0 °C) solution, (C) proteolytic digestion of the sample into peptides, (D) chromatographic separation of the peptides on a UPLC apparatus, (E) analysis of the sample with an ESI-MS, (F) identification of proteolytic peptide ions from previously identified curated list, (G) determination of the centroid of each peptide for each t_{HDX} and the mass increase of each apo- and holo-peptide versus D_2O immersion time.

equilibrated back to the natural protic isotope abundance during digestion and separation. This simplifies data analyses since each amino acid (except proline and arginine) can have a maximum of one deuterium label.

Studies of protein–ligand interactions are conducted with protein in the presence and absence of ligand. The protein–ligand interaction is characterized by examining the differential deuterium uptake of the peptides observed from the holo-protein (protein–ligand complex) and the respective apo-protein (protein alone).

3.2 Protein Preparation

To accurately map protein structure and binding interactions, the native protein conformation must be maintained in solution; however, these native conditions cannot interfere with the RP peptide separation or ionization.

Most current HDX-MS work flows utilize online enzymatic digestion after which the peptides are trapped on a C18 column (Fig. 4). This set-up allows for physiological, high salt buffers to be used for protein preparation, equilibration, and exchange since the salts are washed away prior to gradient elution and ESI-MS. Compounds that interact with the stationary phase, such as surfactant or phospholipids, should be removed prior to online analyses. Surfactants can be retained through interactions with the C18 stationary phase of the trap column, causing these molecules to accumulate and elute during the peptide separation at high concentrations of acetonitrile (CH_3CN), leading to fouling of the analytical column and ion suppression. Though detergents can affect the separation and ionization of peptides, HDX-MS analyses have been performed in the presence of 10% glycerol or low concentrations of dodecyl maltoside, a nonionic detergent.

The ideal protein concentration for an HDX experiment is determined by the sensitivity of the mass spectrometer. During an experiment, the protein is diluted into exchange buffer (D_2O, pD 7.4) and quench buffer (H_2O, pH 2.5). However, following enzymatic digestion, the peptides are concentrated on a trap column. This step allows for low concentrations of protein to be utilized in the exchange reaction since they can be detected following concentration and peptide separation. Alternatively, highly concentrated proteins can be problematic if they are not fully eluted from the proteolytic, trap, or RP column within a single run. Peptides retained on the columns and eluted during later runs experience greater back exchange, which can complicate data analysis. For current instrumentation, 10–25 pmol of protein per time point is sufficient to detect the peptides following labeling, digestion, and separation without significant peptide carry over between runs.

To study protein–ligand binding, the protein and ligand should be present at concentrations that promote interaction after the protein–ligand mixture is diluted into exchange buffer (D_2O). A comparison of the holo-complex to the apo-protein is best achieved when $\approx$100% of the protein is bound to ligand in the holo-state and a single protein population is present. To estimate the optimum ratio and concentrations of protein and ligand, the dissociation constant (ref. Eq. 6) for the interaction must be known. For computational convenience all activity coefficients are assumed equal to 1; thus, K_d has units of concentration. A convenient equation for estimating the optimum ratio is given by Mandell et al.:

$$\%\text{bound} = \frac{L + P + K_d + \sqrt{(L + P + K_d)^2 - 4LP}}{2P} \tag{12}$$

where *P* and *L* are the total quantities of the protein and ligand, respectively (Mandell et al., 1998). For tightly binding interactions ($K_d \approx 1$ nmol/L), the protein and ligand can be mixed at a molar ratio near 1:1. When $K_d = 10–100$ nmol/L, the ligand must be present at a molar excess to approach 100% binding to the protein. When the ligand is a peptide or protein, there may be a maximum amount of ligand that can be used in the experiment. A protein or peptide ligand will be digested, trapped, separated, ionized, and detected like the other protein. If the ligand is at a high molar excess, the peptides of the ligand may coelute with those of the other protein and suppress ion signal from the less concentrated protein.

3.3 Exchange Reaction

The buffer used for exchange should match the solution conditions (buffering agent, additives, acidity) of the equilibration buffer and be prepared with D_2O solvent. The measured pD of this buffer must be corrected for an artifact associated with monitoring the deuterium isotope with a glass electrode. Equation (13) describes the correction for pD when pD_{read} is the pD read directly from the pH meter (Glasoe & Long, 1960).

$$pD_{corrected} = pD_{read} + 0.4 \tag{13}$$

To accurately measure the exchange rate, the amount of hydrogen and deuterium present in the exchange mixture should be known. HDX is often initiated by diluting protein, prepared in equilibration buffer (H_2O), into exchange buffer (D_2O) with D_2O at 80–95% of the final solution volume. This D_2O excess decreases the probability that labeled sites will be replaced by hydrogen if a second exchange event occurs at the same position during the exchange time. Alternatively, if diluting the protein sample will decrease the protein concentration below that which can be effectively detected by MS, protein can be introduced to D_2O using a small gel filtration spin column; however, this method requires that the shortest exchange time point be no less than 1–2 min (Engen & Smith, 2000).

The exchange buffer should be prepared using high purity D_2O (99%) and the pD should be adjusted with deuterated acid (deuterium chloride solution, DCl) or base (sodium deuteroxide solution, NaOD). Alternatively, a solution of known pD can be made using an acid and conjugate base pair. The dissociation of a monoprotic acid is shown in Eq. (14) where HA and A^- are the acid and conjugate base, respectively.

$$HA + H_2O \rightleftharpoons A^- + H_3O^+ \tag{14}$$

The Henderson–Hasselbalch equation (Eq. 15) can be used to calculate the ratio of acid and conjugate base to combine to make a buffer with a particular pH or pD (Harris, 2003). In the equation below, [HA] and [A$^-$] are the molar (mol/L) concentrations of the acid and conjugate base, respectively.

$$\mathrm{pH} = \mathrm{p}K_\mathrm{a} + \log\frac{[\mathrm{A}^-]}{[\mathrm{HA}]} \tag{15}$$

Since the volume of the acid and conjugate base are the same in the buffer, the volume component of the concentrations cancel and the equation can be rewritten to determine the mole ratio of the conjugate base and acid needed to generate a buffer with a particular pH or pD (Eq. 16).

$$\frac{\mathrm{mol\ A}^-}{\mathrm{mol\ HA}} = 10^{(\mathrm{pH}-\mathrm{p}K_\mathrm{a})} \tag{16}$$

If a diprotic or polyprotic acid is needed to make the buffer, the same general equation can be used, but the pK_a should be replaced by the appropriate dissociation constant (pK_{a1}, pK_{a2}, etc.) and the acid and conjugate base should represent the species associated with that equilibrium.

3.4 Automated Versus Manual HDX-MS

During the mid-2000s, the laborious task of manually conducting continuous-labeling HDX-MS experiments was automated (Chalmers et al., 2006). By 2010, automated commercial instruments became available from the Waters Corp. and ThermoFisher Scientific. The emergence of these instruments has increased the use of HDX-MS methods within the biopharmaceutical industry for studying the effect of process changes on biotherapeutic properties and for characterizing protein–ligand interactions (Chow & Ju, 2013; Houde, Arndt, Domeier, Berkowitz, & Engen, 2009; Houde, Berkowitz, & Engen, 2011; Iacob et al., 2013; Kaltashov, Bobst, Abzalimov, Berkowitz, & Houde, 2010). The automated instruments comprise a liquid handling robotic platform that immerses a sample in D_2O, waits for the exchange reaction time to expire, injects the labeled sample into a quench buffer, and then injects the quenched sample into a series of valves for digestion, solvent exchange, and reverse-phase, ultra-performance liquid chromatography (RP-UPLC). Due to the precise timing of the exchange, quench, digestion, and peptide separation steps within temperature-controlled environments, the results obtained with automated sample handing platforms can exhibit excellent repeatability and intermediate measurement precision (Chalmers, Busby, Pascal, Southern, & Griffin, 2007; Hudgens, Huang, & D'Ambro, 2016). Robots can interweave experiments of varying length (t_{HDX}), which shortens the total duration of an experiment containing multiple exchange time points. Each HDX-MS apparatus comes with specialized processing and analysis software.

Though automated platforms do exist, it is too early to declare dead the age of manual measurements. When armed with appropriate temperature and timing controls, investigators can manually perform HDX-MS experiments with highly reproducible results. For "one-off" investigations designed to augment a larger project, manual HDX-MS may be the most effective, economical strategy for adding compelling results to a biochemical study.

3.5 Reaction Quenching

After the exposure time of the sample to D_2O has expired, the exchange reaction is quenched by decreasing the solution to 0 °C and pH 2.5. From this step forward, the labeled protein sample is exposed to a high concentration of H_2O; thus, the quench conditions are necessary to minimize back exchange of labeled backbone amides.

Solution pH is best controlled using a buffer. An acid and conjugate base pair have a high buffering capacity when the solution pH is ±1 pH unit from the pK_a. Phosphate is a triprotic acid with three dissociation constants of pK_a=2.15, 7.20, and 12.15 (Harris, 2003). Phosphate is regularly used for buffering the quench buffer at pH 2.5 and preparing the equilibration and exchange buffers at pH 7.4 and pD 7.4, respectively. The buffering capacity of the quench buffer must be sufficient to decrease the pH of the exchange reaction to pH 2.5, often requiring the concentration of the buffering species to be 10 times higher in the quench solution than in the equilibration and exchange solutions. Quenching can also be accomplished by adding strong acid, e.g., formic acid, to the exchange reaction. However, when the equilibration/exchange buffer has a single dissociation constant with a pK_a near physiological pH, following quenching, the solution will have a pH significantly below the pK_a and may not have sufficient buffering capacity to maintain a reproducible pH≈2.5, resulting in increased or variable back exchange in replicate analyses.

After quenching, samples can be analyzed immediately or rapidly frozen in liquid nitrogen and stored at −80 °C for a day. Sample storage is more common when using a manual HDX-MS platform. To minimize loss of deuterium label, samples should be frozen quickly after quenching and thawed rapidly (<30 s) prior to digestion.

3.6 Enzymatic Digestion

Following quenching of the exchange reaction, the protein sample is rapidly digested into peptides. To achieve higher digestion efficiencies at this rapid time scale, chaotropes and reducing agents are often added to the quench solution to aid in protein denaturation. The concentrations of these additives are limited since they must not degrade the immobilized-protease columns commonly used for digestion. Chaotropes, such as urea or guanidinium hydrochloride (Gdn-HCl), are often included at a molar concentration as high as 3 mol/L. High concentrations of these additives accelerate denaturation, minimizing the amount of time that labeled protein undergoes back exchange in the aqueous analysis solution. The optimum concentration for each reagent is protein dependent and should be determined prior to HDX-MS experiments.

For proteins containing disulfide bonds, reducing agents, such as 3,3′,3″-phosphanetriyltripropanoic acid (i.e., tris(2-carboxyethyl)phosphine hydrochloride or TCEP), are added at high millimolar concentrations (>200 mmol/L). The reducing agent breaks the disulfide bonds, which helps to unfold the protein and improve digestion efficiency. Since TCEP solutions oxidize with time (Burns, Butler, Moran, & Whitesides, 1991), solutions (pH < 5) are best prepared immediately before initiating a series of measurements. Prepared solutions are stored at ≈0 °C and used the same day. Alternately, an electrochemical cell that is designed to reduce disulfide bonds can be added to the fluidic path prior to the proteolytic column (Mysling, Salbo, Ploug, & Jørgensen, 2014).

Denaturation unfolds the protein, and the amide N–H's along the protein backbone become exchange-competent, resulting in back-exchange rates equal to k_{ch} (Reaction 2). It becomes imperative to maintain the protein sample in cold, acidic conditions. For this reason, online digestions utilize a mobile phase of water with 0.1% formic acid to maintain acidic pH ≈ 2.5, and temperature is kept at ≈0 °C by submerging all valves and tubing in an ice bath (manual experiment) or by refrigerating these components in a compartment (automated system).

Most HDX-MS workflows utilize online columns containing immobilized protease, which can digest samples within one to three minutes (Chalmers et al., 2006), due to the large proteolytic enzyme to protein sample molar ratio. Alternatively, some HDX-MS experiments have employed in-solution digestion. In-solution digestion typically employs a 1:1 ratio of protein to solution-phase enzyme and reaches completion after a digestion time of ≤5 min at 0 °C (Brier et al., 2012; Sarkar & Wintrode, 2011).

In-solution methods may combine enzymatic digestion with additional sample preparation steps, such as removal of additives that are incompatible with UPLC-ESI-MS.

The spatial resolution of deuterium uptake is determined by peptide length. Porcine pepsin is an acidic protease that is commonly used for HDX-MS applications due to the high activity of the enzyme in the pH range between 2 and 4. This enzyme has also been successful at digesting proteins at reduced temperatures. Pepsin digestion generally results in peptides comprising (3–30) amino acids (Zhang & Smith, 1993) with an average length of 14 amino acids (Ahn, Cao, Yu, & Engen, 2013). The sequence of these peptides often overlap, which can improve amide resolution. Pepsin is nonspecific in the sense that cleavage positions cannot be predicted from a protein sequence. However, pepsin shows strong cleavage trends, such as pepsin preferentially cleaves after phenylalanine or leucine, but rarely cleaves after a charged residue (arginine, lysine, or histidine) or proline (Hamuro, Coales, Molnar, Tuske, & Morrow, 2008).

Although pepsin is the most common acidic protease used in HDX-MS experiments, other acidic proteases have been utilized. Each enzyme exhibits a unique cleavage trend (Ahn et al., 2013). Because each protease exhibits distinct site cleavage propensity, the use of two or more proteases during an HDX-MS measurement campaign can increase the sequence coverage of the protein sample and improve amide resolution (Hu et al., 2013; Mayne et al., 2011). Alternative proteases include protease type XIII from *Aspergillus saitoi*, protease type XVIII from *Rhizhopus* species (Cravello, Lascoux, & Forest, 2003; Rey, Man, Brandolin, Forest, & Pelosi, 2009; Zhang et al., 2008), and plant aspartic proteases, nepenthesin I (Rey et al., 2013) and nepenthesin II (Yang et al., 2015) from *Nepenthes gracilis*. Pepsin, protease type XIII, and protease type XVIII are commercially available and can be used to prepare enzyme-immobilized columns following the protocol described by Wang, Pan, and Smith (2002) or Rey et al. (2009). Additionally, because of the prevalence of pepsin in HDX-MS, enzyme-immobilized particles and packed columns can be purchased. Nepenthesin I and nepenthesin II require expression and purification prior to enzyme immobilization and use in packed columns.

3.7 Chromatography

After the sample protein is digested, the peptide effluent passes through a C18 guard column that immobilizes the peptides. Peptides trapped on the guard column are washed with additional mobile phase (usually water with 0.1% formic acid) for 30–60 s, which removes salts and additives (e.g., chaotropes, reducing agents, etc.) present in the equilibration, exchange, or quench buffers. Subsequently, the peptides are eluted and separated on a RP-UPLC column. Table 1 lists column configurations and solution gradients that investigators have used to accomplish HDX-MS measurements.

Following buffer exchange, the peptides are eluted using a gradient of water with 0.1% formic acid (solvent A) and a mixture of 80/20/0.1 (v/v/v) acetonitrile/water/formic acid (solvent B). To minimize back exchange at backbone amides, the RP separation is completed within 10 min. Although this rapid separation does not fully resolve the peptides, it does decrease the number of coeluting peptides, which increases the probability that more peptides will be ionized and detected. Furthermore, the mass spectrometer performs a second dimension separation based on m/z. The UPLC separation also ensures that the side chain residues, with more rapidly exchanging hydrogen sites, back exchange to the protic natural abundance because the mobile phase contains 100% H_2O and 0% D_2O. This loss of deuterium label on the side chains (except for arginine) decreases the complexity of the MS data since each residue within the peptide can have a maximum of one exchange site.

3.8 Mass Spectrometry

The peptides eluting from the upstream UPLC apparatus pass directly into the electrospray source of an MS, which measures ion intensity versus m/z as a function of retention time. Typical fluidic flow rates of the UPLC match the required input flow rate into the ESI source. However, traditional HPLC apparatus operating at higher flow rates may need a splitter that shunts off excess effluent, though this may diminish overall sensitivity.

Numerous HDX-MS laboratories have reported results obtained using ESI tandem mass spectrometers of architectures spanning quadrupole-reflectron-time-of-flight (qTOF/qRTOF), ion trap-Fourier transform (ion cyclotron resonance (ICR) or orbitrap), and quadrupole-orbitrap configurations. These systems feature resolutions ($[(m/z)/\Delta m/z]\cdot z$) from 40,000 to 750,000 at 400 m/z, where m/z is the ion mass-to-charge ratio and $z=$ion charge. To avoid corrupted measurements of deuterium content within peptide ions, Fourier transform instruments must be set to an appropriate resolution, which is usually somewhat lower than the advertised maximum performance (Burns, Rey, Baker, & Schriemer, 2013).

Table 1 HDX-MS Fluidic Configurations used to Conduct UPLC Separations of Peptides Prior to Their Injection into a Mass Spectrometer

Guard Column [Column Description (Particle Size) and Vendor, Bore (mm): Dia.×Length], Desalting Procedure	Analytical Column [Model (Particle Size), Vendor, Bore (mm): Dia. × Length]	Solvent Gradients (Flow Rate, Gradient Cycle Duration, Solvents)	References
C18 (5 μm) Grace Discovery Sciences, Carnforth, UK *Bore (mm)*: 1.0 × 50. *Desalt*: 50 μL/min, 100% solvent A; 3 min	C18 Hypersil GOLD (1.9 μm), ThermoFisher Scientific, Waltham, MA *Bore (mm)*: 1.0 × 50	*Flow rate*: 50 μL/min *Gradient sequence*: 5–35% solvent B, 3 min; 35–70% solvent B; 5 min; 70–100% solvent B, 0.5 min; 100% solvent B, 0.5 min; 5% solvent B, 0.5 min *Solvent A*: H_2O, 0.1% formic acid *Solvent B*: 80% CH_3CN, 20% H_2O, 0.1% formic acid	Li et al. (2014)
ACQUITY UPLC BEH C18 VanGuard Pre-column (1.7 μm); Waters Corp., Milford, MA *Bore (mm)*: 2.1 × 5 *Desalt*: 200 μL/min, 100% solvent A; 2.5 min	ACQUITY UPLC BEH C18 (1.7 μm); Waters Corp., Milford, MA *Bore (mm)*: 1.0 × 100	*Flow rate*: 40 μL/min *Gradient sequence*: linear gradient 15–45% solvent B; 6.8 min *Solvent A*: H_2O, 0.1% formic acid *Solvent B*: 90% CH_3CN, 10% H_2O, 0.1% formic acid	Malito et al. (2013)
Protein Micro Trap column; Michrom BioResources, Inc. *Desalt*: 220 μL/min; 5% buffer A, 2 min	C12 Jupiter Proteo C12 (4 μm, 90 Å); Phenomenex, Torrance, CA *Bore (mm)*: 1.0 × 50	*Flow rate*: 50 μL/min *Gradient*: Linear gradient of 5–60% solvent B; 12 min *Solvent A*: H_2O, 0.1% formic acid *Solvent B*: 90% CH_3CN, 10% H_2O, 0.1% formic acid	Brier et al. (2012)

Continued

Table 1 HDX-MS Fluidic Configurations used to Conduct UPLC Separations of Peptides Prior to Their Injection into a Mass Spectrometer—cont'd

Guard Column [Column Description (Particle Size) and Vendor, Bore (mm): Dia.×Length], Desalting Procedure	Analytical Column [Model (Particle Size), Vendor, Bore (mm): Dia. × Length]	Solvent Gradients (Flow Rate, Gradient Cycle Duration, Solvents)	References
C8, Microm Bioresources, Auburn, CA	C18 Hypersil GOLD (3 μm), ThermoFisher Scientific, Waltham, MA *Bore (mm)*: 2.1 × 50	*Flow rate*: 200 μL/min *Gradient*: 4–40% CH_3CN with balance H_2O, 0.3% formic acid; 20 min	Chalmers et al. (2007)
ACQUITY UPLC BEH C18 VanGuard Pre-column trap (1.7 μm); Waters Corp., Milford, MA *Bore (mm)*: 2.1 × 5 *Desalt solution*: 100 μL/min; 100% H_2O, 0.05% formic acid; 2 min	ACQUITY UPLC BEH C18 (1.7 μm), Waters Corp., Milford, MA *Bore (mm)*: 1 × 100	*Flow rate*: 40 μL/min *Gradient*: 8–40% linear CH_3CN, 0.1% formic acid; 6 min	Houde et al. (2009)
Peptide concentration and desalting Microtrap, Bruker-Michrom, Auburn, CA *Desalt*: 100% solvent A; 5 min	ZORBAX 300SB-C18 (3.5 μm), Agilent Tech., Santa Clara, CA *Bore (mm)*: 1 × 50	*Flow rate*: 200 μL/min Gradient sequence: 15–40% solvent B, 5 min *Cleaning trap and column*: 40–90% solvent B, 1 min; 5–90% solvent B, 1 min; 90% solvent B, 1 min *Solvent A*: H_2O, 0.1% formic acid *Solvent B*: 90% CH_3CN, 10% H_2O, 0.1% formic acid	Majumdar et al. (2012)

For HDX-MS, the linearity of the ion detector of the mass spectrometer is also of key importance, as it can limit the accuracy of the intensity measurement. During the analysis of HDX data, scientists should manually search for peculiar ion intensity envelopes that may evidence ion suppression induced by the coincidence of one or more abundant coeluting peptide ions.

For HDX-MS experiments *m/z* accuracy of the MS is important, particularly, for measurement campaigns spanning days or weeks. HDX-MS experimentalists should know the maximum deviation drift rate of their instrument and develop a calibration schedule that maintains the instrument accuracy within the instrument specification over each anticipated instrument run.

To reduce proton–deuteron scrambling in the ESI source, the electrospray capillary temperature must be carefully controlled. It must be high enough to promote desolvation of the peptide ions, but low enough that the deuterium label is not corrupted by back exchange or H/D scrambling via intramolecular migration of protons and deuterons. Experiments have found that H/D scrambling is minimized when the desolvation temperature in the ESI source is set to 100 °C (Coales, Tomasso, & Hamuro, 2008; Walters, Ricciuti, Mayne, & Englander, 2012), rather than the 200–300 °C that is common for proteolytic studies.

Tandem ESI-MS instruments containing an electron transfer dissociation (ETD) or electron capture dissociation (ECD) stage can resolve proteolytic fragmentation HDX-MS data at the unit amide limit. In the middle-down version of this experiment, selected peptide ions are isolated and fragmented by ETD or ECD, which produces c_k and z_k ion fragments. The ion intensity versus *m/z* spectra of the secondary ion fragments are recorded. Each c_k and z_k ion fragment reports the nascent deuterium content of the associated fraction of the parent peptide ion sequence. Simultaneous equations describing data observed for the c_k and z_k ion sets can be solved to determine the D occupancy of the peptide ion sequence with resolution approaching and equal to the unit amide limit. Examples of top-down and middle-down HDX-MS of proteins using ETD (Abzalimov, Kaplan, Easterling, & Kaltashov, 2009; Huang & Hudgens, 2013; Rand, 2013; Rand, Zehl, Jensen, & Jørgensen, 2009; Sterling & Williams, 2010) and ECD (Pan & Borchers, 2014; Pan, Han, & Borchers, 2012; Pan, Han, Borchers, & Konermann, 2009) have appeared in the literature. The emergence of software tools (Table 2) designed for data analysis of ETD/ECD data will only accelerate the use of these techniques that can resolve the HDX of protein structures with single amino acid resolution.

The ETD/ECD produces fragment ions by transferring a low energy electron to the parent peptide cation, typically of charge $z=+2$ or $+3$. Fragment ions resulting from ETD and ECD processes are accompanied by little or no H/D scrambling and no structural rearrangements. However, as parent ion charge is increased, collisions induced by the ion optics increasingly

Table 2 Software Packages and Programs Available for Analyzing HDX-MS Data

HDX-MS Analysis Software	Features	References
AUTOHD	*Public software*: Computer program that takes in a list of mass spectra, defines isotopic clusters, and identifies them using a list of candidate peptides generated from enzymatic fragmentation or dissociation URL: http://www.ms-utils.org/autohd.html	Palmblad, Buijs, and Håkansson (2001)
DynamX	*Commercial product*: Allows simultaneous visualization of raw MS data, identified peptides, deuterium uptake graphs, and comparability plots. Multifunctional user interface. Integrates ion mobility separation and ETD fragmentation data URL: http://www.waters.com/	Waters Corp. (2015)
ExMS	*Public software*: Scans through high-resolution MS data to find the individual isotopic peaks and isotopic envelopes, verifies peptide assignment, extracts deuterium incorporation information URL: http://HX2.med.upenn.edu/download.html	Kan, Mayne, Chetty, and Englander (2011)
HDExaminer	*Commercial product*: Supports various MS data formats including high-resolution MS data and ETD data URL: http://www.massspec.com/HDExaminer.html	HDExaminer, Sierra Analytics, Inc. (2009)
HDsite	*Public software*: An iterative optimization program that integrates multiple peptide acquisitions together with isotopic envelope-shape information and a site-resolved back-exchange correction URL: http://HX2.med.upenn.edu/download.html	Kan, Walters, Mayne, and Englander (2013)

HeXicon 2	*Noncommercial software*: An automated pipeline suite for analysis and visualization of HDX-MS data. Improvements in this version include a peak detection algorithm, chromatogram alignment, and a robust peptide sequence assignment routine URL: http://hx2.mpimf-heidelberg.mpg.de/	Lindner et al. (2014)
HXExpress2	*Public software*: Semi-automatic data analysis software that generates deuterium uptake and peak width plots. Improves the speed of analyzing HDX-MS data URL: http://www.hxms.com/HXExpress	Guttman, Weis, Engen, and Lee (2013)
HDX Workbench	*Noncommercial software*: An integrated desktop program that facilitates automation, management, visualization, and statistical cross-comparison of large HDX data sets. This is a significant upgrade of the previous programs, The Deuterator and HD Desktop (Pascal, Chalmers, Busby, & Griffin, 2009; Pascal et al., 2007) URL: http://hdx.florida.scripps.edu	Pascal et al. (2012)
HDXAnalyzer	*Public software*: HDXAnalyzer is a software package that supports statistical analyses of HDX mass spectrometry data. Program examines raw LC-MS hydrogen exchange data and finishes with complete data visualization. All information for usage and installation is available in the package URL: http://people.tamu.edu/~syuan/hdxanalyzer/	Liu et al. (2011)
HDXFinder	*Public software*: This Web tool application (previously named HDXAnalyzer) relies on high resolution of mass spectrometry to	Miller, Prasannan, Villar, Fenton, and Artigues (2012)

Continued

Table 2 Software Packages and Programs Available for Analyzing HDX-MS Data—cont'd

HDX-MS Analysis Software	Features	References
	extract all isotopic envelopes before correlating these envelopes for individual peptides. The software includes a variety of graphical tools for verifying correlations and rankings of the isotopic peptide envelopes URL: http://www.kumc.edu/mspc/links.html	
Hydra	*Public software*: Software that processes H/D exchange data from MS or tandem MS analyses URL: http://omictools.com/hydra-hydrogen-deuterium-exchange-s7731.html	Slysz et al. (2009)
MagTran	*Public software*: This application calculates centroids from a user-selected m/z window. This utility program uses the ZSCORE algorithm URL: http://magtran.software.informer.com/	Zhang and Marshall (1998)
Mass Spec Studio	*Public software*: High-capacity HXD-MS and HDX-MS^2 mode of operation; supports high structural resolution of labeling chemistries; supports analyses involving ETD and CID fragmentation; can investigate peptides modified by covalent labeling; integrates structural mass spectrometry data with the *H*igh *A*mbiguity *D*riven bimolecular *DOCK*ing (HADDOCK) program URL: https://github.com/MassSpecStudio/MassSpecStudio	Rey et al. (2014)
MS Tools	*Public software*: Suite of Web tools for planning, executing, visualizing, and presenting HX MS data URL: http://ms.biomed.cas.cz/MSTools/	Kavan and Man (2011)

heat the parent peptide ions, which can result in H/D scrambling across the amide sequence. When adjusting ion optics prior to a measurement campaign, control measurements must be performed with defined model peptides to verify that scrambling is not occurring at the conditions used for the HDX experiment (Zehl, Rand, Jensen, & Jørgensen, 2008).

Attempts to enhance the amide resolution of HDX-MS by employing collision-induced dissociation (CID) to produce secondary fragments have proved disappointing. CID is unsuitable for HDX-MS studies because this fragmentation method heats the ions, increasing the vibrational energy and leading to significant H/D scrambling (Ferguson et al., 2007; Jørgensen, Gårdsvoll, Ploug, & Roepstorff, 2005; Rand & Jørgensen, 2007).

3.9 Generating a Proteomic Map

Prior to conducting any HDX-MS experiments, a map comprising the curated list of proteolytic peptide ions must be generated from the protein sample. The peptide map is the filter through which mass spectral data are selected for HDX-MS analyses. The curated list of proteolytic peptide ions includes the verified peptide sequence, ion score, charge (z), and retention time for the UPLC apparatus. Although pepsin and the other acidic proteases described above are nonspecific in that cleavage positions cannot be predicted from a protein sequence, these cleavage trends are highly reproducible when the same conditions (pH, temperature, time, and protein concentration) are used. Under controlled conditions each unique peptide elutes from the analytical chromatography column at a consistent retention time, regardless of its deuterium content.

During generation of the peptide map, the digestion and running parameters are optimized, including the temperature of the enzymatic column and the flow rate, which in combination with the column dimensions, determines the digestion time. Proteases function most effectively at physiological temperatures; however, this efficiency must be balanced against the need to reduce back exchange of the labeled backbone amide N–D within the protein, which is decreased at reduced temperature. Additionally, the faster the flow rate, the less time the protein will be in the presence of water, minimizing back exchange; yet, decreasing the flow rate increases the time the protein is in contact with the protease, potentially increasing digestion efficiency. Interestingly, pepsin more effectively digests model proteins when the protease column is maintained at high pressures (Ahn, Jung, Wyndham, Yu, & Engen, 2012; Jones, Zhang, Vidavsky, & Gross, 2010;

López-Ferrer et al., 2011). However, the construction of the column housing and stationary phase support may limit the pressure and flow rate that can be applied to the column.

Digestion efficiency is determined by observing the UPLC effluent using a tandem MS (MS/MS) instrument that employs CID to fragment the parent peptide ions. CID is an acceptable fragmentation method since these studies are performed prior to exchange, and H/D scrambling is not a factor. These experiments allow digestion conditions (temperature and flow rate), solution conditions (chaotropic and reducing agent concentrations), and the UPLC gradient to be optimized. By searching a database, such as MASCOT, the m/z values of the detected peptides and their fragments can be correlated to specific peptides within the protein sequence. Good digestions yield peptides covering 70–100% of the protein sequence and a large number of overlapping peptides. The elution time and exact mass for each peptide from the database search are then utilized by the HDX-MS software to find labeled peptides and measure the mass increase associated with deuterium uptake.

As described in the chromatography section, sample carry over complicates HDX-MS data analyses. One way to test for carry over is to repeat the tandem MS analysis immediately following peptide mapping using only water as the "sample." Since no protein is present, peptides that are detected in a database search must have eluted from the digestion, RP-trap, or RP-analytical column. To minimize carry over, the peptide mapping can be repeated with a lower protein concentration (protein concentration affects acidic protease digestion efficiency) or additional column washes can be incorporated into the HDX-MS experiment between protein analyses (Majumdar et al., 2012).

3.10 Data Analysis

Manual analysis of an HDX-MS experiment is very time consuming; however, software programs that calculate deuterium uptake for each detected peptide greatly decrease analysis times (Table 2). This reduction of analysis time has encouraged more groups to use HDX-MS for studies of protein structure, dynamics, and binding interactions. Most tools determine average deuterium uptake using the peptide centroid of the isotopic envelope, $\langle m \rangle_{\text{centroid}}$, which is defined as:

$$\langle m \rangle_{\text{centroid}} = z \cdot \left(\frac{\sum_{i=1}^{n} (m/z)_i \cdot I_i}{\sum_{i=1}^{n} I_i} - m_{\text{H}^+} \right) \tag{17}$$

where z is the ion charge, n is the number of isotopic m/z features in the mass spectrum of the ion, $(m/z)_i$ and I_i are, respectively, the measured mass-to-charge ratio and intensity of the ith ion feature, and m_{H^+} is the mass of the charge carrier (usually a proton). The centroid of the isotopic envelope is a derived measurand that reflects the average mass of an isotopic envelope originating from the same peptide. It contains no information about the measurement system or about the properties of the analyzed ion. The centroid calculation applies well for peptides that uptake deuterium under EX2 conditions. However, centroid calculations do not apply for peptides that exchange under EX1 conditions, nor can automated analysis programs detect EX1 signatures in centroid data (Zhang, Ramachandran, Kumar, & Gross, 2013). Thus, it is incumbent upon the experimentalist to verify manually the actuality of EX2 kinetics for each peptide ion. Verification is accomplished by inspecting the isotopic peak intensity distribution envelope of each peptide ion. In practice, peptide ions exhibiting EX1 kinetics are infrequently observed. Often, EX1-like signatures indicate practical problems in the data acquisition process, e.g., carryover due to incomplete column washes (Fang, Rand, Beuning, & Engen, 2011).

The isotopic envelope for each detected peptide may indicate the type of exchange occurring. Peptides that exchange with EX1 kinetics can have bimodal distributions (Fang et al., 2011; Guttman et al., 2013; Kreshuk et al., 2011; Weis, Wales, Engen, Hotchko, & Ten Eyck, 2006). Under EX1 conditions $k_{ch} \gg k_{cl}$; therefore, multiple exchange events occur simultaneously in the unfolded protein. The two populations represent the exchanged and nonexchanged forms of the protein. If the two populations are well separated on the m/z axis, then two distinct peaks will be visible, though if the peaks are not well separated, then a broad isotopic distribution will be apparent.

For peptides that exchange with EX2 kinetics, deuterium uptake results in a binomial distribution. Following quenching, each backbone amide experiences deuterium to hydrogen back exchange with a rate, k^i_{ch} (ref. Eq. 1; superscript i denotes the sequence index). Peptides exiting the UPLC column have a distribution of masses with the average mass of this population representing the average number of deuterium that were incorporated and maintained in the peptide. For most backbone amides, the loss of deuterium due to back exchange is between 20% and 30% (Walters et al., 2012). However, just as the exchange rate is dependent on the primary structure of the protein or peptide, the back-exchange rate is sequence dependent and can vary by up to 30-fold depending on the neighboring side chains. At a specific pH and temperature, the back-exchange rate can be readily estimated for each amide hydrogen within an unfolded peptide sequence (Bai et al., 1993; Connelly et al., 1993; Englander, 2015).

For specific peptides, the amount of back exchange may deviate from its estimate due to interactions with the chromatographic columns (Sheff, Rey, & Schriemer, 2013; Walters et al., 2012). If an accurate value of the exchange rate is needed, then additional control measurements must be performed to determine the amount of deuterium that is lost due to back exchange. The corrections may be applied during data analysis. However, for most ligand-binding experiments, a relative analysis of deuterium uptake is measured between the apo- and holo-protein states with identical experimental conditions used for both proteins. This comparison is sufficient to determine regions of the protein that have different amounts of deuterium uptake in the presence and absence of ligand if the experiments are performed within a short time frame.

Data analysis for deuterium uptake can be complicated in arginine-rich peptides (Morgan & Engen, 2009). Though the backbone amide has the lowest rate of exchange at pH 2.5 and that rate is less than any other functional group, the guanidinium group of arginine has the next slowest exchange rate at this pH (Brier & Engen, 2008). During solvent exchange and UPLC, the presence of H_2O ensures that the side chain residues exchange back to H prior to detection by MS. Since the exchange rate of the guanidinium group is also slow at this pH, with short wash times and UPLC gradients, some deuterium label may be retained in arginine side chains upon ionization.

3.11 Uncertainty Evaluations: Which Deuterium-Uptake Differences Are Meaningful?

The determination of dataset quality requires an uncertainty evaluation for each peptide ion that is computed from two or more replicate measurements at each t_{HDX}. For a few peptides, the determination of deuterium content and uncertainty can be improved by including centroid data from two or three charge states. For HDX-MS data analysis, most software packages (Table 2) automatically evaluate measurement uncertainties for the centroids. Alternately, measurement uncertainties can be evaluated manually.

Pairwise comparisons of peptides common to the apo- and holo-protein data sets can signal differences arising from a protein–ligand interaction. However, knowledge of the uncertainties of these measurements is required to identify those peptides exhibiting a significant difference. A powerful comparison method uses pairwise *t*-tests of each peptide across the t_{HDX}

time points. The *t*-tests presume that apo- and holo-peptide ion data follow a Student's *t*-distribution (Press, Teukolsky, Vetterling, & Flannery, 2007). The threshold for statistical significance between pairs is usually set at $P \leq 0.05$. To conduct simultaneous statistical comparisons of the same peptide across multiple HDX-MS data sets, Tukey's range test can be used (Tukey, 1949). An assortment of *t*-tests is available from numerous commercial and freeware statistics packages (e.g., *R*-language, TR Foundation, 2015).

3.12 Data Display

Deuterium uptake can be visualized in multiple ways. First, data for each peptide that is common in the apo- and holo-species can be plotted together with their measurement uncertainties on the same semi-logarithmic graph. The *y*-axis can show average mass increase in Daltons or it can show normalized mass increase, where the normalization is for the number of exchangeable amide hydrogens within the peptide, excluding the N-terminal amide hydrogens (Huang, Wen, Blankenship, & Gross, 2012). The *x*-axis is plotted as the logarithm of immersion time of the protein in D_2O. Peptide ions exhibiting high levels of deuterium uptake at all-time points indicate regions that are solvent-exposed, unstructured (loops) in the protein. Peptides that pass from low to high deuterium uptake over the time course of the experiment indicate that the peptides are present in dynamic regions of the protein. Peptides that have low levels of deuterium uptake at all-time points represent portions of the protein that are protected from exchange due to secondary structures, such as within α-helices and β-sheets.

Second, a difference map of the deuterium uptake observed in the apo- and holo-protein forms can be presented in a one dimensional, linear array with the detected peptides represented by bars under the protein sequence. Customarily regions absent statistically different deuterium uptake are shown in gray. Regions showing statistical differences are exhibited as a colored heat map and referenced to a separately displayed color scale.

Third, when a crystal structure is known for the protein sample, the differential deuterium uptake can be plotted on the Protein Data Bank (PDB) image (Berman et al., 2000) to give a better indication of the sites where structural and dynamic changes are present. Plotting these data on a three-dimensional structure allow visualization of the binding interaction. This visualization can help distinguish whether the binding site is in a linear sequence or associated with multiple discontinuous regions of the protein.

Plotting the data onto the 3D structure can give significant insight into the protein–ligand interaction; however, it should be realized that the crystal structure is one model of the protein, and the HDX-MS data may represent a different solution-based equilibrium structure of the protein. Additionally, experiments should also be performed to verify that the site is the protein–ligand interface rather than a portion of the protein sequence affected by an allosteric effect.

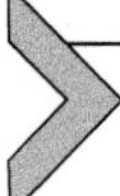

4. INTERPRETING HDX-MS DATA TO DETERMINE PROTEIN–LIGAND INTERACTION MAPS

Proteins can interact by engaging either continuous or discontinuous sequences of amino acids (Barlow, Edwards, & Thornton, 1986; Thornton, Edwards, Taylor, & Barlow, 1986). Continuous, or linear, epitopes, such as short linear motifs (SLiMs), are short lengths of the protein sequence that mediate a protein–protein interaction. The majority of annotated SLiMs consist of 3–11 contiguous amino acids, with an average of just over 6 residues (Diella et al., 2008; Dinkel et al., 2013; Neduva & Russell, 2006). On the other hand, discontinuous or conformational protein–ligand interactions may involve several amino acid sequences that are distant in primary structure; however, conformational folding has placed the active residues proximal in tertiary structure. For example, antigen–antibody interactions often involve interactions of discontinuous epitopes (Pandit et al., 2012; Zhang et al., 2011).

This section outlines procedures that have successfully characterized a linear epitope and a discontinuous motif. For brevity, the presentation focuses on the HDX-MS portion of each study; however, these studies used auxiliary measurements, e.g., sodium dodecyl sulfate–polyacrylamide gel electrophoresis (SDS-PAGE), size exclusion chromatography, assays, etc., to assure purity, structural invariance, and biochemical activity of the protein–ligand complex under HDX-MS conditions. These auxiliary data carry additional value, as they help define properties of the protein–ligand complex.

4.1 Example: Protein–Ligand Interactions Involving Continuous Amide Contacts

In the archaeon, *Thermococcus kodakarensis*, three proliferating cell nuclear antigen (PCNA) monomers can organize into a homotrimeric ring that encircles duplex DNA. The PCNA homotrimer tethers other enzymes in its role as the processivity factor for DNA polymerases. During laboratory investigations of *T. kodakarensis*, Li et al. (2014), discovered a small, ≈8.5 kDa protein that binds to *T. kodakarensis* PCNA1 and PCNA2 (29.3 kDa) *in vitro* (Li, Santangelo, Čuboňová, Reeve & Kelman, 2010). SDS-PAGE data showed that the 8.5 kDa protein inhibits PCNA-dependent activities, likely by preventing the formation of the homotrimer. Hence, this small protein is designated as *Thermococcales* inhibitor of PCNA (TIP). Tests demonstrated that the PCNA-TIP interaction was novel, as mutagenesis of the PCNA protein sequence in locations known to induce unclamping by other ligands did not suppress PCNA activity.

An HDX-MS study was initiated with the objective of determining the interacting regions of PCNA2 and TIP that drive the dissociation of the PCNA homotrimer (Li et al., 2014). Protein stock solution was diluted in phosphate buffer (20 mmol/L sodium phosphate, 500 mmol/L NaCl, pH 7.6) to prepare a 15 μmol/L final analytical concentration and equilibrated at 4 °C for 2 h. The PCNA-TIP complex was prepared by mixing PCNA and TIP at a molar ratio of 1:1 (PCNA monomer:TIP monomer) and equilibrated at 4 °C for 16 h.

Prior to the conduct of HDX-MS investigations, separate proteomic studies were conducted on the native PCNA2 and TIP, comprising pepsin digestion, separation of the peptic peptides by UPLC, and ESI-MS/(CID)MS of the UPLC effluent. Application of MASCOT (Matrix Science, Oxford, UK) and MassMatrix database searches (Xu, 2010) of ion signals observed by standard and tandem mass spectrometry (ESI-MS/(CID)MS) yielded a master list of identified peptide ions. Subsequently, each identification was verified manually. The ion lists provided coverage for 98% of the PCNA2 backbone (85 peptide (+1, +2, +3) ions) and 99% of the TIP backbone (13 peptide (+1, +2, +3) ions).

Robotic HDX-MS measurements were conducted by immersing each protein in buffered D_2O (pD 7.6; 3 °C) for $t_{HDX}=0.5$, 5, 20, 40, 60 min. The robot dispensed protein solution (5 μL) into 21 μL D_2O buffer (20 mmol/L sodium phosphate, 500 mmol/L NaCl, pD 7.6) at 3 °C. This temperature was chosen to slow the H/D exchange rates so that the observed rates better fit the dynamic range of the instrument.

The HDX-MS data were analyzed using HDX Workbench (Pascal et al., 2012). HDX-MS data for the apo-PCNA2 (PCNA2 homotrimer alone) exhibited a range of deuterium uptake rates. After an hour of exposure to

D_2O, the interior-helical regions near the trimeric interfaces exhibited little deuterium content. Exterior to the homotrimer ring, long-loop regions of each PCNA2 also required hours to achieve significant D_2O exchange. On the other hand, backbone amides in some outer helices and loop regions were more labile, as these amides exhibited substantial deuterium uptake within 20 min. In contrast to the structured PCNA2 protein, HDX-MS experiments on apo-TIP (TIP alone) found no evidence of stable secondary structure. After immersion in D_2O solution, amide groups of apo-TIP, especially in the C-terminal region, rapidly exchanged to contain over 80% deuterium within 30 s.

Variations in deuterium uptake rates were observed for various peptides in apo-PCNA2 (PCNA2 homotrimer alone) and holo-PCNA2 (i.e., PCNA2 peptides from the PCNA2-TIP complex). Deuterium uptake plots for peptide ions 243–259 and 187–204 (Fig. 5B and C) of apo-PCNA2 and holo-PCNA2 indicate that the holo-PCNA gains deuterium more slowly. (This conclusion is also supported by pairwise *t*-tests on the D-uptake versus time data, which yielded $P = 0.03$ and $P = 0.01$, respectively, inferring that the apo and holo data differ.) Since peptide 243–259 contains mostly artificial histidine residues, this decreased deuterium uptake likely arose from nonspecific interactions. Alternatively, peptides 104–110 and 34–38 of holo-PCNA2 exhibit greater deuterium uptake than apo-PCNA2 (Fig. 5D and E). These peptides reside in the trimer interfaces. The observations suggested that binding with TIP facilitates the dissociation of PCNA2 homotrimer. Such dissociation may occur by altering the $\Delta_d G_i^\circ$ of these residues, so that the local amide hydrogen bonding becomes labile (K_{op} is increased), which enables greater protic exchange with water. Such behavior is in accord with the type 2 Sowole–Konermann scenario (Sowole & Konermann, 2014).

To confirm that region 187–204 of PCNA2 binds TIP and accounts for the diminished deuterium uptake, three residues, K197, Y200, and Y204, were mutated to alanine and glutamine (ref. Fig. 5F) to make PCNA2-A and PCNA2-E, respectively. Relative to the native PCNA2-TIP complex, the PCNA2-A was expected to bind strongly with TIP, and the PCNA2-E was expected to bind weakly with TIP. Importantly, as determined by functional assays, the mutations forming PCNA2-A or PCNA-E did not alter the PCNA activity.

Prior to conducting HDX-MS studies with PCNA2-A and PCNA2-E, ion lists providing nearly complete sequence coverage were measured and manually verified. HDX-MS experiments involving PCNA2-A,

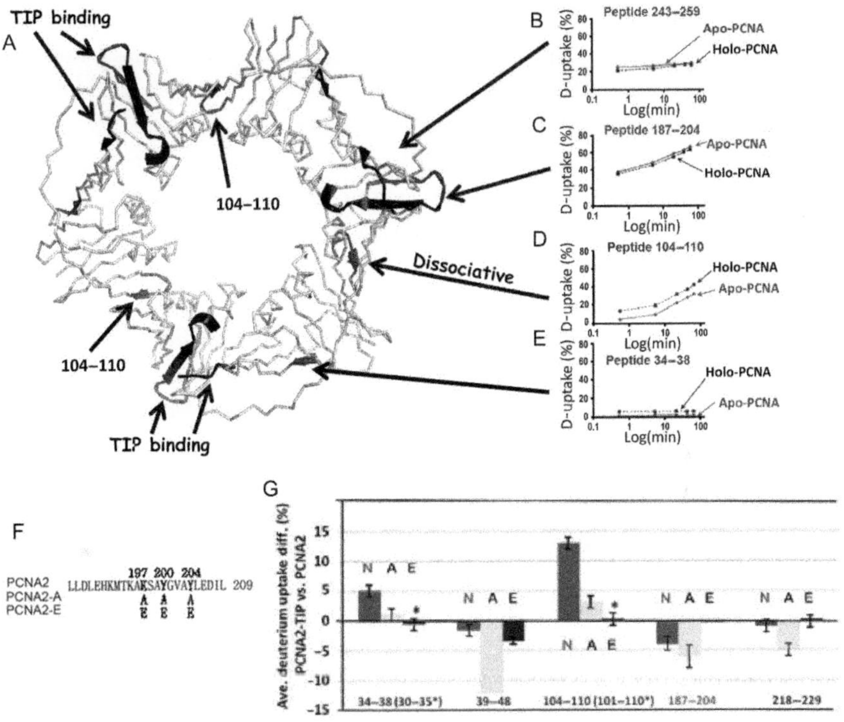

Figure 5 HDX-MS of the protein–ligand system, PCNA2-TIP. (A) Difference D-uptake map between native apo- and holo-PCNA2 plotted on the crystal structure of homotrimeric PCNA2 (PDB: 3LX2) (Ladner, Pan, Hurwitz, & Kelman, 2011). Blue signifies no statistical difference, red denotes a decreased D-uptake rate in native holo-PCNA2, and yellow denotes increased D-uptake rate in native holo-PCNA2. (B) Temporal D-uptake of native PCNA2 peptide 243–259. (C) Temporal D-uptake of native PCNA2 peptide 187–204. (D) Temporal D-uptake of native PCNA2 peptide 104–110. (E) Temporal D-uptake of native PCNA2 peptide 34–38. (F) Partial sequence of PCNA2 showing the mutations of PCNA2-A and PCNA2-E. (G) Effects of mutations on selected peptides of PCNA2. Each colored bar plots the difference ratio (%) between the deuterium uptake by the PCNA2-TIP complex relative to its uncomplexed PCNA2. Colors correspond to the native PCNA2 (N, red), PCNA2-A mutant (A, yellow), and PCNA2-E mutant (E, blue). Each bar was computed from the average deuterium uptake found at five HDX times points, and a bracket (thin lines) denotes its standard deviation (1σ). Asterisks on the colored bars of PCNA2-E indicate data originating from slightly shorter peptides. The 187–204 peptide (red) contains sites subjected to mutagenesis. Each plotted peptide group exhibits statistically significant differential peptide data, as inferred by paired *t*-tests. *Panels (B)–(G) are used with permission of Li et al. (2014).* (See the color plate.)

PCNA2-E, and TIP were conducted. Relative to apo-PCNA2-A, holo-PCNA2-A mutant showed significantly decreased deuterium uptake, not only in region 187–204, but also in regions 218–229 and 39–48 (Fig. 5G). Interestingly, these regions are composed of short loops and are geometrically close to each other, suggesting potential binding interfaces. As expected for a dissociative interface, holo-PCNA2-A shows enhanced deuterium uptake for region 104–110, but the effect is less than observed for the native holo-PCNA2. Overall, the HDX-MS results for holo-PCNA2-A were consistent with the prediction that alanine mutations strengthen PCNA2-TIP interactions. In accord with an expectation for weaker binding, the holo-PCNA2-E mutant responded weakly to the presence of TIP. Decreased deuterium uptake was observed only in region 39–48 (Fig. 5G) and essentially no increase in deuterium uptake was observed in region 101–110. The absence of a deuterium uptake difference in region 187–204 of native holo-PCNA-E indicated the importance of this region in modulating the PCNA2-TIP interactions.

Examination of the differential deuterium uptake of apo- and holo-TIP found that peptides 22–28 and 55–58 of holo-TIP (Fig. 6A) show reduced deuterium uptake rates in the native PCNA2-TIP complex, evidencing a Sowole–Konermann type 1 scenario. These diminished D-uptake rates

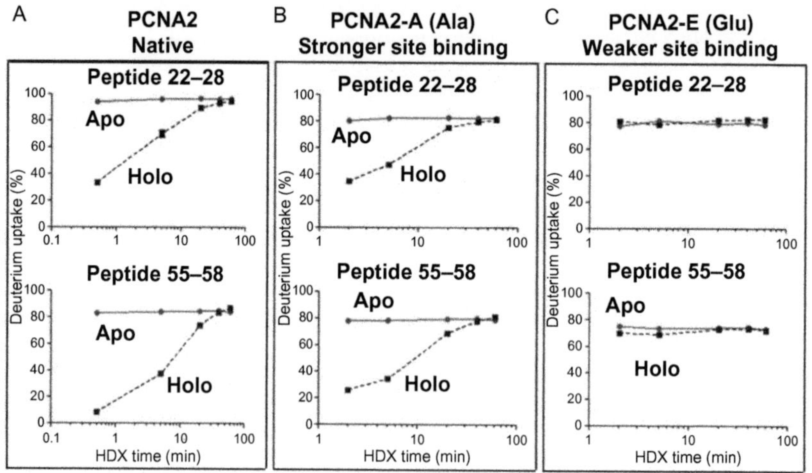

Figure 6 Mutations in PCNA2 affect the interactions with TIP. Panels display the deuterium uptake of peptides 22–28 and 55–58 in apo-TIP (Solid red (dark gray in the print version) lines; TIP alone) and holo-TIP (dashed black lines; TIP in solution with PCNA) for (A) native PCNA2, (B) PCNA2-A mutant, and (C) PCNA2-E mutant. *Figures used with permission of Li et al. (2014).*

suggested that TIP engages with native PCNA2, likely through interactions with peptide region 187–204 of PCNA2. Moreover, the HDX kinetics data show that the deuterium uptake of peptide regions 22–28 and 55–58 of TIP in the presence of native PCNA2 eventually reaches the same levels as that observed in the apo form (TIP alone), thus, evidencing the flexible nature of the TIP structure and its weaker binding with PCNA2.

The PCNA2-A mutant-TIP complex exhibited nearly the same trend of HDX kinetics in regions 22–28 and 55–58 of TIP (Fig. 6B) as compared to the native PCNA2-TIP complex. This similarity is in accord with the formation of the stable PCNA2-A-TIP complex. In contrast, apo-TIP and holo-TIP for the PCNA2-E mutant-TIP complex exhibit no significant difference in deuterium uptake behaviors (Fig. 6C), which is in accord with the weaker binding expected in the PCNA2-E-TIP complex. These data taken together with the behaviors of holo-PCNA2-A and holo-PCNA2-E support assignment of the TIP binding region as residing within residues 187–204 of PCNA2.

4.2 Example: Mapping a Discontinuous Protein–Protein Interaction

Mapping of discontinuous epitopes recognized by functional mAbs is essential for understanding the nature of immune responses and designing improved vaccines, therapeutics, and diagnostics. B-cell paratopes have complex, 3D structures often composed of residues that are discontinuous in the primary structure. Therefore, immunization with an isolated synthetic peptide that mimics a single linear antigen is unlikely to elicit high titers of neutralizing antibodies (Malito et al., 2013).

Important epitopes include those of bacterial meningitis (*Neisseria meningitidis*) that target human complement factor H (fH) protein. Human fH protein is a 155 kDa, 20 domain, complement control protein (CCP) that binds to the glycosaminoglycans (GAGs) that are present on the surface of host cells (Józsi & Zipfel, 2008). The resulting fH glycoprotein decoration marks human host cells as "self," protecting them from lysis by complement. Pathogenic cells and viruses not decorated with fH are effectively "unself" cells that are vulnerable to complement-mediated lysis. To evade complement-mediated killing, bacterial meningitis recruits fH by sequestering it to fHbp, a 27 kDa lipoprotein that resides on the pathogenic cell surface (Schneider et al., 2009). A crystal structure model of the fHbp:fH complex reveals that fHbp folds to form two intermingled N- and

C-terminal β-barrel domains that span amino acids 1–137 and 138–255, respectively (Malito et al., 2013; Schneider et al., 2009).

The fHbp protein on the pathogenic cell surface can serve as a vaccine antigen of bacterial meningitis. Malito et al. developed mAb 12C1, which specifically recognized fHbp (*variant* 1), inhibited fHbp binding to human fH, and exhibited bactericidal activity (Malito et al., 2013).

Malito et al. (2013) used HDX-MS to characterize the interaction between fHbp and mAb 12C1. During the HDX-MS study separate sets of deuterium labeling experiments were conducted on solutions of fHbp alone (apo-fHbp) and mAb 12C1 with fHbp (holo-fHbp). The antibody–antigen complex was formed by adding 225 pmol fHbp to mAb 12C1 using a molar ratio of 1:1.5, followed by incubation for 30 min at ≈20 °C. During the experiment, samples were immersed in deuterated phosphate-buffered saline (PBS) solution (pD 7.4, 25 °C) for specific exposure times ($t_{HDX}=0.5$, 2, 10, 20, 30 min). After each immersion period expired, a 30 μL sample was removed and the H/D exchange reaction was quenched with 30 μL ice-cold 200 mmol/L sodium phosphate (pH 2.4) buffer. The quenched aliquots were immediately frozen in liquid nitrogen and stored at −80 °C for less than 24 h.

When analyzing frozen aliquots, labeled samples were thawed rapidly to ≈0 °C and injected into a Waters nanoACQUITY UPLC. The UPLC injector, switching valve, columns, solvents, and all associated tubing were maintained ≈0 °C (ref. Table 1). The front end of this apparatus digested protein samples into peptides with a Poroszyme Immobilized Pepsin Cartridge. This digestion step required 2.5 min at 20 °C with a solvent flow rate (H_2O, 0.1% formic acid) of 200 μL/min and resulted in desalted peptides that were trapped on a guard column. Subsequently, the peptides were eluted off the guard column, chromatographically separated, and electrosprayed into a Waters tandem SynaptG2 mass spectrometer operating in resolution mode (m/z 100–2000). Mass accuracy was ensured by continuously infusing a Glu-1-Fibrinopeptide B solution [600 fmol/μL in 50% (vol/vol) CH_3CN, 0.1% formic acid] or Leu-Enk (2 ng/μL in 50% CH_3CN, 50% H_2O, 0.1% formic acid; Waters Corp.) through the reference probe of the electrospray ionization source. DynamX software (Waters Corp., 2015) determined the centroid mass of each peptide as a function of labeling time. Only the peptic peptides present in at least three repeated digestions of the unlabeled proteins were considered for the analysis.

Proteomics measurements and application of the ProteinLynx Global Server proteomics search program to apo-fHbp mass spectrum data identified 19 peptide ion fragments that covered 97% of the fHbp sequence.

Similar HDX-MS measurements of holo-fHbp solutions revealed that the presence of mAb 12C1 suppressed deuterium uptake for seven of the nineteen peptide ion fragments of fHbp (Fig. 7). The diminished D-uptake evidences a Sowole–Konermann type 1 interaction scenario induced by binding between 12C1 and fHbp. The peptides define four discontinuous segments (L34-L48, I89-F96, T107-137, and Y214-L251) encompassing 92 residues distributed in the N- and C-terminal domains of fHbp. These four discontinuous segments are clustered in a distinct region (Fig. 7) and comprise the discontinuous antigenic epitope of fHbp.

The HDX-MS results are in accord with other epitope mapping techniques. Most importantly, all four peptides identified by HDX-MS are also present in the fHbp:12C1 epitope determined from crystal structure analysis. Scanning with a library of synthetic fHbp peptides suggested that the C-terminal residues, A238-I249, comprise the primary target recognized by mAb 12C1. Screening a phage-display library of fHbp peptides identified that C-terminal peptide, L224-G250, is also a target of mAb 12C1.

Although both peptides reside in the epitope identified with HDX-MS, neither peptide accounted for the binding affinity found by surface plasmon resonance (SPR). However, the HDX-MS approach not only confirmed the involvement of peptide A238-I249 in binding 12C1 but also revealed additional fHbp N-terminal regions that mediate the binding interaction. These results accounted for the strong binding constant derived from SPR data, and they also supported evidence of a conformational epitope spanning the N- and C-termini of fHbp as the target of mAb 12C1.

Malito et al. concluded "…our results highlight the importance of using sensitive epitope mapping techniques rather than relying on the identification of linear peptides when seeking to fully understand the details of B-cell epitopes mediating antigen–antibody interactions" (Malito et al., 2013). The technique of proteolytic fragmentation HDX-MS is a sensitive platform that can rapidly provide these nuanced maps of protein–ligand interactions.

DISCLAIMER

Certain commercial materials and equipment are identified in order to adequately specify the experimental procedure. Such identification neither implies recommendation or endorsement by the National Institute of Standards and Technology nor does it imply that the material or equipment identified is the best available for the purpose.

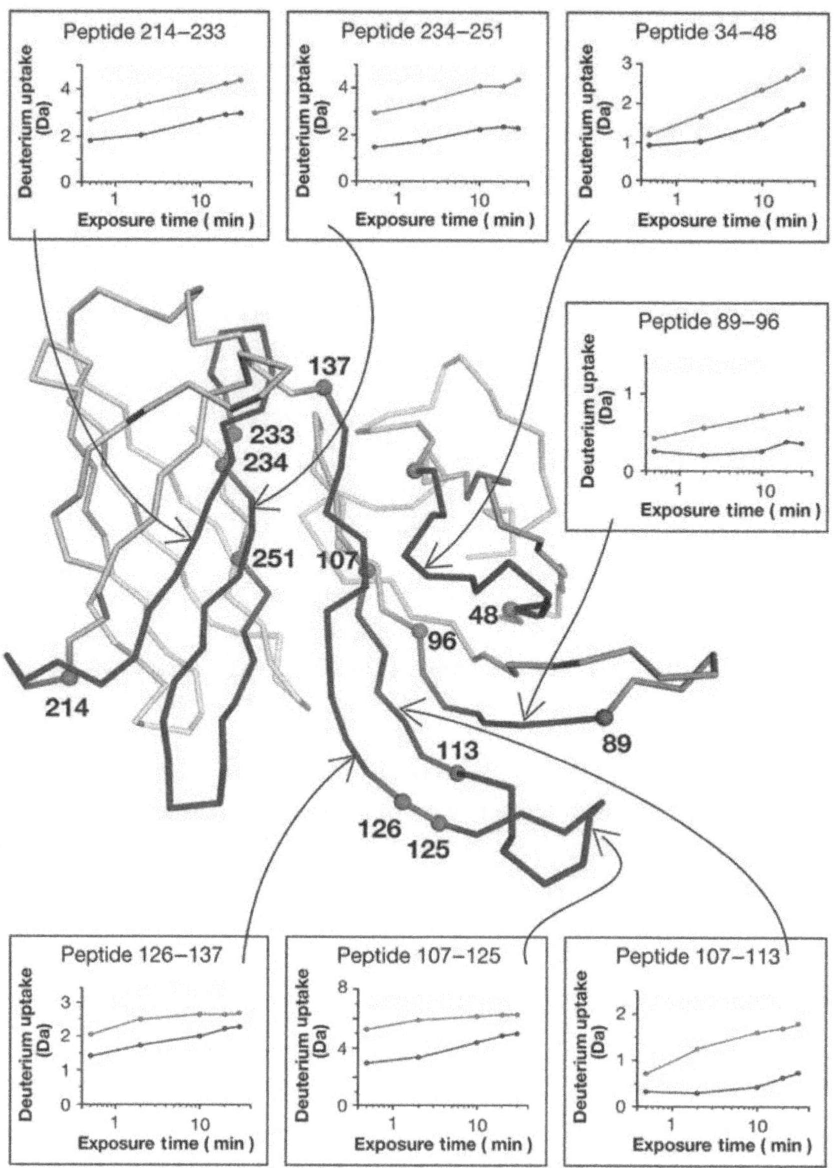

Figure 7 Map of the antigenic epitope of fHbp during interaction with mAb 12C1, as determined by HDX-MS. The peptides protected by mAb 12C1 are highlighted in red on the fHbp structure. Numbered spheres label the start and end of each affected fragment. Subpanels show the deuterium uptake observed in peptides of apo-fHbp (red, always upper line) and holo-fHbp (blue, always lower line), showing discordant behavior during 30 min immersion in D_2O. The relative arrangement of the lines is characteristic of Sowole–Konermann type 1 behavior. *Used with permission of Malito et al. (2013).* (See the color plate.)

REFERENCES

Abzalimov, R. R., Kaplan, D. A., Easterling, M. L., & Kaltashov, I. A. (2009). Protein conformations can be probed in top-down HDX MS experiments utilizing electron transfer dissociation of protein ions without hydrogen scrambling. *Journal of The American Society for Mass Spectrometry*, *20*(8), 1514–1517. http://dx.doi.org/10.1016/j.jasms.2009.04.006.

Ahn, J., Cao, M.-J., Yu, Y. Q., & Engen, J. R. (2013). Accessing the reproducibility and specificity of pepsin and other aspartic proteases. *Biochimica et Biophysica Acta—Proteins and Proteomics*, *1834*(6), 1222–1229. http://dx.doi.org/10.1016/j.bbapap.2012.10.003.

Ahn, J., Jung, M. C., Wyndham, K., Yu, Y. Q., & Engen, J. R. (2012). Pepsin immobilized on high-strength hybrid particles for continuous flow online digestion at 10000 psi. *Analytical Chemistry*, *84*(16), 7256–7262. http://dx.doi.org/10.1021/ac301749h.

Artimo, P., Jonnalagedda, M., Arnold, K., Baratin, D., Csardi, G., de Castro, E., et al. (2012). ExPASy: SIB bioinformatics resource portal. *Nucleic Acids Research*, *40*, W597–W603.

Bai, Y. W., Milne, J. S., Mayne, L., & Englander, S. W. (1993). Primary structure effects on peptide group hydrogen exchange. *Proteins*, *17*(1), 75–86. http://dx.doi.org/10.1002/prot.340170110.

Balasubramaniam, D., & Komives, E. A. (2013). Hydrogen-exchange mass spectrometry for the study of intrinsic disorder in proteins. *Biochimica et Biophysica Acta—Proteins and Proteomics*, *1834*(6), 1202–1209. http://dx.doi.org/10.1016/j.bbapap.2012.10.009.

Baldwin, R. L. (2011). Early days of protein hydrogen exchange: 1954–1972. *Proteins: Structure, Function, and Bioinformatics*, *79*, 2021–2026.

Barlow, D. J., Edwards, M. S., & Thornton, J. M. (1986). Continuous and dicontinuous protein antigenic determinants. *Nature*, *322*, 747–748.

Berman, H. M., Westbrook, J., Feng, Z., Gilliland, G., Bhat, T. N., Weissig, H., et al. (2000). The protein data bank. *Nucleic Acids Research*, *28*, 235–242. http://www.rcsb.org.

Bissantz, C., Kuhn, B., & Stahl, M. (2010). A medicinal chemist's guide to molecular interactions. *Journal of Medicinal Chemistry*, *53*(14), 5061–5084. http://dx.doi.org/10.1021/jm100112j.

Brier, S., & Engen, J. R. (2008). Hydrogen exchange mass spectrometry: Principles and capabilities. In M. Chance (Ed.), *Mass spectrometry analysis for protein-protein interactions and dynamics* (pp. 11–36). Hoboken: Wiley.

Brier, S., Fagnocchi, L., Donnarumma, D., Scarselli, M., Rappuoli, R., Nissum, M., et al. (2012). Structural insight into the mechanism of DNA-binding attenuation of the Neisserial adhesin repressor NadR by the small natural ligand 4-hydroxyphenylacetic acid. *Biochemistry*, *51*(34), 6738–6752. http://dx.doi.org/10.1021/bi300656w.

Burns, J. A., Butler, J. C., Moran, J., & Whitesides, G. M. (1991). Selective reduction of disulfides by tris(2-carboxyethyl)phosphine. *The Journal of Organic Chemistry*, *56*(8), 2648–2650. http://dx.doi.org/10.1021/jo00008a014.

Burns, K. M., Rey, M., Baker, C. A. H., & Schriemer, D. C. (2013). Platform dependencies in bottom-up hydrogen/deuterium exchange mass spectrometry. *Molecular & Cellular Proteomics*, *12*(2), 539–548.

Chalmers, M. J., Busby, S. A., Pascal, B. D., He, Y., Hendrickson, C. L., Marshall, A. G., et al. (2006). Probing protein ligand interactions by automated hydrogen/deuterium exchange mass spectrometry. *Analytical Chemistry*, *78*(4), 1005–1014. http://dx.doi.org/10.1021/ac051294f.

Chalmers, M. J., Busby, S. A., Pascal, B. D., Southern, M. R., & Griffin, P. R. (2007). A two-stage differential hydrogen deuterium exchange method for the rapid characterization of protein/ligand interactions. *Journal of Biomolecular Techniques*, *18*(4), 194–204.

Chen, Y.-B., Chattopadhyay, A., Bergen, P., Gadd, C., & Tannery, N. (2007). The Online Bioinformatics Resources Collection at the University of Pittsburgh Health Sciences

Library System—A one-stop gateway to online bioinformatics databases and software tools. *Nucleic Acids Research*, *35*, D780–D785. http://www.hsls.pitt.edu/obrc/index.php?page=protein_protein_interactions.

Chow, S.-C., & Ju, C. (2013). Assessing biosimilarity and interchangeability of biosimilar products under the Biologics Price Competition and Innovation Act. *Generics and Biosimilars Initiative Journal*, *2*(1), 20–25. http://dx.doi.org/10.5639/gabij.2013.0201.004.

Coales, S. J., Tomasso, J. C., & Hamuro, Y. (2008). Effects of electrospray capillary temperature on amide hydrogen exchange. *Rapid Communications in Mass Spectrometry*, *22*(9), 1367–1371. http://dx.doi.org/10.1002/rcm.3512.

Connelly, G. P., Bai, Y. W., Jeng, M. F., & Englander, S. W. (1993). Isotope effects in peptide group hydrogen exchange. *Proteins*, *17*(1), 87–92. http://dx.doi.org/10.1002/prot.340170111.

Cravello, L., Lascoux, D., & Forest, E. (2003). Use of different proteases working in acidic conditions to improve sequence coverage and resolution in hydrogen/deuterium exchange of large proteins. *Rapid Communications in Mass Spectrometry*, *17*(21), 2387–2393. http://dx.doi.org/10.1002/rcm.1207.

Das, A. A., Sharma, O. P., Kumar, M. S., Krishna, R., & Mathur, P. P. (2013). PepBind: A comprehensive database and computational tool for analysis of protein-peptide interactions. *Genomics, Proteomics & Bioinformatics*, *11*, 241–246. http://dx.doi.org/10.1016/j.gpb.2013.03.002.

Dempsey, C. E. (2001). Hydrogen exchange in peptides and proteins using NMR spectroscopy. *Progress in Nuclear Magnetic Resonance Spectroscopy*, *39*(2), 135–170. http://dx.doi.org/10.1016/S0079-6565(01)00032-2.

Diella, F., Haslam, N., Chica, C., Budd, A., Michael, S., Brown, N. P., et al. (2008). Understanding eukaryotic linear motifs and their role in cell signaling and regulation. *Frontiers in Bioscience*, *13*, 6580–6603.

Dinkel, H., Van Roey, K., Michael, S., Davey, N. E., Weatheritt, R. J., Born, D., et al. (2013). The eukaryotic linear motif resource ELM: 10 years and counting. *Nucleic Acids Research*, *42*, D259–D266. http://dx.doi.org/10.1093/nar/gkt1047.

Engen, J. R., & Smith, D. L. (2000). Investigating the higher order structure of proteins: Hydrogen exchange, proteolytic fragmentation & mass spectrometry. In J. R. Chapmann (Ed.), *Methods in Molecular Biology: 146. Protein and peptide analysis: New mass spectrometric applications* (pp. 93–110). New York: Chapmann, J. R.

Englander, S. W. (2006). Hydrogen exchange and mass spectrometry: A historical perspective. *Journal of The American Society for Mass Spectrometry*, *17*(11), 1481–1489. http://dx.doi.org/10.1016/j.jasms.2006.06.006.

Englander, S. W. (2015). *EXCEL Spreadsheets to calculate intrinsic exchange rate of amide protons*. http://hx2.med.upenn.edu/download.html.

Englander, S. W., & Kallenbach, N. R. (1984). Hydrogen-exchange and structural dynamics of proteins and nucleic acids. *Quarterly Reviews of Biophysics*, *16*(4), 521–655.

Englander, S. W., Mayne, L., Bai, Y., & Sosnick, T. R. (1997). Hydrogen exchange: The modern legacy of Linderstrom-Lang. *Protein Science*, *6*(5), 1101–1109.

Englander, S. W., Mayne, L., & Krishna, M. M. G. (2007). Protein folding and misfolding: Mechanism and principles. *Quarterly Reviews of Biophysics*, *40*(4), 287–326. http://dx.doi.org/10.1017/S0033583508004654.

Fang, J., Rand, K. D., Beuning, P. J., & Engen, J. R. (2011). False EX1 signatures caused by sample carryover during HX MS analyses. *International Journal of Mass Spectrometry*, *302*(1–3), 19–25. http://dx.doi.org/10.1016/j.ijms.2010.06.039.

Ferguson, P. L., Pan, J. X., Wilson, D. J., Dempsey, B., Lajoie, G., Shilton, B., et al. (2007). Hydrogen/deuterium scrambling during quadrupole time-of-flight MS/MS analysis of a zinc-binding protein domain. *Analytical Chemistry*, *79*(1), 153–160. http://dx.doi.org/10.1021/ac061261f.

Fleming, P. J., & Rose, G. D. (2005). Do all backbone polar groups in proteins form hydrogen bonds? *Protein Science*, *14*, 1911–1917.

Fournier, T., Medjoubi-N, N., & Porquet, D. (2000). Alpha-1-acid glycoprotein. *Biochimica et Biophysica Acta—Protein Structure and Molecular Enzymology*, *1482*(1–2), 157–171. http://dx.doi.org/10.1016/s0167-4838(00)00153-9.

Gilson, M. K., Given, J. A., Bush, B. L., & McCammon, J. A. (1997). The statistical-thermodynamic basis for computation of binding affinities: A critical review. *Biophysical Journal*, *72*(3), 1047–1069. http://dx.doi.org/10.1016/S0006-3495(97)78756-3.

Glasoe, P. K. L., & Long, F. A. (1960). Use of glass electrodes to measure acidities in deuterium oxide1,2. *The Journal of Physical Chemistry*, *64*, 188–190.

Gregory, R. B., Crabo, L., Percy, A. J., & Rosenberg, A. (1983). Water catalysis of peptide hydrogen isotope exchange. *Biochemistry*, *22*(4), 910–917. http://dx.doi.org/10.1021/bi00273a031.

Gu, Z. Y., Zitzewitz, J. A., & Matthews, C. R. (2007). Mapping the structure of folding cores in TIM barrel proteins by hydrogen exchange mass spectrometry: The roles of motif and sequence for the indole-3-glycerol phosphate synthase from *Sulfolobus solfataricus*. *Journal of Molecular Biology*, *368*(2), 582–594. http://dx.doi.org/10.1016/j.jmb.2007.02.027.

Guttman, M., Weis, D. D., Engen, J. R., & Lee, K. K. (2013). Analysis of overlapped and noisy hydrogen/deuterium exchange mass spectra. *Journal of The American Society for Mass Spectrometry*, *24*(12), 1906–1912. http://dx.doi.org/10.1007/s13361-013-0727-5.

Hamuro, Y., Coales, S. J., Molnar, K. S., Tuske, S. J., & Morrow, J. A. (2008). Specificity of immobilized porcine pepsin in H/D exchange compatible conditions. *Rapid Communications in Mass Spectrometry*, *22*(7), 1041–1046. http://dx.doi.org/10.1002/rcm.3467.

Harris, D. C. (2003). *Quantitative chemical analysis* (6th ed.). New York: Freeman and Company.

HDExaminer, Sierra Analytics, Inc., Modesto, CA. (2009). http://www.massspec.com/HDExaminer.html.

Houde, D., Arndt, J., Domeier, W., Berkowitz, S., & Engen, J. R. (2009). Characterization of IgG1 conformation and conformational dynamics by hydrogen/deuterium exchange mass spectrometry. *Analytical Chemistry*, *81*(7), 2644–2651. http://dx.doi.org/10.1021/ac802575y.

Houde, D., Berkowitz, S. A., & Engen, J. R. (2011). The utility of hydrogen/deuterium exchange mass spectrometry in biopharmaceutical comparability studies. *Journal of Pharmaceutical Sciences*, *100*(6), 2071–2086. http://dx.doi.org/10.1002/jps.22432.

Hu, W., Walters, B. T., Kan, Z.-Y., Mayne, L., Rosen, L. E., Marqusee, S., et al. (2013). Stepwise protein folding at near amino acid resolution by hydrogen exchange and mass spectrometry. *Proceedings of the National Academy of Sciences of the United States of America*, *110*(19), 7684–7689. http://dx.doi.org/10.1073/pnas.1305887110.

Huang, R. Y. C., & Hudgens, J. W. (2013). Effects of desialylation on human α1-acid glycoprotein–ligand interactions. *Biochemistry*, *52*(40), 7127–7136. http://dx.doi.org/10.1021/bi4011094.

Huang, R. Y.-C., Wen, J., Blankenship, R. E., & Gross, M. L. (2012). Hydrogen–deuterium exchange mass spectrometry reveals the interaction of Fenna–Matthews–Olson protein and chlorosome CsmA protein. *Biochemistry*, *51*(1), 187–193. http://dx.doi.org/10.1021/bi201620y.

Hudgens, J. W., Huang, R. Y.-C., & D'Ambro, E. (2016). Method validation and standards in hydrogen/deuterium exchange mass spectrometry. In D. Weis (Ed.), *Hydrogen-deuterium exchange mass spectrometry: Fundamentals, techniques and applications* (pp. 40–50). New York: Wiley. http://www.wiley.com/WileyCDA/WileyTitle/productCd-1118616499.html.

Hvidt, A., & Linderstrøm-Lang, K. (1954). Exchange of hydrogen atoms in insulin with deuterium atoms in aqueous solutions. *Biochimica et Biophysica Acta*, *14*, 574–575.

Hvidt, A., & Nielsen, S. O. (1966). Hydrogen exchange in proteins. *Advances in Protein Chemistry*, *21*, 287–386.

Iacob, R. E., Bou-Assaf, G. M., Makowski, L., Engen, J. R., Berkowitz, S. A., & Houde, D. (2013). Investigating monoclonal antibody aggregation using a combination of H/DX-MS and other biophysical measurements. *Journal of Pharmaceutical Sciences*, *102*(12), 4315–4329. http://dx.doi.org/10.1002/jps.23754.

Jones, L. M., Zhang, H., Vidavsky, I., & Gross, M. L. (2010). Online, high-pressure digestion system for protein characterization by hydrogen/deuterium exchange and mass spectrometry. *Analytical Chemistry*, *82*(4), 1171–1174. http://dx.doi.org/10.1021/ac902477u.

Jørgensen, T. J. D., Gårdsvoll, H., Ploug, M., & Roepstorff, P. (2005). Intramolecular migration of amide hydrogens in protonated peptides upon collisional activation. *Journal of the American Chemical Society*, *127*(8), 2785–2793. http://dx.doi.org/10.1021/ja043789c.

Józsi, M., & Zipfel, P. F. (2008). Factor H family proteins and human diseases. *Trends in Immunology*, *29*(8), 380–387. http://dx.doi.org/10.1016/j.it.2008.04.008.

Kaltashov, I. A., Bobst, C. E., & Abzalimov, R. R. (2013). Mass spectrometry-based methods to study protein architecture and dynamics. *Protein Science*, *22*(5), 530–544. http://dx.doi.org/10.1002/pro.2238.

Kaltashov, I. A., Bobst, C. E., Abzalimov, R. R., Berkowitz, S. A., & Houde, D. (2010). Conformation and dynamics of biopharmaceuticals: Transition of mass spectrometry-based tools from academe to industry. *Journal of The American Society for Mass Spectrometry*, *21*(3), 323–337. http://dx.doi.org/10.1016/j.jasms.2009.10.013.

Kan, Z.-Y., Mayne, L., Chetty, P. S., & Englander, S. W. (2011). ExMS: Data analysis for HX-MS experiments. *Journal of The American Society for Mass Spectrometry*, *22*, 1906–1915. http://dx.doi.org/10.1007/s13361-011-0236-3.

Kan, Z. Y., Walters, B. T., Mayne, L., & Englander, S. W. (2013). Protein hydrogen exchange at residue resolution by proteolytic fragmentation mass spectrometry analysis. *Proceedings of the National Academy of Sciences of the United States of America*, *110*(41), 16438–16443. http://dx.doi.org/10.1073/pnas.1315532110.

Katta, V., & Chait, B. T. (1991). Conformational-changes in proteins probed by hydrogen-exchange electrospray-ionization mass spectrometry. *Rapid Communications in Mass Spectrometry*, *5*(4), 214–217. http://dx.doi.org/10.1002/rcm.1290050415.

Kavan, D., & Man, P. (2011). MSTools—Web based application for visualization and presentation of HXMS data. *International Journal of Mass Spectrometry*, *302*(1–3), 53–58.

Kirschke, E., Goswami, D., Southworth, D., Griffin, P. R., & Agard, D. A. (2014). Glucocorticoid receptor function regulated by coordinated action of the Hsp90 and Hsp70 chaperone cycles. *Cell*, *157*(7), 1685–1697. http://dx.doi.org/10.1016/j.cell.2014.04.038.

Kreshuk, A., Stankiewicz, M., Lou, X., Kirchner, M., Hamprecht, F. A., & Mayer, M. P. (2011). Automated detection and analysis of bimodal isotope peak distributions in H/D exchange mass spectrometry using HeXicon. *International Journal of Mass Spectrometry*, *302*(1–3), 125–131.

Krishna, M. M. G., Hoang, L., Lin, Y., & Englander, S. W. (2004). Hydrogen exchange methods to study protein folding. *Methods*, *34*(1), 51–64. http://dx.doi.org/10.1016/j.ymeth.2004.03.005.

Krishna, M. M. G., Maity, H., Rumbley, J. N., Lin, Y., & Englander, S. W. (2006). Order of steps in the cytochrome c folding pathway: Evidence for a sequential stabilization mechanism. *Journal of Molecular Biology*, *359*(5), 1410–1419. http://dx.doi.org/10.1016/j.jmb.2006.04.035.

Ladner, J. E., Pan, M., Hurwitz, J., & Kelman, Z. (2011). Crystal structures of two active proliferating cell nuclear antigens (PCNAs) encoded by *Thermococcus kodakaraensis*. *Proceedings of the National Academy of Sciences of the United States of America*, *108*, 2711–2716.

Li, Z., Huang, R. Y.-C., Yopp, D. C., Hileman, T. H., Santangelo, T. J., Hurwitz, J., et al. (2014). A novel mechanism for regulating the activity of proliferating cell nuclear antigen by a small protein. *Nucleic Acids Research, 42*(9), 5776–5789. http://dx.doi.org/10.1093/nar/gku239.

Li, Z., Santangelo, T. J., Čuboňová, L., Reeve, J. N., & Kelman, Z. (2010). Affinity purification of an archaeal DNA replication protein network. *mBio, 1*(5), e00221–00210. http://dx.doi.org/10.1128/mBio.00221-10.

Liepinsh, E., & Otting, G. (1996). Proton exchange rates from amino acid side chains—Implications for image contrast. *Magnetic Resonance in Medicine, 35*, 30–42.

Lindner, R., Lou, X. H., Reinstein, J., Shoeman, R. L., Hamprecht, F. A., & Winkler, A. (2014). Hexicon 2: Automated processing of hydrogen-deuterium exchange mass spectrometry data with improved deuteration distribution estimation. *Journal of The American Society for Mass Spectrometry, 25*(6), 1018–1028. http://dx.doi.org/10.1007/s13361-014-0850-y.

Liu, S., Liu, L., Uzuner, U., Zhou, X., Gu, M., Shi, W., et al. (2011). HDX-Analyzer: A novel package for statistical analysis of protein structure dynamics. *BMC Bioinformatics, 12*(Suppl. 1), S43. http://dx.doi.org/10.1186/1471-2105-12-s1-s43.

López-Ferrer, D., Petritis, K., Robinson, E. W., Hixson, K. K., Tian, Z., Lee, J. H., et al. (2011). Pressurized pepsin digestion in proteomics: An automatable alternative to trypsin for integrated top-down bottom-up proteomics. *Molecular & Cellular Proteomics, 10*(2). http://dx.doi.org/10.1074/mcp.M110.001479. M110.001479.

Majumdar, R., Manikwar, P., Hickey, J. M., Arora, J., Middaugh, C. R., Volkin, D. B., et al. (2012). Minimizing carry-over in an online pepsin digestion system used for the H/D exchange mass spectrometric analysis of an IgG1 monoclonal antibody. *Journal of The American Society for Mass Spectrometry, 23*(12), 2140–2148. http://dx.doi.org/10.1007/s13361-012-0485-9.

Malito, E., Faleri, A., Lo Surdo, P., Veggi, D., Maruggi, G., Grassi, E., et al. (2013). Defining a protective epitope on factor H binding protein, a key meningococcal virulence factor and vaccine antigen. *Proceedings of the National Academy of Sciences of the United States of America, 110*(9), 3304–3309. http://dx.doi.org/10.1073/pnas.1222845110.

Mandell, J. G., Falick, A. M., & Komives, E. A. (1998). Identification of protein–protein interfaces by decreased amide proton solvent accessibility. *Proceedings of the National Academy of Sciences of the United States of America, 95*(25), 14705–14710.

Mayne, L., Kan, Z.-Y., Chetty, P. S., Ricciuti, A., Walters, B. T., & Englander, S. W. (2011). Many overlapping peptides for protein hydrogen exchange experiments by the fragment separation-mass spectrometry method. *Journal of The American Society for Mass Spectrometry, 22*, 1898–1905. http://dx.doi.org/10.1007/s13361-011-0235-4.

Miller, D. E., Prasannan, C. B., Villar, M. T., Fenton, A. W., & Artigues, A. (2012). HDXFinder: Automated analysis and data reporting of deuterium/hydrogen exchange mass spectrometry. *Journal of The American Society for Mass Spectrometry, 23*(2), 425–429. http://dx.doi.org/10.1007/s13361-011-0234-5.

Morgan, C. R., & Engen, J. R. (2009). Investigating solution-phase protein structure and dynamics by hydrogen exchange mass spectrometry. *Current Protocols in Protein Science, 58*. http://dx.doi.org/10.1002/0471140864.ps1706s58. Unit-17.16. 11–17.16. 17.

Mysling, S., Salbo, R., Ploug, M., & Jørgensen, T. J. D. (2014). Electrochemical reduction of disulfide-containing proteins for hydrogen/deuterium exchange monitored by mass spectrometry. *Analytical Chemistry, 86*(1), 340–345. http://dx.doi.org/10.1021/ac403269a.

Neduva, V., & Russell, R. B. (2006). Peptides mediating interaction networks: New leads at last. *Current Opinion in Biotechnology, 17*, 465–471.

Nelson, D. L., & Cox, M. M. (2012). *Lehninger principles of biochemistry* (6th ed.). New York: W.H. Freeman.

Palmblad, M., Buijs, J., & Håkansson, P. (2001). Automatic analysis of hydrogen/deuterium exchange mass spectra of peptides and proteins using calculations of isotopic distributions. *Journal of The American Society for Mass Spectrometry, 12*(11), 1153–1162.

Pan, J. X., & Borchers, C. H. (2014). Top-down mass spectrometry and hydrogen/deuterium exchange for comprehensive structural characterization of interferons: Implications for biosimilars. *Proteomics, 14*(10), 1249–1258. http://dx.doi.org/10.1002/pmic.201300341.

Pan, J. X., Han, J., & Borchers, C. H. (2012). Top-down hydrogen/deuterium exchange and ECD-stitched FTICR-MS for probing structural dynamics of a 29-kDa enzyme. *International Journal of Mass Spectrometry, 325*, 130–138. http://dx.doi.org/10.1016/j.ijms.2012.06.021.

Pan, J., Han, J., Borchers, C. H., & Konermann, L. (2009). Hydrogen/deuterium exchange mass spectrometry with top-down electron capture dissociation for characterizing structural transitions of a 17 kDa protein. *Journal of the American Chemical Society, 131*(35), 12801–12808. http://dx.doi.org/10.1021/ja904379w.

Pan, L. Y., Salas-Solano, O., & Valliere-Douglass, J. F. (2014). Conformation and dynamics of interchain cysteine-linked antibody-drug conjugates as revealed by hydrogen/deuterium exchange mass spectrometry. *Analytical Chemistry, 86*(5), 2657–2664. http://dx.doi.org/10.1021/ac404003q.

Pandit, D., Tuske, S. J., Coales, S. J., E, S. Y., Liu, A., Lee, J. E., et al. (2012). Mapping of discontinuous conformational epitopes by amide hydrogen/deuterium exchange mass spectrometry and computational docking. *Journal of Molecular Recognition, 25*(3), 114–124. http://dx.doi.org/10.1002/jmr.1169.

Pascal, B. D., Chalmers, M. J., Busby, S. A., & Griffin, P. R. (2009). HD Desktop: An integrated platform for the analysis and visualization of H/D exchange data. *Journal of The American Society for Mass Spectrometry, 20*(4), 601–610. http://dx.doi.org/10.1016/j.jasms.2008.11.019.

Pascal, B. D., Chalmers, M. J., Busby, S. A., Mader, C. C., Southern, M. R., Tsinoremas, N. F., et al. (2007). The Deuterator: Software for the determination of backbone amide deuterium levels from H/D exchange MS data. *BMC Bioinformatics, 8*, 156. http://www.biomedcentral.com/1471-2105/8/156.

Pascal, B. D., Willis, S., Lauer, J. L., Landgraf, R. R., West, G. M., Marciano, D., et al. (2012). HDX Workbench: Software for the analysis of H/D exchange MS data. *Journal of The American Society for Mass Spectrometry, 23*(9), 1512–1521. http://dx.doi.org/10.1007/s13361-012-0419-6.

Pirrone, G. F., Iacob, R. E., & Engen, J. R. (2015). Applications of hydrogen/deuterium exchange MS from 2012 to 2014. *Analytical Chemistry, 87*(1), 99–118. http://dx.doi.org/10.1021/ac5040242.

Press, W. H., Teukolsky, S. A., Vetterling, W. T., & Flannery, B. P. (2007). *Numerical recipes. The art of scientific computing* (3rd ed.). Cambridge, U.K.: Cambridge University Press.

TR Foundation (2015). *The R project for statistical computing.* Retrieved 10 May 2015, from, http://www.r-project.org/.

Rand, K. D. (2013). Pinpointing changes in higher-order protein structure by hydrogen/deuterium exchange coupled to electron transfer dissociation mass spectrometry. *International Journal of Mass Spectrometry, 338*, 2–10. http://dx.doi.org/10.1016/j.ijms.2012.08.010.

Rand, K. D., & Jørgensen, T. J. D. (2007). Development of a peptide probe for the occurrence of hydrogen (H-1/H-2) scrambling upon gas-phase fragmentation. *Analytical Chemistry, 79*(22), 8686–8693. http://dx.doi.org/10.1021/ac0710782.

Rand, K. D., Zehl, M., Jensen, O. N., & Jørgensen, T. J. D. (2009). Protein hydrogen exchange measured at single-residue resolution by electron transfer dissociation mass spectrometry. *Analytical Chemistry*, *81*(14), 5577–5584. http://dx.doi.org/10.1021/ac9008447.

Rey, M., Man, P., Brandolin, G., Forest, E., & Pelosi, L. (2009). Recombinant immobilized rhizopuspepsin as a new tool for protein digestion in hydrogen/deuterium exchange mass spectrometry. *Rapid Communications in Mass Spectrometry*, *23*(21), 3431–3438. http://dx.doi.org/10.1002/rcm.4260.

Rey, M., Sarpe, V., Burns, K. M., Buse, J., Baker, C. A. H., van Dijk, M., et al. (2014). Mass spec studio for integrative structural biology. *Structure*, *22*(10), 1538–1548. http://dx.doi.org/10.1016/j.str.2014.08.013.

Rey, M., Yang, M. L., Burns, K. M., Yu, Y. P., Lees-Miller, S. P., & Schriemer, D. C. (2013). Nepenthesin from monkey cups for hydrogen/deuterium exchange mass spectrometry. *Molecular & Cellular Proteomics*, *12*(2), 464–472. http://dx.doi.org/10.1074/mcp.M112.025221.

Sarkar, A., & Wintrode, P. L. (2011). Effects of glycosylation on the stability and flexibility of a metastable protein: The human serpin [alpha]1-antitrypsin. *International Journal of Mass Spectrometry*, *302*(1–3), 69–75.

Schneider, M. C., Prosser, B. E., Caesar, J. J. E., Kugelberg, E., Li, S., Zhang, Q., et al. (2009). *Neisseria meningitidis* recruits factor H using protein mimicry of host carbohydrates. *Nature*, *458*, 890–893.

Sheff, J. G., Rey, M., & Schriemer, D. C. (2013). Peptide-column interactions and their influence on back exchange rates in hydrogen/deuterium exchange-MS. *Journal of The American Society for Mass Spectrometry*, *24*(7), 1006–1015. http://dx.doi.org/10.1007/s13361-013-0639-4.

Skinner, J. J., Lim, W. K., Bédard, S., Black, B. E., & Englander, S. W. (2012). Protein dynamics viewed by hydrogen exchange. *Protein Science*, *21*, 996–1005.

Slysz, G. W., Baker, C. A. H., Bozsa, B. M., Dang, A., Percy, A. J., Bennett, M., et al. (2009). Hydra: Software for tailored processing of H/D exchange data from MS or tandem MS analyses. *BMC Bioinformatics*, *10*, 162. http://dx.doi.org/10.1186/1471-2105-10-162.

Sowole, M. A., & Konermann, L. (2014). Effects of protein-ligand interactions on hydrogen/deuterium exchange kinetics: Canonical and noncanonical scenarios. *Analytical Chemistry*, *86*(13), 6715–6722. http://dx.doi.org/10.1021/ac501849n.

Sterling, H. J., & Williams, E. R. (2010). Real-time hydrogen/deuterium exchange kinetics via supercharged electrospray ionization tandem mass spectrometry. *Analytical Chemistry*, *82*(21), 9050–9057. http://dx.doi.org/10.1021/ac101957x.

Thevenon-Emeric, G., Kozlowski, J., Zhang, Z., & Smith, D. L. (1992). Determination of amide hydrogen exchange rates in peptides by mass spectrometry. *Analytical Chemistry*, *64*, 2456–2458.

Thornton, J. M., Edwards, M. S., Taylor, W. R., & Barlow, D. J. (1986). Location of 'continuous' antigenic determinants in the protruding regions of proteins. *The EMBO Journal*, *5*, 409–413.

Tukey, J. (1949). Comparing individual means in the analysis of variance. *Biometrics*, *5*(2), 99–114.

Tzeng, S.-R., & Kalodimos, C. G. (2009). Dynamic activation of an allosteric regulatory protein. *Nature*, *462*, 368–372.

Walters, B. T., Ricciuti, A., Mayne, L., & Englander, S. W. (2012). Minimizing back exchange in the hydrogen exchange-mass spectrometry experiment. *Journal of The American Society for Mass Spectrometry*, *23*(12), 2132–2139. http://dx.doi.org/10.1007/s13361-012-0476-x.

Wang, L. T., Pan, H., & Smith, D. L. (2002). Hydrogen exchange-mass spectrometry—Optimization of digestion conditions. *Molecular & Cellular Proteomics*, *1*(2), 132–138. http://dx.doi.org/10.1074/mcp.M100009-MCP200.

Waters Corp. (2015). *DynamX Data Analysis Software v 3.0.* Retrieved 24 July 2015 from, http://www.waters.com/webassets/cms/library/docs/720005145en.pdf.

Weinkam, P., Zimmermann, J., Romesberg, F. E., & Wolynes, P. G. (2010). The folding energy landscape and free energy excitations of cytochrome c. *Accounts of Chemical Research, 43*(5), 652–660. http://dx.doi.org/10.1021/ar9002703.

Weis, D. D., Wales, T. E., Engen, J. R., Hotchko, M., & Ten Eyck, L. F. (2006). Identification and characterization of EX1 kinetics in H/D exchange mass spectrometry by peak width analysis. *Journal of The American Society for Mass Spectrometry, 17*(11), 1498–1509. http://dx.doi.org/10.1016/j.jasms.2006.05.014.

Wooll, J. O., Wrabl, J. O., & Hilser, V. J. (2000). Ensemble modulation as an origin of denaturant-independent hydrogen exchange in proteins. *Journal of Molecular Biology, 301*(2), 247–256. http://dx.doi.org/10.1006/jmbi.2000.3889.

Xu, H. (2010). *MassMatrix Database Search Engine.* Retrieved 20 April 2015 from, http://www.massmatrix.net/mm-cgi/home.py.

Yang, M., Hoeppner, M., Rey, M., Kadek, A., Man, P., & Schriemer, D. C. (2015). Recombinant nepenthesin II for hydrogen/deuterium exchange mass spectrometry. *Analytical Chemistry, 87*, 6681–6687. http://dx.doi.org/10.1021/acs.analchem.5b00831.

Zehl, M., Rand, K. D., Jensen, O. N., & Jørgensen, T. J. D. (2008). Electron transfer dissociation facilitates the measurement of deuterium incorporation into selectively labeled peptides with single residue resolution. *Journal of the American Chemical Society, 130*(51), 17453–17459. http://dx.doi.org/10.1021/ja805573h.

Zhang, H. M., Kazazic, S., Schaub, T. M., Tipton, J. D., Emmett, M. R., & Marshall, A. G. (2008). Enhanced digestion efficiency, peptide ionization efficiency, and sequence resolution for protein hydrogen/deuterium exchange monitored by Fourier transform ion cyclotron resonance mass spectrometry. *Analytical Chemistry, 80*(23), 9034–9041. http://dx.doi.org/10.1021/ac801417d.

Zhang, Z., & Marshall, A. G. (1998). A universal algorithm for fast and automated charge state deconvolution of electrospray mass-to-charge ratio spectra. *Journal of The American Society for Mass Spectrometry, 9*, 225–233.

Zhang, J., Ramachandran, P., Kumar, R., & Gross, M. L. (2013). H/D exchange centroid monitoring is insufficient to show differences in the behavior of protein states. *Journal of The American Society for Mass Spectrometry, 24*(3), 450–453. http://dx.doi.org/10.1007/s13361-012-0555-z.

Zhang, Z., & Smith, D. L. (1993). Determination of amide hydrogen exchange by mass spectrometry: A new tool for protein structure elucidation. *Protein Science, 2*, 522–531.

Zhang, Q., Willison, L. N., Tripathi, P., Sathe, S. K., Roux, K. H., Emmett, M. R., et al. (2011). Epitope mapping of a 95 kDa antigen in complex with antibody by solution-phase amide backbone hydrogen/deuterium exchange monitored by Fourier transform ion cyclotron resonance mass spectrometry. *Analytical Chemistry, 83*(18), 7129–7136. http://dx.doi.org/10.1021/ac201501z.

CHAPTER FIFTEEN

Isotope Labeling of Biomolecules: Structural Analysis of Viruses by HDX-MS

Miklos Guttman[1], Kelly K. Lee[1]
Department of Medicinal Chemistry, University of Washington, Seattle, Washington, USA
[1]Corresponding authors: e-mail address: mguttman@u.washington.edu; klee@u.washington.edu

Contents

Abstract

The structural analysis of viruses is often a complex task. In many cases, the details of the viral architecture, especially for enveloped viruses, are limited to low-resolution techniques such as electron microscopy. These structural proteins and assemblies of viruses often populate multiple conformational states and undergo dramatic structural changes, making them difficult to study by most structural methods. They also frequently include highly dynamic regions that are of key functional importance. Many viruses present large surface glycoproteins, which have also proved to be challenging for structural biology due to the intrinsic flexibility and heterogeneity of the glycan decorations. Over the past two decades, hydrogen deuterium exchange coupled to mass spectrometry (HDX-MS) has provided a wealth of information on many diverse viral proteins, glycoproteins, and complexes, in many cases, in multiple conformational states. Here, we describe the methodology for using HDX-MS to investigate the rich structural dynamics of viral systems, and we briefly review the type of systems that have been examined through this type of approach. Though the technique is relatively simple, several potential pitfalls exist at both the sample preparation and the data analysis stage that investigators should be aware of for obtaining reliable data.

Methods in Enzymology, Volume 566
ISSN 0076-6879
http://dx.doi.org/10.1016/bs.mie.2015.05.021

ABBREVIATIONS

ESI electrospray ionization
FA formic acid
HDX-MS hydrogen deuterium exchange coupled to mass spectrometry
TFA trifluoroacetic acid
UPLC ultra-high-pressure liquid chromatography

1. INTRODUCTION

Hydrogen deuterium exchange utilizes the naturally occurring proton exchange that occurs at amides on proteins in solution. By incubating a protein in its native solution state in a deuterated buffer, the amide protons exchange for deuterium. The rate of this exchange is governed by solution conditions such as temperature and pH and also is highly dependent on the local environment of the amide (Englander, 2006; Marcsisin & Engen, 2010). Amides involved in stable hydrogen bonds, such as in alpha helices and beta-sheet secondary structure, exchange several orders of magnitude slower than exposed amides. The conformational dynamics of the structural elements will also influence the amide exchange rate, as breathing and dynamic fluctuations expose hydrogen-bonded amides for brief periods, allowing their protons to exchange with solvent. The kinetics of amide exchange thus can reveal sequence-specific information about secondary structure as well as the local conformational dynamics throughout a protein.

As a technique hydrogen deuterium exchange has been around since the 1950s, but only in the early 1990s was mass spectrometry applied for analyzing deuterium uptake in intact proteins (Englander, 2006; Katta & Chait, 1991). After exchange under native conditions, the deuterium at backbone amides is locked in place by lowering the temperature and the pH to a regime where amide exchange is minimized or "quenched" (~0 °C and pH 2.5), thus enabling analysis by mass spectrometry. Acid-active proteases, such as pepsin, can be used under such quench conditions to digest the protein while leaving the deuterium label in place, thus providing local information on the exchange throughout the protein (Zhang & Smith, 1993; Fig. 1).

The real limitation on the size or complexity of the protein/complexes that can be analyzed by HDX-MS is the ability to resolve and detect the large number of peptides generated. Over the past two decades, these limits have continual expanded thanks to developments in liquid chromatography and

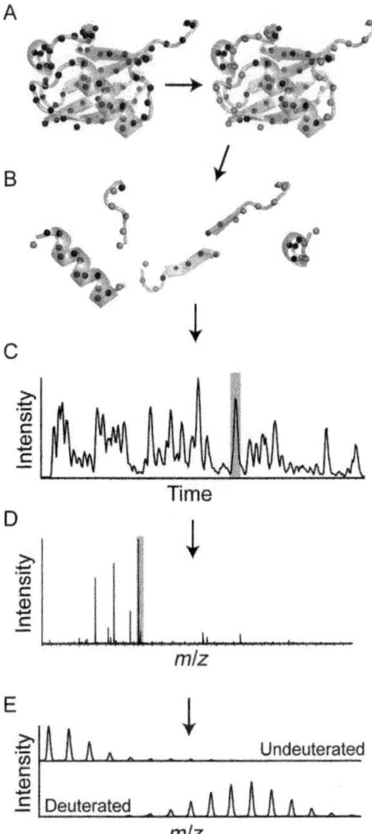

Figure 1 Overview of HDX-MS. (A) The protein is incubated in deuterated buffer under native conditions to allow the amide protons to exchange for deuterium. (B) The exchange is then slowed to lock the deuterium in place and an acid protease is used to rapidly digest the protein into peptides. (C) The peptides are resolved by fast reverse phase liquid chromatography, which is directly infused into the mass spectrometer. (D) The mass spectra are recorded for each eluting peptide where the mass shift due to deuterium incorporation for each deuteration time is quantified (E). (See the color plate.)

mass spectrometry instrumentation. Ultra-high-pressure liquid chromatography (UPLC) offers increased chromatographic resolution that greatly increases the complexity of samples that can be analyzed, more effectively resolving the hundreds of peptides in a typical digest reaction (Wales, Fadgen, Gerhardt, & Engen, 2008). The continual improvement in the sensitivity and resolution of mass spectrometers also expands the limits on sample complexity while also helps to lower the limit of detection. With modern mass spectrometers, complete HDX analysis can be accomplished with as little as ~100 pmol, or around 5 μg of a 50-kDa protein. Software for HDX-MS data processing has also greatly facilitated the analysis of experimental data, though manual validation remains a critical part of generating reliable, complete datasets (Iacob & Engen, 2012).

Because HDX-MS provides a sensitive method for probing local structural dynamics, it has become widely used for mapping protein–protein or protein–ligand interactions, monitoring conformational changes induced by ligand binding or changes in solution conditions, and comparing variants of proteins such as mutants that may differ in stability (Chalmers et al., 2011; Englander, 2006; Marcsisin & Engen, 2010). The method has also been adopted by the biopharmaceutical industry for assessing conformational purity and comparing biosimilars (Houde, Berkowitz, & Engen, 2011). While HDX-MS has primarily been employed to study soluble proteins, with additional steps for detergent or lipid removal prior to LC–MS, it has become possible to investigate the structure of membrane proteins as well (Hebling et al., 2010; Kim et al., 2011; Rey et al., 2010; Zhang, Chien, et al., 2010).

A major focus of structural virology is aimed at obtaining an understanding of construction and structural basis for stability of the protein capsids that encapsulate the viral genomic information. While the structures of many intact capsids have been determined by crystallography and electron microscopy, due to their large size, monitoring the structural or dynamic changes during assembly and subsequent maturation can be challenging. Large-scale changes also take place during receptor binding, entry, and uncoating or delivery of the viral genome. HDX-MS provides unique insights into structural changes in viral capsids and can be used to compare capsids that are in different conformational states, helping to pinpoint the regions that undergo significant local reorganization (Tuma, Coward, Kirk, Barnes, & Prevelige, 2001). HDX-MS has been used to measure local conformational dynamics and gain insight into the mechanism of assembly and capsid maturation for various bacteriophages (Domitrovic et al., 2013; Fu & Prevelige, 2006; Gertsman, Fu, Huang, Komives, & Johnson, 2010; Gertsman et al., 2009; Kang, Hawkridge, Johnson, Muddiman, & Prevelige, 2006; Morton et al., 2010; Veesler et al., 2014), human immunodeficiency virus (HIV) (Lanman et al., 2003, 2004; Monroe, Kang, Kyere, Li, & Prevelige, 2010), Rous sarcoma virus (Bush et al., 2014), Adenovirus (Snijder et al., 2014), Hepatitis B (Bereszczak, Watts, Wingfield, Steven, & Heck, 2014), Brome mosaic virus (Wang, Lane, & Smith, 2001), and Rhinovirus (Wang & Smith, 2005).

One strength of HDX-MS is the ability to detect bimodal spectra, reflecting distinct conformations that a given region of a protein may populate in a sample. Under most exchange conditions, amides are only exposed for short times through fast local structural fluctuations and take up deuterium gradually ("EX2" kinetics shown in Fig. 2A). However, in some cases, proteins can undergo large motions that expose several amides for long periods leading to the complete exchange of this region ("EX1" kinetics shown in Fig. 2B). The result is a bimodal mass spectrum with two populations whose relative intensities shift over the time course of deuteration (Zhang, Post, & Smith, 1996). Such EX1 behavior has been observed in Hepatitis B capsids providing a more detailed level of information about their conformational dynamics (Bereszczak et al., 2014). Alternatively, a bimodal will also result if there are two distinct noninterchanging

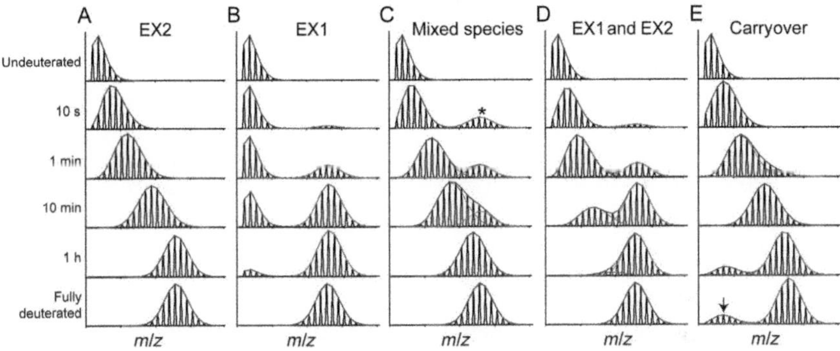

Figure 2 Examples of different exchange behaviors during HDX-MS. (A) Most commonly peptides uptake deuterium gradually by exposing amides through fast local structural fluctuations (EX2). (B) Large protein motions occurring on relatively slow timescales will expose large segments for long periods of time, whereby all amides become fully exchanged in a correlated manner (EX1). This is evidenced by a bimodal distribution where the population shifts to the exchanged species over the course of deuterium exchange. (C) Samples may contain populations of alternate conformational states. This is evidenced by a bimodal where the relative intensities of the two species remain constant over the course of deuteration. In this case, there is approximately 20% contribution to the signal from an unfolded population that exchanges fully by the earliest time point (denoted by *). (D) Proteins can undergo motions on various timescales resulting in mixed EX1/EX2 kinetics. In this case, the peptide is gradually deuterated (EX2) while also shifting to the fully deuterated state via EX1. (E) Sample carryover during LC–MS analysis can cause artifactual bimodals. In this case, the 1 min and 1 h time points show an additional component corresponding to ~15% carryover from a previous injection. The fact that it is also evident in the fully deuterated spectrum (depicted by an arrow), which should be uniformly and completely deuterated, is strong evidence that carryover is responsible for the bimodal behavior. (See the color plate.)

conformations present within an ensemble (Fig. 2C). This type of behavior was observed in HK-97 bacteriophage capsids, where HDX-MS revealed two conformational states that correlated well with the known quasi-equivalent states within the capsid assembly (Gertsman, Komives, & Johnson, 2010). This type of spectra can also reveal conformational heterogeneity in a sample, for example, the presence of some misfolded or aggregated protein (Garcia, Guttman, Ebner, & Lee, 2015; Kaltashov, Bobst, Abzalimov, Berkowitz, & Houde, 2010).

Many viruses contain proteins that interact with host proteins during infection and may be in part or largely unstructured on their own. Unlike many structural techniques, HDX-MS can provide information on even intrinsically disordered proteins in their solutions state (Balasubramaniam & Komives, 2013). Using short timescale deuterium exchange, even weak residual secondary structure can be mapped (Keppel & Weis, 2015). HDX-MS has been used to elucidate the residual structure present in HIV Nef (Hochrein et al., 2006) and nonstructural proteins from Dengue virus (Zhao et al., 2015).

An important question for virology and vaccinology is how the viral surface proteins mediate viral infectivity while at the same time protecting themselves from immune surveillance. In many cases, the surface proteins are highly glycosylated as a means to protect the protein from the immune system (Scanlan, Offer, Zitzmann, & Dwek, 2007). Even for HDX-MS, viral glycoproteins present a challenge not only because material is often only available in very limited quantities, but also because the glycosylation on particular N-linked or O-linked sites tends to exhibit varying degrees of heterogeneity. This results in a multiplicity of ions that all may share the same underlying peptide sequence but have different glycan decorations (microheterogeneity) (Varki et al., 2009). The signal for glycopeptides thus can be relatively low, especially for glycopeptides that have multiple glycan chains. Furthermore, glycopeptides may be too hydrophilic to be retained by the reverse phase columns commonly used for HDX-MS. Together these factors can dramatically limit the detection and sequence coverage of glycosylated segments.

Despite these challenges, in recent years, HDX-MS has been used to examine Hepatitis C E2 glycoprotein (Khan et al., 2014), influenza hemagglutinin (Garcia et al., 2015), HIV envelope glycoprotein (Davenport et al., 2013; Guttman et al., 2014; Guttman et al., 2015; Kim et al., 2011; Kong et al., 2010), and Ebola GP1/GP2 (Bale et al., 2011), revealing their conformational dynamics in solution and providing insight into mechanisms of viral glycoprotein recognition and virus neutralization by antibodies and small molecule inhibitors. The continued progress in the advancement of LC and MS technologies should eventually enable examination of viral proteins on their intact, infectious virions. The first steps have been taken toward applying HDX-MS to analyze surface glycoproteins *in situ* on whole enveloped viruses (Garcia et al., 2015).

2. METHODS

2.1 Initial Considerations for HDX-MS and Sample Requirements

For an HDX experiment, it is important to get suitable deuteration of the protein resulting in a final volume amenable to LC–MS analysis and minimizing deuterium loss (back-exchange) once the exchange is quenched. All samples within a series must be treated identically to ensure there are no deviations in percent deuterium content, buffer conditions, or extent of back-exchange between samples. The buffers should be selected so that the protein retains its native conformation throughout the deuterium exchange, and it is useful to perform native and SDS-PAGE before and after HDX to ensure that longer incubations in deuterium do not result in aggregation or degradation, both of which would confound the analysis.

The starting protein solution should be made to allow a 5- to 20-fold dilution into deuteration buffer, followed by at least a two-fold dilution into the quenching buffer. The final quenched sample should yield around 10–100 pmol of protein per injection on LC–MS, depending instrument sensitivity. The final volume is often limited by the size of the injection loop, typically 100 or 200 μL. It is generally preferable to inject on the higher side to get better data for peptides of weaker intensity, thereby gaining more complete sequence coverage. However, it is important to be aware that too much signal in the MS may cause detector saturation, which can distort the mass envelope, leading to inaccurate deuteration measurements. For aggregation-prone proteins, the starting concentration may need to be lower, and final volumes of up to 1 mL can be used, but will require longer injection times.

Buffers for HDX, especially the quench solution, should be made with LC–MS grade water. Using MilliQ water is not advised as it often contains polymeric species that can interfere with MS analysis. For studies that aim to examine binding of small molecules, including a cosolvent such as 1–2% dimethyl sulfoxide may be required for compound solubility. In this case, it is also important to include the cosolvent in the unliganded protein exchanges, as the solvent may subtly perturb the structure of the protein and thereby alter the deuterium exchange kinetics. It is also important to note that when the ligand is in large excess, the compound may give rise to a large peak in the LC gradient that can affect the detection of the peptides.

2.2 Deuterium Exchange Reaction

The exchange reaction is initiated by diluting the protein stock into deuterated buffer. Ideally, the deuteration times should be selected to get an adequate measure of the exchange rate for all amides. With many proteins, unprotected amides will be exchanged within milliseconds, while very highly ordered amides may require weeks. It is generally best to sample a broad timescale from seconds to several hours to maximize the temporal sampling. It is also important to consider that both pH and temperature have a large effect on the intrinsic exchange rate (the rate that an unprotected, completely accessible amide exchanges with solvent deuterium). Increasing the pH by 1 or the temperature by ~24 °C will increase the exchange rate 10-fold (Bai, Milne, Mayne, & Englander, 1993). One approach for expanding the temporal range covered by HDX-MS has been to run parallel experiments under different conditions and combine the exchange data (Coales et al., 2010).

Whenever working with D_2O (>99.9%), use freshly opened containers and keep samples well sealed as the D_2O will exchange with moisture in the air. If the HDX data are to be used for thermodynamic measurements, then it is important measure the pD during deuterium exchange (Bai et al., 1993; Hamuro et al., 2003). The pH* (the pH measured in D_2O) should be recorded for the buffers, which is used to calculate the pD (Krezel & Bal, 2004). It is also important to record the exact temperature during the deuterium exchanges. The downstream LC–MS should be compatible with nearly all buffer systems commonly used for proteins (with the exception detergents). For neutral pH samples, phosphate buffer is preferable as it also contributes to buffering capacity at the pH used for the subsequent quench step, pH 2.5.

For experiments that aim to compare two or more sets of data by HDX, it is advisable to include in an internal standard to provide an exact measure of the exchange conditions. This is best accomplished by spiking the starting protein stock solution with the tetrapeptide HN-Pro-Pro-Pro-Ile-OH

(PPPI), which contains only a single, slow-exchanging amide (Zhang, Zhang, & Xiao, 2012) (~10 pmol per LC–MS injection works well in our hands). This provides a means to assess whether the exchange conditions were identical in all samples under comparison. Alternatively, the standard's exchange profile can be used to normalize data sets performed under different buffer conditions. Small peptides can also be spiked and preequilibrated in the buffered D_2O solution prior to the addition of the protein, to be used to test and correct for any deviations in back-exchange between samples (Zhang et al., 2012). When performing studies with protein–ligand complexes, it is necessary to use a sufficient excess of ligand and preequilibrate the complex to ensure that the complex is fully formed and remains intact during deuterium exchange.

2.2.1 Quenching the Exchange

After deuteration for a specified time, the amide exchange is slowed (or "quenched") by lowering the temperature to ~0 °C and the pH to 2.5. This decreases the exchange rate by an order of about 10,000 where the average half-life of the deuterium label is ~1 h. The protein can then be either left intact for global deuterium examination or rapidly digested under quench conditions with an acid-active protease to generate peptides that retain the deuterium labels. In either case, there will still be a small amount of deuterium loss (back-exchange) that should be kept as consistent as possible between samples. Quenching is typically performed by mixing deuteration reactions 1:1 into ice-cold acidification buffer (e.g., 0.1% trifluoroacetic acid [TFA]). Mock samples (containing buffers but not protein) are used to empirically adjust the pH of the quench solution to obtain a final pH of 2.5 after mixing. It is best to use thin walled PCR tubes for more efficient heat transfer during the quench step.

2.2.2 Digestion Optimization and Peptide Identification

Unless the deuteration is to be monitored on the intact protein level, a major consideration is whether enough peptides can be observed throughout the protein by LC–MS to comprehensively monitor the entire sequence (a metric referred to as the "sequence coverage"). An early step in HDX-MS studies is optimizing quench and digest conditions to provide the highest possible observable peptides to ensure maximal sequence coverage. Since the sample needs to be maintained at low pH, the availability of proteases is limited. Most commonly pepsin, which is very robust and works under quench conditions, is used. Alternative acid proteases such as protease XVII from Rhizhopus, protease type XIII from Aspergillus Saitoi, and Nepenthesin protease have also been used with HDX-MS to provide complementary sequence coverage due to their differences in sequence specificity (Cravello, Lascoux, & Forest, 2003; Kadek et al., 2014). Due to the nonspecific nature of the proteases used, multiple partially overlapping peptides are often observed. This is beneficial as it provides multiple data points for a given region of the protein providing a higher level of redundancy.

Many proteins are sufficiently destabilized under such acidic conditions to enable subsequent digestion, though some may require additional denaturants to increase accessibility of the peptide backbone to the protease. In this case, the quench solution can be supplemented with urea of guanidine to help unfold the protein. For disulfide bond-containing proteins, tris-2-carboxy-ethylphosphine (TCEP) is included in the quench to reduce disulfides during protease digestion (Zhang, McLoughlin, et al., 2010). The amount of denaturant/TCEP, digestion time, and the protein:enzyme ratio should be varied when determining optimal quench and digest conditions. Pepsin will still retain activity in 8 *M* urea, up to 2 *M* GndHCl, or up to 1 *M* TCEP. Some hyper-stable proteins may be so protease resistant that even with high concentrations of denaturants under quench conditions the digestion will produce very few peptides, severely limiting the sequence coverage.

Some proteins may require detergent for solubility or stability that can be included during deuterium exchange. It is possible to supplement buffers with low concentrations of certain nonionic detergents (e.g., 0.1% *n*-dodecyl maltoside or β-octyl glucoside) that are compatible with LC–MS. For higher concentrations of detergents, an additional organic extraction step can be used for detergent removal after quenching (Rey et al., 2010). For membrane-embedded proteins, zirconium affinity has been shown to be an effective approach for phospholipid depletion under quench conditions (Hebling et al., 2010).

Digestion can be carried out manually with soluble or immobilized protease. Alternatively, in many cases, an online protease column is incorporated into the LC–MS system. With in-solution digestion, a method of rapidly mixing the protease into the quenched solution is to add the small volume of protease solution into the cap of the microcentrifuge tube. This way after the protein is added to the quench it is possible to immediately vortex and efficiently combine the protease with the quenched sample. Immobilized proteases or protease columns have the advantage of generating a digest with a lower background (no autolysis fragments, or other impurities from the protease). Immobilized pepsin on high-pressure resistant particles is now also commercially available (Enzymate column, Waters). This enables digestions at pressures up to 10,000 psi, which can greatly aid in digestion efficiency by disrupting the protein's residual structure, without contributing to back-exchange (Ahn, Jung, Wyndham, Yu, & Engen, 2012). One disadvantage of using online digestion columns is that they add another source of carryover (discussed below) during LC, which can confound HDX-MS data.

Another choice for HDX sample preparation is whether to freeze the sample or proceed directly to LC–MS. Since it is best to match all conditions as closely as possible for all deuteration time points, many labs perform all deuterium reactions side by side and flash freeze in liquid nitrogen. Deuterated samples can be stored at −80 °C for months with no detectable loss of the deuteration. Modern automated HDX systems that utilize robots to perform exchanges immediately inject the quenched samples onto LC–MS. This approach has automated the mixing, quenching, and injection steps and has enabled HDX as a tool for high-throughput analysis and screening (Chalmers et al., 2011; Englander et al., 2003). Only pepsin columns compatible with HDX-MS should be used, as some pepsin columns have been observed to cause severe deuterium loss (Wu, Kaveti, & Engen, 2006).

2.2.3 Additional HDX Controls

An undeuterated control is made by identical dilution and subsequent digestion of the protein in purely aqueous buffer. This sample is used for peptide identification and also serves as a reference for deuterium shift analysis. A fully or maximally deuterated control is helpful to include as they serve as a reference for whether deuterium exchange went to completion at the longest time points. For simple comparative studies (i.e., free protein vs. ligand bound), the quantitative differences at each time point will already highlight the locations where changes occur, and the fully deuterated control may not be necessary.

Preparation of a fully deuterated sample may not always be trivial. It has to achieve full deuteration, but the buffer should match the other time points to ensure consistent back-exchange conditions. With many proteins unfolding a starting protein stock with denaturant (e.g., heating in 3–6 *M* GndHCl), diluting into the buffered D_2O and extensive heating (65 °C for 1 h) should provide a fully deuterated sample that approximately matches the buffer composition of the other samples. Performing the deuteration step in high concentrations of denaturant (i.e., GndHCl) may significantly offset the intrinsic exchange rate (Zhang et al., 2012), and thereby lead to altered back-exchange. Since the fully deuterated sample is fully denatured (and reduced), it can also have a different digestion profile (due to increased digestion efficiency). Using a lower concentration of protease can help generate digestion profiles to match closer to the other deuterated samples while keeping the same level of back-exchange. In some cases, it may be necessary to digest the protein first and deuterate the peptides. For this approach, the exchange and quench buffers must be carefully matched to the other time points to ensure identical back-exchange.

During the digestion step, the sample is exposed to the residual deuterium content after the quench mixing. This can cause artifactual "in-exchange," especially with longer digestion times, where the unfolded and digested protein slowly takes up deuterium in the quenched solution (Hoofnagle, Resing, & Ahn, 2003). Some peptides will appear to have become partially deuterated when in fact they did not exchange at all during the deuterium incubation step. For this reason, many labs perform a zero time point correction by premixing the quench with the D_2O buffer, and subsequently adding the stock protein, thereby only sampling the in-exchange.

2.2.4 Liquid Chromatography–Mass Spectrometry

Once the sample is digested, it is then analyzed by mass spectrometry to measure incorporation of deuterium. It is critical that quench conditions be maintained throughout the analysis to avoid further back-exchange. Both matrix-assisted laser desorption ionization (MALDI) and electrospray ionization (ESI) have been used for HDX-MS studies (Mandell, Falick, & Komives, 1998; Zhang & Smith, 1993); however, the coupling of ESI to reverse-phase UPLC has made it the dominant method for analysis, in general providing greater sequence coverage and more complete structural analysis. In a typical setup, the quenched sample is desalted and preconcentrated on a trapping column, and subsequently the peptides are resolved with a relatively short (10 min) gradient, though for highly complex samples longer gradients can be used (Walters, Ricciuti, Mayne, & Englander, 2012). LC gradients should be empirically optimized for each sample, but a good starting condition is 5–40% B over 10 min (A: 0.1% formic acid (FA), 5% acetonitrile with ~0.02% TFA for pH 2.5; B: 0.1% FA, 100% acetonitrile). The syringe, injection loop, lines, and columns are kept at or near 0 °C to minimize back-exchange. Systems for maintaining cold temperature LC runs can be constructed rather inexpensively (Keppel, Jacques, Young, Ratzlaff, & Weis, 2011). Alternatively, commercial systems are now also available that maintain low temperatures and automate the injection process (HDX Manager, Waters).

To avoid any deviations in the data that could arise from day-to-day variation, it is always preferable to collect data on all samples on the same day with identical conditions (Houde et al., 2011). A major concern during LC–MS data collection is residual sample from a previous injection appearing in the subsequent run (sample "carryover"). Significant carryover will result in a convoluted mass envelope leading to an inaccurate measure of deuterium incorporation (Fig. 2E). Therefore, minimizing such carryover during a sequence of injections is critical for reliable HDX-MS data. Elaborate washing protocols have been established to minimize this (Fang, Rand, Beuning, & Engen, 2011); however, it is still advisable to run periodic blanks to ensure that carryover is not significant (<5%). Incorporation of an online protease column adds another source of carryover, which requires an additional set of wash cycles (Majumdar et al., 2012). In this case, it is also important to keep in mind that some of the harsher wash solvents for cleaning the trapping column should not be passed over the protease column.

High-resolution quadrupole time-of-flight mass (Q-TOF) analyzers have been popular for HDX-MS as they provide the necessary resolution for observing each isotopic peak. More recent FT-based instruments offer even higher resolution, but suffer from potential destructive interference effects that can result in inaccurate deuteration measurements (Burns, Rey, Baker, & Schriemer, 2013). Another consideration with the mass spectrometer is that the heating during ESI will also result in some back-exchange (Katta & Chait, 1993). For this reason, HDX-MS on most instruments is conducted using a lower source temperature. MS instrumentation with an ion mobility stage adds an additional dimension for the gas-phase separation of peptides, thereby expanding the complexity of samples that can be analyzed (Iacob, Murphy, & Engen, 2008). The limits on sample size and complexity for HDX-MS are expected to expand in parallel with the continual advancement of LC and MS technologies.

2.2.5 Peptide Identification

Due to the characteristic proteolytic activity of pepsin, the peptides observed in HDX-MS vary greatly in length, ranging from 3 to 30 residues, and identification is accomplished using exact mass and fragmentation data from an undeuterated sample. Due to the inherent hydrogen scrambling by collision-induced dissociation (CID), MS/MS of deuterated peptides does not yield additional information regarding deuterium localization (Bache,

Rand, Roepstorff, Ploug, & Jorgensen, 2008; Ferguson & Konermann, 2008). However, MS/MS by electron capture (Abzalimov, Kaplan, Easterling, & Kaltashov, 2009) and electron transfer dissociation (Rand, Pringle, Morris, Engen, & Brown, 2011) provide gas-phase fragmentation without inducing hydrogen scrambling, thus enabling deuterium exchange measurements with single-residue resolution (Landgraf, Chalmers, & Griffin, 2012).

Identification of glycosylated peptides, which are common on viral surface proteins, is more challenging as conventional CID MS/MS yields little fragmentation of the peptide backbone (Medzihradszky, 2005). Electron transfer dissociation is often used as an effective alternative MS/MS tool for identification of glycopeptides (Syka, Coon, Schroeder, Shabanowitz, & Hunt, 2004). Samples can also be reneutralized after digestion, treated with endoglycosidases to remove glycans, and then used to obtain CID fragmentation data for the deglycosylated peptides. With regard to glycopeptides, it is also important to note that amido groups within the glycan chains also take on and retain deuterium during HDX-MS analysis (Guttman, Scian, & Lee, 2011).

2.2.6 HDX Data Analysis

Once the datasets have been collected, the deuterium incorporation needs to be extracted from the mass spectra. Most often, determination of deuterium uptake for a given peptide involves measuring the geometric center (centroid) of each mass envelope for each time point for each observable peptide, which can be hundreds to thousands of spectra. To this end, countless software tools have been developed to provide automation and turn weeks of manual analysis into days. Today, there are both commercial and academic software packages including: DynamX (Waters); HD-Examiner (Sierra Analytics), Mass Spec Studio (Rey et al., 2014), HD-Workbench (Pascal et al., 2012), EX-MS (Kan, Mayne, Chetty, & Englander, 2011), and HX-Express (Weis, Engen, & Kass, 2006), to name only a few.

Techniques have also been developed for utilizing overlapping peptides, commonly generated by the nonspecific proteases used for HDX-MS, for obtaining higher resolution exchange information (Fajer, Bou-Assaf, & Marshall, 2012; Zhang et al., 2012). However, unpredictable back-exchange due to residual secondary structure or column interactions adds a complication that can confound this approach to gaining higher resolution information from the data (Sheff, Rey, & Schriemer, 2013). It should be appreciated that a simple centroid analysis is insufficient for interpreting the full information content within HDX data (Zhang, Ramachandran, Kumar, & Gross, 2013). Slow correlated protein motions, or mixtures of conformational states in a sample, can give rise to bimodal mass spectra, which can sometimes be subtle and not readily evident (Fig. 2; Weis, Wales, Engen, Hotchko, & Ten Eyck, 2006). Recent software developments have begun to address detection and even deconvolution of such bimodal mass spectra (Chik, Vande Graaf, & Schriemer, 2006; Guttman, Weis, Engen, & Lee, 2013; Kan et al., 2011).

HDX-MS is often used for mapping protein–protein and protein–ligand interactions (Coales, Tuske, Tomasso, & Hamuro, 2009; Mandell, Baerga-Ortiz, Falick, & Komives, 2005). However, ligand binding can cause long-range allosteric effects, which are also readily observed by HDX-MS (Chalmers et al., 2011). When comparing the effects of ligand binding, it is also important to consider the pathway of deuterium exchange. For some protein–ligand complexes, the predominant exchange pathway actually occurs through the unliganded state during transient ligand dissociations (Wildes & Marqusee, 2005). These caveats should be considered during the interpretation of HDX-MS comparisons, especially in the absence of other structural information.

3. EXAMPLE HDX-MS EXPERIMENT WITH BOVINE RIBONUCLEASE A

Prepare buffers[1]: wherever possible use LC–MS grade water for all solutions;

20 × phosphate-buffered saline (PBS): 400 m*M* sodium phosphate, pH 7.4, 150 m*M* sodium chloride, 1% sodium azide, 20 m*M* ethylenediamine-tetraacetic acid (EDTA);

Ribonuclease A (RNAseA) stock: 0.1 mg/mL RNAseA in 1 × PBS;

Pepsin solution: 1 mg/mL pepsin (3 × recrystallized, Worthington Biochemicals) in 100 m*M* sodium phosphate, pH 4.0;

Quench buffer: 0.2% FA, 100 m*M* TCEP (Sigma-Aldrich) in LC–MS grade water;

Buffered D_2O[2]: Freshly opened D_2O (>99.9%, Cambridge isotope labs), mixed with 20 × PBS for 1 ×.

[1] Optional: include 0.5 μg/mL of the tetrapeptide: Pro-Pro-Pro-Ile (PPPI).

[2] Keep this solution sealed as the D_2O will exchange with ambient moisture.

Prepare mock samples for optimizing the quench pH:

90 μL buffered D_2O,

10 μL 1 × PBS,

100 μL quench buffer,

10 μL pepsin buffer (100 m*M* sodium phosphate, pH 4.0).

Vortex and test the pH. Make small additions of 1 *M* hydrochloric acid or 1 *M* sodium hydroxide to the master quench solution and continue to prepare mock samples until the final pH is 2.5. Approximately 2 μL of 10 *M* sodium hydroxide per milliliters of quench solution will be needed to achieve this final pH. Have a container of liquid nitrogen ready for freezing-quenched samples.

3.1 Deuteration and Quenching

To initiate deuterium exchanges add 10 μL of the RNAseA stock into 90 μL of buffered D_2O, pipette up and down to mix, seal the tube. For the sake of consistency, it is best to use the same pipettor and identical pipette tips for all deuterium exchanges. For incubation times longer than 1 h, it is advisable to wrap the tube cap with parafilm to avoid exchange with ambient moisture. About 1 min before the desired deuteration time, place the quench tube with 100 μL of quench solution on ice. Open the tube and place 10 μL of pepsin solution in the cap, leave on ice with the cap open. At the desired deuteration time, pipette the entirety of the deuteration mixture and inject into the quench tube. Quickly close and vortex the tube to mix and initiate digestion. Leave the tube on ice for 5 min to reduce and digest, and flash freeze in liquid nitrogen. For incubations less than 15 s, place 10 μL RNAseA in new tube. Using a 100 μL pipette transfer 90 μL of buffered D_2O to the protein, mix rapidly and with the same pipette tip, transfer the full volume into the quench tube. For incubations greater than 8 h, it is best to initiate the exchange the evening before and leave sealed overnight, so it can be quenched side by side with the shorter incubations. Samples can be stored at −80 °C for months.

Undeuterated sample

Prepared as all deuterium incubations but with buffered H_2O instead of buffered D_2O.

Fully deuterated sample

Prepare a denatured solution of RNAseA at 0.1 mg/mL:

0.1 mg/mL RNAseA,

1 × PBS,

4 *M* guanidine hydrochloride,

10 m*M* dithiothreitol (DTT),

Heat at 90 °C for 30 min.

Mix 10 μL denatured RNAseA into 90 μL buffered D_2O, seal the cap with parafilm, and heat to 65 °C for 1 h. Allow the sample to cool and quench, digest, and freeze as described above.

Optional: zero time point

Prepare a quench tube with quench buffer and pepsin as described above. In quick succession, add 90 μL of buffered D_2O and then 10 μL of RNAseA stock. Digest and freeze as described above.

ACKNOWLEDGMENTS

We wish to thank Elizabeth A. Komives, David D. Weis, and John R. Engen for insightful discussions, Matthew Honaker for assistance with data analysis, and Dale Whittington and Ross Lawrence for keeping our mass spectrometers in excellent condition. This work was supported by NIH grants R01-GM099989 and F32-GM097805 (M.G.), as well as the Bill and Melinda Gates Foundation Collaboration for AIDS Vaccine Discovery (CAVD) grant OPP1033102.

REFERENCES

Abzalimov, R. R., Kaplan, D. A., Easterling, M. L., & Kaltashov, I. A. (2009). Protein conformations can be probed in top-down HDX MS experiments utilizing electron transfer dissociation of protein ions without hydrogen scrambling. *Journal of the American Society for Mass Spectrometry*, *20*(8), 1514–1517.

Ahn, J., Jung, M. C., Wyndham, K., Yu, Y. Q., & Engen, J. R. (2012). Pepsin immobilized on high-strength hybrid particles for continuous flow online digestion at 10,000 psi. *Analytical Chemistry*, *84*(16), 7256–7262.

Bache, N., Rand, K. D., Roepstorff, P., Ploug, M., & Jorgensen, T. J. (2008). Hydrogen atom scrambling in selectively labeled anionic peptides upon collisional activation by MALDI tandem time-of-flight mass spectrometry. *Journal of the American Society for Mass Spectrometry*, *19*(12), 1719–1725.

Bai, Y., Milne, J. S., Mayne, L., & Englander, S. W. (1993). Primary structure effects on peptide group hydrogen exchange. *Proteins*, *17*(1), 75–86.

Balasubramaniam, D., & Komives, E. A. (2013). Hydrogen-exchange mass spectrometry for the study of intrinsic disorder in proteins. *Biochimica et Biophysica Acta*, *1834*(6), 1202–1209.

Bale, S., Liu, T., Li, S., Wang, Y., Abelson, D., Fusco, M., et al. (2011). Ebola virus glycoprotein needs an additional trigger, beyond proteolytic priming for membrane fusion. *PLoS Neglected Tropical Diseases*, *5*(11), e1395.

Bereszczak, J. Z., Watts, N. R., Wingfield, P. T., Steven, A. C., & Heck, A. J. (2014). Assessment of differences in the conformational flexibility of hepatitis B virus core-antigen and e-antigen by hydrogen deuterium exchange-mass spectrometry. *Protein Science*, *23*(7), 884–896.

Burns, K. M., Rey, M., Baker, C. A., & Schriemer, D. C. (2013). Platform dependencies in bottom-up hydrogen/deuterium exchange mass spectrometry. *Molecular & Cellular Proteomics*, *12*(2), 539–548.

Bush, D. L., Monroe, E. B., Bedwell, G. J., Prevelige, P. E., Jr., Phillips, J. M., & Vogt, V. M. (2014). Higher-order structure of the Rous sarcoma virus SP assembly domain. *Journal of Virology*, *88*(10), 5617–5629.

Chalmers, M. J., Pascal, B. D., Willis, S., Zhang, J., Iturria, S. J., Dodge, J. A., et al. (2011). Methods for the analysis of high precision differential hydrogen deuterium exchange data. *International Journal of Mass Spectrometry*, *302*(1–3), 59–68.

Chik, J. K., Vande Graaf, J. L., & Schriemer, D. C. (2006). Quantitating the statistical distribution of deuterium incorporation to extend the utility of H/D exchange MS data. *Analytical Chemistry*, *78*(1), 207–214.

Coales, S. J., E, S. Y., Lee, J. E., Ma, A., Morrow, J. A., & Hamuro, Y. (2010). Expansion of time window for mass spectrometric measurement of amide hydrogen/deuterium exchange reactions. *Rapid Communications in Mass Spectrometry*, *24*(24), 3585–3592.

Coales, S. J., Tuske, S. J., Tomasso, J. C., & Hamuro, Y. (2009). Epitope mapping by amide hydrogen/deuterium exchange coupled with immobilization of antibody, on-line proteolysis, liquid chromatography and mass spectrometry. *Rapid Communications in Mass Spectrometry*, *23*(5), 639–647.

Cravello, L., Lascoux, D., & Forest, E. (2003). Use of different proteases working in acidic conditions to improve sequence coverage and resolution in hydrogen/deuterium exchange of large proteins. *Rapid Communications in Mass Spectrometry*, *17*(21), 2387–2393.

Davenport, T. M., Guttman, M., Guo, W., Cleveland, B., Kahn, M., Hu, S. L., et al. (2013). Isolate-specific differences in the conformational dynamics and antigenicity of HIV-1 gp120. *Journal of Virology*, *87*(19), 10855–10873.

Domitrovic, T., Movahed, N., Bothner, B., Matsui, T., Wang, Q., Doerschuk, P. C., et al. (2013). Virus assembly and maturation: Auto-regulation through allosteric molecular switches. *Journal of Molecular Biology*, *425*(9), 1488–1496.

Englander, S. W. (2006). Hydrogen exchange and mass spectrometry: A historical perspective. *Journal of the American Society for Mass Spectrometry*, *17*(11), 1481–1489.

Englander, J. J., Del Mar, C., Li, W., Englander, S. W., Kim, J. S., Stranz, D. D., et al. (2003). Protein structure change studied by hydrogen-deuterium exchange, functional labeling, and mass spectrometry. *Proceedings of the National Academy of Sciences of the United States of America*, *100*(12), 7057–7062.

Fajer, P. G., Bou-Assaf, G. M., & Marshall, A. G. (2012). Improved sequence resolution by global analysis of overlapped peptides in hydrogen/deuterium exchange mass spectrometry. *Journal of the American Society for Mass Spectrometry*, *23*(7), 1202–1208.

Fang, J., Rand, K. D., Beuning, P. J., & Engen, J. R. (2011). False EX1 signatures caused by sample carryover during HX MS analyses. *International Journal of Mass Spectrometry*, *302*(1–3), 19–25.

Ferguson, P. L., & Konermann, L. (2008). Nonuniform isotope patterns produced by collision-induced dissociation of homogeneously labeled ubiquitin: Implications for spatially resolved hydrogen/deuterium exchange ESI-MS studies. *Analytical Chemistry*, *80*(11), 4078–4086.

Fu, C. Y., & Prevelige, P. E., Jr. (2006). Dynamic motions of free and bound O29 scaffolding protein identified by hydrogen deuterium exchange mass spectrometry. *Protein Science*, *15*(4), 731–743.

Garcia, N. K., Guttman, M., Ebner, J. L., & Lee, K. K. (2015). Dynamic changes during acid-induced activation of influenza hemagglutinin. *Structure*, *23*(4), 665–676.

Gertsman, I., Fu, C. Y., Huang, R., Komives, E. A., & Johnson, J. E. (2010). Critical salt bridges guide capsid assembly, stability, and maturation behavior in bacteriophage HK97. *Molecular & Cellular Proteomics*, *9*(8), 1752–1763.

Gertsman, I., Gan, L., Guttman, M., Lee, K., Speir, J. A., Duda, R. L., et al. (2009). An unexpected twist in viral capsid maturation. *Nature*, *458*(7238), 646–650.

Gertsman, I., Komives, E. A., & Johnson, J. E. (2010). HK97 maturation studied by crystallography and H/2H exchange reveals the structural basis for exothermic particle transitions. *Journal of Molecular Biology*, *397*(2), 560–574.

Guttman, M., Cupo, A., Julien, J. P., Sanders, R. W., Wilson, I. A., Moore, J. P., et al. (2015). Antibody potency relates to the ability to recognize the closed, pre-fusion form of HIV Env. *Nature Communications*, *6*, 6144.

Guttman, M., Garcia, N. K., Cupo, A., Matsui, T., Julien, J. P., Sanders, R. W., et al. (2014). CD4-induced activation in a soluble HIV-1 Env trimer. *Structure*, *22*, 974–984.

Guttman, M., Scian, M., & Lee, K. K. (2011). Tracking hydrogen/deuterium exchange at glycan sites in glycoproteins by mass spectrometry. *Analytical Chemistry*, *83*(19), 7492–7499.

Guttman, M., Weis, D. D., Engen, J. R., & Lee, K. K. (2013). Analysis of overlapped and noisy hydrogen/deuterium exchange mass spectra. *Journal of the American Society for Mass Spectrometry*, *24*(12), 1906–1912.

Hamuro, Y., Coales, S. J., Southern, M. R., Nemeth-Cawley, J. F., Stranz, D. D., & Griffin, P. R. (2003). Rapid analysis of protein structure and dynamics by hydrogen/deuterium exchange mass spectrometry. *Journal of Biomolecular Techniques*, *14*(3), 171–182.

Hebling, C. M., Morgan, C. R., Stafford, D. W., Jorgenson, J. W., Rand, K. D., & Engen, J. R. (2010). Conformational analysis of membrane proteins in phospholipid bilayer nanodiscs by hydrogen exchange mass spectrometry. *Analytical Chemistry*, *82*(13), 5415–5419.

Hochrein, J. M., Wales, T. E., Lerner, E. C., Schiavone, A. P., Smithgall, T. E., & Engen, J. R. (2006). Conformational features of the full-length HIV and SIV Nef proteins determined by mass spectrometry. *Biochemistry*, *45*(25), 7733–7739.

Hoofnagle, A. N., Resing, K. A., & Ahn, N. G. (2003). Protein analysis by hydrogen exchange mass spectrometry. *Annual Review of Biophysics and Biomolecular Structure*, *32*, 1–25.

Houde, D., Berkowitz, S. A., & Engen, J. R. (2011). The utility of hydrogen/deuterium exchange mass spectrometry in biopharmaceutical comparability studies. *Journal of Pharmaceutical Sciences*, *100*(6), 2071–2086.

Iacob, R. E., & Engen, J. R. (2012). Hydrogen exchange mass spectrometry: Are we out of the quicksand? *Journal of the American Society for Mass Spectrometry*, *23*(6), 1003–1010.

Iacob, R. E., Murphy, J. P., 3rd., & Engen, J. R. (2008). Ion mobility adds an additional dimension to mass spectrometric analysis of solution-phase hydrogen/deuterium exchange. *Rapid Communications in Mass Spectrometry*, *22*(18), 2898–2904.

Kadek, A., Mrazek, H., Halada, P., Rey, M., Schriemer, D. C., & Man, P. (2014). Aspartic protease nepenthesin-1 as a tool for digestion in hydrogen/deuterium exchange mass spectrometry. *Analytical Chemistry*, *86*(9), 4287–4294.

Kaltashov, I. A., Bobst, C. E., Abzalimov, R. R., Berkowitz, S. A., & Houde, D. (2010). Conformation and dynamics of biopharmaceuticals: Transition of mass spectrometry-based tools from academe to industry. *Journal of the American Society for Mass Spectrometry*, *21*(3), 323–337.

Kan, Z. Y., Mayne, L., Chetty, P. S., & Englander, S. W. (2011). ExMS: Data analysis for HX-MS experiments. *Journal of the American Society for Mass Spectrometry*, *22*(11), 1906–1915.

Kang, S., Hawkridge, A. M., Johnson, K. L., Muddiman, D. C., & Prevelige, P. E., Jr. (2006). Identification of subunit-subunit interactions in bacteriophage P22 procapsids by chemical cross-linking and mass spectrometry. *Journal of Proteome Research*, *5*(2), 370–377.

Katta, V., & Chait, B. T. (1991). Conformational changes in proteins probed by hydrogen-exchange electrospray-ionization mass spectrometry. *Rapid Communications in Mass Spectrometry*, *5*(4), 214–217.

Katta, V., & Chait, B. T. (1993). Hydrogen/deuterium exchange electrospray ionization mass spectrometry: A method for probing protein conformational changes in solution. *Journal of the American Chemical Society*, *115*(14), 6317–6321.

Keppel, T. R., Jacques, M. E., Young, R. W., Ratzlaff, K. L., & Weis, D. D. (2011). An efficient and inexpensive refrigerated LC system for H/D exchange mass spectrometry. *Journal of the American Society for Mass Spectrometry*, *22*(8), 1472–1476.

Keppel, T. R., & Weis, D. D. (2015). Mapping residual structure in intrinsically disordered proteins at residue resolution using millisecond hydrogen/deuterium exchange and residue averaging. *Journal of the American Society for Mass Spectrometry*, *26*(4), 547–554.

Khan, A. G., Whidby, J., Miller, M. T., Scarborough, H., Zatorski, A. V., Cygan, A., et al. (2014). Structure of the core ectodomain of the hepatitis C virus envelope glycoprotein 2. *Nature*, *509*(7500), 381–384.

Kim, M., Sun, Z. Y., Rand, K. D., Shi, X., Song, L., Cheng, Y., et al. (2011). Antibody mechanics on a membrane-bound HIV segment essential for GP41-targeted viral neutralization. *Nature Structural & Molecular Biology*, *18*(11), 1235–1243.

Kong, L., Huang, C. C., Coales, S. J., Molnar, K. S., Skinner, J., Hamuro, Y., et al. (2010). Local conformational stability of HIV-1 gp120 in unliganded and CD4-bound states as defined by amide hydrogen/deuterium exchange. *Journal of Virology*, *84*(19), 10311–10321.

Krezel, A., & Bal, W. (2004). A formula for correlating pK_a values determined in D_2O and H_2O. *Journal of Inorganic Biochemistry*, *98*(1), 161–166.

Landgraf, R. R., Chalmers, M. J., & Griffin, P. R. (2012). Automated hydrogen/deuterium exchange electron transfer dissociation high resolution mass spectrometry measured at single-amide resolution. *Journal of the American Society for Mass Spectrometry*, *23*(2), 301–309.

Lanman, J., Lam, T. T., Barnes, S., Sakalian, M., Emmett, M. R., Marshall, A. G., et al. (2003). Identification of novel interactions in HIV-1 capsid protein assembly by high-resolution mass spectrometry. *Journal of Molecular Biology*, *325*(4), 759–772.

Lanman, J., Lam, T. T., Emmett, M. R., Marshall, A. G., Sakalian, M., & Prevelige, P. E., Jr. (2004). Key interactions in HIV-1 maturation identified by hydrogen-deuterium exchange. *Nature Structural & Molecular Biology*, *11*(7), 676–677.

Majumdar, R., Manikwar, P., Hickey, J. M., Arora, J., Middaugh, C. R., Volkin, D. B., et al. (2012). Minimizing carry-over in an online pepsin digestion system used for the H/D exchange mass spectrometric analysis of an IgG1 monoclonal antibody. *Journal of the American Society for Mass Spectrometry*, *23*(12), 2140–2148.

Mandell, J. G., Baerga-Ortiz, A., Falick, A. M., & Komives, E. A. (2005). Measurement of solvent accessibility at protein-protein interfaces. *Methods in Molecular Biology*, *305*, 65–80.

Mandell, J. G., Falick, A. M., & Komives, E. A. (1998). Measurement of amide hydrogen exchange by MALDI-TOF mass spectrometry. *Analytical Chemistry*, *70*(19), 3987–3995.

Marcsisin, S. R., & Engen, J. R. (2010). Hydrogen exchange mass spectrometry: What is it and what can it tell us? *Analytical and Bioanalytical Chemistry*, *397*(3), 967–972.

Medzihradszky, K. F. (2005). Characterization of protein N-glycosylation. *Methods in Enzymology*, *405*, 116–138.

Monroe, E. B., Kang, S., Kyere, S. K., Li, R., & Prevelige, P. E., Jr. (2010). Hydrogen/deuterium exchange analysis of HIV-1 capsid assembly and maturation. *Structure*, *18*(11), 1483–1491.

Morton, V. L., Burkitt, W., O'Connor, G., Stonehouse, N. J., Stockley, P. G., & Ashcroft, A. E. (2010). RNA-induced conformational changes in a viral coat protein studied by hydrogen/deuterium exchange mass spectrometry. *Physical Chemistry Chemical Physics*, *12*(41), 13468–13475.

Pascal, B. D., Willis, S., Lauer, J. L., Landgraf, R. R., West, G. M., Marciano, D., et al. (2012). HDX workbench: Software for the analysis of H/D exchange MS data. *Journal of the American Society for Mass Spectrometry*, *23*(9), 1512–1521.

Rand, K. D., Pringle, S. D., Morris, M., Engen, J. R., & Brown, J. M. (2011). ETD in a traveling wave ion guide at tuned Z-spray ion source conditions allows for site-specific hydrogen/deuterium exchange measurements. *Journal of the American Society for Mass Spectrometry*, *22*(10), 1784–1793.

Rey, M., Mrazek, H., Pompach, P., Novak, P., Pelosi, L., Brandolin, G., et al. (2010). Effective removal of nonionic detergents in protein mass spectrometry, hydrogen/deuterium exchange, and proteomics. *Analytical Chemistry*, *82*(12), 5107–5116.

Rey, M., Sarpe, V., Burns, K. M., Buse, J., Baker, C. A., van Dijk, M., et al. (2014). Mass spec studio for integrative structural biology. *Structure*, *22*(10), 1538–1548.

Scanlan, C. N., Offer, J., Zitzmann, N., & Dwek, R. A. (2007). Exploiting the defensive sugars of HIV-1 for drug and vaccine design. *Nature*, *446*(7139), 1038–1045.

Sheff, J. G., Rey, M., & Schriemer, D. C. (2013). Peptide-column interactions and their influence on back exchange rates in hydrogen/deuterium exchange-MS. *Journal of the American Society for Mass Spectrometry*, *24*(7), 1006–1015.

Snijder, J., Benevento, M., Moyer, C. L., Reddy, V., Nemerow, G. R., & Heck, A. J. (2014). The cleaved N-terminus of pVI binds peripentonal hexons in mature adenovirus. *Journal of Molecular Biology*, *426*(9), 1971–1979.

Syka, J. E., Coon, J. J., Schroeder, M. J., Shabanowitz, J., & Hunt, D. F. (2004). Peptide and protein sequence analysis by electron transfer dissociation mass spectrometry. *Proceedings of the National Academy of Sciences of the United States of America*, *101*(26), 9528–9533.

Tuma, R., Coward, L. U., Kirk, M. C., Barnes, S., & Prevelige, P. E., Jr. (2001). Hydrogen-deuterium exchange as a probe of folding and assembly in viral capsids. *Journal of Molecular Biology*, *306*(3), 389–396.

Varki, A., Cummings, R., Esko, J., Freeze, H., Stanley, P., Bertozzi, C., et al. (2009). *Essentials of glycobiology*. Plainview, NY: Cold Spring Harbor Laboratory Press.

Veesler, D., Khayat, R., Krishnamurthy, S., Snijder, J., Huang, R. K., Heck, A. J., et al. (2014). Architecture of a dsDNA viral capsid in complex with its maturation protease. *Structure*, *22*(2), 230–237.

Wales, T. E., Fadgen, K. E., Gerhardt, G. C., & Engen, J. R. (2008). High-speed and high-resolution UPLC separation at zero degrees celsius. *Analytical Chemistry*, *80*(17), 6815–6820.

Walters, B. T., Ricciuti, A., Mayne, L., & Englander, S. W. (2012). Minimizing back exchange in the hydrogen exchange-mass spectrometry experiment. *Journal of the American Society for Mass Spectrometry*, *23*(12), 2132–2139.

Wang, L., Lane, L. C., & Smith, D. L. (2001). Detecting structural changes in viral capsids by hydrogen exchange and mass spectrometry. *Protein Science*, *10*(6), 1234–1243.

Wang, L., & Smith, D. L. (2005). Capsid structure and dynamics of a human rhinovirus probed by hydrogen exchange mass spectrometry. *Protein Science*, *14*(6), 1661–1672.

Weis, D. D., Engen, J. R., & Kass, I. J. (2006). Semi-automated data processing of hydrogen exchange mass spectra using HX-Express. *Journal of the American Society for Mass Spectrometry*, *17*(12), 1700–1703.

Weis, D. D., Wales, T. E., Engen, J. R., Hotchko, M., & Ten Eyck, L. F. (2006). Identification and characterization of EX1 kinetics in H/D exchange mass spectrometry by peak width analysis. *Journal of the American Society for Mass Spectrometry*, *17*(11), 1498–1509.

Wildes, D., & Marqusee, S. (2005). Hydrogen exchange and ligand binding: Ligand-dependent and ligand-independent protection in the Src SH3 domain. *Protein Science*, *14*(1), 81–88.

Wu, Y., Kaveti, S., & Engen, J. R. (2006). Extensive deuterium back-exchange in certain immobilized pepsin columns used for H/D exchange mass spectrometry. *Analytical Chemistry*, *78*(5), 1719–1723.

Zhang, X., Chien, E. Y., Chalmers, M. J., Pascal, B. D., Gatchalian, J., Stevens, R. C., et al. (2010). Dynamics of the beta2-adrenergic G-protein coupled receptor revealed by hydrogen-deuterium exchange. *Analytical Chemistry*, *82*(3), 1100–1108.

Zhang, H. M., McLoughlin, S. M., Frausto, S. D., Tang, H., Emmett, M. R., & Marshall, A. G. (2010). Simultaneous reduction and digestion of proteins with disulfide bonds for hydrogen/deuterium exchange monitored by mass spectrometry. *Analytical Chemistry*, *82*(4), 1450–1454.

Zhang, Z., Post, C. B., & Smith, D. L. (1996). Amide hydrogen exchange determined by mass spectrometry: Application to rabbit muscle aldolase. *Biochemistry*, *35*(3), 779–791.

Zhang, J., Ramachandran, P., Kumar, R., & Gross, M. L. (2013). H/D exchange centroid monitoring is insufficient to show differences in the behavior of protein states. *Journal of the American Society for Mass Spectrometry*, *24*(3), 450–453.

Zhang, Z., & Smith, D. L. (1993). Determination of amide hydrogen exchange by mass spectrometry: A new tool for protein structure elucidation. *Protein Science*, *2*(4), 522–531.

Zhang, Z., Zhang, A., & Xiao, G. (2012). Improved protein hydrogen/deuterium exchange mass spectrometry platform with fully automated data processing. *Analytical Chemistry*, *84*(11), 4942–4949.

Zhao, Y., Soh, T. S., Zheng, J., Chan, K. W., Phoo, W. W., Lee, C. C., et al. (2015). A crystal structure of the Dengue virus NS5 protein reveals a novel inter-domain interface essential for protein flexibility and virus replication. *PLoS Pathogens*, *11*(3), e1004682.

AUTHOR INDEX

Note: Page numbers followed by "*f*" indicate figures and "*t*" indicate tables.

C

G

SUBJECT INDEX

Note: Page numbers followed by "*f*" indicate figures, "*t*" indicate tables and "*s*" indicate schemes.

E

F

G

H

M

N

T

U

V

W

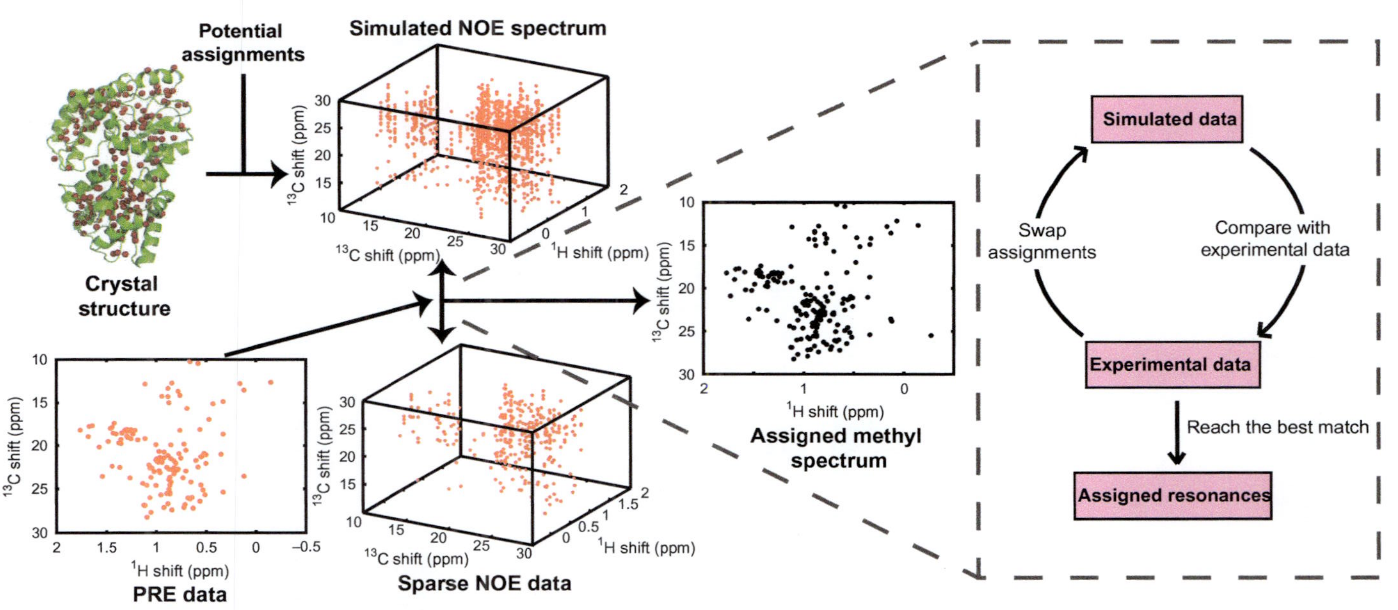

Jonggul Kim *et al.*, Figure 5 Overview of the FLAMEnGO algorithm. An input structure and the [^{1}H,^{13}C]-HMQC spectrum are used to simulate a NOE spectrum. This spectrum is compared with the experimental NOE data, the assignments swapped and the process is repeated until a best match is found. Other experimental restraints, such as PRE data, are used. *Adapted from Chao et al. (2012).*

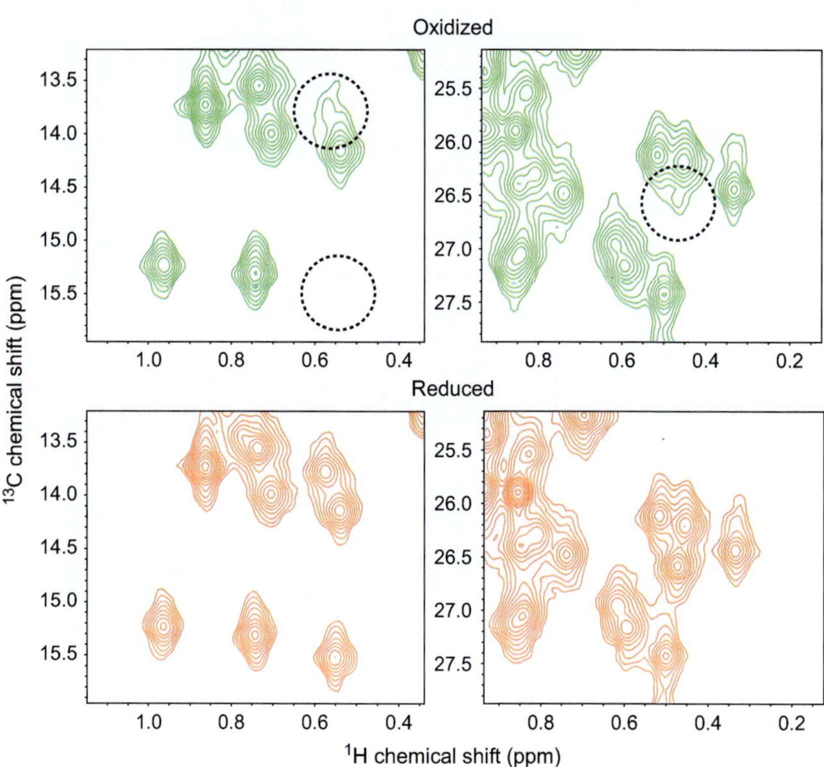

Jonggul Kim *et al.*, Figure 7 Spectra of oxidized (green) and reduced (red) PKA-C with a spin label on residue 244. Note the spectrum should be nearly identical before and after the reduction of the spin label.

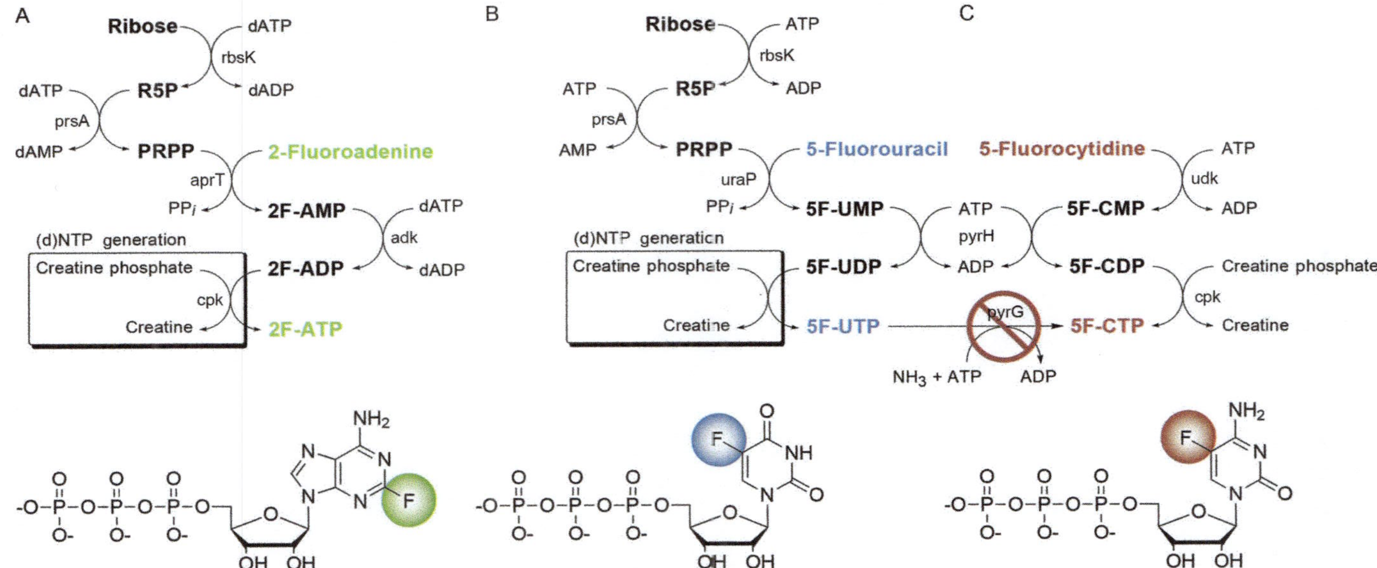

Lincoln G. Scott and Mirko Hennig, Figure 3 Enzymatic synthesis of fluorinated nucleotides. (A) Synthesis of 2-fluoroadenosine-5′-triphosphate (2F-ATP) from 2-fluoroadenine and ribose. (B) Synthesis of 5-fluorouridine-5′-triphosphate (5F-UTP) from 5-fluorouracil and ribose. (C) Synthesis of 5-fluorocytidine-5′-triphosphate (5F-CTP) from 5-fluorocytidine. Phosphorylation of 5-fluorocytidine by uridine kinase (udk) proceeds to produce 5F-CMP, which is then converted into 5F-CTP. In contrary, 5F-UTP proved not to be an acceptable substrate for the enzyme CTP synthetase (pyrG).

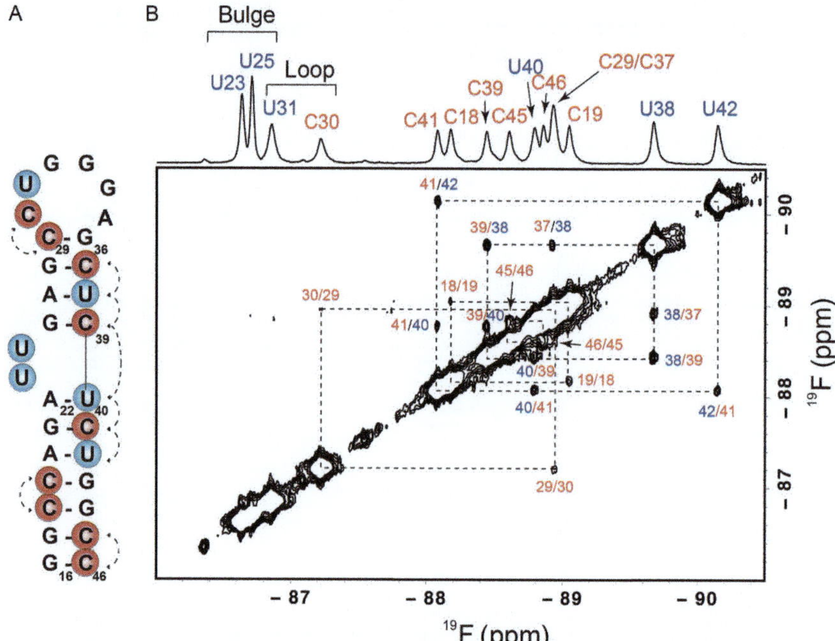

Lincoln G. Scott and Mirko Hennig, Figure 8 (A) Sequence and secondary structural representation of HIV-2 TAR with 5-fluorocytidine and 5-fluorouridine substitutions highlighted by circles. Dashed lines indicated observable cross-peaks between stacked 5F-Ura/Cyt-modified nucleobases. (B) Phase-sensitive ^{19}F,^{19}F-NOESY experiments (^{19}F resonance frequency of 470.5 MHz) showing fluorine connectivities between stacked bases in 5F-Ura/Cyt-substituted TAR. Assignments of individual homonuclear ^{19}F-NOESY cross-peaks are given. The NOESY mixing time τ_{mix} was 400 ms.

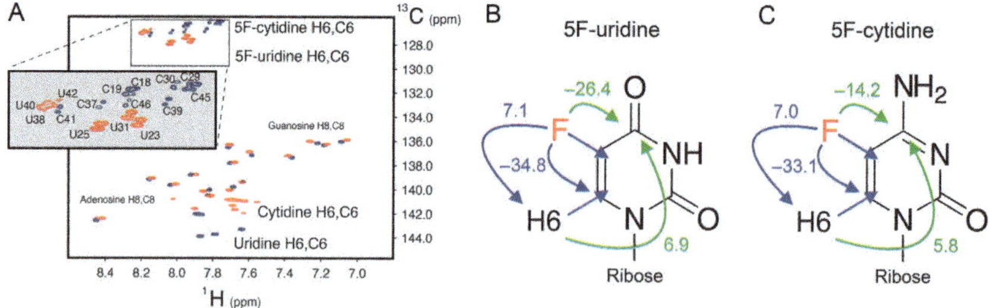

Lincoln G. Scott and Mirko Hennig, Figure 11 (A) ^{19}F-coupled ^{1}H,^{13}C HSQC experiments (^{1}H resonance frequency of 600 MHz) recorded for 5F-Ura- and 5F-Cyt-substituted TAR. Assignments of individual HSQC cross-peaks between aromatic H6 protons and C6 carbons (split in ω_1 by ^{2}J(C6,F) in 5-fluoropyrimidine-substituted TAR are given in the inset. Scalar *J*-couplings within (B) 5-fluorouridine and (C) 5-fluorocytidine obtained in d_4-methanol at 600 MHz and 293 K.

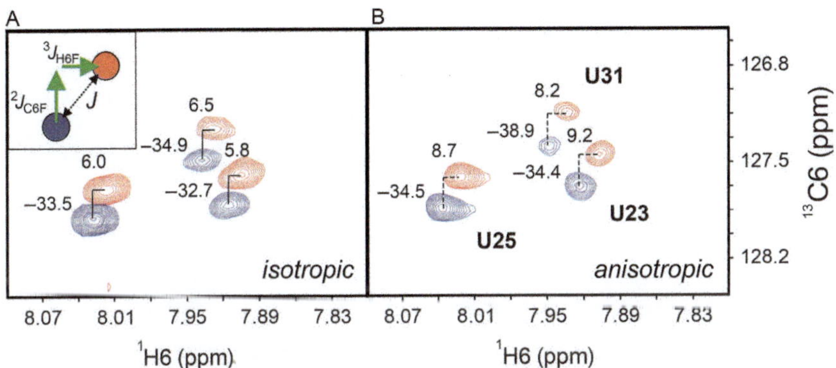

Lincoln G. Scott and Mirko Hennig, Figure 13 (S^3E)-^{19}F-α,β-edited ^{1}H6,^{13}C6-HSQC experiment for (A) 5F-Ura- and (B) 5F-Ura-substituted TAR partially aligned using ~25 mg mL^{-1} Pf1 phage. Individual subspectra shown are superimposed and correspond to the passive ^{19}F spins adopting α- and β-states, respectively. Assignments of $U^{5F}23$, $U^{5F}25$, and $U^{5F}31$ ^{1}H6,^{13}C6 cross-peaks are given and extracted *J* (*isotropic*) and *J*+*D* (*anisotropic*) couplings are reported. The inset shows a schematic superposition of the two subspectra (^{1}H6,^{13}C6 HSQC cross-peak) with the ^{19}F spin in either α- (red) and β-states (blue). The displacement vector *J* between the submultiplets consists of the ^2J(C6,F) coupling in ω_1 and the ^3J(H6,F) coupling in ω_2.

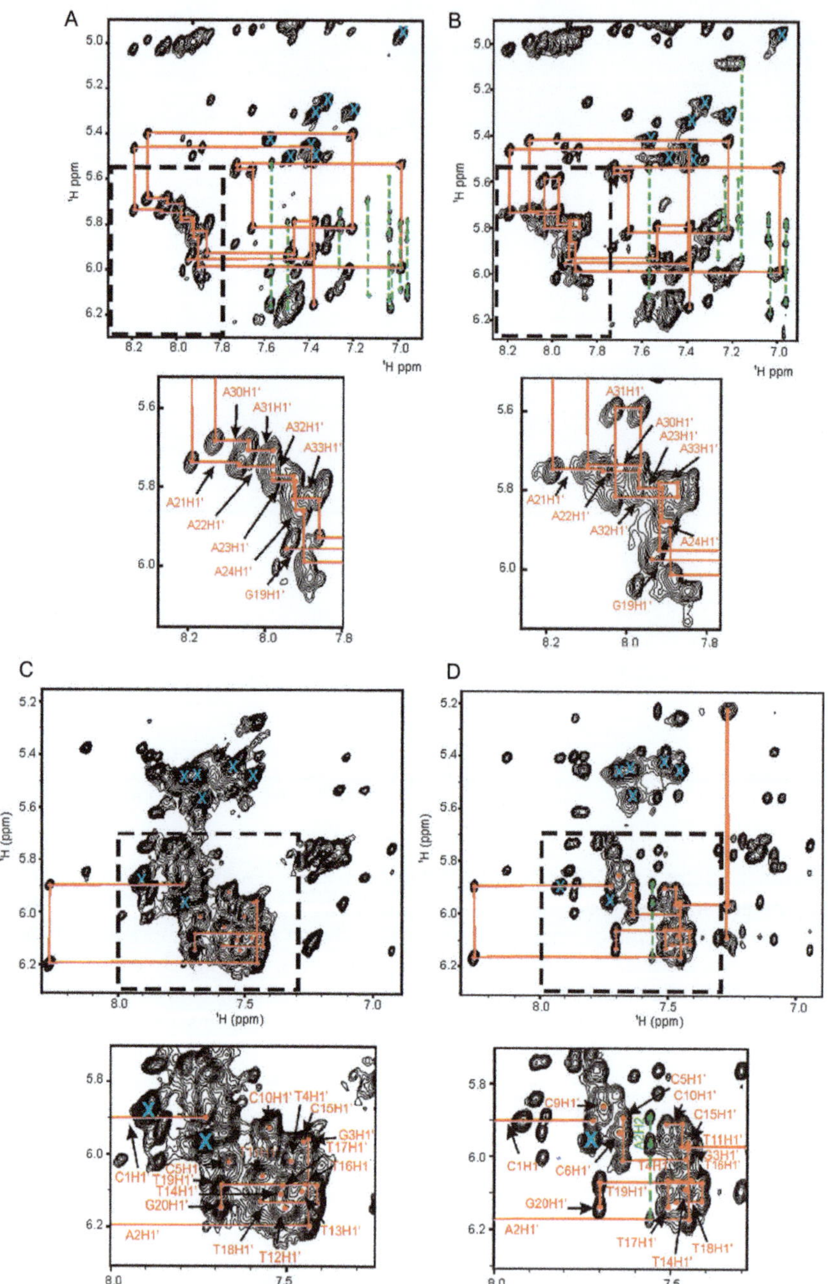

Robert G. Brinson *et al.*, Figure 4 Expansion of the anomeric (H1′) to aromatic region of a NOESY ($\tau_{NOE} = 250$ ms) experiment applied to a 1.0 m*M* sample of the (A) unmodified A-tract DNA duplex, (B) A-tract dF5 DNA duplex, (C) unmodified HIV, and (D) HIV dF12 duplex. The local dispersion of peaks due to dF substitution is emphasized by the expansions in the dotted boxes below each spectrum. The sequential assignment "NOE walk" for residues G19 through C36 is shown in red lines for the A-tract and C1 though G20 for the HIV spectra. For clarity, assignment labels were removed from the full spectrum but retained in the expansions. Dashed green lines are used to annotate correlations of adenosine H2 protons with 5′ sequential and 3′ cross-strand anomeric (H1′) protons. An "X" in cyan indicates a cytosine H5–H6 correlation.

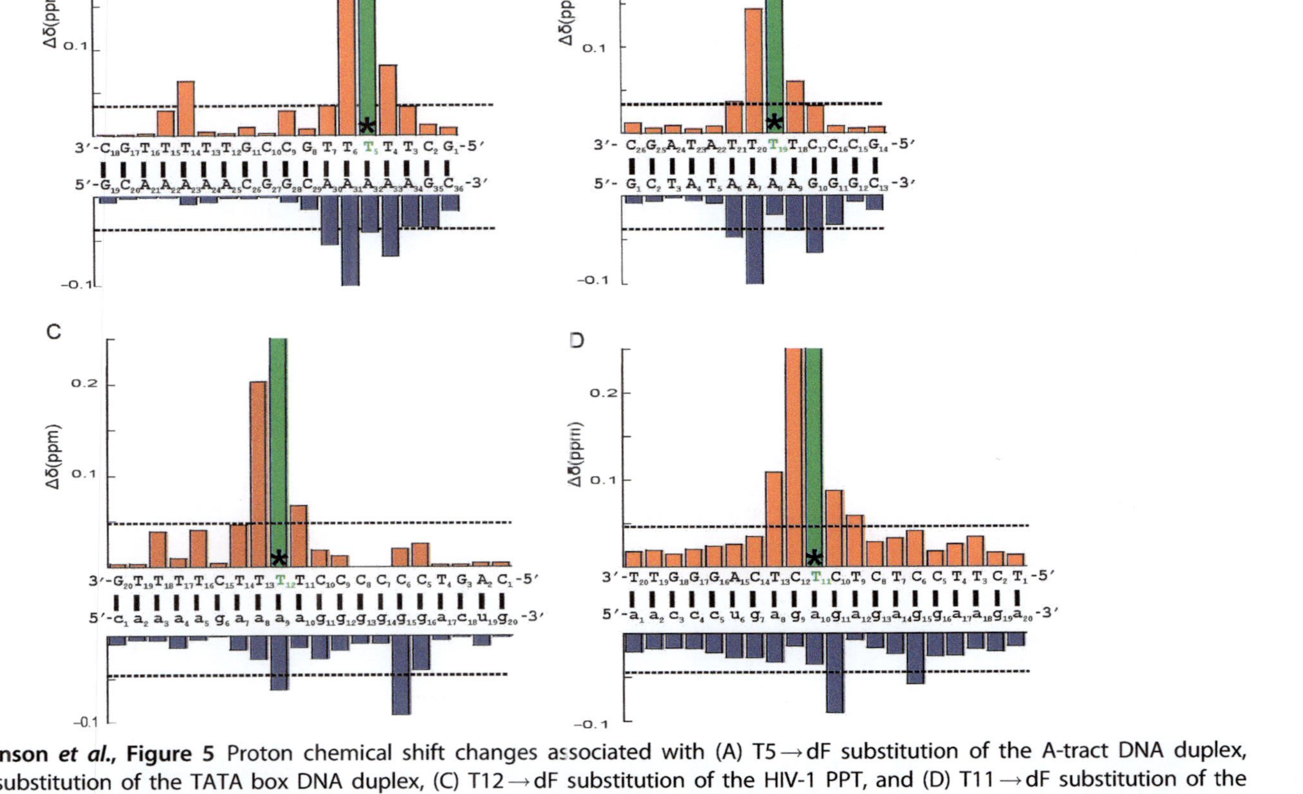

Robert G. Brinson *et al.*, Figure 5 Proton chemical shift changes associated with (A) T5 → dF substitution of the A-tract DNA duplex, (B) T19 → dF substitution of the TATA box DNA duplex, (C) T12 → dF substitution of the HIV-1 PPT, and (D) T11 → dF substitution of the Ty3 PPT. DNA and RNA sequences of each hybrid are shown in upper case and lower case letters, respectively. Positions of dT → dF substitution are highlighted in green and an asterisk "*." Each bar represents the average difference in normalized ppm values obtained for equivalent nucleotides within dF-substituted and unmodified duplexes. All aromatic and H1′ resonances were included in the average. The dotted gray line designates the threshold of a significant chemical shift difference. Chemical shift changes (ppm) for dF-substituted DNA strands are shown above the sequence schematics (red bars), while unmodified strands are depicted below (blue bars).

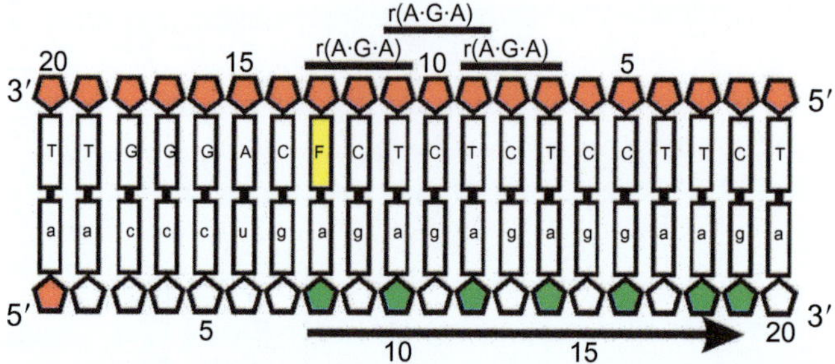

Robert G. Brinson *et al.*, Figure 6 A schematic representation of the Ty3 dF13 PPT hybrid sequence. dF13 substitution is in yellow. Wild-type ribose sugars in mixed C3′/C2′-endo conformation are shown in red. The symmetric 5′-r(A·G·A)-3′ steps are labeled. Highlighted collectively in green are ribose sugars for which a C3′-endo to C3′/C2′-endo sugar pucker switch is observed when the PPT hybrids, and arrow indicates the propagation of the sugar pucker switch to the 3′ end of the sequence. *Figure was modified from published data Yi-Brunozzi et al. (2008).*

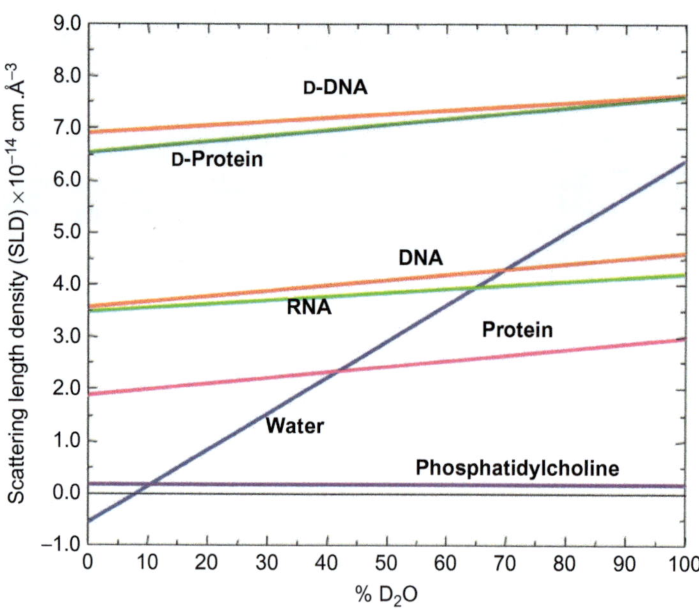

Michael Haertlein *et al.*, Figure 1 Scattering length densities (SLDs) plotted as a function of D_2O/H_2O solvent composition for various types of biological macromolecule. The points at which these lines intersect that for the isotopically mixed water defines the match points at which that biological component becomes invisible in terms of its coherent scattering.

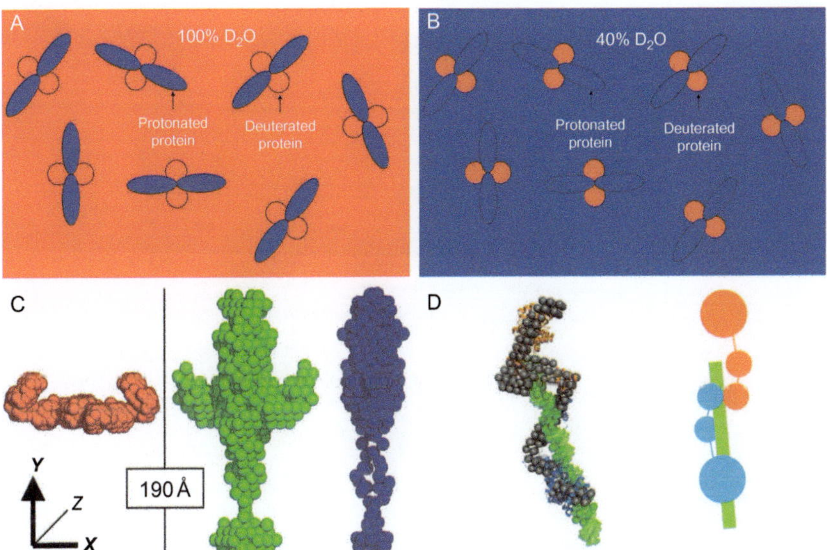

Michael Haertlein *et al.*, Figure 2 Graphical representation of the use of solvent contrast variation and some examples. (A) Schematic representation of the matchout of the deuterated protein in a complex in 100% D_2O solvent, (B) analogous representation of matchout of hydrogenous parts of the same complex in 40% D_2O, (C) restriction modification system studied by contrast variation in conjunction with selective deuteration (Callow, Sukhodub, Taylor, & Kneale, 2007), and (D) contrast variation study of a mariner transposase system using deuterated protein in complex with hydrogenated DNA (Cuypers, Trubitsyna, et al., 2013).

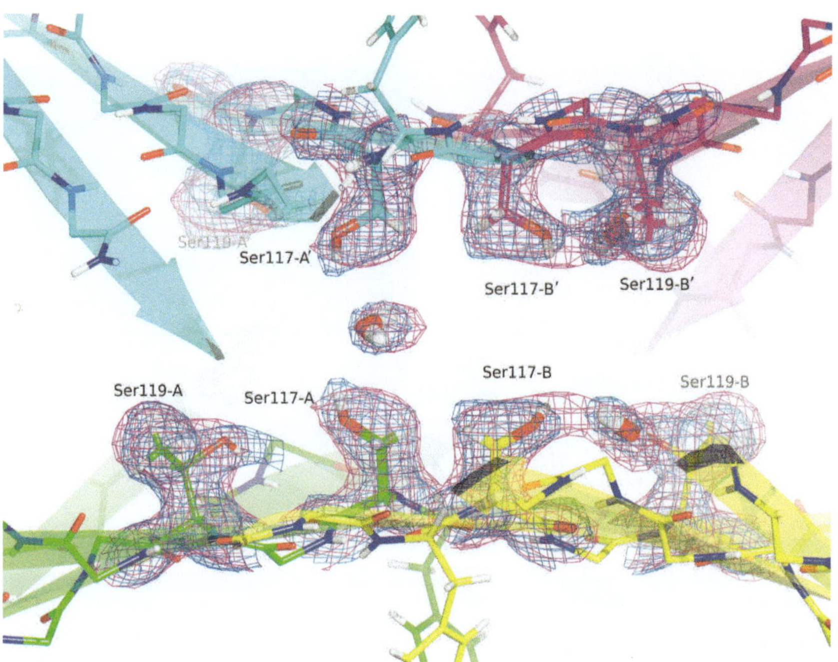

Michael Haertlein *et al.*, Figure 4 View of the central part of the bivalent binding pocket of tetrameric human transthyretin (TTR). The neutron study used perdeuterated TTR, with data recorded from both LADI-III and D19 diffractometers at the ILL. The neutron scattering density maps show a key water molecule trapped between hydroxyl groups of serine residues in adjacent monomers. This imparts an intrinsic asymmetry to the TTR tetramer in a region that is very important for the binding of thyroxine and of ligands that have been designed to prevent tetramer dissociation and subsequent amyloid formation. *Picture from Haupt et al. (2014) with permission of the International Union of Crystallography.*

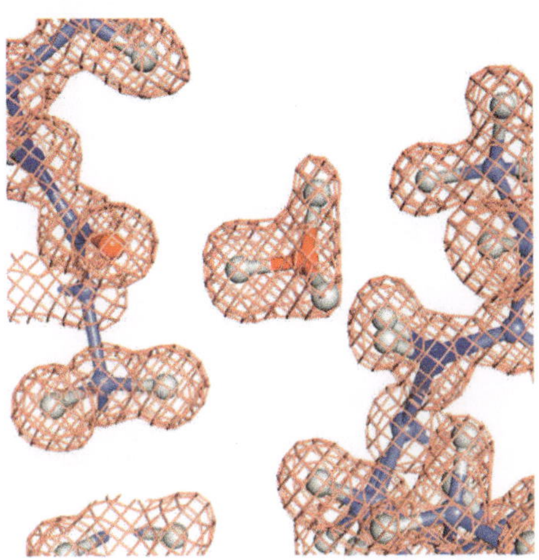

Michael Haertlein *et al.*, Figure 5 High-resolution studies (current limit 0.88 Å) of both oxidized and reduced forms of perdeuterated *Pyrococcus furiosus* rubredoxin have been carried out on the D19 diffractometer at the ILL and show unparalleled detail. A tautomeric shift is noted following the change from oxidized to reduced forms (Cuypers, Mason, et al., 2013). Highly ordered water networks are observed that provide definitive evidence of hydronium ions that may play a role in charge transport in the redox protein.

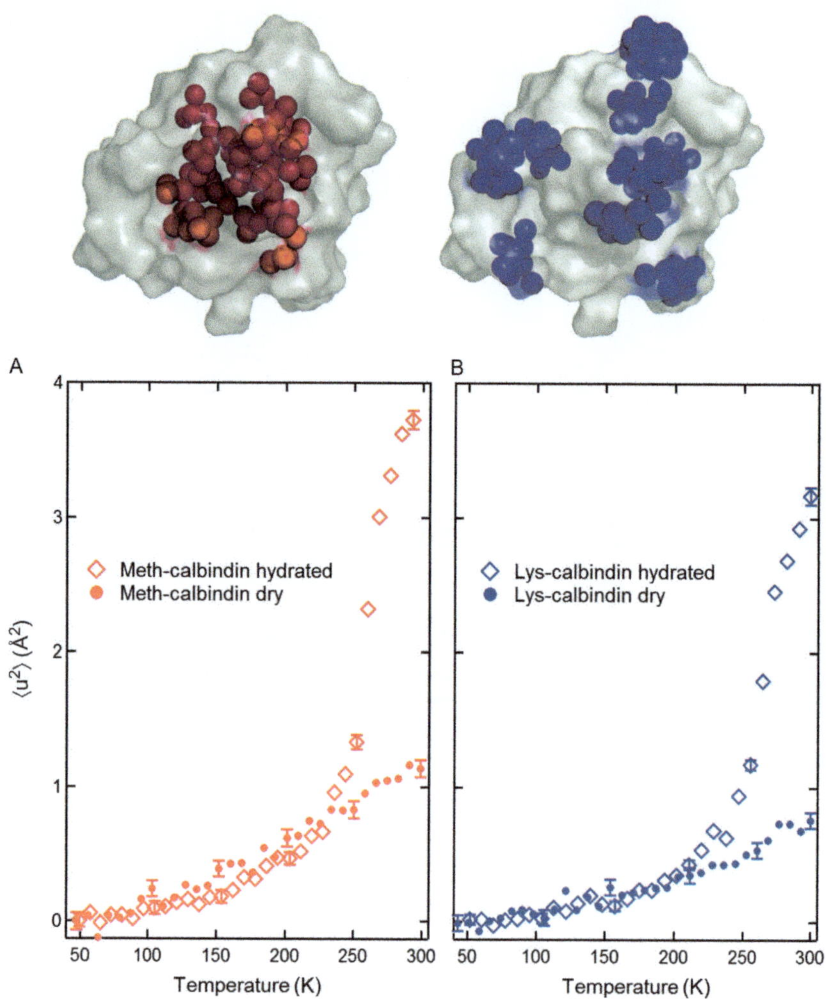

Michael Haertlein *et al.*, Figure 6 Coupling of protein and hydration water nanosecond dynamics as measured from the interior and exterior of calbindin by neutron elastic incoherent scattering (see Wood et al., 2013).

B

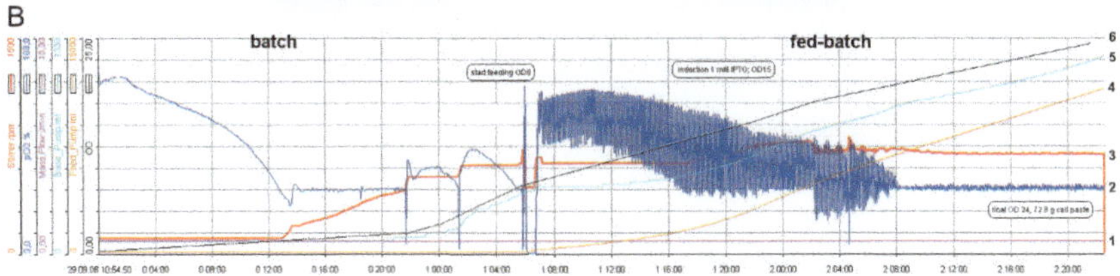

Key: 1 (pink) - airflow; 2 (blue) - pO2; 3 (red) - stirring speed; 4 (orange) - feeding; 5 (turquoise) - NaOD; 6 (black) - OD

Michael Haertlein *et al.*, Figure 8 (A) Fermentation equipment used for high cell density cultures and (B) tracking of fermentation parameters for a typical deuterated HCDC run, showing stirrer speed, dissolved oxygen concentration, air flow volumes, as well as feed and base additions. Following inoculation, cellular growth (oxygen consumption) is monitored by a decrease in dissolved oxygen concentration (pO_2, in blue) at constant stirrer speed. When the set-point for pO_2 of 30% is reached, the stirrer speed (rpm in red) increases to compensate for the increased oxygen demand of the growing cells. After about 30 h in the batch phase at an OD_{600} of about 4–5, the deuterated glycerol (5 g L^{-1}) is consumed by the growing cells, as demonstrated by the sudden increase in pO_2. d_8-glycerol feeding initiates the fed-batch phase. Cells are typically induced with 1 m*M* IPTG at an OD_{600} of 15 and growth stopped when cells reach an OD_{600} of $\sim$20.

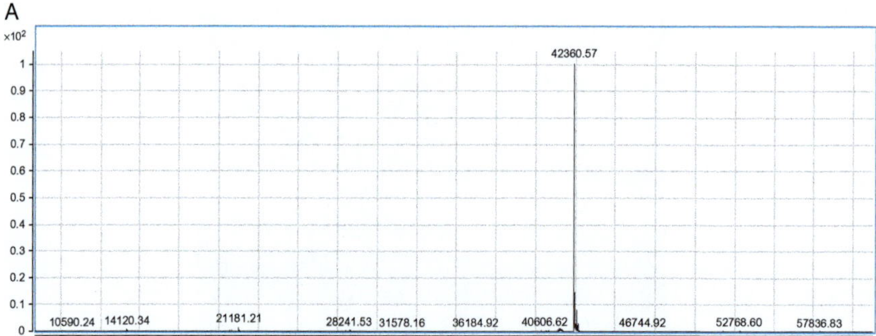

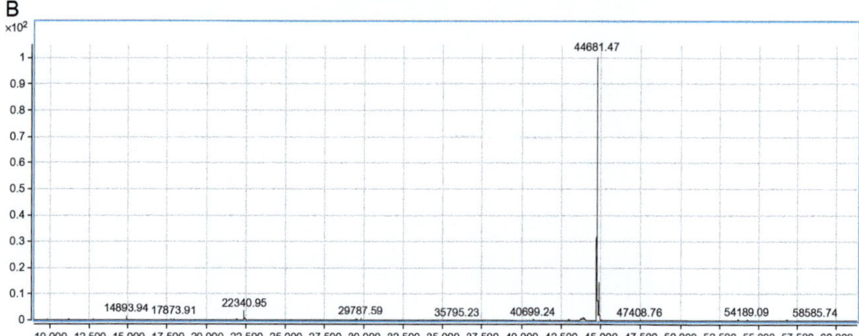

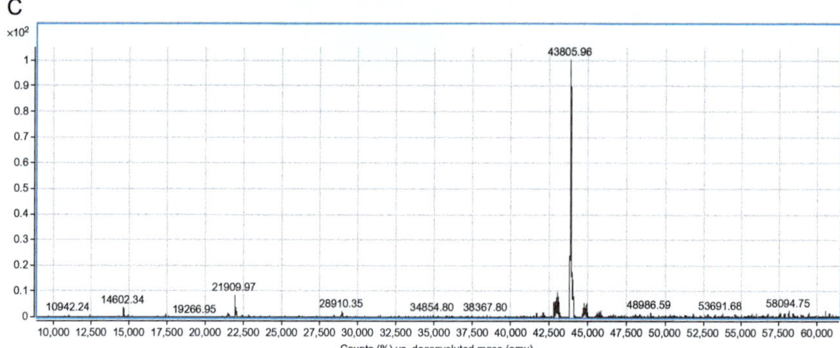

Michael Haertlein *et al.*, Figure 9 Mass spectrometry analysis (LC-TOF) for three differently labeled samples of maltose-binding protein (MBP): (A) hydrogenated (unlabeled) MBP, (B) perdeuterated MBP, and (C) "matchout labeled" MBP. Mass spectrometry measurements were carried out in hydrogenated buffers; the mass differences between the various samples arise from the non-exchangeable deuterium atoms incorporated into the samples.

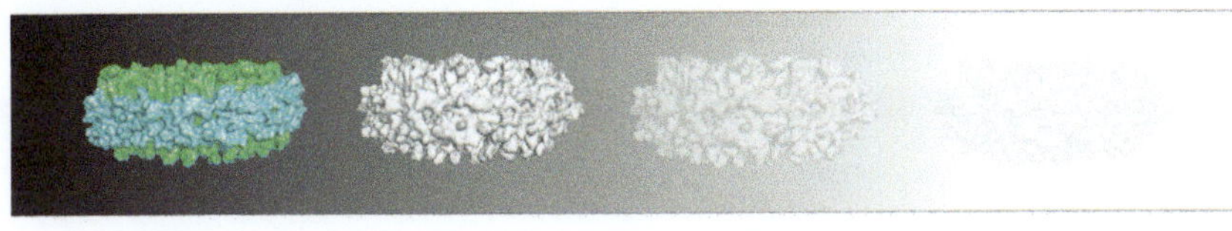

Michael Haertlein *et al.*, Figure 10 Schematic representation of the visibility by SANS of the "stealth" nanodisc in buffer solution with a gradually increasing levels of solvent D_2O. (A) Hydrogenated nanodisc consisting of a phospholipid bilayer (green) and membrane scaffold protein (MSP) (light blue). (B) Stealth nanodisc comprised of deuterated lipid and deuterated MSP at 60% D_2O. (C) Stealth nanodisc at 80% D_2O. (D) Stealth nanodisc at 100% D_2O. *Picture reproduced from Maric et al. (2014) with permission of the International Union of Crystallography.*

Michael Haertlein *et al.*, Figure 11 Photobioreactor system (Biostat PBR2S, Sartorius) being used for the production of biomass from the cyanobacterium *Anabaena*. This system allows control of temperature, pH (pD), recirculation rate, and light intensity.

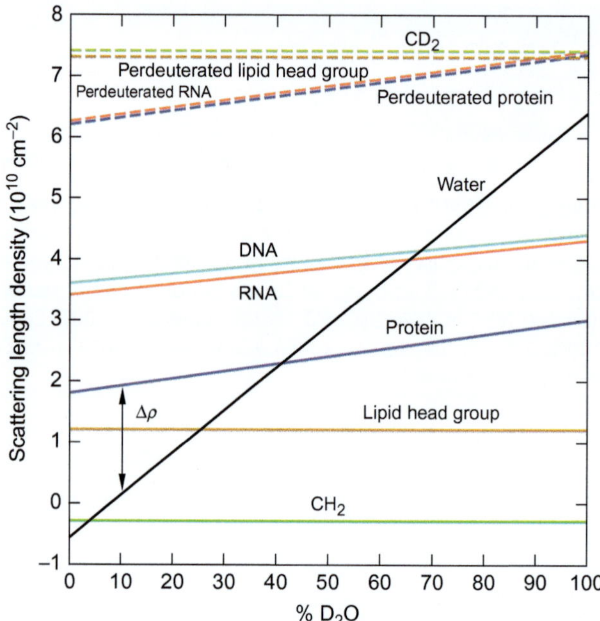

Nathan R. Zaccai *et al.*, Figure 1 Plot of the neutron scattering length density versus % D_2O in the solvent for water compared to those for protein, RNA, DNA, and the components of lipids (lipid head group and CH_2), along with their perdeuterated counterparts.

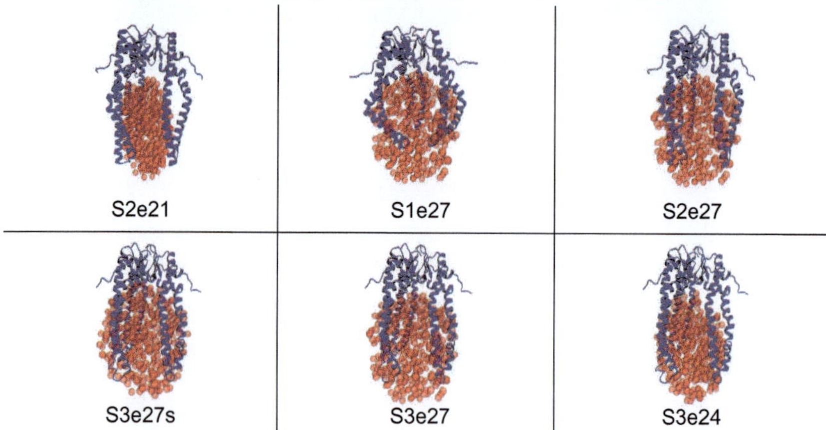

Nathan R. Zaccai *et al.*, Figure 7 Skp–OmpW model structures tested against the SANS data as described in Section 3. The ball representation of the ellipsoids is for clarity. The ellipsoids were represented by nonoverlapping points for SANS curve calculations, as described in Section 3.

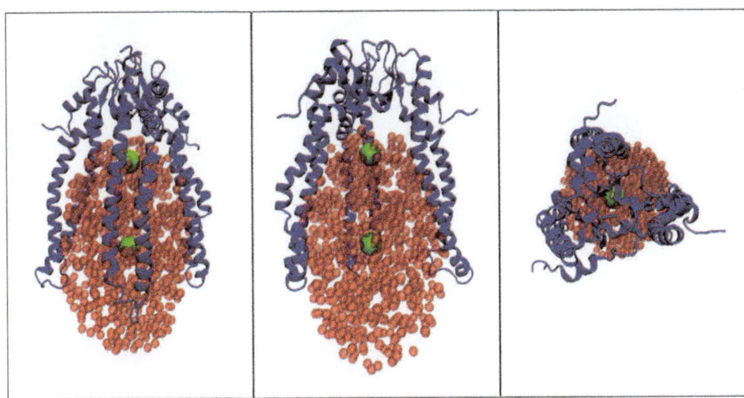

Nathan R. Zaccai *et al.*, Figure 9 Several views of Skp–OmpW model S3e24, which was the best-fit model for this complex. The larger green balls are at the CM locations of S3 Skp and the e24 ellipsoid. The ball representation of the ellipsoids is for clarity. The ellipsoids were represented by nonoverlapping points for SANS curve calculations, as described in Section 3.

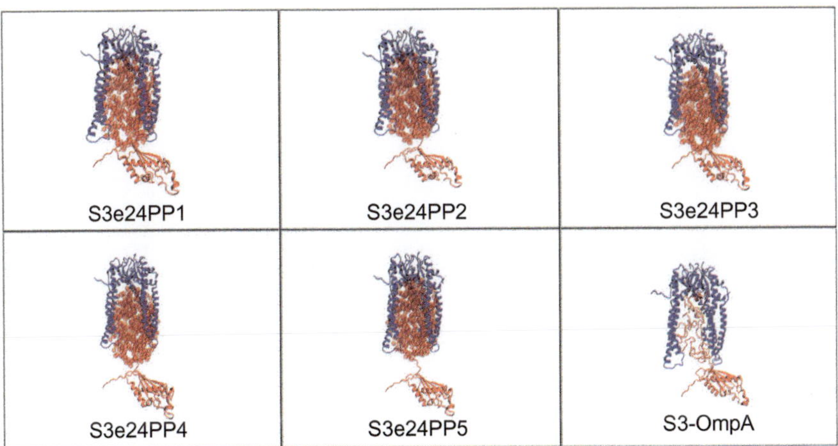

Nathan R. Zaccai *et al.*, Figure 11 Skp–OmpA starting structure models for SASSIE runs as described in Section 3. The ball representation of the ellipsoids is for clarity. The ellipsoids were represented by nonoverlapping points for SANS curve calculations, as described in Section 3.

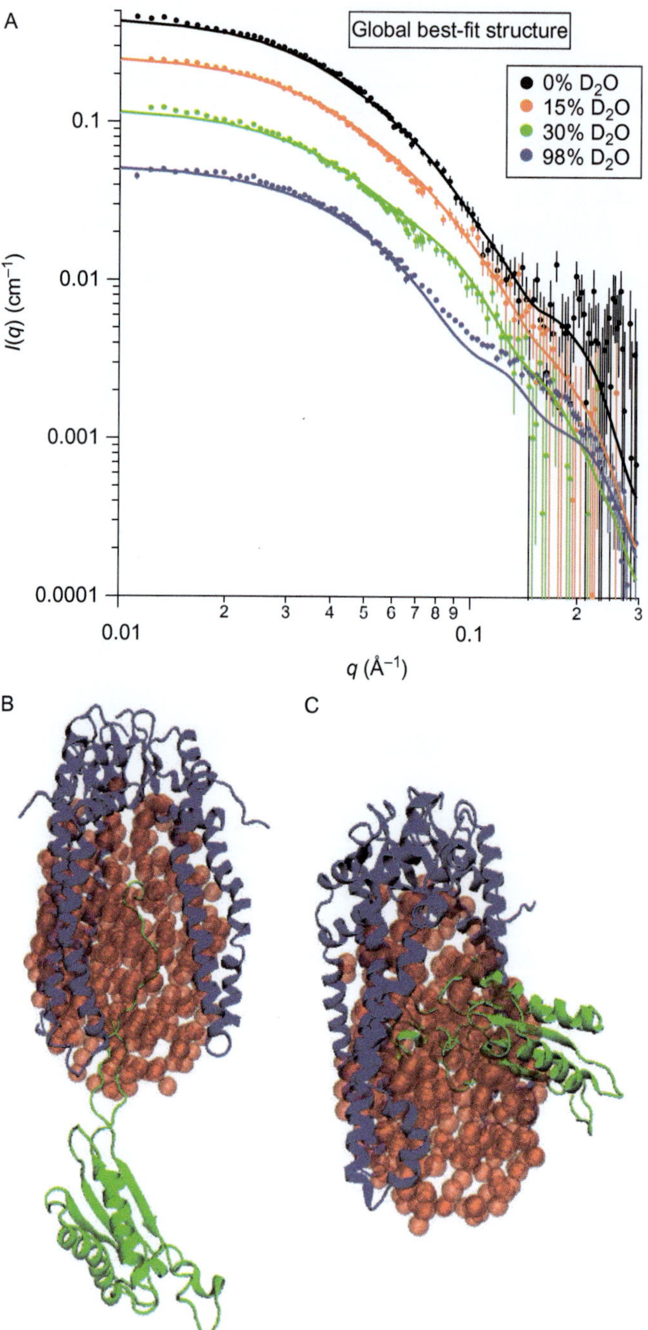

Nathan R. Zaccai *et al.*, Figure 18 (A) Skp–OmpA SANS data on an absolute scale along with model SANS curves from the global single best-fit S3e24PP5 structure. Error bars represent the standard error of the mean with respect to the number of pixels used in the data averaging. The (B) best-fit and (C) worst-fit model structures are also shown for comparison. The ball representation of the ellipsoids is for clarity. The ellipsoids were represented by nonoverlapping points for SANS curve calculations, as described in Section 3.

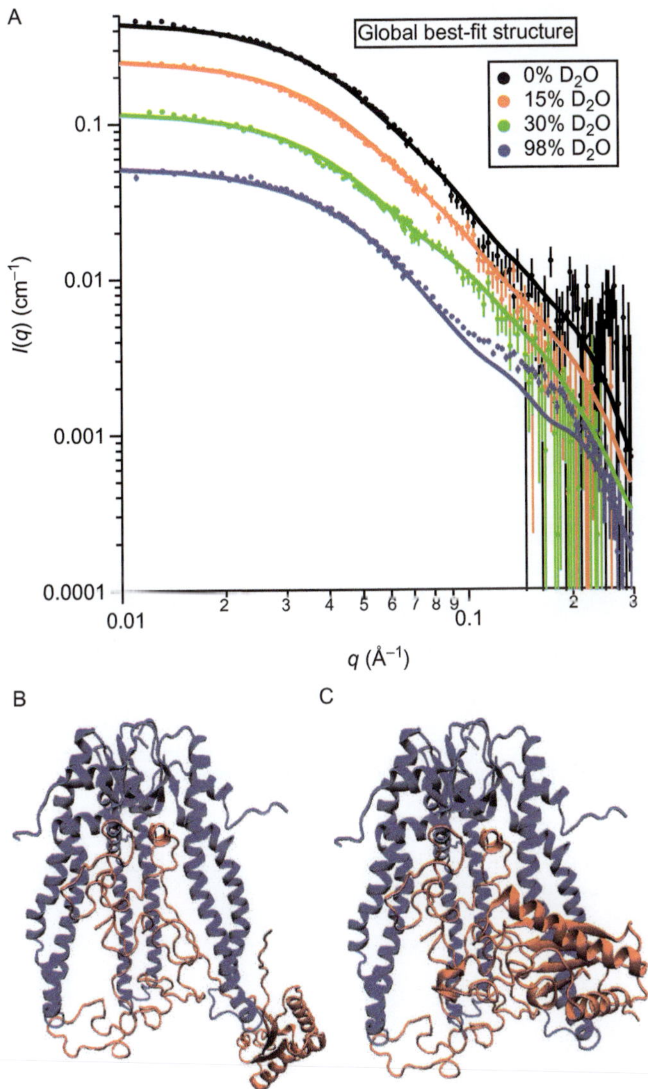

Nathan R. Zaccai *et al.*, Figure 19 (A) Skp–OmpA SANS data on an absolute scale, along with model SANS curves from the global single best-fit S3-OmpA structure. Error bars represent the standard error of the mean with respect to the number of pixels used in the data averaging. The (B) best-fit and (C) worst-fit model structures are also shown for comparison.

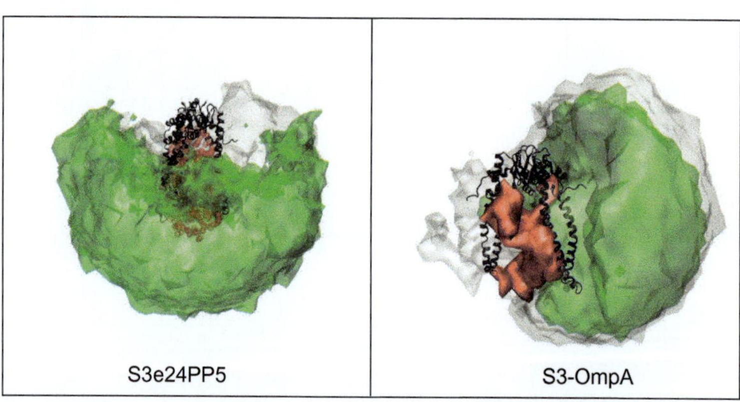

Nathan R. Zaccai *et al.*, Figure 20 Structure density plots representing all of conformation space (gray) and the best-fit conformation space within the rectangles in Figs. 13 and 15 (green) explored by the OmpA PP domain using the S3e24PP5 and S3-OmpA models. The OmpA TM domain is shown in red and Skp is shown in black.

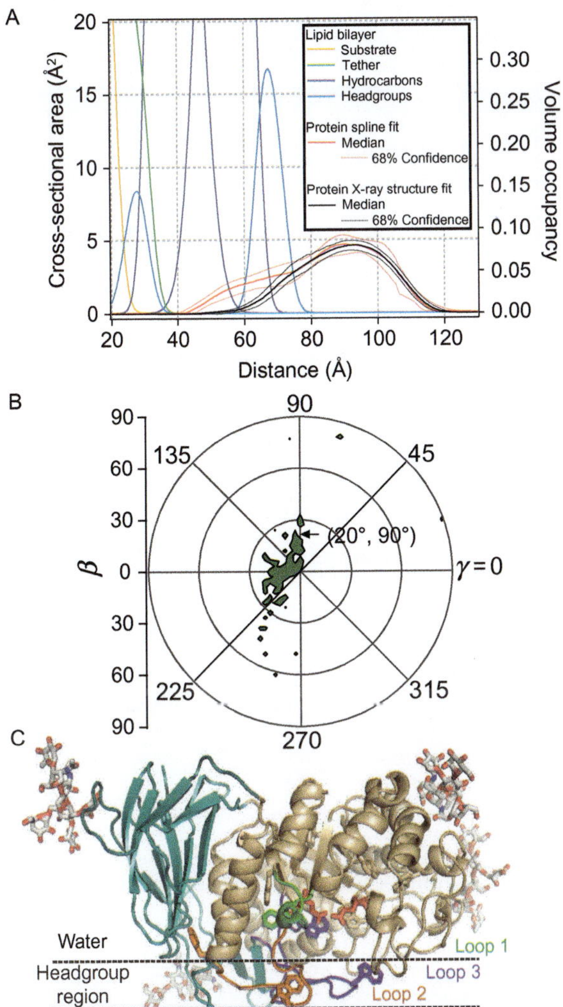

Frank Heinrich, Figure 2 Example of a rigid body modeling of a protein (GCase) associated with a tethered lipid membrane composed of 50:50 molar fraction POPC:POPS (Yap et al., 2015). (A) Composition space modeling result. A difference between results obtained from a free-form modeling approach using a Hermite spline and an integrative modeling approach using an available X-ray crystal structure is apparent at the substrate–distal headgroup–hydrocarbon interface. It is likely that the protein adopts a different configuration when membrane-bound compared to its crystalline form with a deeper penetration of flexible loops into the hydrocarbon region of the substrate–distal lipid leaflet. (B) Probability plot of two Euler angles (β and γ) defining the protein orientation at the interface. Robust methods to determine modeling uncertainties are essential for complex modeling approaches. Here, a Monte Carlo Markov Chain algorithm was employed (Kirby et al., 2012). (C) The orientation of the protein at the membrane in agreement with the NR data and further supported by additional fluorescence data (Yap et al., 2015). *This research was originally published in the Journal of Biological Chemistry (Yap et al., 2015), © The American Society for Biochemistry and Molecular Biology.*

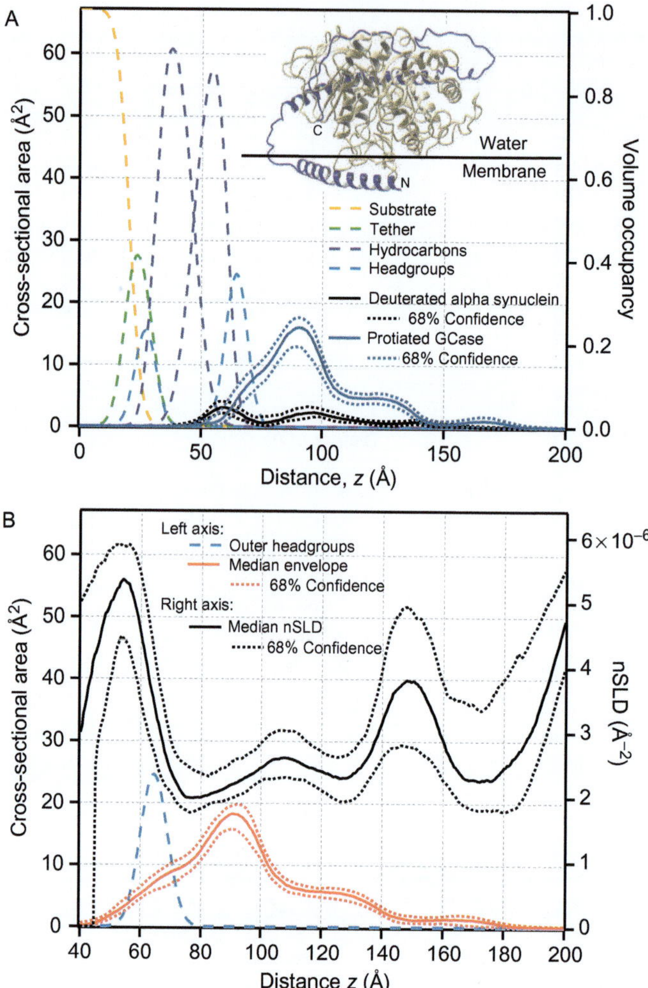

Frank Heinrich, Figure 4 Characterization of a membrane-bound protein–protein complex of α-synuclein and GCase (see Fig. 2) at the lipid membrane using NR and specific deuteration (Yap et al., 2015). (A) Modeling result for the lipid bilayer using a continuous distribution model and for the volume occupancy profile of the protein complex using Hermite splines. Within 68% confidence limits, the volume occupancy profile of deuterated α-synuclein in the complex shows a bimodal envelope, which is in contrast to membrane-bound-free α-synuclein that exhibits only one maximum at the headgroup/hydrocarbon interface. Inset: A molecular model of the complex that satisfies the constraints set by NR and auxiliary data from fluorescence spectroscopy. In order to fit the NR profile, the second α-helix has to lift off the bilayer and is most likely closely associated with the GCase protein. (B) Implementation of the free-form Hermite spline modeling for protein–protein complexes with specific deuteration. Two Hermite splines model the volume occupancy profile and the nSLD profile independently, sharing the position of control points that are in this example on average 20 Å apart. The volume occupancy profile can vary between 0% and 100% volume filling, the nSLD profile can vary between values for protiated GCase (1.67×10^{-6} Å^{-2}) and deuterated α-synuclein (6.12×10^{-6} Å^{-2}). Individual volume occupancy profiles for the two components as shown in (A) are computed from the two Hermite splines and confidence limits are propagated. *This research was originally published in the Journal of Biological Chemistry (Yap et al., 2015), © The American Society for Biochemistry and Molecular Biology.*

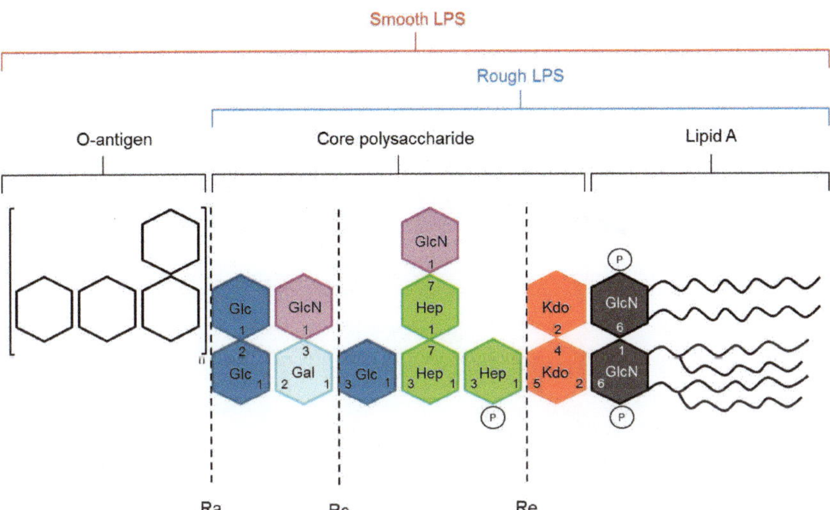

Anton P. Le Brun *et al.*, Figure 1 The general structure of LPS from B strains of *E. coli* showing the R3 core. The numbers in the colored hexagons indicate the carbons through which the glycosidic bonds are formed between each sugar moiety and the abbreviations are Kdo, 2-keto-3-deoxyoctonic acid; Hep, L-glycero-D-manno heptose; Glc, glucose; Gal, galactose; GlcN, 2-amino-2-deoxyglucose; P (in circle), phosphate. The lipid A tails are represented by the wavy lines and consist of four (R)-3-hydroxy-mystic acids, one myristic acid, and one lauric acid.

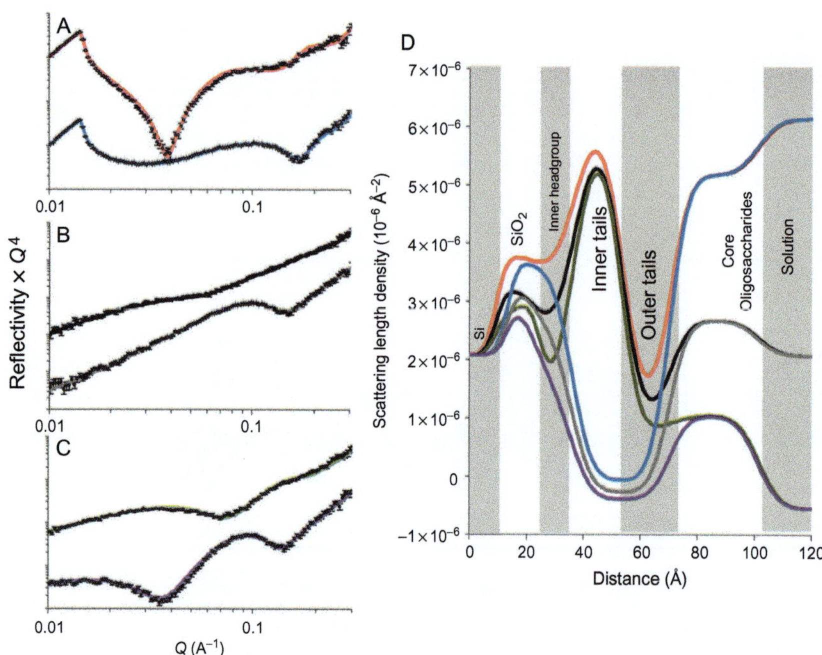

Anton P. Le Brun *et al.*, Figure 8 Neutron reflectometry profile and model data fits (A–C) and the scattering length density profiles these fits describe (D) for asymmetrically deposited DPPC (inner leaflet): Ra-LPS (outer leaflet) bilayer in the presence of 20 m*M* HEPES, pH 7.2 buffer with 5 m*M* $CaCl_2$. The six simultaneously fitted isotopic contrasts shown are (A) d-DPPC/Ra-LPS in D_2O (red line), h-DPPC/Ra-LPS in D_2O (blue line); (B) d-DPPC/Ra-LPS in 38% D_2O (black line), h-DPPC/Ra-LPS in 38% D_2O (gray line); and (C) d-DPPC/Ra-LPS in H_2O (green line), h-DPPC/Ra-LPS in H_2O (purple line). *Reproduced from Clifton et al. (2015) with permission, copyright © 2014 American Chemical Society.*

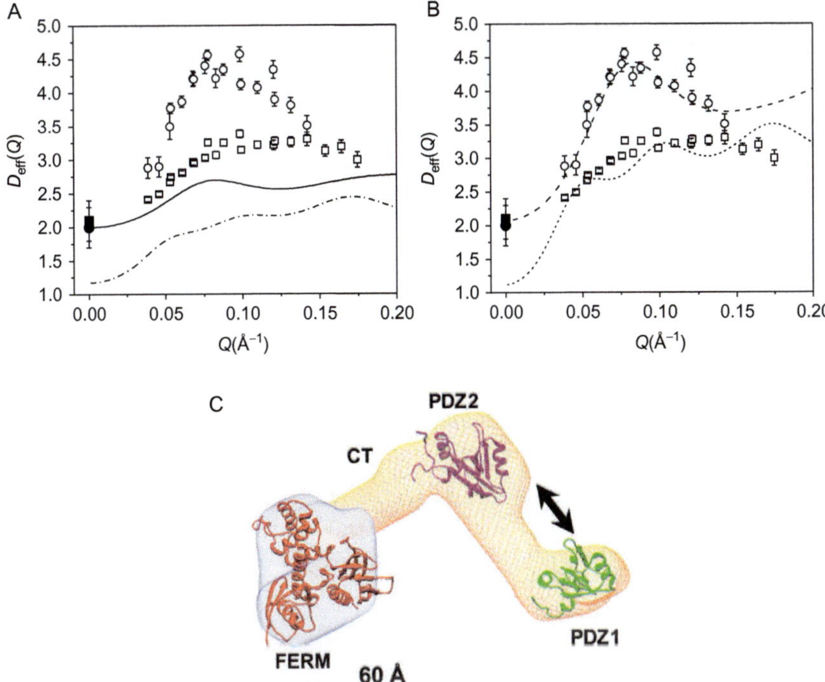

David J.E. Callaway and Zimei Bu, Figure 3 Activation of interdomain motion in NHERF1 upon binding to the FERM domain of Ezrin. (A) Comparing experimental $D_{eff}(Q)$ of NHERF1·dFERM and NHERF1·hFERM with rigid-body calculations. Open circle are the NSE data from NHERF1·dFERM. Open squares are the NSE data from NHERF1·hFERM. Solid circle and square are the self-diffusion constants D_0 of NHERF1·dFERM and NHERF1·hFERM obtained from PFG NMR, respectively. The solid line is from rigid-body model calculations of the NHERF1·dFERM complex. The dot dash line is from rigid-body model calculations of the NHERF1·hFERM complex. (B) Comparing experimental $D_{eff}(Q)$ of deuterated complex NHERF1·dFERM and hydrogenated complex NHERF1·hFERM with calculations incorporating interdomain motion between PDZ1 and PDZ2. The symbols for the experimental data are the same as in (A). The dashed curve is calculated from model incorporating domain motion between PDZ1 and PDZ2 for the NHERF1·dFERM complex. The short dashed curve is calculated from model incorporating domain motion between PDZ1 and PDZ2 for the NHERF1·hFERM complex. The comparisons in (A) and (B) show that deuteration of the FERM domain amplifies the effects of protein internal motions detected by NSE. (C) A model representing domain motion between PDZ1 and PDZ2 in the complex. The 3-D shape of the complex is reconstructed from SANS (Li et al., 2009). The known high-resolution structure fragments of PDZ1, PDZ2, and the FERM domain (PDB code: 1NI2) are docked into the envelope using UCSF chimera (Pettersen et al., 2004). The arrows represent translational motion between PDZ1 and PDZ2. A length-scale bar of 60 Å is shown.

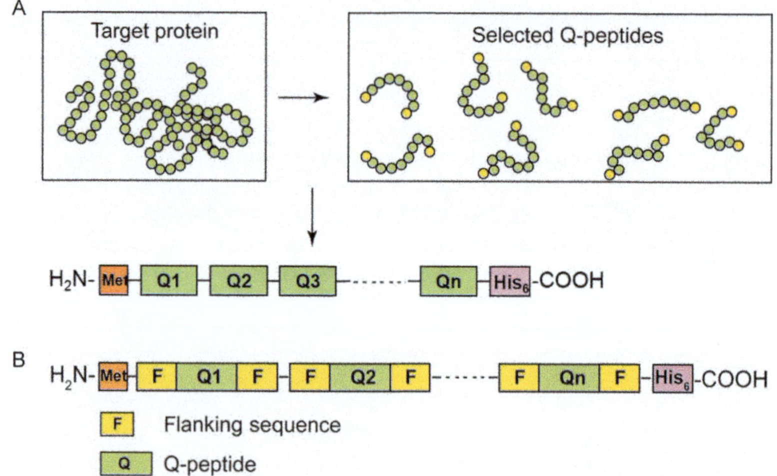

Kerry Bauer Scott *et al.*, Figure 1 QconCAT schematic. (A) Selected peptides from a protein of interest concatenated into an artificial protein. (B) QconCAT with peptides carrying flanking sequences made up of their natural amino acid sequences on both sides of the peptides. *Adapted from Chen and Turko (2014). Reprinted with permission.*

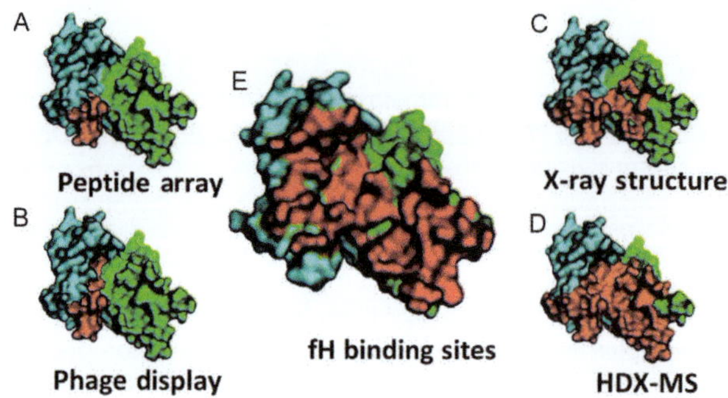

Elyssia S. Gallagher and Jeffrey W. Hudgens, Figure 1 Binding maps of the fHbp antigenic epitope plotted by different techniques (Malito et al., 2013). The N- and C-terminal domains of fHbp are presented in green and cyan, respectively. (A) Red denotes the fHbp epitope found by synthetic peptide arrays. (B) Red denotes the fHbp epitope found by phage display. (C) Red denotes the 23 residues of fHbp contacted by the Fab of mAb 12C1, as determined from the X-ray structure of the fHbp:Fab 12C1 complex. (D) Red denotes the peptide regions of fHbp contacted by the Fab of mAb 12C1 within the fHbp:12C1 complex, as determined by HDX-MS measurements. (E) Red denotes 52 fHbp surface residues bound with surface residues of domains 6 and 7 in human fH, as prepared by Malito et al. by modeling the crystal structure of fHbp:fH complex (Schneider et al., 2009). *Adapted images used with the permission of Malito et al. (2013).*

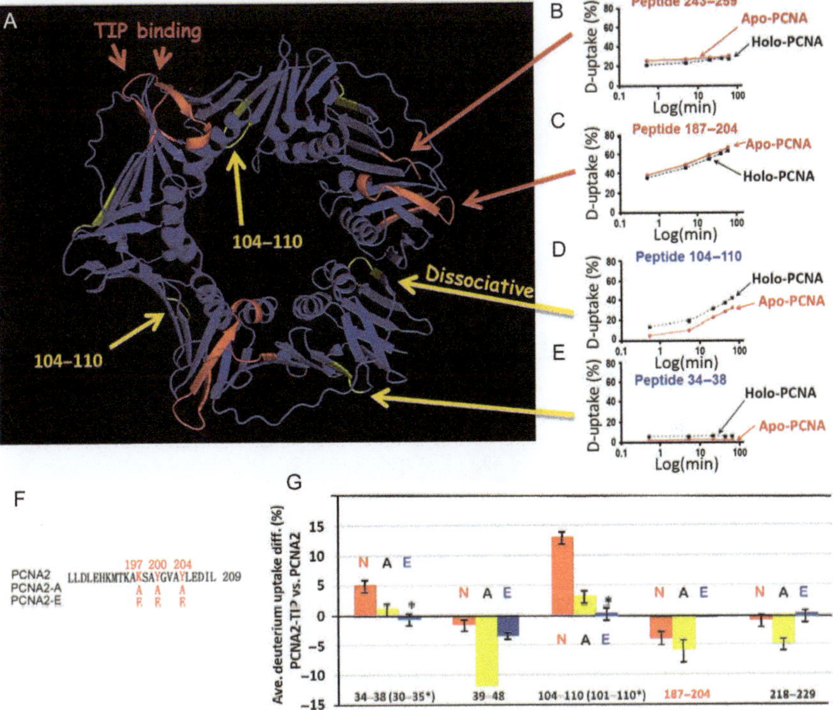

Elyssia S. Gallagher and Jeffrey W. Hudgens, Figure 5 HDX-MS of the protein–ligand system, PCNA2-TIP. (A) Difference D-uptake map between native apo- and holo-PCNA2 plotted on the crystal structure of homotrimeric PCNA2 (PDB: 3LX2) (Ladner, Pan, Hurwitz, & Kelman, 2011). Blue signifies no statistical difference, red denotes a decreased D-uptake rate in native holo-PCNA2, and yellow denotes increased D-uptake rate in native holo-PCNA2. (B) Temporal D-uptake of native PCNA2 peptide 243–259. (C) Temporal D-uptake of native PCNA2 peptide 187–204. (D) Temporal D-uptake of native PCNA2 peptide 104–110. (E) Temporal D-uptake of native PCNA2 peptide 34–38. (F) Partial sequence of PCNA2 showing the mutations of PCNA2-A and PCNA2-E. (G) Effects of mutations on selected peptides of PCNA2. Each colored bar plots the difference ratio (%) between the deuterium uptake by the PCNA2-TIP complex relative to its uncomplexed PCNA2. Colors correspond to the native PCNA2 (N, red), PCNA2-A mutant (A, yellow), and PCNA2-E mutant (E, blue). Each bar was computed from the average deuterium uptake found at five HDX times points, and a bracket (thin lines) denotes its standard deviation (1σ). Asterisks on the colored bars of PCNA2-E indicate data originating from slightly shorter peptides. The 187–204 peptide (red) contains sites subjected to mutagenesis. Each plotted peptide group exhibits statistically significant differential peptide data, as inferred by paired *t*-tests. *Panels (B)–(G) are used with permission of Li et al. (2014).*

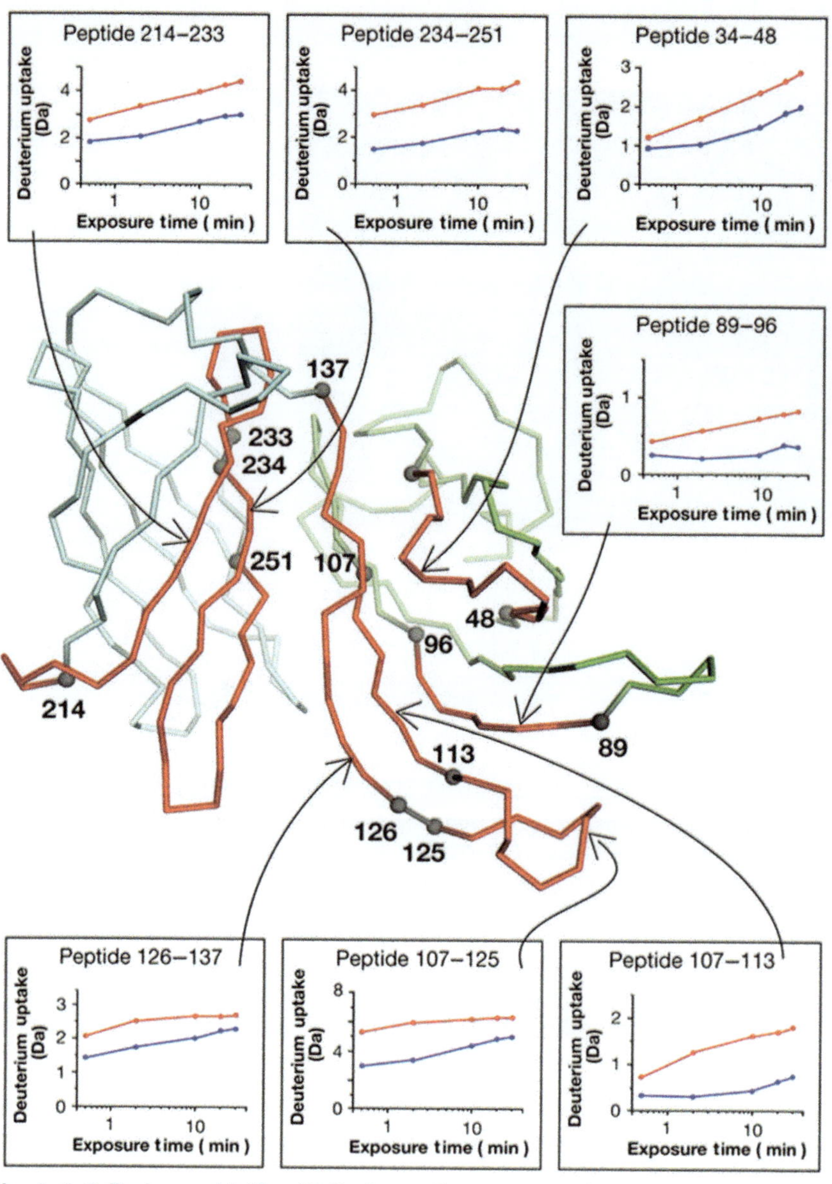

Elyssia S. Gallagher and Jeffrey W. Hudgens, Figure 7 Map of the antigenic epitope of fHbp during interaction with mAb 12C1, as determined by HDX-MS. The peptides protected by mAb 12C1 are highlighted in red on the fHbp structure. Numbered spheres label the start and end of each affected fragment. Subpanels show the deuterium uptake observed in peptides of apo-fHbp (red, always upper line) and holo-fHbp (blue, always lower line), showing discordant behavior during 30 min immersion in D_2O. The relative arrangement of the lines is characteristic of Sowole–Konermann type 1 behavior. *Used with permission of Malito et al. (2013).*

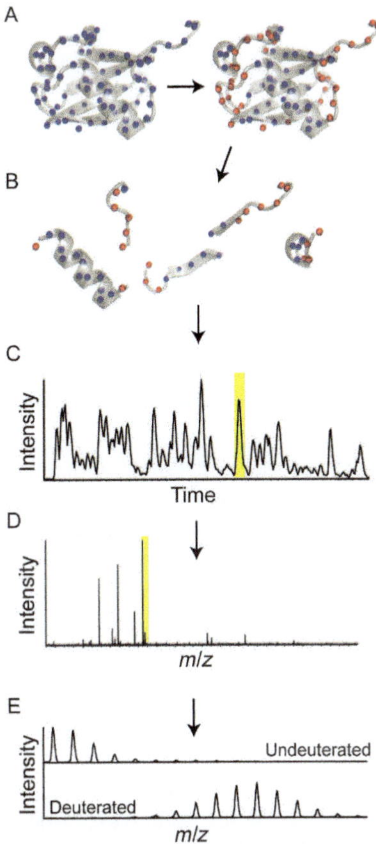

Miklos Guttman and Kelly K. Lee, Figure 1 Overview of HDX-MS. (A) The protein is incubated in deuterated buffer under native conditions to allow the amide protons to exchange for deuterium. (B) The exchange is then slowed to lock the deuterium in place and an acid protease is used to rapidly digest the protein into peptides. (C) The peptides are resolved by fast reverse phase liquid chromatography, which is directly infused into the mass spectrometer. (D) The mass spectra are recorded for each eluting peptide where the mass shift due to deuterium incorporation for each deuteration time is quantified (E).

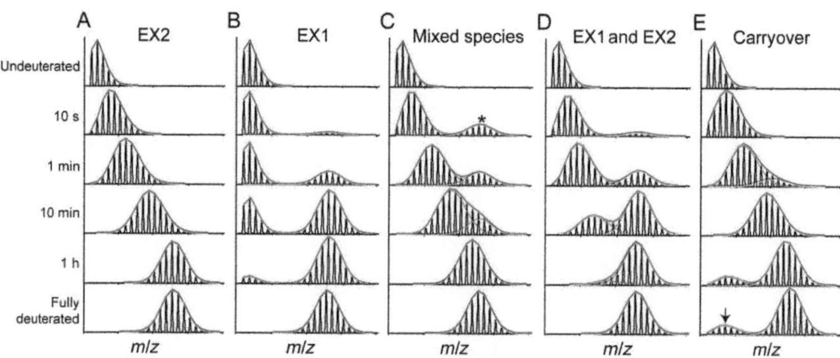

Miklos Guttman and Kelly K. Lee, Figure 2 Examples of different exchange behaviors during HDX-MS. (A) Most commonly peptides uptake deuterium gradually by exposing amides through fast local structural fluctuations (EX2). (B) Large protein motions occurring on relatively slow timescales will expose large segments for long periods of time, whereby all amides become fully exchanged in a correlated manner (EX1). This is evidenced by a bimodal distribution where the population shifts to the exchanged species over the course of deuterium exchange. (C) Samples may contain populations of alternate conformational states. This is evidenced by a bimodal where the relative intensities of the two species remain constant over the course of deuteration. In this case, there is approximately 20% contribution to the signal from an unfolded population that exchanges fully by the earliest time point (denoted by *). (D) Proteins can undergo motions on various timescales resulting in mixed EX1/EX2 kinetics. In this case, the peptide is gradually deuterated (EX2) while also shifting to the fully deuterated state via EX1. (E) Sample carryover during LC–MS analysis can cause artifactual bimodals. In this case, the 1 min and 1 h time points show an additional component corresponding to ~15% carryover from a previous injection. The fact that it is also evident in the fully deuterated spectrum (depicted by an arrow), which should be uniformly and completely deuterated, is strong evidence that carryover is responsible for the bimodal behavior.

CPI Antony Rowe
Chippenham, UK
2016-01-21 02:44